PAPERS PRESENTED AT THE
INTERNATIONAL CONFERENCE ON THE

PLANNING, CONSTRUCTION, MAINTENANCE & OPERATION OF SEWERAGE SYSTEMS

READING, ENGLAND
12-14 SEPTEMBER 1984

Organised and sponsored by
BHRA, THE FLUID ENGINEERING CENTRE

BHRA
THE FLUID ENGINEERING CENTRE

In conjunction with the
WATER RESEARCH CENTRE

Editor: Ms. J. Pounsford

EDITORIAL NOTES

The Organisers are not responsible for statements or opinions made in the papers. The papers have been reproduced by offset printing from the authors' original typescript to minimise delay.

When citing papers from this volume the following references should be used:-

Title, Author(s), Paper No., Pages, International Conference on the Planning, Construction, Maintenance & Operation of Sewerage Systems, Reading, England. Organised and sponsored by BHRA, The Fluid Engineering Centre, Cranfield, Bedford MK43 0AJ, England. (12-14 September, 1984).

British Library Cataloguing in Publication Data

International Conference on the Planning,
Construction, Maintenance and Operation
of Sewerage Systems (1984 : Reading)
International Conference on the Planning,
Construction, Maintenance and Operation of
Sewerage Systems.
1. Sewerage 2. Sewage disposal
I. Title II. Watts, G.A. III. Pounsford, J.
628'.2 TD645

ISBN 0-906085-96-9

Printed and published by
BHRA, The Fluid Engineering Centre
Cranfield, Bedford MK43 0AJ, England

ISBN 0-906085-96-9

ACKNOWLEDGEMENTS

The valuable assistance of the Organising Committee and Panel of Referees is gratefully acknowledged.

ORGANISING COMMITTEE

D.N. Young (Chairman)	North West Water Authority
G. Cox	Thames Water Authority
D. Fiddes	WRC Engineering Centre
C.G. Boyd	Brickhouse Dudley plc
C. Hayward	John Taylor & Sons
E. Porter	Water Research Centre
Dr. R.K. Price	Hydraulics Research Ltd.
D. Rees	Consultant
Dr. H.S. Wheater	Imperial College of Science and Technology
D.A. Crow	BHRA, The Fluid Engineering Centre
Jacqueline Pounsford Conference Organiser	BHRA, The Fluid Engineering Centre

OVERSEAS CORRESPONDING MEMBERS

S. Bendixen	V.B.B. Sweden
R.M. Bremner	Department of Public Works, Toronto, Canada
Dr. Ing. E.L. Kuntze	Freie & Hansestadt Baubehorde, F.R.G.
Dr-Ing. H. Laikari	National Board of Waters, Finland
R.H. Sullivan	American Public Works Association, U.S.A.
Dr. Ir. W.C. Witvoet	DHV Consulting Engineers, Netherlands

Cover picture courtesy of Rees Seergun

International Conference on the

Planning, Construction, Maintenance and Operation of Sewerage Systems

Reading, England: 12-14 September, 1984

CONTENTS

How to prevent sewer survey analysis becoming a drain

When the closed circuit tv survey of the sewers has been completed, you could be swamped by the sheer volume of data collected. Processing and analysing this data could tie up staff for days and drain your resources.

A hundred kilometre sewer survey can provide up to half a million separate items of data.

Given time, Tompkins could eventually find the information you need, select what is most important or urgent, and compare it with information collected in last year's survey. But could he do it all by Monday?

The solution to this and many other problems was invented by Seer. It is called PRISM and it is unique. Designed specifically for closed circuit tv surveying, it is a system of data management with many attractive facilities.

There is a choice of reporting methods to suit your needs.

- **Long form reports, in which detail is recorded and which are legible at a glance.**
- **Short form reports which can highlight areas requiring action, based on the engineer's criteria.**
- **Facility for addition of later survey data.**
- **Facility for comparison of status changes.**
- **Facility for including earlier (pre PRISM) survey data.**
- **Facility to interrogate the data bank (e.g. list all sewer lengths in which infiltration occurs simultaneously with a sewer depth less than 2 metres which are adjacent to buildings).**

There is a range of options available including hard copy, microfiche, video cassette recordings and photographs. Seer's control over fundamental disciplines and methods ensures consistently high standards throughout.

Accurate analysis relies on thorough surveying methods.

Contact us now and ask for more information about PRISM-the answers could be on your desk by next Monday.

SEER T.V. SURVEYS LTD

Seer TV Surveys, Westminster House, Old Woking, Surrey GU22 9LF.
Telephone Woking (04862) 62221.

International Conference on the
PLANNING, CONSTRUCTION, MAINTENANCE & OPERATION OF SEWERAGE SYSTEMS
12-14 September, 1984

PAPER A1

PLANNING AND IMPLEMENTING A SEWERAGE STRATEGY

P.W. Styles

BSc, CEng, MIMunE
Principal Engineer (Strategic Planning)

G.K. Robinson

BSc, CEng, MICE
Principal Regional Officer (Sewerage)

Severn-Trent Water Authority
Birmingham
England

Summary

After considering the "National Assessment" in 1977, the Severn-Trent Water Authority carried out a statistical survey of its sewers to determine the extent of structural decay. Following publication of the results in 1981, the production of a strategy for sewerage in the eighties was commenced.

First a programme of improvement of sewer records was instigated. Methods of hydraulic analysis were investigated resulting in adoption of the Wallingford Procedure. The Authority assisted in setting up national standards for CCTV work and subsequently carried out a systematic census of strategic sewers.

A strategy document has been published which lays down the programmes of action, including "Drainage Area Studies", over a ten year period. These studies, which bring together the hydraulic and structural investigations, are central to the strategy.

The strategy, which is considered cost effective, is now being implemented at local level, in conjunction with the WAA/WRC "Sewerage Rehabilitation Manual". The adoption of national standards of documentation is emphasised.

"...........some seem to think that their first aim in life is to save money; instead of which they are sent to their present positions, not primarily in order to save money, but in order to spend it wisely."

Joseph Chamberlain (1836-1914)
Mayor of Birmingham

Held at The University of Reading, England. Organised and sponsored by BHRA, The Fluid Engineering Centre, Cranfield, Bedford in conjunction with the Water Research Centre.

1. BACKGROUND

1.1 Severn-Trent Water Authority was created by the Water Act of 1973 and serves 8.25m customers in the Midlands of England. Whilst the Authority directly provides a wide range of water services, including water resources, supply, land drainage and water reclamation, the sewerage function is largely carried out by local councils acting as agents. Large scale strategic surveys and exercises therefore require considerable organisation and effort to achieve an equitable spread of financial resources. Approximately £35m per annum is expended on sewerage schemes, this being an increasing proportion of the total capital works programme. The value of the region's sewerage system exceeds £3.5 billion.

2. INTRODUCTION

2.1 At the Institution of Civil Engineer's 'Sewerage 81' Conference Tom King (then Minister for Local Government and Environmental Services) and Peter Stott of the National Water Council both referred to the need to set up a strategic approach to solving the problems of dereliction(1).

2.2 With the economic recession of the 1970s and early 1980s, and the consequent decline in the construction of sewers for housing, there has been a steady increase in the proportion of expenditure on sewer rehabilitation. None the less a House of Lords Select Committee recently estimated that £310m per year is required to maintain the Nation's sewers whereas only £205m per annum is being spent(2). The Select Committee recommended that there should be a greater sense of urgency and more attention paid to strategic planning.

2.3 In its report on its investigations into the Severn-Trent Water Authority et al, the Monopolies and Mergers Commission(3) recommended:

> "that the Authority should press ahead with its engineering survey to establish the physical state of its underground assets ...." "that it should develop a fully integrated strategy, not only to ensure consistency over time in the allocation of financial and physical resources, but also to make trade-offs between standards of service and financial resources more explicit."

2.4 The Authority's Objectives include:

> "To convey, without nuisance, the community's water borne waste to an appropriate site for treatment and/or harmless disposal.
>
> To protect the community from harmful or uneconomic effects of flooding.
>
> To maintain and restore, to agreed objective standards, the quality of rivers, canals and enclosed waters ...."
>
> "To invest in the maintenance of existing, and the construction of new assets where economically justified by clearly identified needs,.........."

2.5 These Objectives provide a conceptual framework on which to base policies and strategies. Much of the Sewerage Strategy(4) is taken up with the detailed identification of those needs and the processes that bring needs forward into concrete actions. In essence the problems of developing a strategy for sewerage are not concerned directly with the construction of new assets. The strategy is concerned with the setting of priorities, data collection, survey and analysis to determine where construction and renovation are appropriate.

3. THE NATIONAL ASSESSMENT ETC

3.1 In 1977 the Standing Technical Committee (STC) for Sewers and Water Mains, set up by Government and the National Water Council, issued its widely publicised National Assessment(5) which suggested that the nation's sewers were in a very neglected state. The report pointed out that there was a serious lack of information about the system and included very high estimates for the cost of renewal.

3.2 The National Assessment states:

> "There is a serious lack of information relating to the lengths, diameters, age, material and condition of both sewers and water mains. This lack of information prevents all except a broad estimate of monetary requirements for the maintenance and renewal of sewers and water mains".

3.3 The figures suggested in the report give an approximate requirement of £260m per annum (Jan 1975 prices - say £750m at Aug 84 prices) to maintain the water and sewerage networks. This would equate to approximately £63m per annum for the sewers in the Severn-Trent region. (Aug 1984 prices)

3.4 At about the same time concern began to grow over the number of sewers which were collapsing under traffic routes. This concern was especially evident in the North West of England and generated considerable publicity. Whilst concern was expressed in the Midlands it was thought to relate to a number of the older town centres and mining areas which have particular problems.

4. THE SEWER RENEWALS SURVEY

4.1 In response to the National Assessment, and political and public concern over the size of the problem, a sample survey was set up to determine the composition of the sewerage system and the scale and distribution of dereliction in the Midlands.

4.2 The results of this survey were reported to the "Sewerage 81" Conference by Messrs Balmer and Meers in their paper "Surveying Severn-Trent's Sewers"(6). Their main findings were that:

(1) the problem is not as bad in the Severn-Trent area as envisaged nationally by the STC,

(2) about 2% of the region's sewers would require attention to prevent collapse over a ten year period,

(3) the esimated cost of dealing with existing dereliction over the ten years was £149m, plus or minus £45m (1980/81 prices).

Thus the low estimate was £104m, or approximately £10m per annum, and the high estimate £194m, or approximately £20m per annum.

4.4 Currently the Strategy is based upon the low estimate to develop the system and bring in the required new technology. After deducting the £5m per annum already expended on renewal and renovation, but often triggered by other purposes, the minimum increase in capital expenditure amounts to approximately £5m per annum which is manageable within current spending levels.

4.5 The sample survey indicated the scale of the problem and the type of sewers most at risk. It provided valuable information at a time when no comparable data was available in the industry. However, because it was only a sample survey, it could not identify individual sewers at risk. For this reason a census of strategic sewers was undertaken and a programme of CCTV survey commenced as part of the sewerage strategy.

5. SEWER RECORDS

5.1 Spending on sewer records has been patchy and sporadic since 1974 when the sewers, but not the records, were transferred to Water Authorities. In 1974 there were over seventy different record systems in use in the region of varying quality. It was estimated that roughly half of these records were adequate for most purposes but very few came up to the standards required to develop a strategy.

5.2 Sewer records, while costly, are considered to be cost effective in terms of managing a complex asset worth over £3.5 billion. Severn-Trent's Director of Finance stated in the Institution of Civil Engineer's debate in March 1983 "... the more we have got to know about the state of underground assets, the more we have reduced the perception of what is needed in renewal terms"[7].

5.3 The programme of Sewer Record improvements, using the NWC Specification [8] & [9], is planned for completion in 1991. The annual cost is £0.5m giving an approximate total of £5m to complete the work. This is based upon about half of the sewers requiring survey and drawing whilst with the other 50%, records are adequate to require redrawing only. Standard overlay sheets have been produced and a standard manhole record card is in use throughout the region (Fig 1).

5.4 Survey and drawing to the NWC specification costs about £1 per head of population in the area to be surveyed. Redrawing only, costs around 10p per head.

5.5 The Authority uses the Model Conditions of Contract[10] and maintains an approved list of contractors capable of working to the NWC specification. Councils carrying out records work financed by Severn-Trent are now required to compete with contractors from the approved list. As the strategy progresses hydraulic data will become an integral part of the records system.

5.6 Early in the development of the strategy it was realised that a common referencing system would be required for all data collection on sewer lengths. In a one-off exercise lasting 12 months, all existing sewer records were labelled with the NWC node references at a total cost of £50,000. These references form the link between the records, the strategic sewer census and the CCTV surveys. Unfortunately the system cannot be used within WASSP thus requiring two referencing systems to be in parallel use.

6. CCTV - IDENTIFYING THE PRIORITY SEWERS

6.1 Approximately 97 per cent of the UK sewerage system is too small for direct inspection by man entry. Hence, until the advent of Closed Circuit Television (CCTV) sewer inspections in the 1960s, very little knowledge of the condition of the system could be made available. In 1970 there were only 3 or 4 CCTV contractors; in 1984 there are over 50 active in the UK.

6.2 The Sewer Renewals Survey concluded by recommending that future CCTV inspection should be concentrated in those areas where the cost of collapse repair was highest and where the risk of collapse (due to age, material etc) was greatest.

6.3 At about the same time, the Water Research Centre was developing a policy for sewer rehabilitation[11] in which they proposed the classification of sewers as Category A, B or C. Category A sewers (some 5% of the system) are those where collapse repair costs would be highest and would be surveyed regularly. Category B sewers (about 15% of the system) are the intermediate group of high relative importance but not as critical as the Category A sewers. They would be surveyed less frequently. Category C sewers comprise the remainder of the system (80%) where the cost of inspection could not be financially justified.

6.4 STWA has combined the WRC proposals with the conclusions of its own survey to produce a plan of action for sewer inspection. Of paramount importance is the identification of those sewers where CCTV inspection will produce a positive return. This was achieved by a strategic sewer census.

7. THE STRATEGIC SEWERS CENSUS

7.1 The census, which followed from the logic of Dennis Peters' paper at "Sewerage 81"(12), sought to identify the strategic sewers and those that have a high risk of collapsing. (The term 'strategic' is largely synonymous with the WRC 'critical' and relates to what were known as 'Category A and B' sewers).

7.2 In order to be cost effective a survey must concentrate on the areas that will be most productive, thus there are two elements to take into account:

(1) What is the effect of failure?

(2) What is the risk of failure?

The first relates to the importance and location of a sewer (Categories A and B) and the second largely to its age and construction.

7.3 A standard form was devised (see Fig 2) and 200,000 distributed to local councils to fill in. Reimbursement was at a standard rate of £1 per form.

7.4 The results were processed on the ICL 2988 computer to give a priority listing of sewers for survey. The ranking formula used was:

LS = (SV + LOC) x (MAT + AGE + SF)

Where LS = Length Score

SV = Strategic Value

LOC = Location Score

MAT = Material Score

AGE = Age Score

SF = Special Factors Score

The individual scores relate to the definitions used in filling in the forms as follows:

Strategic Value	-	High	20
		Med	12
		Low	10
Location	-	Central Commercial Area	40
		Major Traffic Route	25
		Other	10
Material	-	Brick	32
		Other	10
Age	-	Before 1918	44
		1918-1945	12
		Post 1945	10
Special Factors	-	Agressive Effluent	10
		Poor Ground	10
		Public Health Risk	10
		Septicity Problems	10
		Subsidence Area	10

7.5 The need to further refine the scores is not immediate as the system is concerned to identify the order of survey only. Better information will be available in say five years when refinement may be necessary.

7.6 A relatively clear dividing line occurs at around a length score of 1,000 and this figure, which has some 15% of the system above it, has been used to define the sewers requiring structural survey. Sewers with lower scores than 1,000 are not being considered as part of the current programme for survey. The maximum score has been around 6,600. (Fig. 3).

7.7 The census, which was carried out during 1983/84, has cost approximately £150,000 giving information on about 100,000 relevant sewer lengths - about 15% of the total system

7.8 A secondary use has been made of the information to assist in setting priorities for the Drainage Area Studies. Since each pipelength has the Ordnance Survey grid number included within its NWC reference, it is easy to list the pipelengths according to their grid square. The sums of the length scores for each km square, when plotted on a grid, give a clear indication of the priority areas for investigation (Fig. 4).

8 CCTV CONTRACTS

8.1 CCTV contracts let in the STWA region use the Manual of Sewer Condition Classification[13] and the Model Contract Documents[14]. All CCTV work is carried out by contractors from an approved list. The contract documents have been augmented recently to include requirements for quality control.

8.2 When CCTV was used mainly to investigate problems, the contractor's longhand report with photographs and/or videotape were sufficient. With the advent of strategic surveys there has been an increase in the length of sewer surveyed and only a proportion will warrant closer examination. Computer programs become essential, therefore, to screen the mass of information produced to identify that relatively small percentage that should be examined in detail. Some 90% of the length surveyed will have insufficient structural deterioration to warrant further attention.

8.3 The standard form of field sheet devised by water authorities and CCTV contractors has been adopted (Fig. 5). The submission of survey data from the contractor on magnetic tape or over telephone lines for direct input to the mainframe computer is encouraged and a suite of computer programs has been developed to analyse data. Defects are weighted and peak, mean and total defect scores calculated for each sewer length. Very high scoring sewers are identified to the engineer for him to view the videotape and judge for himself the urgency and degree of rehabilitation needed.

8.4 Suggested re-survey dates are derived for the remainder depending upon the defect scores and the sewer category. These range from five years to 30 years on. The programs allow on-line interrogation of the header data files for analysis of strategic sewers by operating division, district council, age, material, defect score etc.

8.5 The Sewerage Strategy sets out a 15 year programme for the inspection of all Category A and B sewers. The associated budget started in 1983/84 at £100,000, rises to £500,000 by 1987/88 and continues at that level to the end of the programme. Larger sewers will be inspected manually and photographic surveys will be preferred for some applications.

9 HYDRAULIC ANALYSIS

9.1 Investigations into hydraulic analysis systems commenced in 1977 and it was rapidly concluded that current design methods were inadequate for simulating conditions within an existing pipe network as surcharge is a frequent occurrence. When the trial versions of the Wallingford Procedure[15] (WASSP) were produced, Severn-Trent assisted as one of the selected trialists. Since that time usage has gradually increased despite a number of minor set backs with the simulation package (WASSP-SIM).

9.2 WASSP-SIM, as a tool for hydraulic simulation, is central to the analysis of the causes of flooding problems and is currently run on an ICL 2988 mainframe computer. The procedure has been adopted as a standard within the region and simulation will become mandatory prior to the approval of rehabilitation schemes in the near future.

9.3 Hydraulic Simulation is estimated to cost between 50p and £1 per head of the population in the area covered. The high cost is due to the extra data and greater effort required in building a realistic model for simulation. This is, of course, a one-off exercise which need not be repeated once the model is complete.

10. MODEL VERIFICATION

10.1 Flow measurement and rainfall gauging can be expensive and difficult, though recent advances in equipment have improved the situation. Against the high costs of measurement the potential benefits arising from a verified model must be weighed. Due to its expensive nature a flexible approach has been adopted to the collection of such data for model verification:

1) first the model is run

2) the results are compared with physical symptoms

3) if the results and symptoms do not match the model is examined for correctness

4) if the model is not sufficiently accurate for the work in hand then flow (or surcharge) and rainfall measurement is undertaken

10.2 This gives a cost effective and reasonable answer in many cases without having to resort to the effort and expense of flow measurement every time.

10.3 Model verification by flow survey typically costs between 50p and £1 per head of population served; the optimum area covering a population of about 30,000 persons. Measurement of surcharge is less expensive.

11. DRAINAGE AREA STUDIES

11.1 The Authority commenced a programme of "Forward Planning Studies" in 1976. These covered all relevant factors within a drainage area with a view to formulating proposals. They were limited in extent and not part of a concerted effort.

11.2 The programme of Drainage Area Studies is planned to cover all areas, with a contiguous development of more than 10,000 population, by 1994. The Studies are based upon complete 'drainage areas' (i.e. the piped area as distinct from 'catchment' which is the natural area drained). They cover both hydraulic simulation and CCTV survey to determine structural problems.

11.3 The first step is programming. Initially priorities have been based upon a subjective assessment of need but the 1985 programme will be based upon a common regional system. This contains three elements:

1) the performance indicator

2) the investment indicator

3) the strategic sewers indicator

11.4 To calculate these, first the urbanised areas are split into drainage areas of between 10,000 and 50,000 population. Areas larger than 50,000 are considered twice; once crudely for the whole area and secondly in more detail with each component drainage area.

11.5 Maps are produced of the areas and problems are plotted along with storm overflows, pumping stations and proposed development. The performance indicator is calculated using weighted scores from this information.

11.6 The investment indicator represents the proposed capital works in the drainage area. It is thus an indication of the possible cost effectiveness of the study if part of the proposed expenditure can be saved through a more effective scheme.

11.7 The strategic sewers indicator is based upon the length scores taken from the strategic sewers census outlined earlier. These three indicators give clear pointers to the priority areas for study. Further derived figures show potential benefit against cost.

11.8 The first step in carrying out a study is to construct a skeleton of the network - this contains details of all the sewers necessary to model the system. Provision is made at an early stage to improve access to sewers that form part of the main skeleton (the 'core sewers') if it is not adequate. The CCTV programme is planned so that all strategic sewers will have been surveyed.

11.9 Hydraulic analysis of the system is carried out using WASSP-SIM and the model verified as previously outlined. After verification a full statement is available on the hydraulic and structural performance of the strategic sewers. This is put alongside the statement of problems and pressures such as development proposals. Only then are proposals made to solve the problems and these, being based on a full understanding of the system operation, are usually more cost effective than simple "re-designs". The proposals, which will consist of a number of options at this stage, are still subject to full scheme appraisal procedures.

11.10 When alterations and extensions to the system are complete the hydraulic data and the sewer records are updated so that further analysis can be quickly carried out in response to new problems or pressures.

12 PROVISION OF RESOURCES AND COST EFFECTIVENESS

12.1 It is important to appreciate the difficulty in providing additional resources to fund work of this nature at a time when staff numbers are reducing and there is great pressure to reduce costs. The strategy must be able to demonstrate that the investment in data and analysis will result in a sufficient reduction in future costs to justify its adoption.

12.2 The major expenses come in sewer records, CCTV, flow measurement and hydraulic simulation. If a proper system already exists the cost of providing information on flooding, collapses and development are minimal. In order to achieve a reasonable resource allocation and programme, it is essential to develop the strategy carefully, over a period of time, after adequate consultation with interested parties. The production of an agreed strategy document and its adoption by the policy makers is essential.

12.3 The cost effectiveness of the Severn-Trent Strategy was demonstrated using a discounted cash flow analysis over a twenty year period (Fig 6).

12.4 The analysis requires a number of assumptions, each related to a specific area, and supported (albeit subjectively) by experience of similar exercises. These are based upon percentage savings in associated operating or capital costs. The analysis shows:

(1) a payback period of 7 years

(2) a NPV over 20 years of £23m

(3) an Internal Rate of Return of 37%

The apparently high NPV of £23m must be related to the projected capital expenditure of £700m over the 20 year period. The spending profile is shown in Fig 7. This differs from the shape in the comparable WRC diagram as it assumes a constant spending pattern (£35m p.a.) as being more probable than an increasing one.

12.5 Whilst in practice the spending limit is likely to be maintained the improvement in effectiveness means that more works can be achieved from the same spending.

13 TRAINING AND IMPLEMENTATION

13.1 What skills are necessary for the new generation of drainage engineers that will succeed us? Most of us in the UK were brought up as project design engineers in times of growth and engineering meant new works. Times have changed and so must the drainage engineer with it.

13.2 Todays sewerage engineers need skills relating to:

- model building
- hydraulic simulation with computers
- closed circuit television
- renovation
- strategic planning
- maintenance

13.3 In addition to these basic skills, modern engineers need to possess the skills of project appraisal so that, in addition to knowing how to design a scheme, they should be able to assess whether a scheme is necessary at all and which are the most economic of the options available.

13.4 Severn-Trent Water has identified specific personnel to co-ordinate these training needs to ensure the necessary specialisms exist at the appropriate level.

13.5 Implementation of the strategy commenced with a series of seminars in October 1983. In May 1984 a further series of seminars were held to release the WRC Rehabilitation Manual[11] which is central to implementation of the strategy.

14 CONCLUSION

14.1 If it is to be worthwhile, a strategy for any service must be based upon levels of service to the customer and the effects of malfunctioning of the system. It is not concerned with levels of activity or with the construction of individual schemes.

14.2 As a service, sewerage is usually only perceived to any extent by the customer, who is totally unaware of design criteria, when it fails. The main causes of failure may be structural or hydraulic, and long or short term. The usual symptoms will be flooding of some description, pollution or collapse.

14.3 Many areas that are, in theory, hydraulically overloaded cause no great problem to customers. The location of the small percentage of structurally inadequate sewers is largely unknown. The Strategy is required to deal with these two main issues in a cost effective manner which gives greatest benefit to the public. This benefit may include: the continuance of the service despite other changes; the provision of a service to new customers without adverse effect on existing customers; and the alleviation of problems.

14.4 In order to achieve this the Severn-Trent Sewerage Strategy has two key aspects:

(1) Renewal and Renovation which concentrates on identifying and bringing up to standard those sewers which are in poor condition and which are also cost effective to renovate or renew,

(2) Hydraulic simulation which gives knowledge of the operation of existing systems. When related to problems (or the lack of them) simulation enables best use to be made of the existing systems and makes capital investment more effective.

14.5 These two aspects are being co-ordinated within a programme of studies - "Drainage Area Studies"; priorities being largely determined on a local basis in conformity with regional guidelines.

14.6 The major problem in progressing a strategy is in obtaining a commitment to the allocation of resources for the investigation and analysis of problems prior to the definition of capital works. The resources required include finance, manpower, information, skills and computer time however, because of its overall effects on the others, the provision of adequate finance is the foremost priority.

15 POSTSCRIPT

The views contained in this paper are those of the authors and not necessarily those of their employer. The Authors are grateful to WRC staff for their help in producing the strategy and this paper.

The figures and costs quoted are included in the paper, not as statements of unalterable fact, but more with a view to eliciting contributions and discussions from others based upon their own experiences.

16 REFERENCES

1 Inst. of Civil Engineers, "Sewerage 81" Conference. "Proceedings of Conference on Maintenance, Repair, Renovation and Renewal of Sewerage Systems, London, June 1981". Opening and closing addresses.

2 House of Lords Select Committee on Science and Technology. "The Water Industry". HMSO 1983.

3 Monopolies and Mergers Commission. "Report on the STWA, EWWCo and SSWCo". HMSO 1981.

4 STWA. "Sewerage - a Strategic Approach". Internal document, August 1983.

5 Standing Technical Committee for Sewers and Water Mains Report No. 4, "Sewers and Water Mains - A National Assessment". NWC/DOE, 1977.

6 Balmer, R. and Meers, K. "Surveying Severn-Trent's Sewers". "Sewerage 81" Conference, London 1981 (as 1 above).

7 Debate on Spending on Water Infrastructure, Inst. of Civil Engineers, March 1983, Unpublished.

8 Standing Technical Committee on Sewers and Water Mains Report No. 25, "Sewer and Water Main Records". NWC/DOE, 1980.

9 Styles, P.W. and Hedderley, H.M. "Sewers and Water Mains - a Standard Record System". "Sewerage 81" Conference, London 1981 (as 1 above).

10 Ad Hoc Group of Water Authority Engineers. "Sewer Survey and Records - Model Contract Documents".

11 Water Research Centre/Water Authorities Association. "Sewerage Rehabilitation Manual". WRC/WAA, April 1984.

12 Peters, D. "The Development of a Policy for Sewer Rehabilitation". "Sewerage 81" Conference, London 1981 (as 1 above).

13 Standing Technical Committee on Sewers and Water Mains Report No. 24, "Manual of Sewer Condition Classification". NWC/DOE, 1980.

14 Ad Hoc Group of Water Authority Engineers. "Model Contract Documents for CCTV Survey".

15 Standing Technical Committee on Sewers and Water Mains Report No. 28, "The Design and Analysis of Urban Storm Drainage : The Wallingford Procedure". NWC/DOE, 1981.

Fig. 1.

MANHOLE DETAILS

MANHOLE REFERENCE

ST. 2571

DISTRICT COUNCIL	CODE	NATIONAL GRID REFERENCE

LOCATION

STATUS	USE	TYPE	OTHER TYPE	YEAR BUILT	DRAINAGE AREA CODE	DATE OF SURVEY	COVER LEVEL

STATUS: PU - Public; 24 - Section 24 applies; PR - Private; HD - Highway Drain; CW - Culverted Watercourse
USE: F - Foul; S - Surface Water; C - Combined; O - Storm. Water Overflow Sewer
TYPE: MH - Manhole; HB - Hatchbox; LH - Lamphole; OT - Other(detail in "Other Type" box)

	INVERT LEVEL	DEPTH	UPSTREAM MANHOLE REF.	SHAPE	SIZE (DIA)	BACK DROP (DIA)	
A							INCOMING SEWERS
B							
C							
D							
E							
F							
			DOWNSTREAM MANHOLE REF.				
X							OUTGOING SEWERS
Y							

SHAPE: C - Circular, E - Egg, R - Rectangular, T - Trapezoidal, O - Other

COVER			MANHOLE					
Shape	Duty	OPENING SIZE (DIA)	Mat'l	Entry	Access	Slab/Taper	SHAFT SIZE (DIA)	CHAMBER SIZE (DIA)

SHAPE: C - Circular; T - Triangular; S - Square; D - Double Triangular; M - Multiple; O - Other
DUTY: H - Heavy; M - Medium; L - Light; N - Not Known
MATERIAL: B - Brick; P - Pre-cast concrete sections; S - Segmental; I - In-situ concrete; O - Other
ENTRY: T - Top entry; S - Side entry
ACCESS: I - Irons; L - Ladder; S - Stairs; O - Other
SLAB/TAPER: R - Reducing Slab; T - Taper; N - None; A - Brick Arch

LOCATION SKETCH	PLAN OF MANHOLE

Rev. 9/82

Fig. 2.

SEWERAGE STRATEGY

CLASSIFICATION OF SEWERS

Line Type **Ø1** [3] ☐☐☐☐☐☐☐☐☐☐☐ NWC PIPELENGTH REFERENCE

FORM No. ☐☐☐☐☐

PLEASE READ EXPLANATORY NOTES ON FRONT COVER BEFORE COMPLETING FORM.

1. COUNCIL — 15 ☐☐☐☐ (USE CODE NUMBER)

2. SIZE — 20 ☐☐☐☐ mm(dia.) x 25 ☐☐☐☐ mm

3. LENGTH — 30 ☐☐☐ metres

4. DEPTH (to invert) — 34 ☐ — 1 = UNDER 2m; 2 = 2-3m; 3 = OVER 3m — 36 ☐☐☐☐ metres — STATE ACTUAL DEPTH IF KNOWN

5. MATERIAL — 41 ☐ — B = BRICK; O = OTHER — 43 ☐☐☐ (USE CODE) — STATE "OTHER" IF KNOWN

6. DATE OF CONSTRUCTION — 47 ☐ — 1 = BEFORE 1918; 2 = 1918-45; 3 = AFTER 1945 — 49 ☐☐☐☐ — STATE ACTUAL DATE IF KNOWN

7. LOCATION — 54 ☐ — C = CENTRAL COMMERCIAL AREA; T = MAJOR TRAFFIC ROUTE; O = OTHER

8. STRATEGIC VALUE — 56 ☐ — H = HIGH; M = MEDIUM; L = LOW

9. SPECIAL FACTORS — Line Type **Ø2** — 3 ☐☐☐ — 7 ☐☐☐ — 11 ☐☐☐ — 15 ☐☐☐

19 ☐☐☐☐☐☐☐☐☐☐☐☐☐☐☐☐☐☐☐☐☐☐

☐☐☐☐☐☐☐☐☐☐☐☐☐☐☐☐☐☐☐☐ — 61 ☐☐ FOR OFFICE USE ONLY

10. HAS THIS LENGTH OF SEWER PREVIOUSLY BEEN INSPECTED BY CCTV? — 64 ☐ Y=YES N=NO

11. IF "YES" ARE CCTV REPORTS COMPILED IN ACCORDANCE WITH NWC "MANUAL OF SEWER CONDITION CLASSIFICATION"? — 66 ☐ Y=YES N=NO

12. IS A VIDEOTAPE RECORD AVAILABLE? — 68 ☐ Y=YES N=NO

ST. 2494

FORMAT PROGRAM SYFPO1

Distribution of Strategic Sewer Lengths by Score Fig. 3.

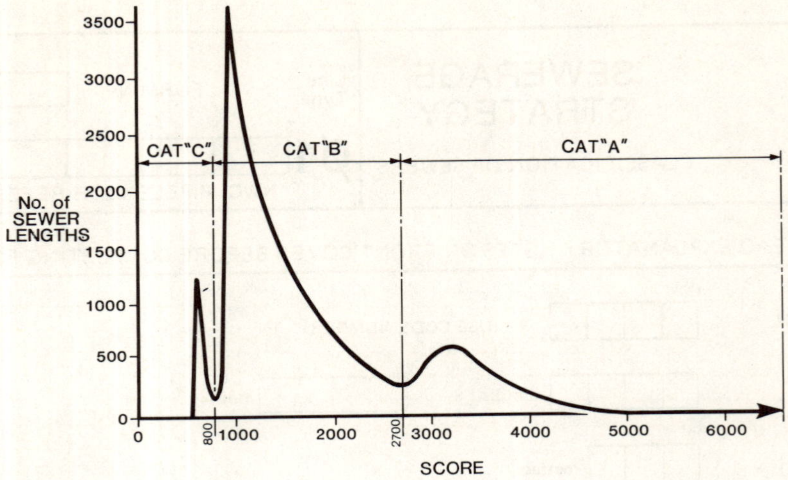

Total Strategic Sewer Length Scores per 1/2 km x 1/2 km Grid Square Fig. 4.

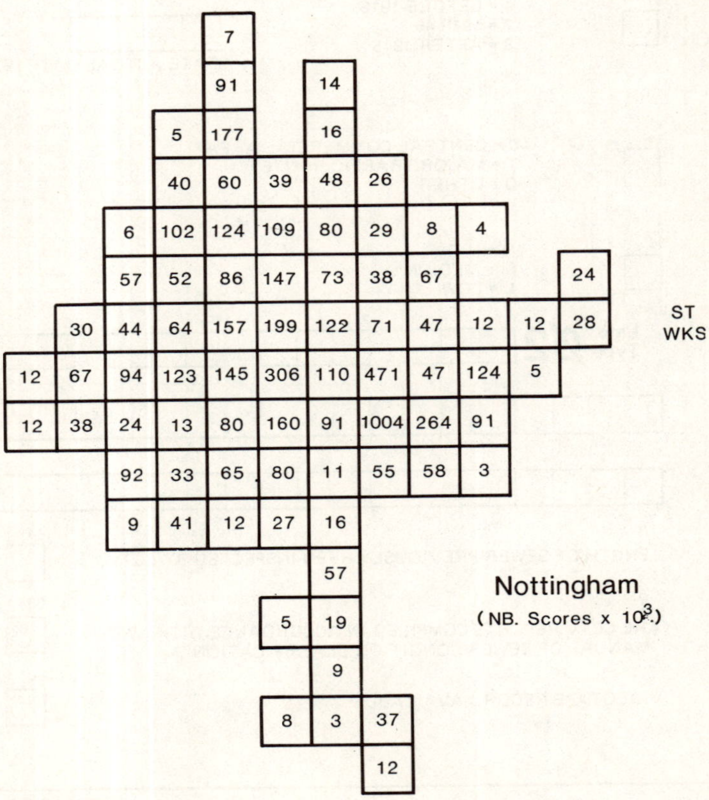

Fig. 5.

CCTV SURVEY

SEVERN-TRENT WATER AUTHORITY

District Council ______ Sheet No. ______ Continuation Sheet No. ______

Contractor | Contract No. | Job No. | Drainage Area Code | Division | District | Pipe Length Reference

Date | Time | Location

START Manhole No. | Depth | Cover L. | Invert L. | FINISH Manhole No. | Depth | Cover L. | Invert L.

Sewer Use | Search Direction | Size (dia) x Size | Shape | Material | Lining | Pipe Length | Total Length | Year Laid

VT. No. | Comments (Videotape Recorder) (Weather) (General)

For Office Use Only

Video No.	Photo No.	Chainage (metres)	Cont Def. &extra lines	LETTERS Code	NUMBERS Dia of JN/CN (mm)	NUMBERS Clock at to	NUMBERS Intrusion Etc. (mm) %	Remarks
1 2 3 4	5 6 7	9 10 11 12 13	15 16	17 18 19 20 21	23 24 25	27 28 29 30	32 33 34 35	36

START A SEPARATE SHEET FOR EACH MANHOLE LENGTH

Pipe Length Reference (Do not Punch)

Sewerage Strategy – Discounted Cash Flow Fig. 6.

Costs in brackets, savings without. All figures in £M

YEAR	0	1	2	3	4	5	6	7	8	9	YEARS 10-19 EACH
COSTS:											
RECORDS	(0.5)	(0.5)	(0.5)	(0.5)	(0.5)	(0.5)	(0.5)	(0.1)	(0.1)	(0.1)	(0.1)
MODELLING	(0.5)	(0.5)	(0.5)	(0.5)	(0.5)	(0.5)	(0.5)	(0.5)	(0.5)	(0.5)	(0.1)
CCTV	(0.1)	(0.2)	(0.2)	(0.3)	(0.4)	(0.5)	(0.5)	(0.5)	(0.5)	(0.5)	(0.5)
STAFFING	(0.1)	(0.2)	(0.3)	(0.3)	(0.3)	(0.3)	(0.3)	(0.3)	(0.3)	(0.3)	(0.3)
BENEFITS:											
RECORDS	0.2	0.2	0.3	0.3	0.4	0.4	0.5	0.5	0.5	0.5	0.5
CAPITAL(HYDRAULICS)	0	0.2	0.5	0.7	1.0	1.2	1.5	1.7	2.0	2.2	2.5
CAPITAL(STRUCTURAL)	0	0.2	0.4	0.5	0.6	0.8	1.0	1.0	1.0	1.0	1.0
DESIGN COSTS	0	0	0.1	0.1	0.2	0.3	0.3	0.4	0.4	0.5	0.5
CASH FLOW	(1.0)	(0.8)	(0.2)	0	0.5	0.9	1.5	2.2	2.5	2.8	3.5
5% DISCOUNT FACTOR	1.00	0.95	0.91	0.86	0.82	0.78	0.75	0.71	0.68	0.64	4.99
DISCOUNTED CASH FLOW	(1.0)	(0.8)	(0.2)	0	0.4	0.7	1.1	1.6	1.7	1.8	17.5

NETT PRESENT VALUE = £22.8M, PAYBACK PERIOD = 7 YEARS, IRR = 37%

Effect of Sewerage Strategy on Capital Expenditure Fig.7.

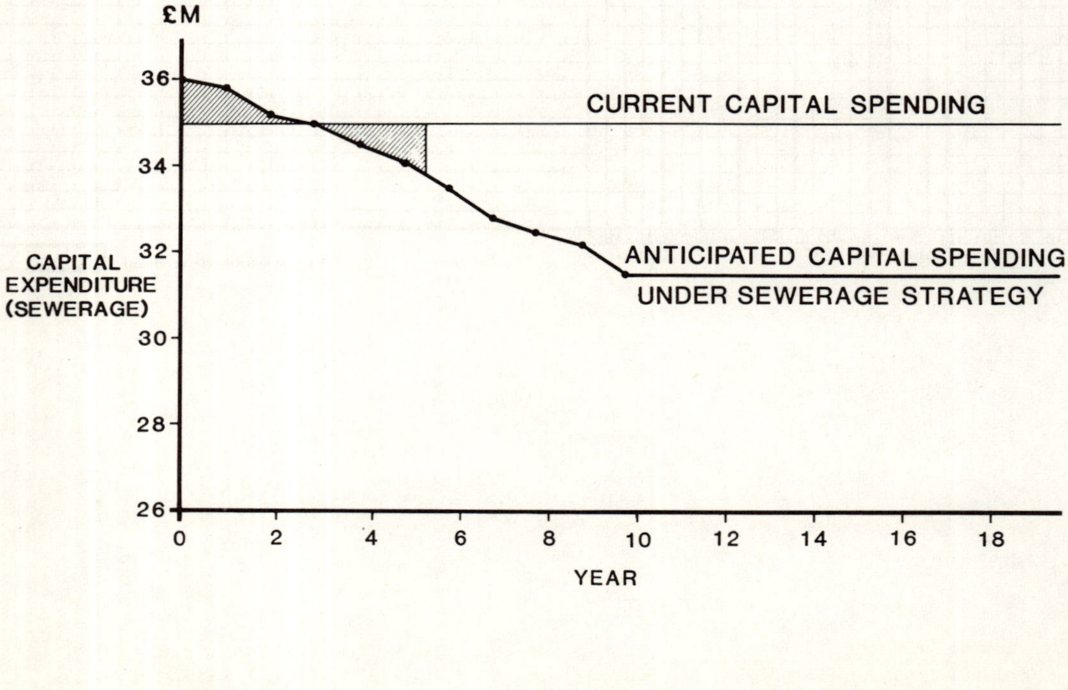

International Conference on the
PLANNING, CONSTRUCTION, MAINTENANCE & OPERATION OF SEWERAGE SYSTEMS
12-14 September, 1984

PLANS AND PRACTICALITY

R. M. Bremner, P. Eng., C. Eng., F.I.C.E.
Commissioner of Public Works & City Engineer,
City of Toronto

and

C. S. Kitchen, P. Eng., M.E.I.C., M.C.S.C.E.
Director of Engineering Systems,
City of Toronto

Summary

This paper describes the work required to carry out a programme of sewer replenishment in an urban area. Matters dealing with inventory, conditions, analyses and staging of the overall plan are discussed. The paper describes the way that the City of Toronto's plan was financed and suggests possible solutions to the predicament of sustained funding. The paper also touches upon the reality of political support that the municipal engineer must harness.

Held at The University of Reading, England. Organised and sponsored by BHRA, The Fluid Engineering Centre, Cranfield, Bedford in conjunction with the Water Research Centre.

1. INTRODUCTION

In urban areas the sewer system provides an essential service which few people think about until failure occurs; the old adage "out of sight, out of mind" applies. Municipalities have large investments in systems, developed over the years and which require positive and secure programmes for maintenance and replacement to ensure an affordable and uninterrupted level of service.

Any plan to replenish a sewer system requires decisions to be made; these decisions not only need a forthright engineering evaluation of what exists but an understanding by those in responsible charge of the level of service required, how the plan is to be carried out and who has to pay. Also, the municipal engineer must generate, through the policy-makers, the political resolve required to carry out such a plan.

The obvious requirements in a systematic approach to putting the plan together, are as follows:

a) Determine the extent and condition of the existing system.

b) Identify deficiencies.

c) Set practical levels of service.

d) Develop a staged plan to eliminate deficiencies and in order of priority.

e) Establish programmes of preventative maintenance for those items which are not deficient.

Putting the plan together requires time and dedicated effort, all of which, although somewhat tedious, is reasonably straightforward. There remains, however, three very important additional steps:

- make sure the public knows about the plan
- propose equitable funding methods to implement the plan
- ensure that the plan is politically palatable and capable of support

2. THE TORONTO PLAN

2.1 Inventory.

In 1964, the City of Toronto was experiencing a most energetic period of redevelopment (over 743,000 square metres of new construction). The problem to be solved was the provision of services for this redevelopment in a City which was frequently experiencing extensive basement flooding due to the hydraulic inadequacies of the existing system. A plan was conceived with the following prime objectives.

(a) A programme of construction of new sewers to eliminate basement flooding conditions during rainfalls.

(b) A programme of construction of new sewers to accommodate continued development and growth of the City and to meet the demands of changing land use.

(c) A rehabilitation programme for the older sewers in the City which had adequate hydraulic capacity but which were in a state of structural disrepair.

(d) A general construction programme conceived and carried out in such a manner to ensure that pollution of the lakefront would be reduced to minimum practical limits thereby protecting the highly desirable recreational facilities.

As serious as the problems were at the time and as in the case of many jurisdictions, there was no factual worthwhile information available on either the physical or hydraulic conditions of the system. Consequently, in 1964, "plain leg work" was required and a first attempt was made by field survey and inspection to determine the inventory and condition of the existing system. The initial inspections were simple and made by lamp from manholes and by walking through sewers of larger diameter. In other sections of the City where sewers were of smaller diameter, camera inspections took place.

From this modest start and with a continuous and sustained effort over 20 years, all of the details of the inventory of the City's sewer system are now captured in one of the segments of the Department of Public Works Master File.

The sewer records in the Master File contain the identification of each sewer as well as its function, physical description, construction and maintenance history and condition. The identification field describes the location, ownership and reference to the original construction plan. The physical description field describes the sewer material, bedding, shape, type, length and diameter, as well as the number of manholes, the overflow chambers and catch basins. Construction and maintenance history fields document the years of construction and repair, reaming, lining and the two most recent cleaning and TV camera inspection dates. The condition field contains a code giving the relative physical conditions of the sewer segment according to a set of criteria.

Sewer records may be accessed for inquiry purposes by staff who require information but records are updated and maintained by a small group of technicians to ensure that the file will retain its integrity.

From its crude beginnings in 1964 the sewer inventory in the Master File is now a formidable factual information tool.

2.2 Condition Rating.

The condition rating of a sewer system requires measurement and quantification of deficiencies. When dealing with an extensive network the authors believe that the subjective approach is more practical; by using experienced staff to assign condition ratings based on established criteria a good consistency can be obtained and maintained.

The criteria used in Toronto for the condition rating of sewers is as follows:

1.	Collapse	6.	Ponding
2.	Irregular line, grade	7.	Calcite
3.	Extensive calcite	8.	Spiderweb cracks
4.	Root penetration	9.	Minor cracks
5.	Protruding laterals	10.	Satisfactory

Condition ratings for the Toronto sewers are contained in the sewer records and it is, therefore, possible to produce reports for work on sewers on the basis of condition, either alone or in combination with other criteria.

Unless an accurate inventory coupled with condition ratings is developed, it is nigh impossible to analyze deficiences and to plan a purposeful and economical replenishment programme.

2.3 Analyses of the System.

In the Toronto case two investigations were required. First the system was examined for structural deficiencies and second it was analyzed with respect to the capacity of the system.

Examination of the system for deficiencies becomes much less difficult once an inventory is in place. The manipulation of data allows priority programmes to be developed on such factors as age, material, failures and condition rating, either alone or in combination.

In 1964, all sewers in the City of Toronto were analyzed using the rational formula, on the basis of prevailing land use and anticipated new development. That analysis showed that about 70% of the existing sewers in the City did not have adequate capacity to accommodate storm runoff. Of the existing sewers, not having sufficient capacity, many were overloaded and surcharged as much as 250% while the average surcharge was in the order of 100%.

Upon completion of the analysis of the hydraulic conditions of the existing sewers and the determination of inadequacies, a programme for the construction of new road storm sewers was set out which would provide the required relief and meet all other objectives of the programme.

In 1972, with the advent of sewer network analysis by computer, the City's system was analyzed using the Dorsch Hydrograph Volume Method (HVM). The principle of HVM analysis is that for each time interval the outflow volume for each segment in the network equals the inflow volume from all sources plus or minus any change in the storage volume for that segment. HVM builds a mathematical model of flow conditions for different rainfall patterns and intensities and produces listings of flow and backwater levels for every sewer segment; the model may also produce flow hydrographs for any and all sections of the network.

The HVM analysis has predicted conditions for runoff flows and surcharge that are very similar to those actually observed in the field. Although it took almost five years of continuous effort to model and analyze the 1800 kilometers of sewers in the network, the results of the analysis allowed the City to complete its 25-year sewer programme four years ahead of schedule and with net savings of approximately 25% of the original 1964 projected costs.

The work of properly analyzing a sewer system is expensive but in relation to the overall cost of the programme, it is vital and truly justified.

2.4 Staging the Plan.

In 1964, two fundamental issues were involved:

(a) The relief of basement flooding; and

(b) The provision of new services and capacity for the City's extensive redevelopment.

A good balance had to be maintained in staging the plan to meet the right of the taxpayer for protection of property from flooding and the desire of City Council to encourage good redevelopment. These two issues could only be dealt with by judicious staging of the plan over a period of years. The factors at play here are evident for no municipality can afford massive expenditures concentrated over a short period of time; also, the disruptive effect of the actual construction programme on the general mobility of an urban area would be catastrophic.

With a staged plan, the taxpayers, although they may continue to experience difficulties, will accept the fact that relief is promised at a particular point in time. You must keep faith, respond to this understanding and make sure the taxpayers have the confidence in the commitments for relief made by the municipality.

The City's programme is now nearing completion with private property in residential areas receiving protection in excess of a two-year return storm; in industrial and commercial districts protection in excess of a five-year return storm is provided. It is worth noting that since 1964 and while the replenishment programme was under way, some six thousand, three hundred million dollars ($6,300,000,000) worth of redevelopment construction has been completed in the City of Toronto.

2.5 Funding the Plan.

1.) Sewer Impost Charge

In 1962, the City of Toronto obtained through the Government of the Province of Ontario, legislation which:

> "would impose upon the owners of highrise and other buildings that impose or may impose a heavy load on the sewer system or water supply system or both by reason of which expenditures are or may be required to provide additional sanitary or storm sewer or water supply capacity which in the opinion of the Council would not otherwise be required a special charge or charges over and above all other rates and charges to pay for all or part of the cost of providing the additional capacity."

Toronto's taxpayers faced a substantial cost for an extensive programme for sewer replenishment and consideration had to be given to a method of direct charging which would assist in and would apportion the cost of the construction programme in an equitable manner. Toronto was no longer involved in the servicing of raw land but was engaged in the reservicing of lands whose current owners had every right to expect an adequate sewer service. Therefore, in a reservicing plan it is quite inequitable for the taxpayer to bear the full brunt of the cost of the entire programme.

When considering the basis for a proposed formula and charge, the sanitary sewer requirements, even at peak flow periods, did not entail more than 5% of the ultimately required sewer capacity for storm runoff. Sanitary flow, therefore, represented an insignificant provision compared to the capacity required for storm runoff; of the two types of flow, storm runoff was considered the practical determinant of the capacity required within the system and was the prime factor in apportioning the responsibility that new construction had in the need for updating the existing system. Having established this direction, it followed that whereas a building use may be residential, commercial or industrial, the formula and charge, based on increased "waterproof" surfaces, would be quite independent of these classifications.

In developing the formula, it was necessary to acknowledge and recognize the fact that an allowance had to be made for the sewer services to the structures existing on the property prior to redevelopment. Since much of the redevelopment was occurring in areas where single family residences predominated, an allowance of 3,300 square feet (306.57 square metres) of floor area was deducted from the calculation of the gross floor area of the proposed new building.

An examination of the gross floor area of new building construction which took place in the years 1962, 1963, 1964, together with the technical study, showed that new building construction could be held responsible to an extent of as much as but not more than 50% of the total capacity required in the sewer system. With the authority and legislation obtained from the Province of Ontario, a charge of 40 cents per square foot (4.31 dollars per square metre) was proposed on all new building construction; this charge would raise funds to pay for about one-half of the annual cost of the construction of the City's new sewer programme.

City Council, when considering the ramifications of this proposed charge (representing an increase of between 1.5% to 4% on building construction costs) was concerned that the levy would slow down or, worse still, drive developers away from Toronto; also, the City would be opposed by developers at the Ontario Municipal Board and the resultant hearing and technical defence of the proposal would be protracted and costly. Consequently, in 1965, Council decided to enact a charge of 20 cents per square foot (2.15 dollars per square metre); this decision was accepted by the development industry.

While at the time the proposers of the 40 cent charge felt that they had received a defeat at the hands of the politicians, in retrospect it can be regarded as an "accommodation for the moment". Later actions, through recommendations to City Council, brought the 20 cent charge to 40 cents in 1978 and to 50 cents in 1981.

To date, construction of Toronto's sewer plan has cost $195,000,000 gross. The Province of Ontario, through statutory subsidies, has contributed $67,000,000, leaving a net cost of $128,000,000. The sewer impost charge has raised $26,000,000 and accordingly, the direct cost to the City of Toronto taxpayers for the plan is $102,000,000.

While the monies raised from the sewer impost fund represent 20.3% of the net cost of the programme and despite the disappointment of the "political reality", a key principle was sustained in that "those who need additional capacity in a system should pay for that requirement."

2.) Sewage Service Rate

In many jurisdictions, funds required for the maintenance of the sewer system are raised directly from property taxes. In hard times or where there are more politically attractive projects, the municipal engineer is generally faced with substantial cuts to the maintenance budget. Programmed maintenance to essential services should not be subject to this vicissitude.

In the Statutes of Ontario, a sewage service rate is defined as a charge for the operation, repair and maintenance of sewage works and include a charge for depreciation, deferred maintenance or a reserve fund for any such purpose and allows:

> "The Council of a municipality, by by-law, to provide for imposing upon owners or occupants of land who use sewage works, a sewage service rate. Also, the municipality may, by by-law, establish a sewage service rate structure upon which sewage service rates shall be based and calculated, and in establishing the rate structure the Council of the municipality shall have regard to differentiating between classes of users, nature, volume and frequency of use and all other relevant matters to ensure that sewage service rates are imposed upon a basis that is equitable and just."

In 1964, the replacement cost for Toronto's existing system was estimated at $311,000,000; its depreciated or unexpired value was estimated at $137,400,000. The average life of a sewer properly maintained was taken as 100 years and a charge of $1,374,000 or 1% of the unexpired value of the system was considered a reasonable basis for reserve funding. In 1964 the annual cost of operation of the City's existing system was $750,000 and, therefore, a total of $2,124,000 ($1,374,000 + $750,000) could be charged on a sewage service rate.

The most equitable and practical way to levy such a sewage service rate was on the basis of water consumption; the City's water revenue in 1964 amounted to $9,356,658. To raise a sewage service rate in the amount of $2,124,000, a surcharge of 22.7% ($2,124,000 ÷ $9,356,658 x 100)% was applied in separate billing on the water bill. This charge represented a surcharge of $4.00 annually on the water bill of the average householder in the City or an increase of 1.7 cents on the metered rate of 7.48 cents per 1,000 litres.

In order to distinguish between classes of users, nature, volume and frequency of use and to ensure that the sewage service rate imposed was equitable and just, provision was made in the by-law to take into account those who use water from the City's system but do not use the sewer system; those who, through consumptive use of water, discharge less into the sewer system than obtained from the City's water system and those who augment water from the City's water system and, therefore, discharge more than obtained into the City's sewer system.

The Sewage Service Rate By-law was passed by City Council in 1966 and since that time an amount has been raised to match the cost to the City of Toronto over the same period for the maintenance and upkeep of the sewer system. This cost to date amounts to some $67,000,000.

Accordingly, the vital work of planned maintenance has not been subject to "political tampering" when Council decides on overall City needs and establishes the annual tax rate.

3. CONCLUSION

A cursory examination of the Water Act of 1973, suggests to the authors that the levying of charges, similar in spirit to the sewer impost and sewage service rate, is not inconsistent with Section 30 of the Act. The authors do not intend to venture into the thickets of legal interpretation of the various powers contained in the Act but we are suggesting that specific legislation, recognizing the principle of user charges, might be obtained thus enabling some financial muscle to be brought to bear on the difficult task of sewer replenishment.

We suspect that the major problems in the United Kingdom may not be ones of capacity but rather those of maintenance and restoration. In this respect, we commend the sewage service rate legislation. With maintenance removed from the tax rate, through a user charge, and the consequent provision for deferred maintenance and reserve funds, some financial stability is obtained. In Toronto it was possible, using the fund, to develop a method of reaming and lining small diameter sewers and to carry out and pay for the extensive HVM analysis. We are sure that there are many more opportunities for innovative work in the United Kingdom, and at the local level, given stability of funding.

For 20 years, the authors have been privileged to take part in the enterprise of translating a plan to practicality. Over these years we have experienced changes in political desires and societal direction. We have "explained" to many new elected members through ten changes of City Council, the whys and wherefores of Toronto's sewer replenishment programme. Our relationship with the public who have been "flooded out" has, at times, been volatile and stormy, especially at public meetings two or three days after heavy rain. Those who have mooted new methods of approaching problems and their theoretical solution have not always been kind to us; numerous studies published by those who are not in the municipal "front line", have led to queries about effectiveness of the Toronto plan as it related, for example, to the abatement of pollutional discharges. Despite all of the hurly-burly, what we have learned over the years boils down to something like this:

Have the temerity to start no matter how uncertain and simple the start may be. Bring together a small group of energetic colleagues and be single minded in direction and purpose. Endure the tedium and long hours of gathering, sorting and analyzing the field facts. Decide on an affordable level of service and be mindful that it is the taxpayer who foots the bill. Put the plan together, stage it sensibly and propose fair methods for funding. Broadcast the plan to the public and media and harness political support. Don't tell fairytales about the plan - tell the truth no matter how unpleasant. Be ready to accept criticism and keep abreast of changes in technology. Don't lose heart in the face of difficulties and above all, keep a sense of humour.

Acknowledgement

The authors wish to acknowledge the help and assistance of Mr. D. N. Young of the North West Water Authority.

International Conference on the
PLANNING, CONSTRUCTION, MAINTENANCE & OPERATION OF SEWERAGE SYSTEMS
12-14 September, 1984

PLANNING FOR SEWER SURVIVAL

Jan Adamsson
Master of Science in Civil Engineering
Gothenburg Water and Sewage Works

Bernt Persson
Master of Science in Civil Engineering
Gothenburg Water and Sewage Works

SUMMARY

A master plan for sewerage in Gothenburg has been developed. The Water and Sewage Works serves 420.000 inhabitants in a 140 km^2 area. The sewers have a total length of 2.000 km.

The first Gothenburg master plan for sewerage was adopted in 1866. Combined sewers were laid until 1957. Thereafter separate storm sewers have been laid in all new suburban areas and in clearance areas. This has brought about considerable areas with non-active separated sewers.

An important objective for the new master plan was to avoid harmfull effects to the receiving waters. A complete rebuilding of the combined system into a separated system was calculated to cost 4.200 million Swedish kronor (SEK).

An investigation of the need of renewal based on compilation of reported disturbances in operation, CCTV-inspections, flow control and detailed field investigations has been performed. The capacity of the existing network was analysed by computer modelling. A study of optimal design criteria for combined sewers has also been performed. The evaluation of alternative solutions for every catchment area set priorities for potential projects.

The results from the field investigations have been used as a base for forming a master plan for the future sewer system. The proposed plan recommends that most of the combined sewer system should be retained. Necessary investments for completing the plan are estimated to cost 400 million SEK.

Held at The University of Reading, England. Organised and sponsored by BHRA, The Fluid Engineering Centre, Cranfield, Bedford in conjunction with the Water Research Centre.

NOMENCLATURE

F_c	=	average damage cost for a basement flooding (Swedish kronor, SEK)
I_c	=	investment cost (SEK)
P_p	=	present value of investment cost plus the sum of difference in costs for basement floodings after and before measurements (SEK)
$S_{d/i}$	=	decrease/increase of storm water volume to the treatment plant per year (m^3)
W_o	=	reduction overflow volume, sanitary sewage, per year into the receiving water from overflow (SEK/m^3 year)
Vw	=	cost benefit ratio
k_p	=	factor of present value
n_1, n_2	=	number of basement floodings in average per year before resp. after measurement
u_c	=	unit cost for pumping and treatment of stormwater at the treatment plant (SEK)
ΔH	=	distance between level of basement floor and head of pipe in the street

1. INTRODUCTION

The Water and Sewage Works of Gothenburg serves 420.000 inhabitants in a 140 km^2 area.

The first master plan for sewerage was adopted in 1866. Already in these early days there was a discussion whether separated or combined system should be chosen. The effects of discharges on receiving waters were thoroughly discussed. The combined system came out as the most economic and reliable system. Combined sewers were laid until 1957. Thereafter separate storm sewers have been laid in all new suburban areas and in clearance areas in central parts of the city. About 30 % of the catchment areas are today served by combined sewers. The different sections are linked together by a rock tunnel system leading to a central treatment plant.

The policy of the National Environment Protection Board (EPB) during the golden days of the 60- and 70-ies was that all combined sewers should be abandoned. In 1976,the City of Gothenburg was imposed to replace the combined sewer system with a separated system and take others measures to reduce pollution in the receiving waters.

Later the EPB changed their policy and declared in 1978 that the impending requirements for separation of combined sewers could be revised provided the local authority presented a rehabilitation plan.

2. SEWER SYSTEM 1980

In 1979 a research project started in cooperation with Chalmers University of Technology whose objective was to model the flow to the central treatment plant (Ref. 1)

The Water and Sewage Works started the work with the master plan at the same time. In the first phase the study included the following steps:

* Inventory and description of the tunnel and sewer trunk system and the catchment areas

- sanitary sewers 770 km
- combined sewers 530 km
- stormwater sewers 770 km
- sewer rock tunnels 74 km
- stormwater rock tunnels 12 km
- overflows 120
- retention basins 5 (11.000 m^3)
- retention volume in the tunnel system 250.000 m^3

* Flow analysis with the NIVA-model developed by the Norwegian Institute of Water Research

* Flow measurements in the sewer network. The equipment consist of 6 portable Manning ultrasonic recorders. Calibrations, shown in Fig 1, are made with lithium chloride.

* Evaluation of operation-disturbances

The results from the inventory is documented in The Sewarage Plan 1980. Every catchment area is described with regard to

- catchment area - size, calculated flows and results from flow measurements
- capacity, profile and critical basement levels for the trunk sewers
- deficiencies in the sewer network
- alternative solutions to solve the problems
- planned measurements

It is intended to keep the plan updated with regard to operation-disturbances and performance measurements. The function of the sewer system is shown in Fig 2.

A separate Stormwater Plan 1980 describes the stormwater flow and the overflows.

3. NEED OF RENEWAL

The main part of the sewer system consists of gravity pipes. Transportation of sanitary water without disturbances demands a minimum slope to be selfcleaning. Ground subsidences affect the slope and tightness of the pipes. The subsidences can even result in negative slope and give rise to serious disturbances, for instance:

- ground and basement floodings
- stench and gas explosion
- infiltration/loss of capacity
- overflow for dry weather conditions

The Swedish Water and Wastewater Works Association (VAV) have collected information on the interruption of service in sewers and water mains from 27 locations in Sweden (Ref 2). Some of the results are shown in Fig 3.

In Gothenburg the water distribution system and the sewerage system were briefly reviewed with the objective of finding out the lowest acceptable level of renewal for the next 10 years. The investigations were concentrated on areas where disturbances in service were frequent and in areas with a subsidence over 5 mm/year and with pipes more than 60 years old. The results are shown in Fig 4 and 5. The frequency of interruptions in sewers is shown in Fig 6.

Fig 7 shows that the real renewal problem in Gothenburg would arise at about year 2000 if the age of the pipes is taken as a parameter. If a lifetime of 100 years is possible then the renewal peak will occur about the year 2040. Accurately predicting the serviceable life of a sewer is at best difficult. One point is clear, new sewers should be designed for a lifetime greater than 100 years.

Since the start of the "Sewer System 1980" project methodology for investigations has been successively improved. The reasons behind disturbance occurences are compiled for every case and neccessary counter measures are proposed. During 1982 investigations and solutions were made for 15 areas. 35 km of sewer pipes were inspected by CCTV.

The experiences from this way of tackling the problem - "getting to know your sewers" - are positive. The question about what will happen in the future if you only deal with acute problems - a successively increasing average life time - can never be answered but the prediction of service life will be better if it is based on a continous compilation of disturbance occurences.

4. ANALYSIS AND OPTIMAL DESIGN OF COMBINED SEWER SYSTEMS

The combined system was designed for a return period of 2 years. The Swedish National Water Supply and Sewage Tribunal have raised the demands upon service level even for the existing system. In practice this means that the Water and Sewage Works has to prove that the system can handle a storm with a return period of 5 years without surcharging otherwise they are responsible for damage caused by flooding.

There was a very intense storm in Gothenburg on July 13, 1983 which caused a lot of basement floodings. The intensity of the storm was equal to a return period of 10-100 years for different duration times and different areas. Analysis of the combined system in the worst affected areas showed that most of the trunk sewers within areas with flooding had not the capacity for storms with a return period of 5 years. The cost for flooding damages for 80 properties in areas were the pipes did not have sufficient capacity amounted to 1.5 million SEK. Distribution of costs for flooding damages is shown in Fig 8.

The Water and Sewage Works in Oslo, Norway and in Stockholm and Gothenburg, Sweden have performed a study of optimal design of combined sewer systems in a joint project between the cities (Ref 3). Some of the results from the study are illustrated in Fig 9 to 11. The study recommends, for areas where combined sewer systems are to be retained, that the sewers should be analysed for a model storm with duration $\geq$ 5 min and a return period of 10 years. If basements are flooded alternative measures that reduce the risk of exceeding the critical pressure line include: percolation of stormwater; storage; pumping of individual properties; increasing capacity at critical sections; overflow; connection to separate system; etc. Design of extension or renewal projects should have a design return period of 10 years without surcharging. The difference between lowest basement floor and the inner head of the main sewer should be at least 0,5 m if gravity connection shall be allowed.

5. PROPOSED FUTURE SEWER SYSTEM

The results from the field investigations have been used as a base for forming a plan for the future sewer system. Every catchment area has been studied with regard to conditions in the receiving waters, existing sewer system network and capacity, types and frequency of malfunctions, need of renewal, planned slum clearance, street renewal and development of residential and industrial areas.

The existing sewer system represents an immense investment. Most of the old sewers fullfill their service requirements. Besides the environmental aspects, a rehabilitation plan must also take into consideration the acceptable level of service for the users and the minimization of costs for operation and maintenance.

The impact on the receiving waters from stormwater and overflows is shown in Table 1. In Table 2 the specific values used are listed. The volumes are taken from the graphs in Fig 12.

Table 1 Annual total pollutant load to the receiving waters from overflows and separate storm sewers (kg/ha impermeable area)

	Urban storm water from separate storm sewers	Overflows in combined system Overflow setting 2 times DWF	4 times DWF	9 times DWF
BOD_7	100	100	50	15
P_{tot}	2,0	3,0	1,5	0,5
SS	1300	700	350	75
Pb	1,3	0,6	0,3	0,06

Table 2 Values for pollutant concentrations (g/m^3)

	Sanitary sewage	Urban storm water Residential area	Industrial areas	Traffic areas	Down town areas
BOD_7	200	10	15	20	15
P_{tot}	7	0,2	0,7	0,4	0,4
SS	300	90	300	500	200
Pb	0,02	0,1	0,2	0,4	0,3

With respect to the pollutant load from overflows in the combined systems, from stormwater and from the treatment plant a rebuilding of the combined system would theoretically reduce the load to half the total value. In practice the reduction would be less. A separate system with 1 % cross connections should imply the same pollutant load as a combined system with an overflow setting of 4 times DWF. The total load on the receiving water does not change much if the overflow setting is varied. Locally the overflow setting is however of great importance.

The master plan is based on the following:

* Combined system in the central parts of the city should be retained if overflows can be discharged to a suitable recipient and/or flow retention can be used

* In areas where renewal of the combined system is necessary an evaluation will be made taking the risk for basement flooding into consideration

* In areas with new housing projects stormwater should be disposed locally

* In areas where the choice of systems is not obvious the need for renewal and the conditions for separation of house connections should be investigated

The proposed system plan is shown in Fig. 2. The following factors have been of importance for the choice of system:

- the regional rock tunnel system, shown in Fig. 13.
- location of possible connection points to discharge sanitary sewage without overflow
- available capacity of trunk sewers that feed the new tunnel system
- retention volume in the tunnel system
- retaining combined system by adding detention basins
- linking overflows and shifting their location to more suitable receiving waters.

In discussing which areas a combined system should be retained or separated, alternative solutions have been analysed and given a priority within each receiving water with the following simple cost/benefit equation.

$$V_w = \frac{I_c \pm u_c \cdot S_d/i \cdot k_p}{W_o}$$

Overflow discharge and stormwater flow to the treatment plant is taken from Fig. 12 (derived from Ref. 4).

For evaluation of alternative measures in order to increase the return period for basement floodings the following equation has been used:

$$P_p = I_c - (n_1 - n_2) \cdot F_c \cdot k_p$$

The discharge of sanitary sewage from overflows will be transferred from weak to strong receiving waters and the volume reduced from 500.000 to 340.000 m^3/year. The stormwater flow to the treatment plant will decrease by 2,2 million m^3/year or 15 %.

A complete rebuilding of the combined system was estimated to cost 4.200 million SEK. The cost of the proposed plan is estimated to be 400 million SEK with the following main expenditures:

Environmental aspects	260	million SEK
Service level-basement flooding problems	40 "	"
Reinvestments in		
combined system	50 "	"
separated system	50 "	"

The proposed plan premises a continous follow-up of troubles in the sewer network. The plan will be revised yearly and function as the base for investment planning.

6. REFERENCES

1. Lyngfelts S. and Svensson G.:
"Urban stormwater runoff from large areas" (Dagvattenavrinning från stora urbana områden. Simuleringsteknik exemplifierat för Göteborgsregionen). The Urban Geohydrology Research Group of Chalmers University of Technology, 1983 (In Swedish). Report No. 68.

2. "Operational malfunctions in water and sewerage systems (Driftstörningar på ledningsnät). The Swedish Water and Waste Water Works Association, 1981 (In Swedish).

3. Adamsson J, Bergström T, Ljunggren O, Persson B and Saltvelt N.:

"Analysis and optimal design of combined sewer systems". The Water and Waste Water Works in Oslo, Norway, Stockholm and Gothenburg, Sweden, 1984 (internal report, unpublished).

4. Arnell V.:

"Rainfall data from Lundby in Gothenburg 1921-1939" (Nederbördens varaktighet och mängd vid Lundby i Göteborg 1921-1939). The Urban Geohydrology Research Group of Chalmers University of Technology, 1979 (In Swedish). Report No 44.

Portable V-notch weir and Palmer-Bowls flume for pipes with diameter ≤ 400 mm

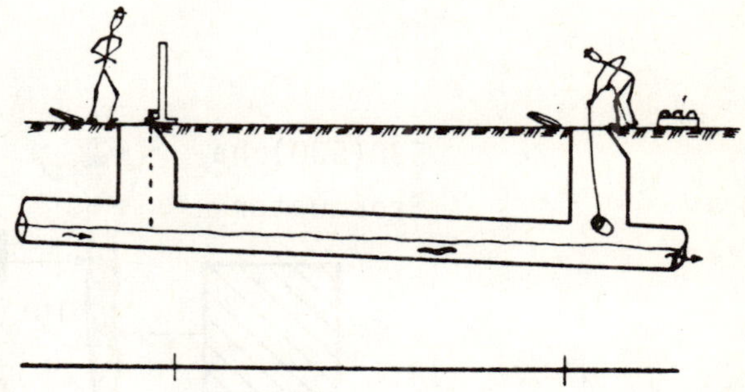

Flow measurement in pipes with diameter > 400 mm is based on depth of water and slope. Calibration with lithium chloride is necessary

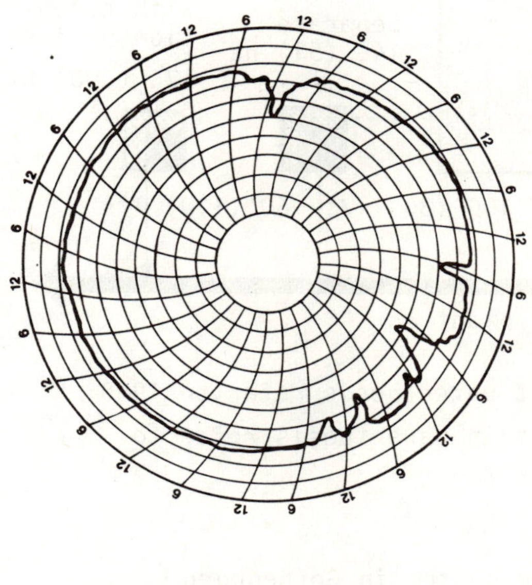

Registrations of water level during one week with a Manning Ultrasonic recorder

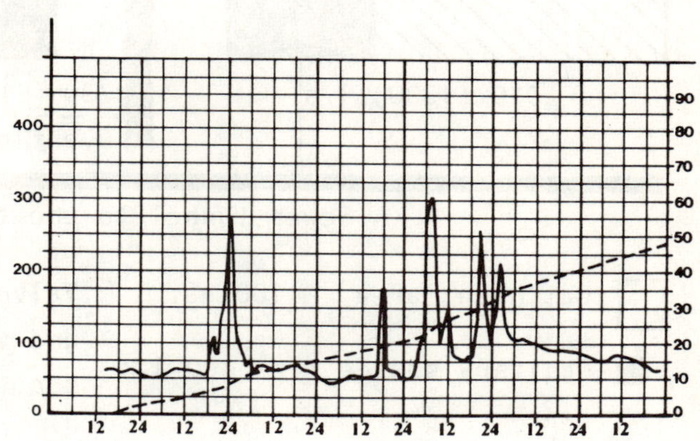

The graph is evaluated with a micro computer

FIG 1 Flow measurements in the sewer network

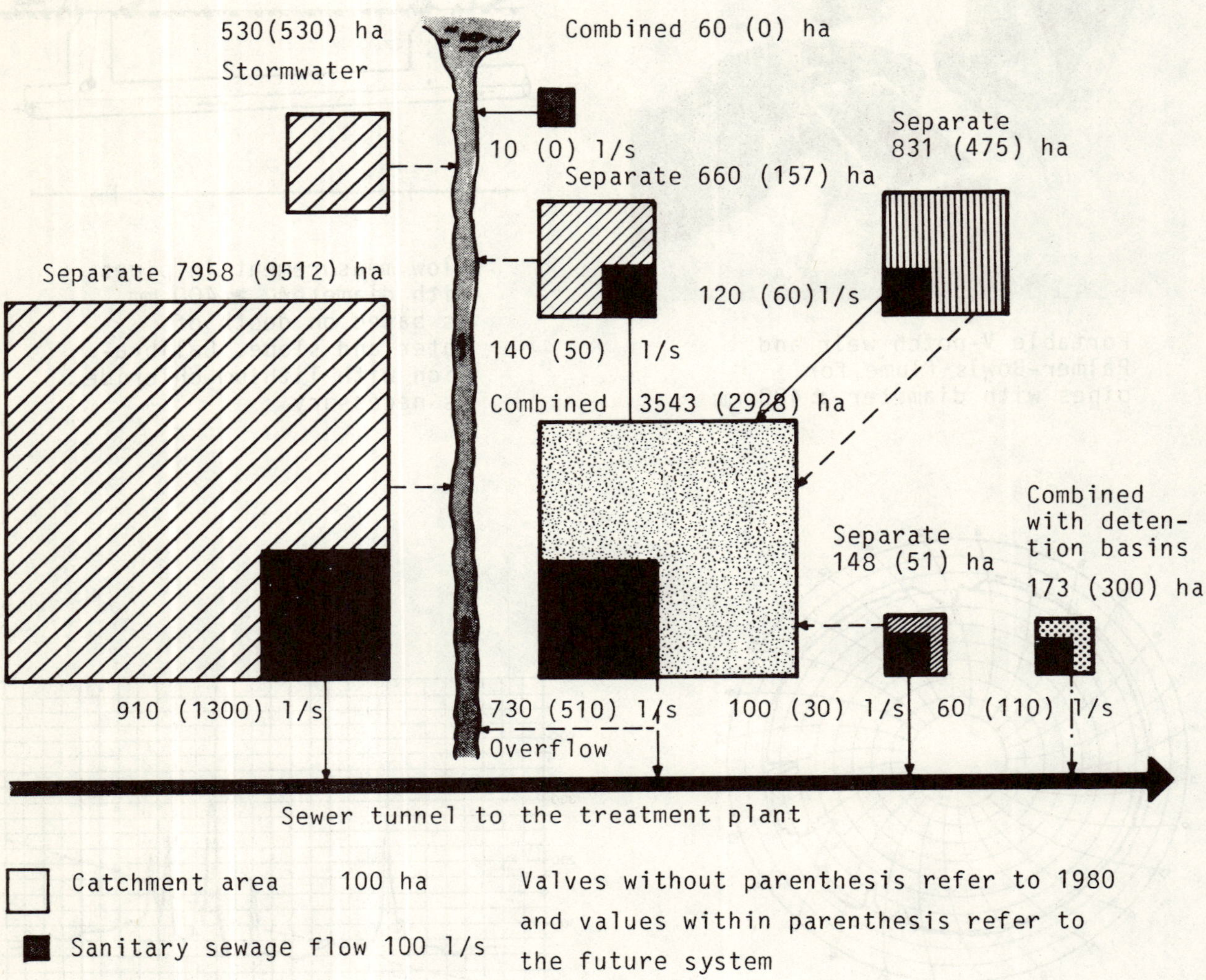

Fig 2 Present and future function of the sewer system in Gothenburg

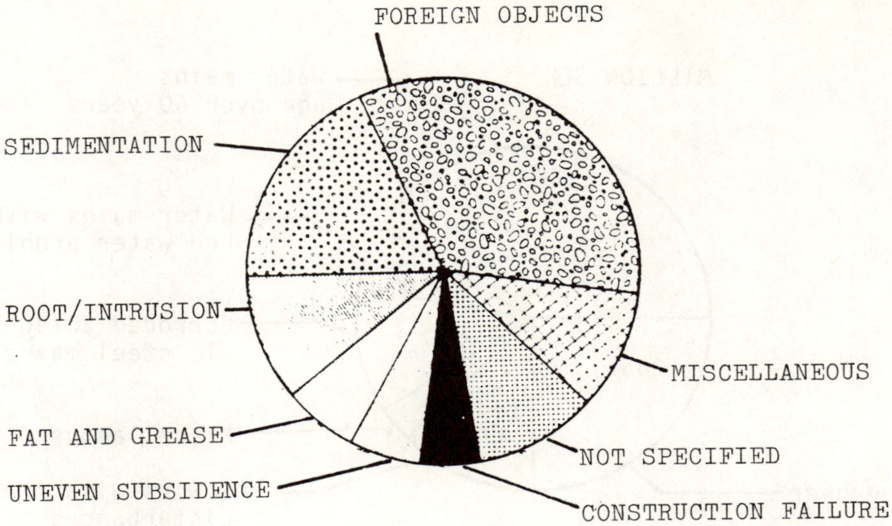

FIG 3 Causes of obstructions in sewers - 425 interuptions in service

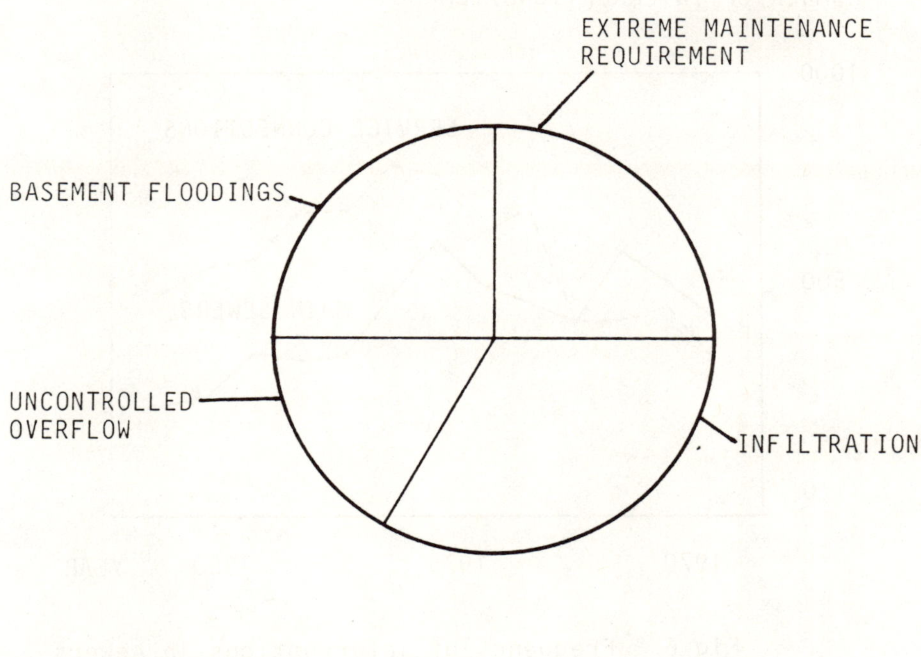

Fig 4 Type of disturbances in problem areas in Gothenburg sewerage system

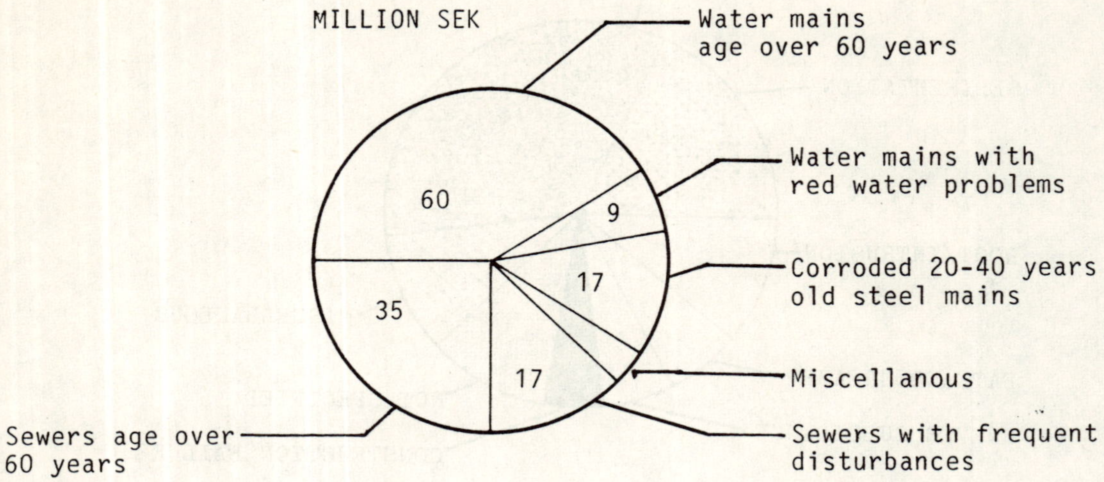

Fig 5 Least acceptable level for renewal of water mains and sewers in Gothenburg for the next 10 years

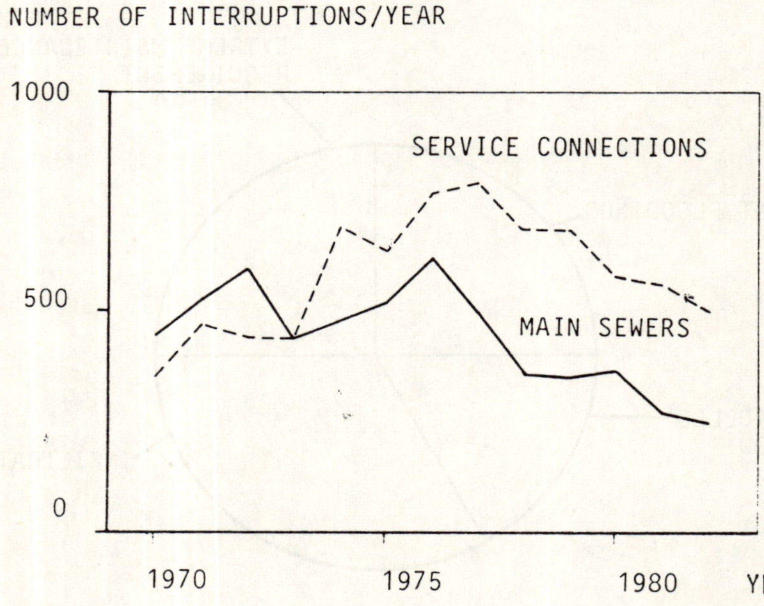

Fig 6 Frequency of interruptions in sewers in Gothenburg

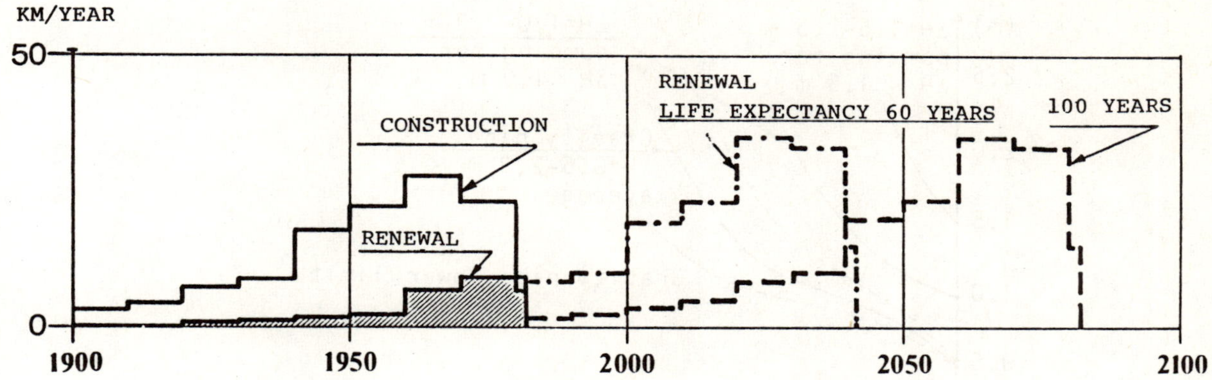

Fig 7. When will the renewal peak occur ?

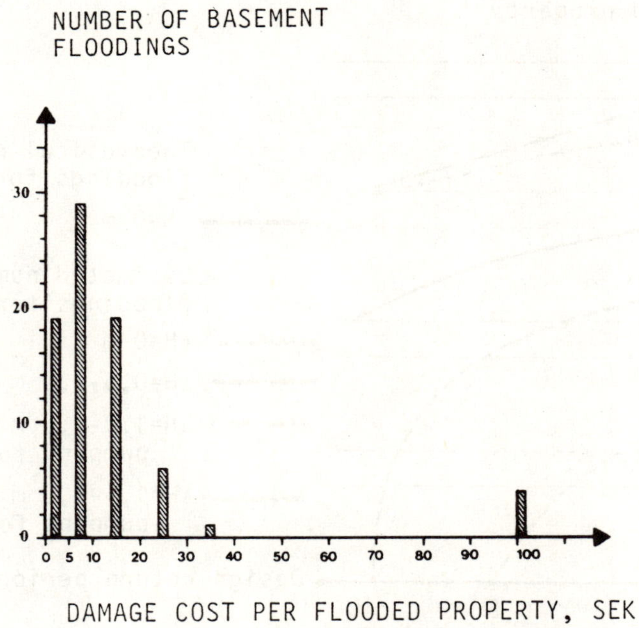

FIG 8 Distribution of costs for flooding damages caused by the intense storm in Gothenburg July 13, 1983

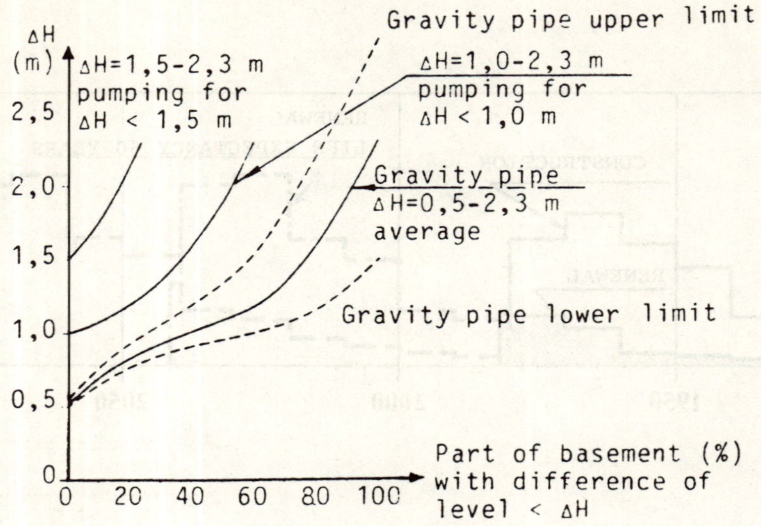

Figure 9 Distribution of the distance between level of basement floor and head of pipe in street according to inventory (ΔH).

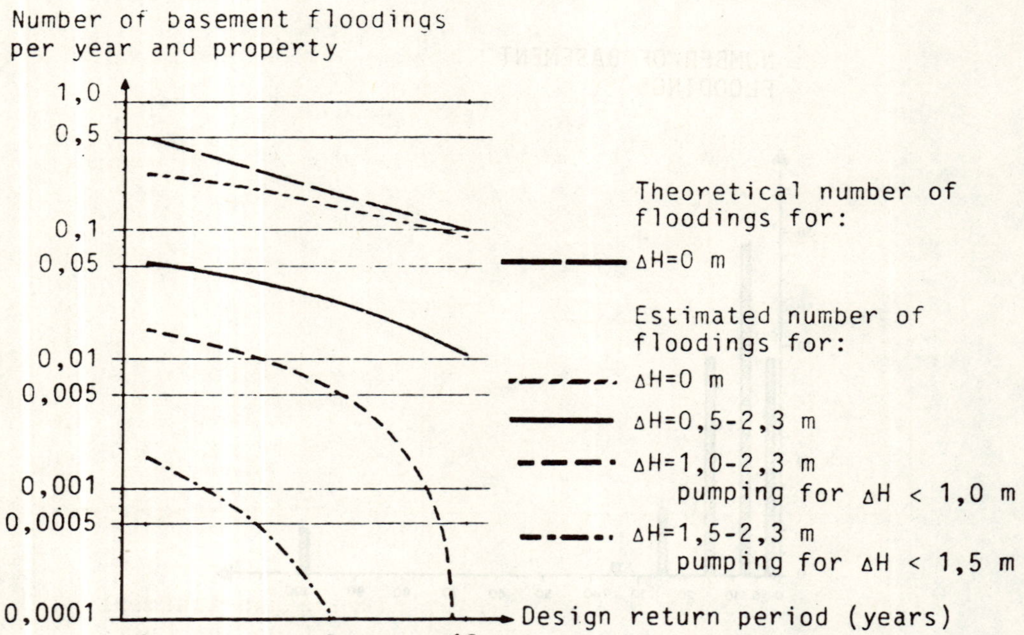

Figure 10 Return design period and number of basement floodings per property for different values of ΔH.
Catchment area of 0,1 km^2. 77 properties
Total length of pipes:1800m.

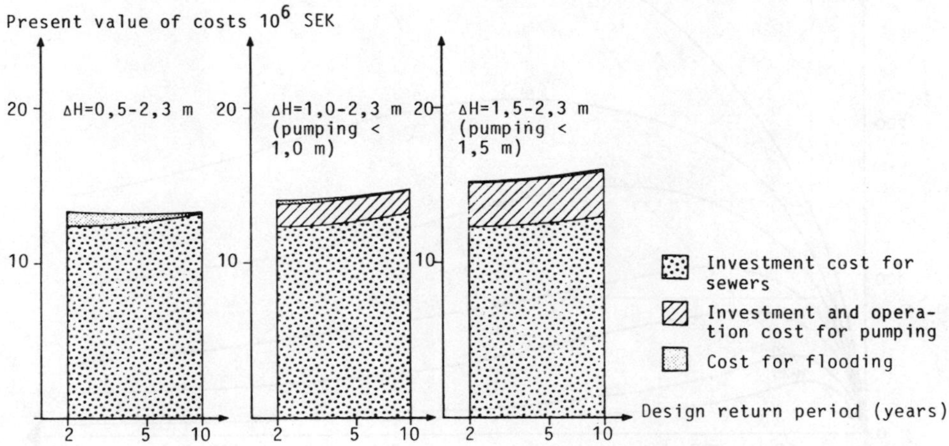

Investment, pumping and basement flooding costs. Damage cost 10.000 SEK per flooded property.

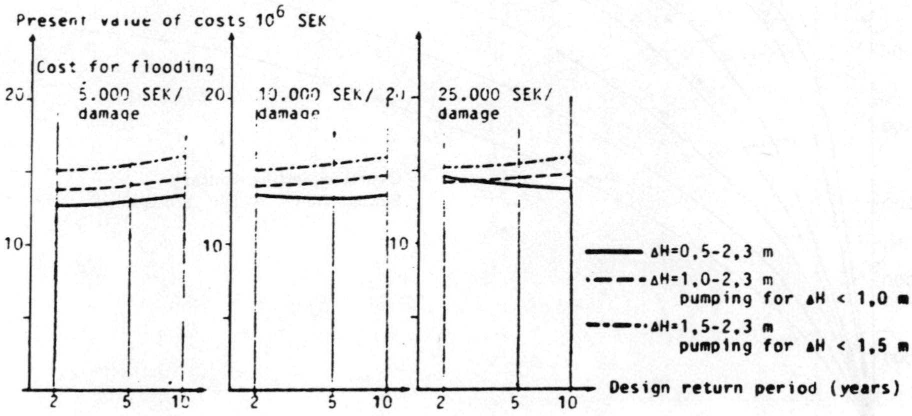

Figure 11 The sum of investment, pumping and basement flooding costs for different flooding cost alternatives.

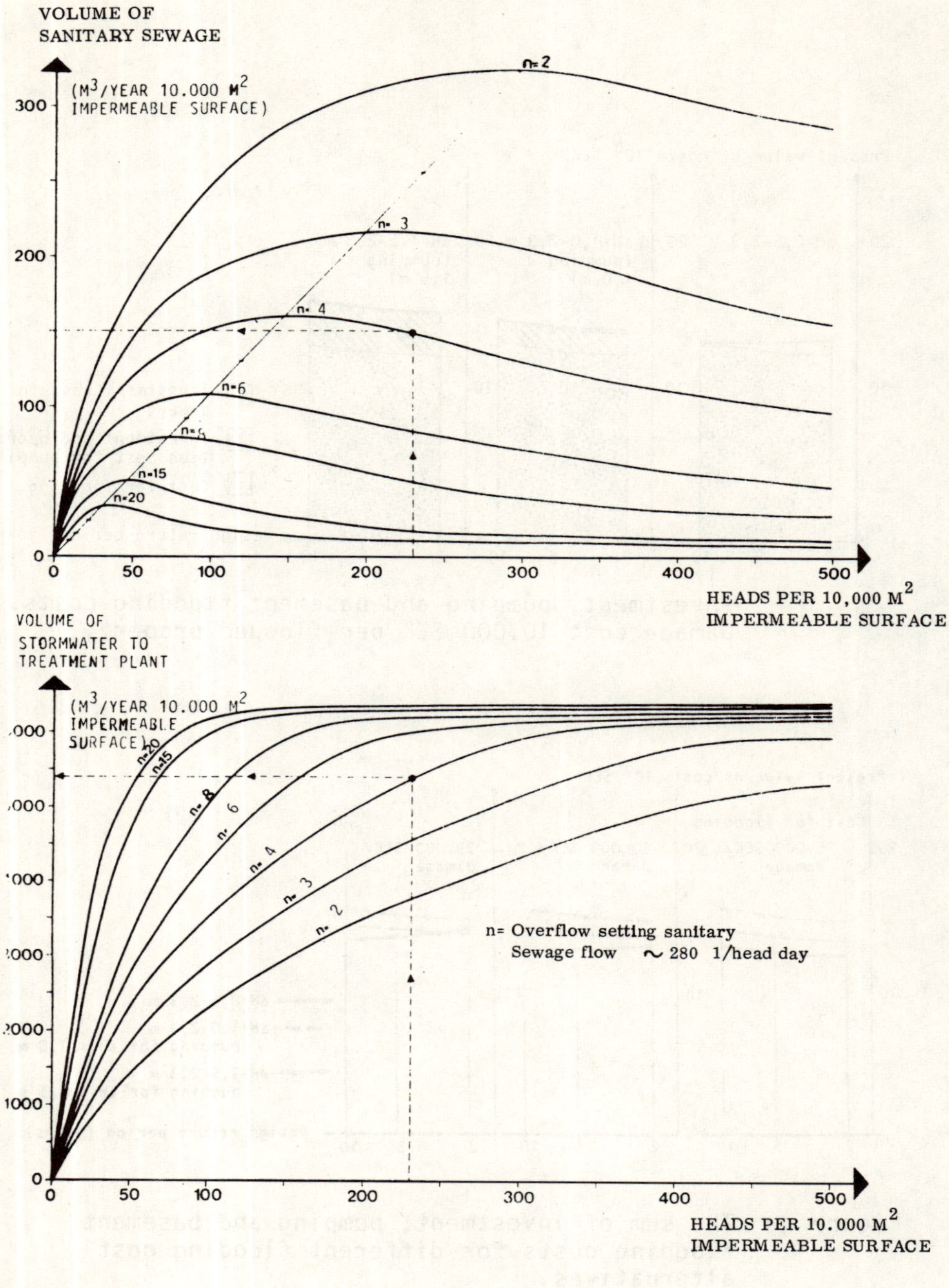

FIG 12 Graphs for estimation of sanitary sewage and stormwater volumes from overflow devices in combined system based on 18-years storm records in Gothenburg

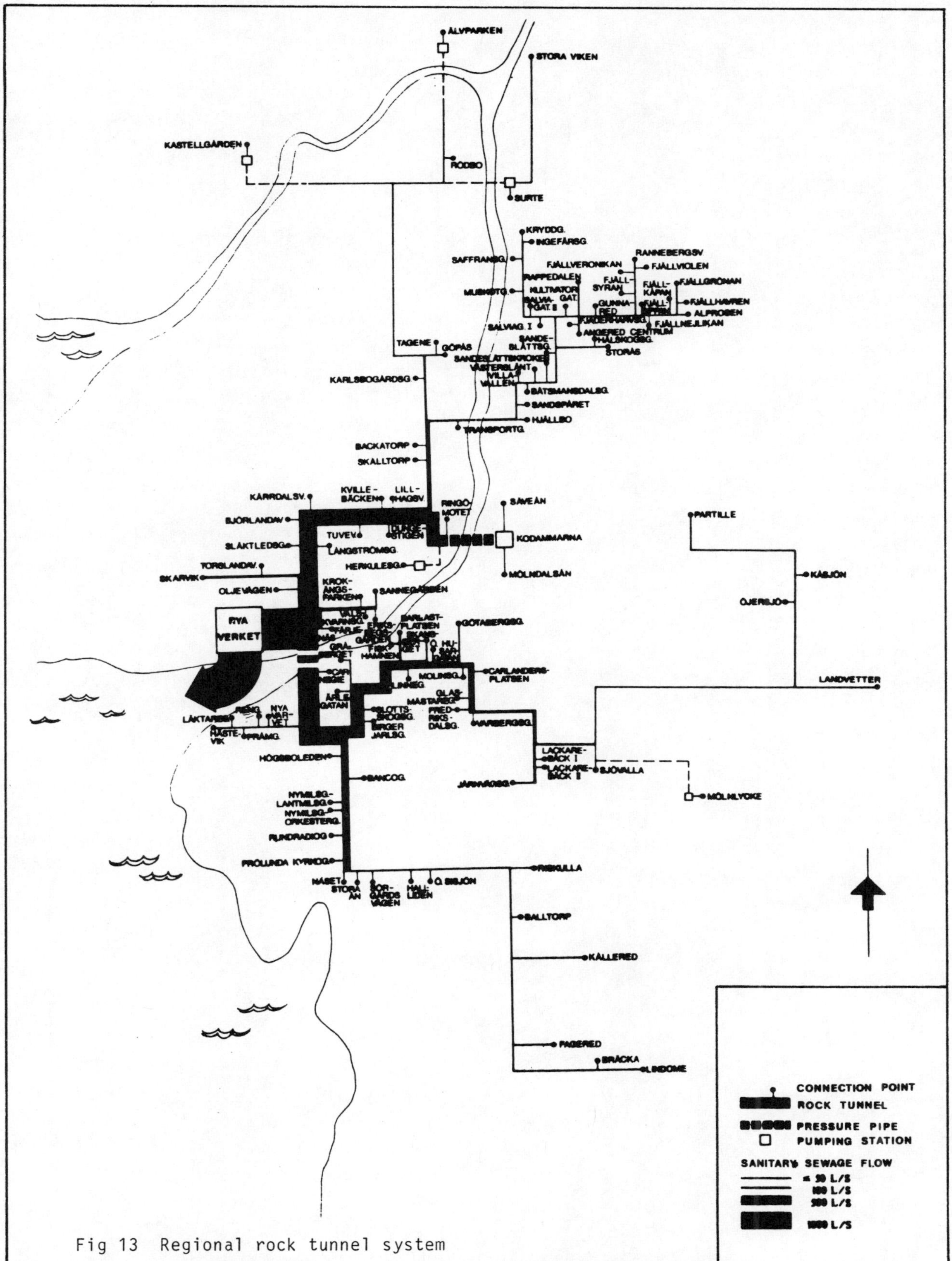

Fig 13 Regional rock tunnel system

International Conference on the
PLANNING, CONSTRUCTION,
MAINTENANCE & OPERATION
OF SEWERAGE SYSTEMS
12-14 September, 1984

PAPER B1

A SEWERAGE REHABILITATION STRATEGY FOR THE UNITED KINGDOM

D FIDDES BSc, MSc, C.Eng, MICE, FIPHE, DIC

WRC ENGINEERING

PO BOX 85

FRANKLAND ROAD

BLAGROVE

SWINDON

WILTSHIRE

SN5 8YR

The paper describes the recent development of a strategy for sewerage rehabilitation which takes into account the problems of dereliction, hydraulic overload and urban runoff pollution. The approach put forward is a selective policy of reactive/planned upgrading. Survey, appraisal and pre-emptive rehabilitation is concentrated on "critical sewers" which make up about 20% of the total network, with a reactive response to failure policy for the remainder.

The procedure for implementing the strategy contains four phases.

i) Identification of the critical sewers.

ii) Assessment of structural and hydraulic performance.

iii) Preparation of integrated upgrading solutions (The Drainage Area Plan).

iv) Undertaking the capital works and updating the Plan.

The benefits and implications of introducing the strategy are reviewed and the programme of implementation described.

Held at The University of Reading, England. Organised and sponsored by BHRA, The Fluid Engineering Centre, Cranfield, Bedford in conjunction with the Water Research Centre.

1. INTRODUCTION

Over the last decade there has been a major change in emphasis in the sewerage capital programme from a concentration on building new sewers to performance upgrading within the existing system. This coincides with a downturn in the new house building programme and an increasing awareness, following the 1974 Water Industry reorganisation, of the problems besetting our ageing drainage networks. These problems tend to be of three main kinds:

Collapses - brought about by fabric deterioration or interference.

Flooding - public health hazards resulting from hydraulic overloading.

Pollution - generally from the inefficient operation of storm overflows.

New technology has become available and a strategy developed which allows the confident and comprehensive dealing with these aspects of performance failure. The approach proposed is contained in the WAA/WRC "Sewerage Rehabilitation Manual" which was published earlier this year. It brings together the available expertise, understanding and technology, developed by the Industry and sets out systematic procedures for applying these to achieve:

- significant cost savings in the necessary rehabilitation works,
- a means of controlling potential cost escalations as the network gets older and
- the quantifying and justifying of the required funding levels for future upgrading programmes.

The Manual is published in loose leaf form to allow updating to incorporate technological developments in hand. It is made up of three volumes.

Volume 1 outlines the recommended procedures and implementation issues.

Volume 2 sets out how to choose the parts of the network appropriate for investigation, describes the planning procedures and advises on how to consider hydraulic and structural upgrading.

Volume 3 enables engineers to decide whether the renovation option is appropriate for structural upgrading, and if so how to prepare the detailed design.

The Manual is aimed principally at meeting the sewerage rehabilitation needs of the UK Water Industry. This country has one of the oldest and most extensive sewer systems in the world, but there are many comparable systems elsewhere some of which were built by our Victorian forefathers and so the methodology developed will also have overseas application.

2. BASIC PHILOSOPHY

The Industry wishes to maintain its high standards of service, but to do so in a period of severe financial stringency it must improve its efficiency. Areas of high current or future spending must be identified, and activities in them optimised.

Applying such arguments to sewerage the following emerge:

Britain spends about £70m per annum dealing with structural deterioration and sewer collapses, and about £120m on overcoming hydraulic overloading. Although problems are relatively well spread over the network the expensive and disruptive incidents in both areas are concentrated in the core of the system and frequently occur together. It is only in these areas (termed critical sewers) that a pre-emptive rehabilitation policy can be justified. The critical sewers make up a relatively small and readily definable part of the network, on average less than one fifth of the total. Elsewhere it can be shown that a reactive approach of dealing with problems after they come to light is the most cost-effective.

Traditionally the solutions for hydraulic and structural problems have been replacement and renewal respectively. These solutions are both expensive and, due to the location of the critical sewers, likely to be very disruptive to the population at large. With the advent of improved hydraulic analysis tools and established renovation methods, alternatives are now available, which are not only significantly cheaper but retain the bulk of the industry's major asset "the hole in the ground". However, to achieve the savings, increased effort must be made in understanding the true existing system performance.

The rationalisation of storm overflows to alleviate pollution of receiving streams puts different and generally increased hydraulic loads onto the sewer network. If this is not to require major sewer replacement, again the performance of the system must be fully understood and optimised.

All aspects of performance tend to be worse in the older parts of the network. Unless adequate levels of performance are achieved in the near future, cost escalation will result or standards of service will deteriorate.

Summarised, the basic philosophy is to provide an acceptable service at least cost and using resources efficiently by:

1. Concentrating survey analysis and pre-emptive rehabilitation on the critical sewers.

2. Considering all aspects of performance concurrently.

3. Retaining as much as practicable of the current "assets".

4. Looking for solutions which can be phased.

3. METHODS OF ACHIEVING COST SAVINGS

It is claimed above that cost savings will result from the simultaneous upgrading of the three aspects of performance and the optimal use of recently developed technology. This is best illustrated with an example of the kind of outline feasibility study now possible.

Consider the sewer network shown in Figure 1. This is the upper section of a much larger system. The diagram shows the hydraulic performance under a typical design storm; a 2 year 60 minute event. The network suffers from recurring flooding problems, many sewers are in a serious structural condition made worse by regular surcharging, and there is a pollution problem due to excessive storm overflow discharges to the river.

The defective and overloaded lengths could be replaced by new sewers of adequate capacity but the cost to the authority would be high and the disruption to the community caused by the works would be considerable. The increased flows would overload downstream sections of the system requiring either continuation of the trunk sewer replacement programme to the outfall, or the building of a new storm overflow at the downstream part of the current works until such time as capacity downstream can be augmented.

An alternative solution could eliminate the flooding and structural problems by a combination of storm overflow rationalisation and renovation as shown in Figure 2. The lower hydraulic roughness following renovation of the defective brick sewers is a major help in improving the hydraulic performance. Note that only one short undersized pipe length which was severely throttling flows, has needed to be replaced. Although the total spill from this "design" storm is only a little reduced, the improved overflow settings do significantly reduced the premature spill from more frequent smaller storms. The pollution can be largely eliminated by incorporating some additional storage into the new overflows as shown in Figure 3.

A comparison of the scheme costs and their hydraulic performance during the assumed two year sixty minute storm are shown in Table 1. Obviously much more detailed analyses, involving a wide range of storms, would be required before the final design could be prepared, but Table 1 illustrates the ease with which savings can be demonstrated and their scale. It demonstrates that all aspects of performance can be concurrently improved without the necessity of abandoning valuable assets.

The financial savings in Table 1 were possible due to the much improved sewer inspection and stability assessment techniques and the availability of reliable design methods for renovation. Renovation can now be confidently used as a permanent structural upgrading option in a wide range of situations. The speed and ease of installation also largely eliminates the very considerable social costs associated with open cut sewer construction.

4. SUMMARISED REHABILITATION PROCEDURE

The investigations involve four distinct phases as will be seen from the flow diagram in Figure 4. The first phase will generally be completed simultaneously on all networks within a region and is largely concerned with identifying the priority for undertaking each individual study, based on the relative lengths of critical sewer and perceived level of performance failure.

In the second phase the problem critical sewer lengths are pinpointed. The structurally defective sewers are identified with the aid of in-sewer colour photography associated, in the smaller sewers, with close circuit television and manual inspection in the larger sewers. The defects found together with associated factors, are considered to give each sewer length a condition rating on a five point scale as shown in Figure 5.

To assess the hydraulic performance a computer model of the network is required. The WASSP-SIM model from the Wallingford Procedure (Ref 1) is recommended in the Manual for this purpose. Data errors can easily be inadvertently incorporated, so it is essential to verify the model before it is used for the detailed design of, possibly, multi-million pounds schemes. The use of short period in-situ flow surveys has been found to be an economic and reliable method of model verification. Relatively common storms are adequate to show up crucial factors such as incorrectly identified contributing areas (Figure 6) or overflows not shown on the record drawings (Figure 7). Other frequently found anomalies are throttles due to partial collapses or smaller diameter pipes introduced during repairs and backing up from drowned outlets.

Having established a trustworthy model it can be used to assess the current level of hydraulic performance by simulating the system response to a range of storms. An example is shown in Figure 8. Each pipe is assessed against two aspects of performance, frequency of surcharge and frequency of flooding. These frequencies can then be compared to the authority's required performance criteria to identify the lengths providing an inadequate level of service. It will be noted on the figure that a 1 or 2 year protection against surcharge does not necessarily give, as frequently assumed, a 20 - 30 year protection against flooding. This demonstrates the need for a simulation model which can accurately model surcharge. Where storm overflows are present the frequency, volume and dilution of spill are computed and again compared to the authority's performance levels.

In phase 3 the results of the two parts of the second phase are considered together and alternative, composite solutions worked up and the most cost effective chosen. Three solutions were possible for the example shown in Figures 5 & 8. These are shown in Figures 9 to 11. They range from the "traditional" solution relying heavily on interceptors to overcome hydraulic overload (Figure 11) to the almost exclusive use of tanks for flow attenuation (Figure 9). In this case the solution of tanks in the upper reaches of the catchment with a limited amount of reinforcement near the outfall gave the best compromise between engineering cost, achievement of target performance, and social costs. This solution was also very amenable to phasing, with each phase giving a substantial system performance improvement.

The fourth phase will see the works progressively built and keep under review the network performance and requirements in the light of any authority policy changes.

5. IMPLEMENTATION AND BENEFITS

Each Regional Water Authority is considering how it wishes to implement the Manual procedures and some have produced (Ref 2), or are currently preparing, strategy documents setting out how the programme will be organised, monies allocated and indicating the performance criteria to be worked to.

There are a number of aspects which are novel and still require refining. These include structural assessment, and the optimisation of hydraulic upgrading. To assist, pilot studies are being set up in each Water Authority area. These will have the advantages that:

a) The Authority can satisfy itself on how it wishes to apply the strategy generally.

b) Any residual bugs or required refinements to the procedure can be efficiently dealt with and resolved prior to the first major update of the Manual in 1986.

c) A body of local expertise is speedily generated.

The pilot studies will allow the financial benefits of using the Manual to be quantified. However, it is already possible to identify the likely scale of savings from case studies where many of the newer components of the procedures have been effectively applied. Three are quoted in the Manual and are listed in Table 2.

From such figures it is expected that overall savings from implementation, on individual schemes, should average 20% to 30% across a region. The benefits will therefore be to make scarce funds go considerably further and reduce the backlog of required rehabilitation works. It will also be possible for the first time to gain sufficient knowledge of the remaining rehabilitation needs to prepare and justify suitable future investment plans.

6. THE FUTURE

The ability to quantify and manipulate performance of sewer systems opens up many opportunities for the future. The advantages of passive forms of flow attenuation, such as tanks, have been demonstrated in the example above. Performance can be further improved by the real time operation of such storages using computer controlled penstocks in association with inflow forecasts from telemetering raingauges or weather radar. Several such schemes are currently being installed on major systems. One can foresee further into the future the use of both passive and active control being available on smaller systems also and the range of flow attenuation and reduction options widening, particularly those applied above the surface. As techniques such as on-line replacement and trenchless pipelaying become available the opportunities for improving system performance further will increase substantially.

It will be possible also to integrate and optimise the joint performance of the natural drainage and sewer (the major and minor) systems. Such investigations will increasingly cause engineers to reappraise conventional urban drainage philosophy. The procedures described in the Manual are adaptable and can be used to investigate and evaluate these possibilities also. Is it too much to claim that we are at the threshold of a revolution in sewerage practice as significant as that during the mid 19th century?

7. REFERENCES

1. Design and Analysis of Urban Storm Drainage - The Wallingford Procedure. NWC/DoE Standing Technical Report No 29. NWC, September 1981.

2. Severn Trent Water Authority. Sewerage - A Strategic Approach. August 1983.

TABLE 1 - EXAMPLE NETWORK - PERFORMANCE DURING 60 MINUTE 2 YEAR STORM

SCHEME	LENGTH SURCHARGED (m)	VOL. OF FLOODING (m^3)	SPILL TO RIVER (m^3)	COST (£k)
1. As existing	6920 total	183	2920	-
2. 'Traditional' solution Replacement of defective sewers + new overflow at d/s end	0	0 (if max continuation flow ≤ 6 dwf	3430	2880
3. Renovation of the defective sewers + overflow rationalisation + replacement of one pipe length	405	0	2665	1260
4. As 3. + storage at new/renovated overflows (Total storage ≃ 2600 m^3)	300	0	404	1490

TABLE 2 - SAVINGS IN MANUAL CASE STUDIES

	CASE STUDY	"TRADITIONAL" SOLUTION	FINAL SOLUTION
1.	York	£9.6m	£2.4 m
2.	Torquay	£6.5m	£2.5m
3.	Crawley	£70k	£45k

FIGURE 1.
PERFORMANCE OF EXISTING SYSTEM
2 year 60 minute storm

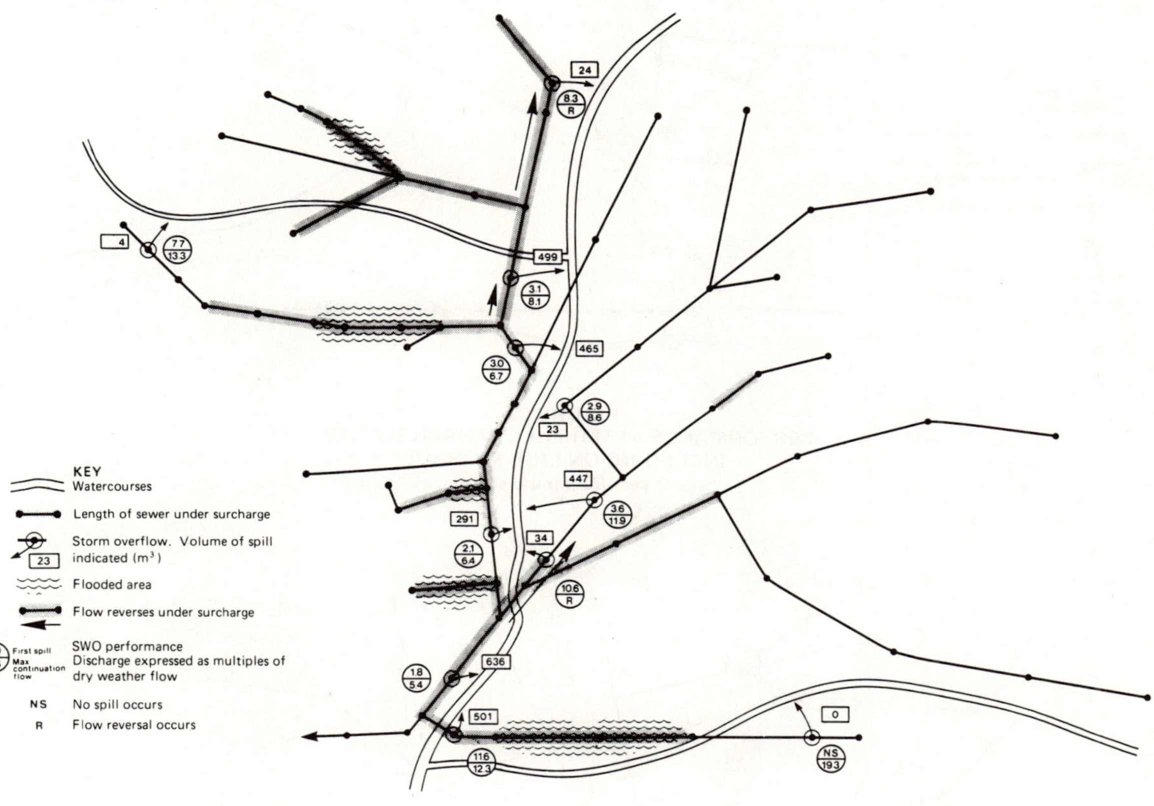

FIGURE 2.
PERFORMANCE AFTER OVERFLOW RATIONALISATION AND RENOVATION OF EXISTING SEWERS
2 year 60 minute storm

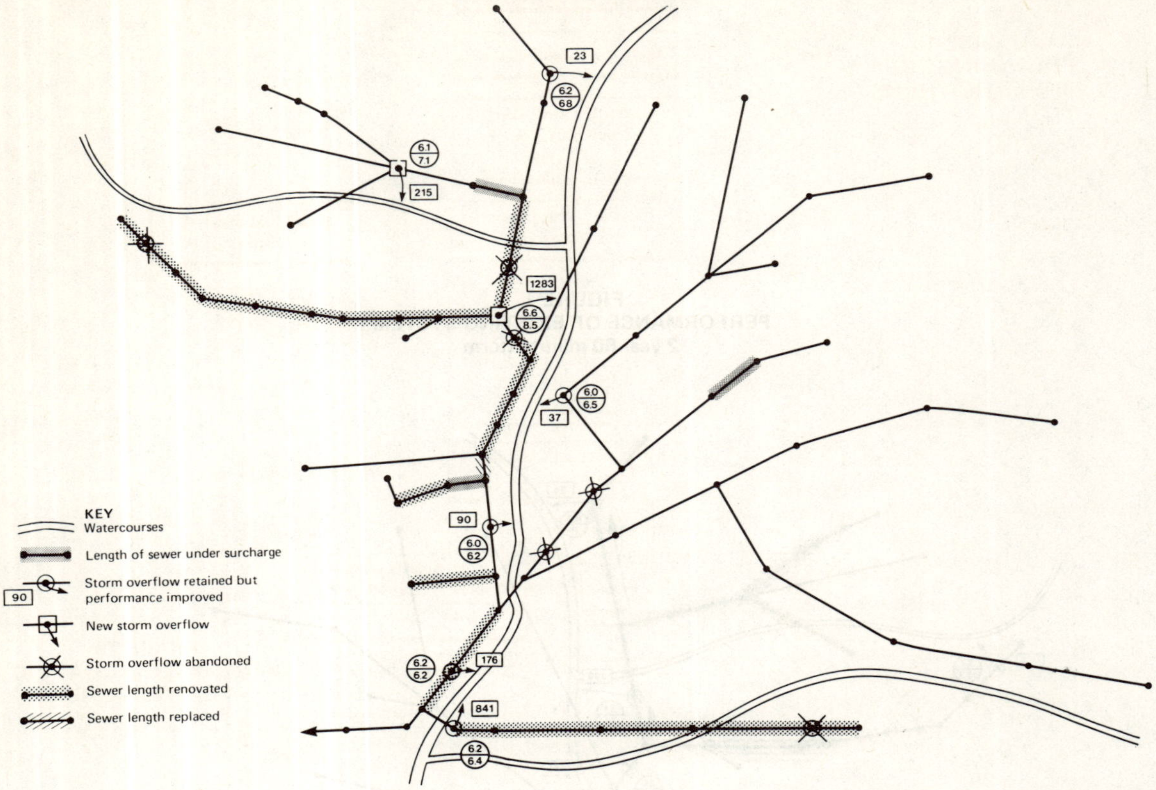

FIGURE 3.
PERFORMANCE AFTER FULL REHABILITATION INCLUDING ON-LINE STORAGE
2 year 60 minute storm

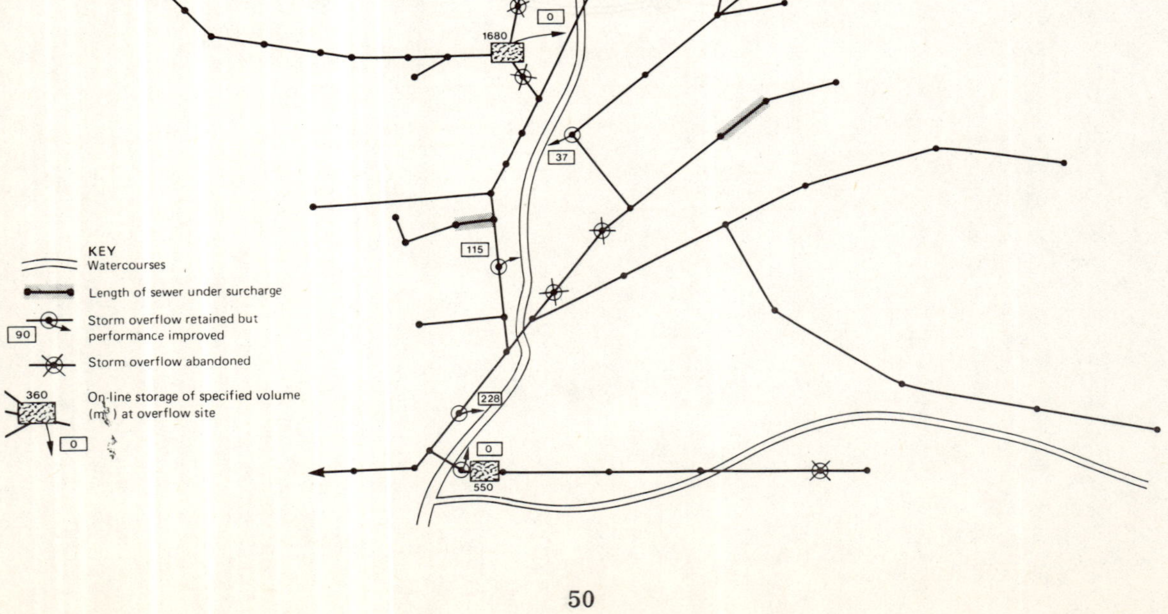

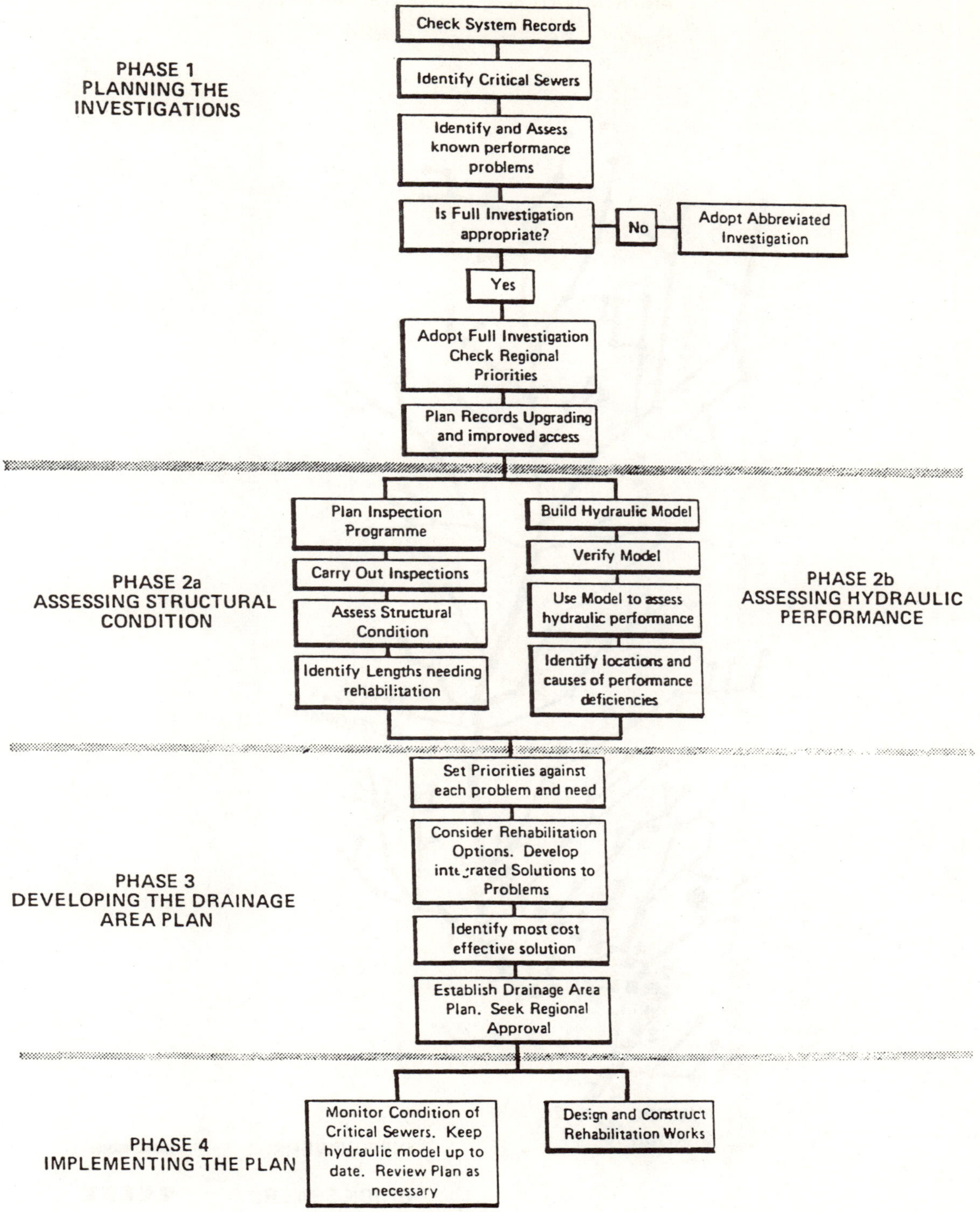

Fig. 4 Full investigation flow chart

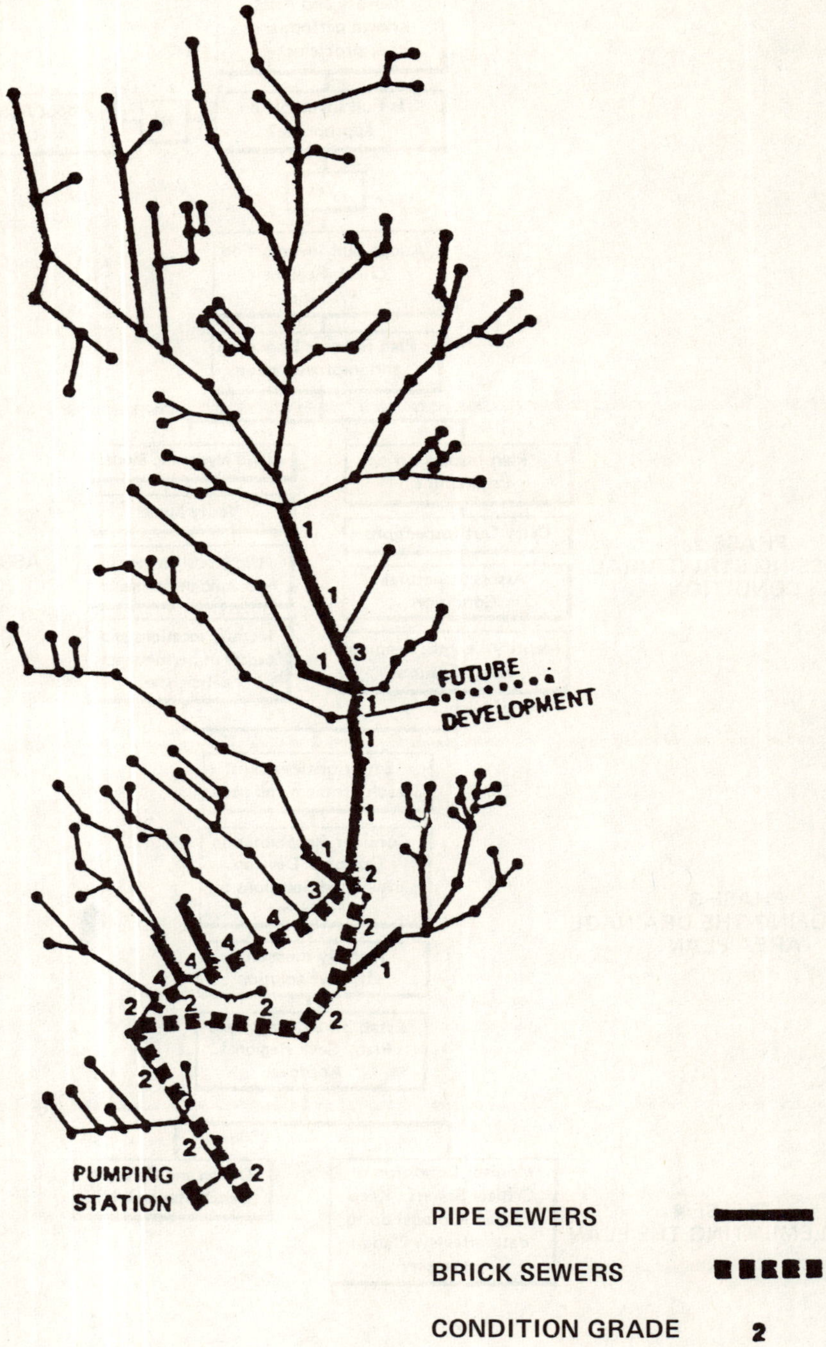

Fig. 5 Sewer Condition Assessments

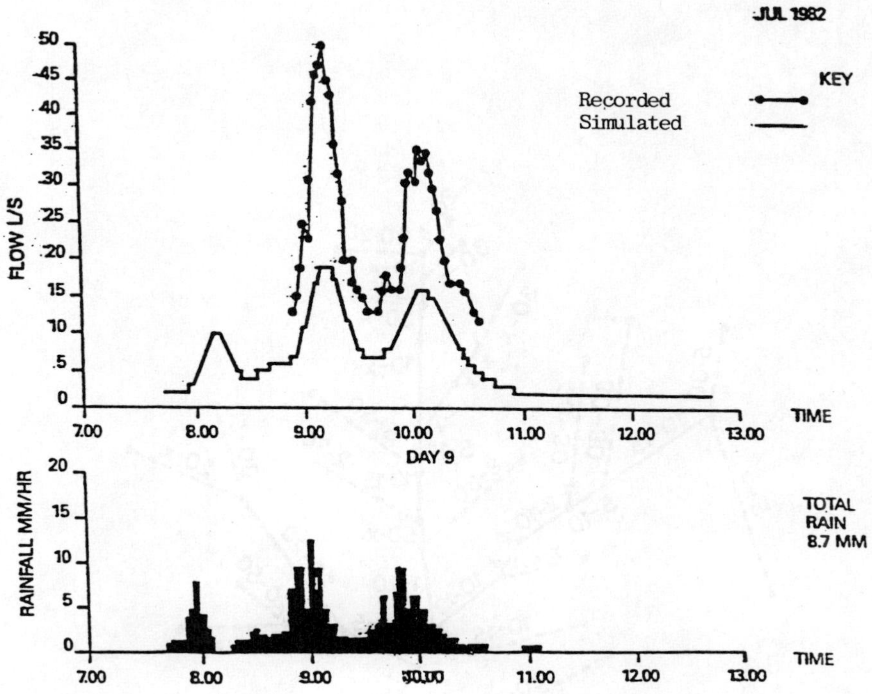

Fig. 6 - Poorly Defined Contributing Areas

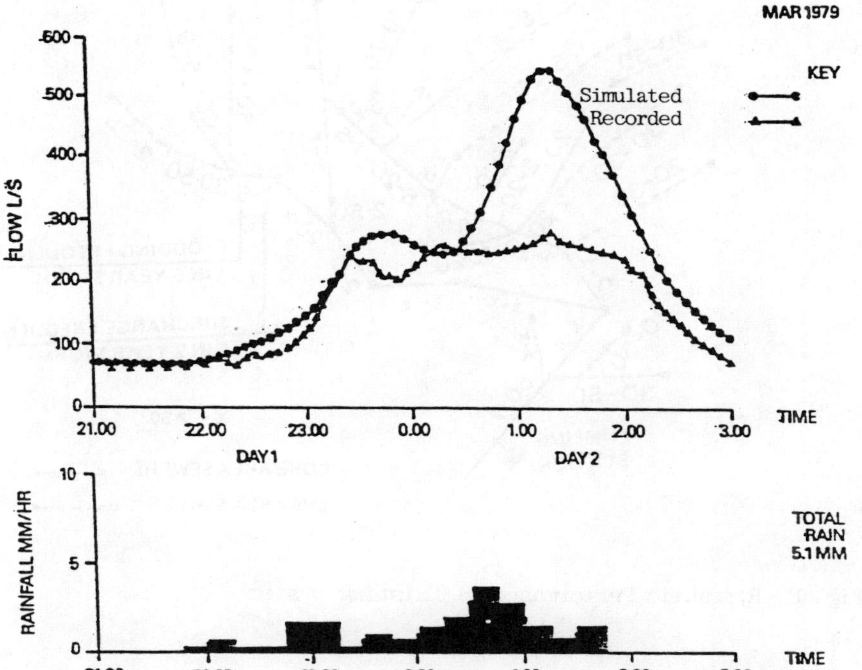

Fig. 7 - Evidence of Unrecorded Overflow

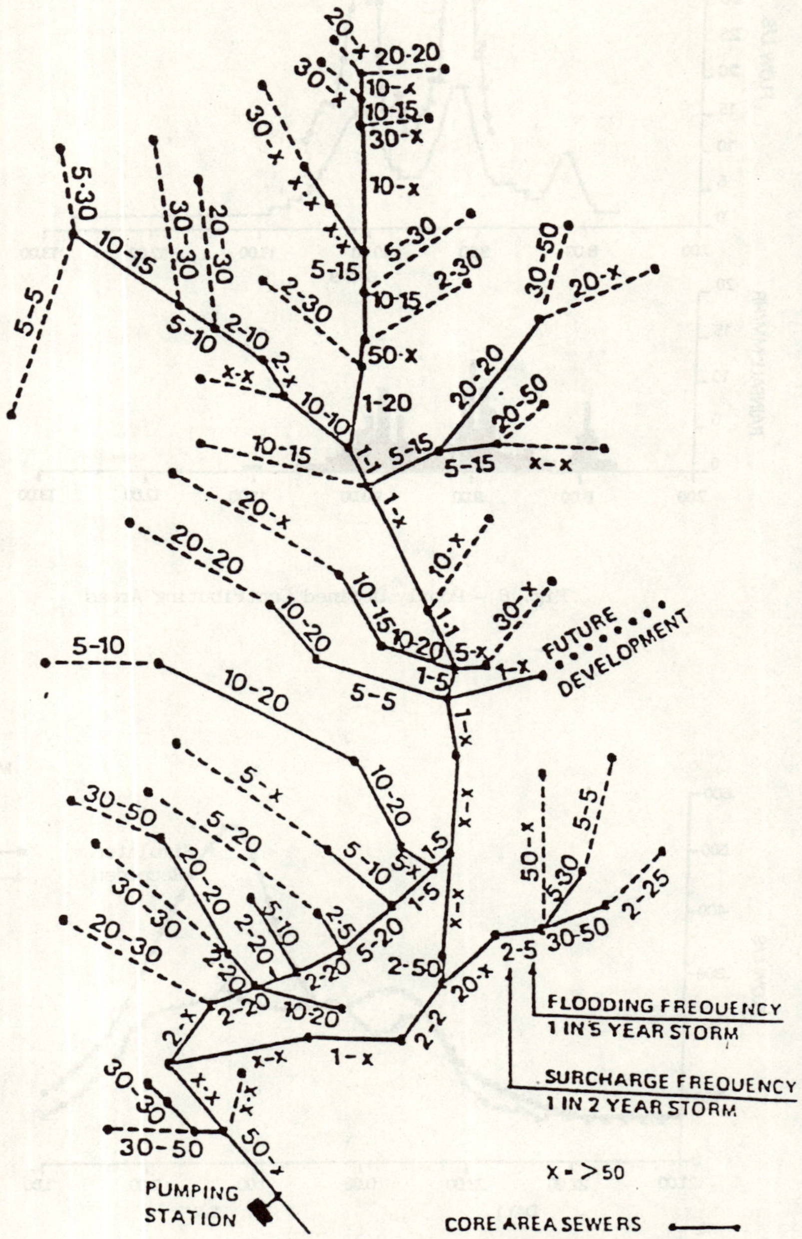

Fig. 8 Hydraulic Performance of Existing System

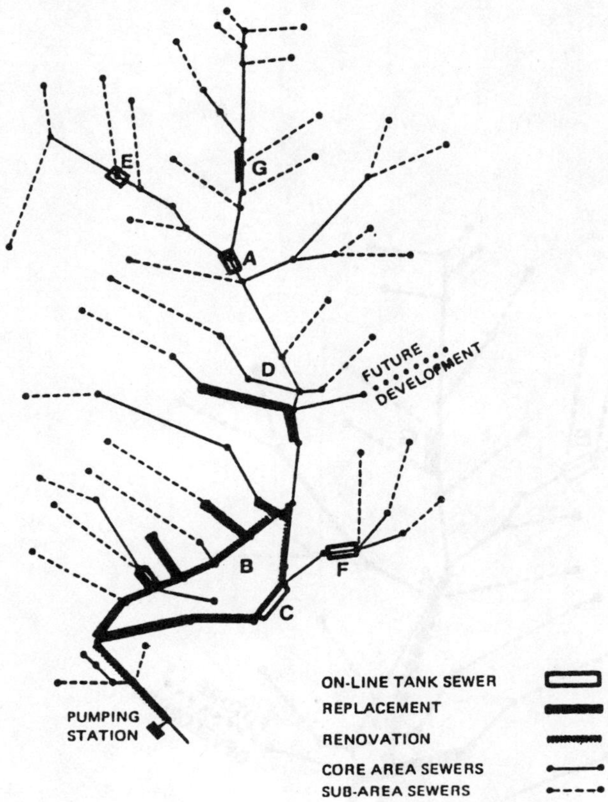

Figure 9 Proposed works — Scheme X

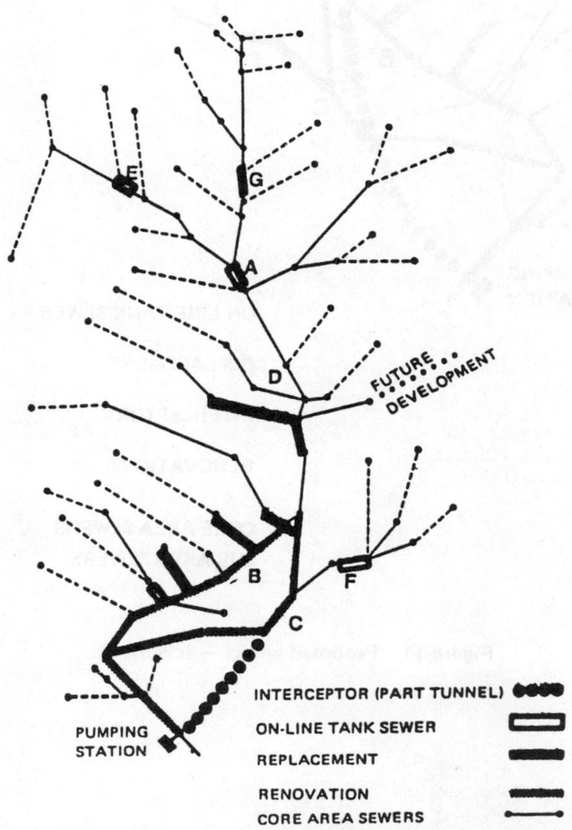

Figure 10 Proposed works — Scheme Y

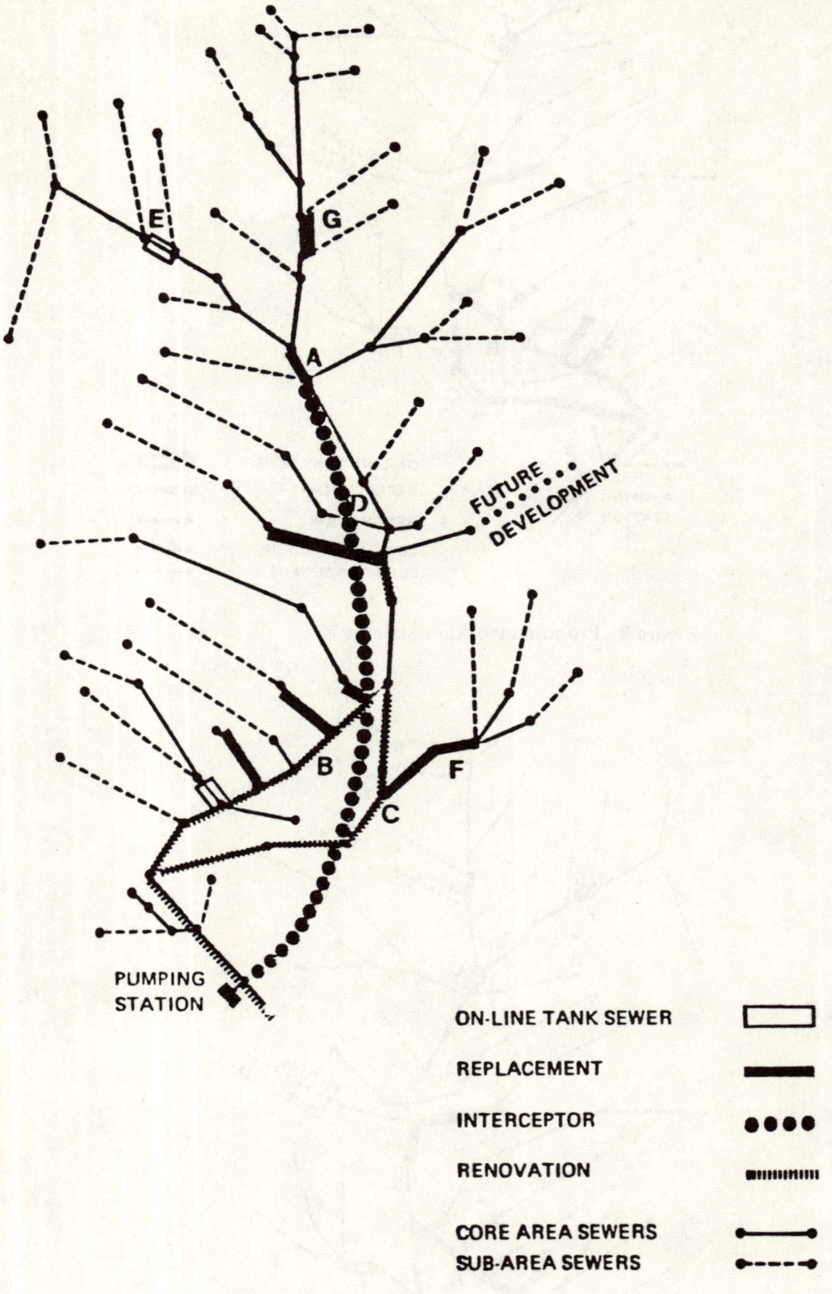

Figure 11 Proposed works – Scheme Z

International Conference on the
PLANNING, CONSTRUCTION, MAINTENANCE & OPERATION OF SEWERAGE SYSTEMS
12-14 September, 1984

PAPER B2

FORMULATION OF A SEWER RELIEF STRATEGY IN EDMONTON, ALBERTA

S. Vatagodakumbura, M.Sc., P.Eng. And K. Chua, M.S.C.E., P.Eng.
Director of Drainage Upgrading — Drainage Modelling Engineer

Edmonton Water and Sanitation
12220 Stony Plain Road
Edmonton, Alberta T5N 3Y4
Canada

Summary

Upgrading of large combined sewer systems to acceptable service levels is a major responsibility of many municipalities in North America. The City of Edmonton, with it's 4,860 ha combined sewer district aged 30 to 80 years, is implementing such a sewer rehabilitation program, costing over $130 million and lasting over 20 years. To maximize the benefits, all proposed relief works have been assessed, compared and priorized into a single implementation scheme. The priority ranking is based on a common set of performance evaluation criteria applicable to each specific drainage system so that the worst affected areas are relieved the earliest, in this long duration program.

This paper presents and discusses the relevant performance parameters and demonstrates their adoption as a set of evaluation criteria. These include indicators such as sewer backup occurrences with different storm conditions; the historical changes in sewer system design criteria as well as comparison of simulated system response to varying hydrologic loading conditions. The simulated responses are obtained from U.S. EPA SWMM RUNOFF and EXTENDED TRANSPORT for a range of hydrologic events. The model capability to predict potential sewer surcharge levels above basement elevations as well as sewer loading ratios (simulated flow/free flow capacity) are incorporated into this systematic evaluation process. The ranking matrix recognizes, through weightage factors, the impact level variation with land use and the frequency of the evaluating rainfall event.

Overall evaluation approach includes: model validation for recorded rainfall events with recorded basement floodings; normalization of flooding records for uniformity over the area; simulation of hydraulic loading tests on the existing sewer networks for a number of design storms within a range of applicable return periods and comparison of response levels of performance characteristics. Thereafter, relief construction is priorized on the basis of relative potential impacts on the study areas on local and on global levels. This evaluation methodology was extremely helpful in completing the priorization process in an objective manner.

Held at The University of Reading, England. Organised and sponsored by BHRA, The Fluid Engineering Centre, Cranfield, Bedford in conjunction with the Water Research Centre.

1. INTRODUCTION

Urban drainage management, as presently practiced for new developments within the City of Edmonton in Alberta, Canada incorporates a 'total concept'. That is, it provides for separated storm and wastewater sewer systems, well defined overland flow patterns for orderly, predictable and safe conveyance of storm runoff for a range of rainfall events up to 100-year level design storms; as well, the servicing standards allow and encourage application of various stormwater management techniques using detention/retention facilities etc. (Ref. 1, 2, 3). Thus, by adopting this comprehensive systems approach, the City can protect the new areas from recurrent drainage related problems inherent to many of its older areas. The 'recent' developments in the fields of urban planning and watershed drainage management as well as the use of computerized mathematical models as analytical tools were instrumental in the adoption of the above up to date policies.

Applying the same techniques of analysis, the City has embarked on an ambitious program to upgrade the service levels of the older combined area sewer systems. The upgrading need of combined sewers is common to all older urban areas of North America and it is a very important responsibility since millions of dollars are spent every year, mainly by municipal governments, for sewer performance upgrading.

This paper presents how a comprehensive implementation strategy was evolved to upgrade the large network of combined sewers in Edmonton. The methodology involved the application of the latest mathematical modelling techniques, the use of a few years basement flooding records for recorded rainfall events and the comparison of relative levels of existing sewer capabilities. Since the upgrading program is extensive and is in high demand over the total area, the objective was to establish a priority order based on reliable criteria so that the worst affected areas are upgraded first. Review of existing literature and past efforts did not reveal any existing special methodology. Therefore, the methodology briefly presented in this paper was adopted, making use of all available relevant information.

2. HISTORY AND BACKGROUND

2.1 Study Area

The combined sewer district of the City of Edmonton comprises about 4860 ha of the central core. In area it is only about 7% of the total City. As shown in Fig. 1, the North Saskatchewan River divides this area into two parts, one on the north side and the other on the south side. The local, shallow (less than 10 m deep) combined sewers are connected by drop structures to the deep (15 to 50 m depth range) interceptor trunk sewers, constructed mostly as tunnels. These trunk sewers from the north and south sides converge at the Gold Bar Wastewater Treatment Plant located on the south bank of the river just outside the eastern boundary of the combined district. There are twelve combined sewer overflows located along the banks of the river.

2.2 Servicing Practices

Some of the salient sewer characteristics and the servicing practices associated with this combined district and relevant to this study are, in brief, as follows:

(i) The sewers are aged from 30 to over 80 years and are constructed of different materials such as clay, cement mortar, concrete, brick linings, etc.

(ii) The local sewers are laid at least 2.5 meters below ground level to protect from frost/freezing conditions since winter temperatures go down to -40°C level.

(iii) Basements (as utility areas) with floors about 2 m below ground are an integral part of all buildings; these are used to house furnaces, water heaters and other utility related appliances etc. Thus, backfills around houses could be subject to heavy percolations during wet weather.

(iv) Each property has a single combined sewer service draining by gravity to the combined sewer on the street, as the street sewers are adequately deep.

(v) The weeping tile system draining the foundations and the backfill area around the house is directly connected to the service sewer.

(vi) The roof leader downspouts are also directly connected to these services sewers.

(vii) In most buildings, at the time of construction, backwater prevention devices were installed to protect against sewer backups. However, with time, many of these units have been removed due to various reasons such as maintenance, change of ownership, remodelling basements etc.

(viii) The sewers were designed using the Rational formula and the 5-year level design standard as per Intensity-Duration-Frequency (I.D.F.) curves available at the time.

Fig. 2 shows the sewer connection practices in Edmonton for combined and separated sewer systems. Thus, the combined sewer systems were designed and constructed, as the most economical at that time, using the design criteria and standards including protection from sewer backups.

However, with time, as with any other urban environment, many changes have occurred which cumulatively resulted in overloading of these sewers more frequently.

2.3 Changes in Criteria

The changes in the major criteria that affect the sewer system performance and the public's perception of the expected sewer service level are critical to the municipality when carrying out proper maintenance and rehabilitation works. In Edmonton, the following major changes have occurred:

(i) Land Use: Over the years, due to redevelopments and paving of lanes, driveways and sidewalks, the effective imperviousness has increased significantly, sometimes doubling the initial percentages. Also, through the peripheral expansion of the City and the resulting population growth, the wastewater flow component to the trunk sewer network has increased with lesser overall impact as compared to that due to increases in paved areas.

(ii) Design Criteria and Standards: As shown in Fig. 3, the design storm for sewer sizing has been upgraded for the new areas based on accumulated historical records. Similarly, the newer areas are serviced by separated storm and wastewater sewer systems; as well, the latest Standards provide for comprehensive drainage management to provide protection up to 100-year storm level in the separated areas only. These upgraded service levels in the newer areas have created greater expectations in the combined area residents, for their sewer systems. In general, this conforms with the trend among citizens to demand higher and more reliable service levels from public utilities.

(iii) Basement Area Use: In keeping with the new trends, especially in separated areas, many residents in the combined areas also have upgraded their basements from "utility spaces" to "living and recreational uses" such as bedroom suites, family rooms involving heavy investments in carpets, hi-fi systems, furniture, decorations etc. Thus, they are vulnerable to suffer significant damages in case of sewer backups. In fact, during the past 10-15 years many such sewer backup related damages have been recorded in these areas especially by residents who had not maintained or installed backwater prevention devices.

2.4 Rehabilitation Strategy

To alleviate the flooding problems associated with sewer backups, initially in 1968, the City started a sewer separating program for these areas. However, due to the extremely high costs of this effort, this approach was modified to a combined sewer upgrading program (Ref. 4). At present, the estimates are that this program will cost about $130 million (1983$). The work consists of local shallow sewer upgrading (Ref. 5, 6, 7, 8), temporary detentions (Ref. 9) and new deep tunnels (Ref. 10, 11).

The sewer network service capacity will be upgraded to the 5-year design storm level and for the approved land use zoning conditions, being the most cost-effective alternative as established through engineering reviews (Ref. 12, 13). The above references show that the methodologies used for the engineering analyses are the most up to date as they are based on systems analysis approach and computerized mathematical models such as USEPA SWMM and EXTRAN models (Ref. 14, 15, 16). The models were modified to suit specific needs of the City's sewer networks and validated using real rainfall and basement flooding data within these areas.

The nature of the rehabilitation effort, the extent of work with a price tag of over $130 million as well as the limited annual resources available, made it imperative that this work be programmed into a 22-year schedule. This meant that an acceptable priority ranking be adopted for implementing these works.

3. IMPLEMENTATION STRATEGY

3.1 General Criteria

The general criteria used in the process of defining the proposed works into a priorized implementation strategy were that:

(i) The worst affected areas be relieved first thus reducing the cumulative damages due to sewer backups during the implementation period. In other words, the capital expenditures would provide the most benefits in the shortest time.

(ii) The priority ranking methodology be based on a set of evaluation criteria which are objective; it also should depend on data relevant to these systems and their responses to differing hydraulic loading conditions.

3.2 Priority Ranking Scheme

3.2.1 Overall Concept and Methodology

The combined sewer network has been constructed over a few decades up to the end of 1950's. Hence, the original capabilities of each system varied with the design standards in use at the time. Therefore, the only acceptable approach was to test the response of each of these existing sub-systems for the same hydraulic loading conditions making use of their existing land uses and service connections; then comparing their relative levels of loadings/overloadings. For this purpose, a series of design storm events with assigned frequencies, as well as observed floodings for recorded storm conditions were used.

In these efforts, the rainfall and sewer flow data collected through the extensive data acquisition system maintained by the City were of great benefit (Ref. 17).

3.2.2 Local vs. Trunk Sewers

As indicated earlier, the trunk sewer network is very deep and is connected by drop structures to the comparatively shallow local sewers. Analyses indicated both systems to be overloaded for the 5-year service level needs. After studying the results and the flooding complaints, it was decided that relieving the local sewer network was the more urgent of the two. Therefore, immediate priority was given to relieving the local systems. But, since the majority of the works were local sewer upgrading spread within the 4860 ha area, still the need for priorization was very high for these proposed works.

4. LOCAL PRIORITY RANKING METHODOLOGY

4.1 General

The local priority ranking required the establishing of relative service levels of each sub-area sewer system on the basis of their comparative responses to increasing hydraulic loadings. For this purpose, two distinct but similar scenerios were used. They are:

(i) Simulated comparative response of each sewer network for a range of applicable hydraulic loadings, and

(ii) Observed comparative response of these networks for some loading ranges. These are explained below.

4.2 Simulated Responses

Under a given hydraulic loading, a sewer system's performance can be measured by two major response characteristics; i.e. the extent of sewer surcharging indicated by the maximum hydraulic grade line and the extent of maximum sewer loading indicated by peak flows in comparison to the designed free flow capacity.

For a given sewer network, if these responses can be adequately defined, for a range of rainfall events with increasing impacts, then they can be interpreted for service level ranking purposes. In this study, this was achieved as described below.

4.2.1 Simulation Tools

For defining the above responses, the computer models USEPA SWMM II RUNOFF (Ref. 14) and WRE EXTRAN (Ref. 15, 16) are eminently suitable hydrologic and sewer network hydraulic routing models. These models are well used, capable of handling complete dynamic network routings including surcharge simulation; the documentation is also excellent as given in above references. Moreover, the City has extensive experiences with them and therefore were used in this analyses.

4.2.2 Simulation Events

The range of required and applicable hydraulic loadings were obtained by imposing three high frequency design rainfall events; (i.e. 1-year, 2-year and 5-year level design storms) which are based on long (over 50 years) meterological records in Edmonton and the "Chicago distribution" (Ref. 18). The 2-year level storm was an actual event recorded in Edmonton. These rainfall histograms and the Edmonton I.D.F. curves are given in Fig. 4.

4.2.3 Watershed and Sewer Network Models

For analyzing the local watershed and its sewer network, the area was discretized into 33 large subcatchments, 21 on the northside and 12 on the southside as per SWMM II RUNOFF needs (Fig. 5). For further detailed evaluation, each of these subcatchments were further divided into a number of sub-areas totalling 111.

The sewer network sub-models for these areas, modelled as per EXTRAN needs, constituted of inflow points, sewer elements, junctions and outflow points. As discussed earlier, on the basis of previous trunk sewer system analyses results (Ref. 10), these simulations assumed that there will be no backwater effects from the deep trunk sewers.

4.3 Observed Responses

Historical records of observed basement floodings, street floodings and sewer flows/depths within the combined sewer district of Edmonton were available for certain recorded summer rainfall events. Although these data were not considered complete and accurate representation of the sewer network response for these events, they were of sufficient merit to be used as reliable indicators. Therefore, with certain modifications to ensure a degree of uniformity and consistency in keeping with the changes inherent to the incident rainfall characteristics over the area and the durations, these data were used as an evaluation criterion in conjunction with the simulated responses.

4.4 Ranking Procedure

The ranking of each sub-area was done at two levels; the first level was the event based ranking when the ratings of each area as per simulated and observed responses for a given event were added and compared using the selected set of criteria and weightage factors. The second level was a composited one using the total of the ratings for the three given events for each area. These are described in the next section.

5. PRIORITY RANKING CRITERIA AND WEIGHTAGE FACTORS

5.1 Criteria for Simulated Responses

These criteria and the methodology are schematically shown in Fig. 6 and explained below.

5.1.1 Sewer Backwater Level

In Edmonton, in general, the basement weeping tile system is about 2 meters below street level as indicated in Section 2.2. Therefore, if the simulations indicated hydraulic grade lines above this level, the affected areas were classified as subject to flooding. From this, the number of residential units so affected as compared to the total residential units in the area and the proportion of street frontages so affected in comparison to the total street frontage covered by the commercial, industrial and institutional areas were used as the measuring criteria for these land use categories.

5.1.2 Sewer Loading Level

Sewers are designed as gravity conveyance elements flowing full at free flow capacity for design loadings. Therefore, the extent of their overloading for a given event is a good indication of their capacity shortfall. In this study, this was used as an evaluating parameter by aggregating points depending on the level of overloading of each sewer as follows:

Loading Ratio (r) (Peak flow/Free flow Capacity)	Evaluation Points
$0 < r \leq 1.00$	1
$1.00 < r \leq 1.25$	2
$1.25 < r \leq 1.50$	3
$1.50 < r \leq 2.00$	4
$r \geq 2.00$	6

From these, all points for $r > 1.25$ were classified as unacceptable and expressed as a fraction of the total points to indicate the level of overloading for each event.

5.2 Criteria for Observed Responses

Recorded basement floodings were the main historical data available. These flooding records were correlated with the intensity of the recorded event by normalizing them to the 30-minute average recorded maximum rainfall intensity of the event incident on each sub-area. The 30-minute duration was selected as the average time of concentration for the sub-systems modelled. A further correction factor was introduced to relate these normalized floodings with simulated potential floodings for each evaluation event so that a composite and correlated value is obtained. This value was then expressed as a fraction of the total houses within the study area to obtain the observed flooding measurement criterion.

5.3 Weightage Factors

Each of the above criteria was assigned a different weightage factor to ensure that following conditions are met.

(i) The more damaging residential area floodings and sewer overloadings are given higher priority at the event based evaluation level. The five weightage factors taken together form unity and are described later.

(ii) The more frequent storm related floodings and sewer overloadings are given greater weightage during composited ranking using the three event based ranking; these are also described later.

5.6 Event Based Ranking (Re)

The above criteria and weightage factors were used in the form of two equations for ranking each area on the basis of accumulated points, for each of the evaluation events, as follows:

$$Re_k = (a_k\ w_a + b_k\ w_b + c_k\ w_c + d_k\ w_d + e_k\ w_e) \times 100 \text{ ------ } (1)$$

$$1.0 = w_a + w_b + w_c + w_d + w_e \text{ ---------------------------- } (2)$$

where:

k = Evaluation event (1, 2, or 5 year design event)

Re= Event based ranking of the area

$$a = \frac{\text{No. of Simulated basement floodings in the given area}}{\text{Total No. of houses in the given area}}$$

$$b = \frac{\text{Length of Simulated frontage flooded for commercial/industrial institutional properties in the area}}{\text{Length of total frontages in the area}}$$

$$c = \frac{\text{Unacceptable sewer overloading related points in the residential parts of the area}}{\text{Total Sewer loading points for the area}}$$

d = Similar to c for commercial, industrial and institutional parts of the area

$$e = \frac{\text{normalized and corrected floodings reported in the area}}{\text{Total no. of houses in the area}}$$

w_a, w_b, w_c, w_d, and w_e are the fractional weightage factors associated with each of the above criteria a, b, c, d and e. In this study, the values assigned for the weightage factors were 0.37, 0.06, 0.37, 0.16 and 0.04 respectively.

It can be seen that the higher the value of Re, the greater is the comparative priority for relief.

5.7 Composite Ranking (Rc)

The ultimate ranking was based on the overall comparative performance of each network for the storm frequency range of 1 to 5 years so that behavior changes in the sewer network with rainfall event is also addressed to some extent. The range of events was kept at this lower intensity level in keeping with the catchbasin inflow capacities as well as the design standards adopted for these relief works. The equations used are as follows:

$$Rc = Re_1\ w_1 + Re_2\ w_2 + Re_5\ w_5 \text{ ----------------- } (3)$$

$$1.0 = w_1 + w_2 + w_5 \text{ ---------------------------- } (4)$$

where:

Rc = composite comparative ranking of the given area

Re_1, Re_2 and Re_5 are the 1, 2 and 5 year event based rankings of the area. w_1, w_2 and w_5 are the weightage factors for the three events 1, 2 and 5 year design storms, respectively and were assigned values 0.63, 0.25 and 0.12 respectively.

Similar to event based ranking, the higher the value of Rc, the greater is the relative priority for relief on a composite basis of the three evaluation events. This composite ranking is, therefore, the final measuring criterion to establish the priorized implementation schedule.

5.8 Relative System Capacity(S)

As an added refinement, on the basis of event based ranking points, relative system capacities in terms of derived services levels could be assigned for different areas. This needs certain tedious efforts of trial and error to arrive at the ranking level that is comparable to the service level represented by the return frequency of the evaluation event (eg. 1 year design storm). This was attempted and the results are presented.

6. RESULTS OF ANALYSIS

The above analysis was conducted in 1982 to repriorize majority of the combined areas after establishing their detailed upgrading requirements.

6.1 Detailed Priority Ranking

The results of the detailed ultimate priority ranking of sub-areas is given in Table 1. The results are tabulated in the order of priority with the highest priority for sub-area 16 of the study area 11S as it scored the maximum ranking point of 72.061. It can be seen that the relative priorities change from one to the other of the three evaluation events. This change is to be expected in view of the non-linear dynamic processes that occur in this hydrologic and hydraulic routing phenomenon.

Also, for this ranking order, the data points are plotted as per their assigned service levels in Fig. 7. For these study areas, the data points can be approximately represented by an exponential curve of the form:

$$Rc = \frac{68.2275}{(S + 0.6523)}\ 1.6362 \qquad \text{----------------} \quad (5)$$

where:

Rc = (composited) Ultimate Ranking of the Study area
and S = (Assigned) Relative System capacity of the study area.

These are used in the actual implementation scheduling of the works program.

6.2 Study area Priority Ranking

Table 2 and Fig. 8 give the results for 'lumped' areas, using the larger study areas as the basis. These were produced for the purposes of overall policy oriented presentations. The level of accuracy is lower than that of the above.

In this case, the data points are matched by a curve of the form:

$$Rc = \frac{31.6974}{(S + 0.1528)^{0.9758}} \quad \text{----------------} \quad (6)$$

6.3 System Capacity(S)

The ranges of comparative system capacities in Table 1 and 2 and the Figures 7 and 8 show that majority of the areas have assigned service levels of 1 to 2 years; also some detailed small areas have service levels lower than the 1 year level. Since these system service levels are established on a comparative basis only, their absolute correctness is open to question. This aspect does not affect the ranking order of the works.

6.4 Implementation Schedule

Table 3 shows, in summary form, the finalized implementation schedule for the above works. In formulating these final schedules, the above results were used as the primary guidelines. Modifications were only made to accommodate external constraints such as staging in conjunction with roadway rehabilitation programs for cost optimization, meeting extensive redevelopment needs, accommodating specific policy directives from City Council etc.

7.0 CONCLUSIONS

The simple but effective methodology presented here was greatly helpful in the orderly and meaningful use of the extensive data available for the analyses. It also provided a systems approach that minimized impacts due to individual limitations as the criteria were objectively defined with due consideration to the hardships and damages associated with sewer backup related floodings. Furthermore, by helping to layout a well defined long term schedule of activities, it has assisted in better coordination with other city agencies such as roadway improvements, rezoning and redevelopments etc.

8.0 ACKNOWLEDGEMENTS

The authors acknowledge the contributions of the staff of the Consultant firms of Associated Engineering Services Ltd., MacLaren Plansearch Services Ltd. and Proctor Redfern Butler and Krebes Ltd., and of the Edmonton Water and Sanitation staff during these studies. In particular, that of Mr. A.T. Davies, P. Eng. Chief, Drainage Engineering for the support and approval of the project and that of Mr. M. Diebel, Drainage Modelling Engineer, who wrote the computer program for final priority ranking evaluation. The assistance of the Edmonton Water and Sanitation staff in the preparation of this manuscript is also gratefully acknowledged.

9.0 REFERENCES

1. Vatagodakumbura, S. and Fok, A.T.K.: "A review of Methodologies and Techniques of Urban Hydrologic Analysis" In: Proc. 5th Canadian Hydrotechnical Conference (Fredericton, New Brunswick, Canada: May 26-27, 1981) Pg. 817-837.

2. Fok, A.T.K., Vatagodakumbura, S. et. al.: "Application of Stormwater Management Concepts for a Master Drainage Planning Study" In: Pro. International Symposium on Urban Storm Runoff (University of Kentucky, Lexington, Kentucky, U.S.A.: July 23-26, 1979) Pg. 275-282.

3. City of Edmonton, Alberta, Canada: "The City of Edmonton Servicing Standards Manual", 1982.

4. City of Edmonton Water and Sanitation Department: Technical Report, "Capital Expenditures and Financing for Upgrading the combined sewer Systems", May, 1978.

5. Associated Engineering Services Ltd.: Technical Report, "The City of Edmonton - Combined Relief Sewer Design Study, Areas 11S and 9S", June 1983.

6. MacLaren Plansearch Services Ltd.: Technical Report, "The City of Edmonton - Combined Sewer District Relief Design Study in Areas 3S, 4S and 5S", July 1983.

7. Proctor Redfern Butler and Krebes Consultants Ltd.: Technical Report, "The City of Edmonton - Combined Sewer Relief Study Areas 1N, 3N, 4N and 17N", July 1983.

8. Water and Sanitation Department, City of Edmonton: "Internal Technical Reports on Various Combined Sewer Relief Study Areas" 1978 to 1983.

9. Vatagodakumbura, S. and Chua, K.: "Use of Underground Storage Unit for Combined Trunk Sewer System Relief" In: Proc. Stormwater and Water Quality Model Users Group Meeting (Montreal, Quebec, Canada,: September 8-9, 1983).

10. MacLaren Engineers Planners and Scientists, Inc.: Technical Report, "Combined Trunk Sewer System Relief and Pollution Abatement Study", January, 1981.

11. Water and Sanitation Department, City of Edmonton: Technical Report, "South Side Combined Trunk Sewer System Functional Relief Study", August, 1983.

12. Water and Sanitation Department, City of Edmonton: Technical Report, "Report on combined Sewer Backup During July 14, 1981 Storm", January, 1982.

13. Vatagodakumbura, S. and Choudhry, R.: "Review of Combined Sewer Relief Service in Edmonton" In: Proc. 1982 Annual Conference, Canadian Society of Civil Engineering (Edmonton, Alberta: May 27-28, 1982).

14. Huber, Wayne et. al.: "Storm Water Management Model User's Manual-Version II (U.S. Environmental Protection Agency, Cincinnati, Ohio), March, 1975.

15. Kibler, D.F.; Monser, J.R. and Roesner L.A.: "San Francisco Stormwater Model, User's Manual and Program Documentation", prepared for the City and County of San Francisco U.S.A. (Water Resources Engineers, Walnut Creek, California).

16. Roesner, Larry A; Shubinski, Robert P. and Aldrich, John A.: "Stormwater Management Model User's Manual Version III, Addendum I Extran" (U.S. Environmental Protection Agency, Cincinnati, Ohio), Final Draft - November, 1981.

17. Water and Sanitation Department, City of Edmonton.: Technical Report (Draft) "The City of Edmonton Raingauge Network Siting - A Technical Review" - September, 1983.

18. Keifer, C.J. and Chu, H.H.: "Synthetic Storm Pattern for Drainage Design" Journal of the Hydraulic Division, ASCE, August 1957, Paper #1332.

Table 1 Ranking by Sub-Area

ULTIMATE RANKING OF SUB AREAS

STUDY AREA	SUB AREA	AREA (ACRE) *	APPROX COST OF RELIEF $	1-YEAR RANKING	2-YEAR RANKING	5-YEAR RANKING	ULTIMATE RANKING	SYSTEM CAPACITY
11S	16.	42.0	546453.	1.	2.	2.	72.061	.31
11S	15.	18.0	138959.	2.	3.	3.	58.775	.39
11N	1.	57.0	1322936.	3.	1.	8.	53.956	.53
11S	14.	162.0	3577955.	5.	5.	4.	43.222	.65
3S	3.	101.0	3844307.	4.	10.	16.	43.369	.60
11N	4.	61.0	891824.	6.	9.	11.	42.039	.67
12N	8.	89.0	2862568.	8.	6.	5.	41.569	.78
6S	6.	7.0	296009.	33.	11.	1.	33.523	1.18
5S	2.	230.0	3328511.	14.	14.	13.	33.515	1.01
11S	2.	182.0	491734.	15.	16.	15.	32.056	1.02
9N	1.	36.0	430732.	10.	19.	30.	31.923	.84
12N	10.	28.0	481585.	34.	7.	6.	31.840	1.17
4N	1.	424.0	6799695.	11.	20.	31.	31.271	.85
11S	17.	30.0	236782.	53.	4.	10.	31.022	1.20
9N	2.	56.0	755204.	13.	23.	21.	30.534	.97
12N	5.	73.0	1607182.	9.	43.	28.	28.688	.81
7N	8.	17.0	212899.	7.	47.	52.	28.261	.69
6S	7.	8.0	478169.	41.	8.	19.	26.109	1.22
20N	3.	33.0	1315726.	28.	18.	27.	25.943	1.22
6S	3.	52.0	856151.	32.	27.	20.	23.852	1.36
8N	3.	45.0	1280812.	40.	26.	14.	23.522	1.40
6S	5.	55.0	906625.	69.	15.	7.	23.506	1.37
12N	2.	36.0	2941X2.	77.	12.	12.	23.268	1.35
7N	4.	81.0	1954994.	31.	25.	35.	23.162	1.30

* NOTE: 1 ACRE = 0.405 ha.

PARTIAL LIST ONLY.

Table 2 Ranking by Study Area

ULTIMATE RANKING OF STUDY AREAS

STUDY AREA	SUB AREA	AREA (ACRE) *	APPROX COST OF RELIEF $	1-YEAR RANKING	2-YEAR RANKING	5-YEAR RANKING	ULTIMATE RANKING	SYSTEM CAPACITY
4N	1.	534.0	7261724.	1.	3.	8.	27.318	1.01
5S	1.	507.0	7081764.	3.	2.	3.	25.894	1.24
3S	1.	368.0	6578057.	6.	1.	4.	25.502	1.25
9N	1.	183.0	1660339.	2.	8.	1.	25.497	1.28
12N	1.	527.0	8078069.	7.	6.	6.	22.676	1.34
8N	1.	237.0	4694469.	9.	12.	2.	22.506	1.48
17N	1.	340.0	315937.	5.	7.	10.	22.230	1.32
9S	1.	671.0	1223545.	10.	9.	7.	21.269	1.42
10N	1.	200.0	2163149.	16.	4.	5.	20.779	1.41
7N	1.	256.0	3957046.	4.	11.	13.	20.602	1.42
11N	1.	498.0	4658389.	8.	13.	12.	19.636	1.49
20N	1.	76.0	1315726.	15.	5.	9.	19.496	1.43
6S	1.	316.0	4350609.	17.	10.	11.	16.796	1.58
1N	1.	374.0	1234738.	13.	15.	16.	16.422	1.70
2N	1.	307.0	3859899.	12.	16.	15.	16.048	1.86
11S	1.	1456.0	12804421.	11.	18.	17.	15.396	2.99
21N	1.	163.0	2009479.	19.	14.	14.	14.882	1.67
3N	1.	256.0	3609820.	14.	17.	18.	14.476	1.90
4S	1.	470.0	3638847.	18.	19.	19.	10.717	4.09
1S	1.	211.0	269444.	20.	20.	20.	9.012	4.97
13N	1.	247.0	2080099.	21.	21.	21.	7.890	5.10
2S	1.	219.0	0.	22.	22.	22.	4.917	5.10
15N	1.	83.0	0.	23.	23.	23.	1.922	5.10

* Note: 1 Acre = 0.405 ha

Table 3 Annual Construction Schedule and Yearly Expenditure

Table 3a. Annual Construction Schedule and Estimated Yearly Expenditure - North Side

Study Area No.	Drainage Areas (Acres) ▲	Predominant Land Use	Relief Cost in Thousands (1983 $)																	
			1983	1984	1985	1986	1987	1988	1989	1990	1991	1992	1993	1994	1995	1996	1997	1998	1999	2000
1	374	Single Family Residence												2120			1235			
2	307	" " "										482	1200	1011						
3	256	" " "												727		2883				
4	534	Industrial						732	3220	2848								462		
*5	370	Single Family Residence																		
*6	310	" " "																		
7	256	" " "								197	3287					473				
8	237	Multiple Res. & Comm.									1163	3476								
9	183	" " "						430	1230											
10	221	Single Family Residence										492	1671							
11	498	Multiple Res. & Comm.			890	1330	515						1267					505		
12	526	Commercial					2572	3288									752		1626	
13	246	Single Family Residence												132						1171
*14	359	" " "																		
15	83	Commercial																		0
*16	382	Single Family Residence																		
17	340	" " "													316					
+18	770	Municipal Airport																		
+19	110	Exhibition Grounds																		
20	76	Single Family Residence								1404										
21	163	High Rise Residence												821	1189					
Total	6,601		0	0	890	1330	3087	4450	4450	4449	4450	4450	4138	4811	1505	3356	1987	967	1626	1171

Note: * Study AREAS 5,6,14 AND 16 RELIEF SEWER CONSTRUCTION COMPLETED
+ Study AREAS 18,19 LOCAL RELIEF MAY NOT REQUIRED

Table 3b. Annual Construction Schedule and Estimated Yearly Expenditure - South Side

Study Area No.	Drainage Areas ▲ (Acres)	Predominant Land Use	Relief Cost in Thousands (1983 $)																	
			1983	1984	1985	1986	1987	1988	1989	1990	1991	1992	1993	1994	1995	1996	1997	1998	1999	2000
1	221	Single Family Residence																	270	
2	219	" " "																		0
3	368	" " "	567	5008	841															
4	470	" " "	1536	815	1287															
5	507	" " "	1846	3132	1942															
6	316	" " "	3181														703			
+7	220	" " "																		
*8	317	" " "																		
9	316	" " "																	1224	
*10	460	Multiple Res. & Comm.																		
11	1456	Industrial		575		3100	1317								2944	1347	344	3177		
12	87	Single Family Residence																		500
Total	4,957		7130	9530	4070	3100	1317	0	0	0	0	0	0	0	2944	1347	1047	3177	1494	500

Note: * Study Areas 8 & 10 Relief Sewer Construction Completed
+ Study Area 7 Local Relief May Not Be Required

Table 3c Combined Sewer District Estimated Yearly Expenditure

Description	Relief Cost in Thousands (1983 $)																		
	1983	1984	1985	1986	1987	1988	1989	1990	1991	1992	1993	1994	1995	1996	1997	1998	1999	2000	TOTAL 1983-2000
North Side	0	0	890	1330	3087	4450	4450	4449	4450	4450	4138	4811	1505	3356	1987	967	1626	1171	47117
South Side	7130	9530	4070	3100	1317	0	0	0	0	0	0	0	2944	1347	1047	3177	1494	500	35656
Combined Trunk Relief	0	0	0	890	1551	1951	2051	2051	2051	2051	2362	1740	2051	1797	3466	2439	3000	4549	34000
Total	7130	9530	4960	5320	5955	6401	6501	6500	6501	6501	6500	6551	6500	6500	6500	6583	6120	6220	116773

▲ Note: 1 Acre = 0.405 ha

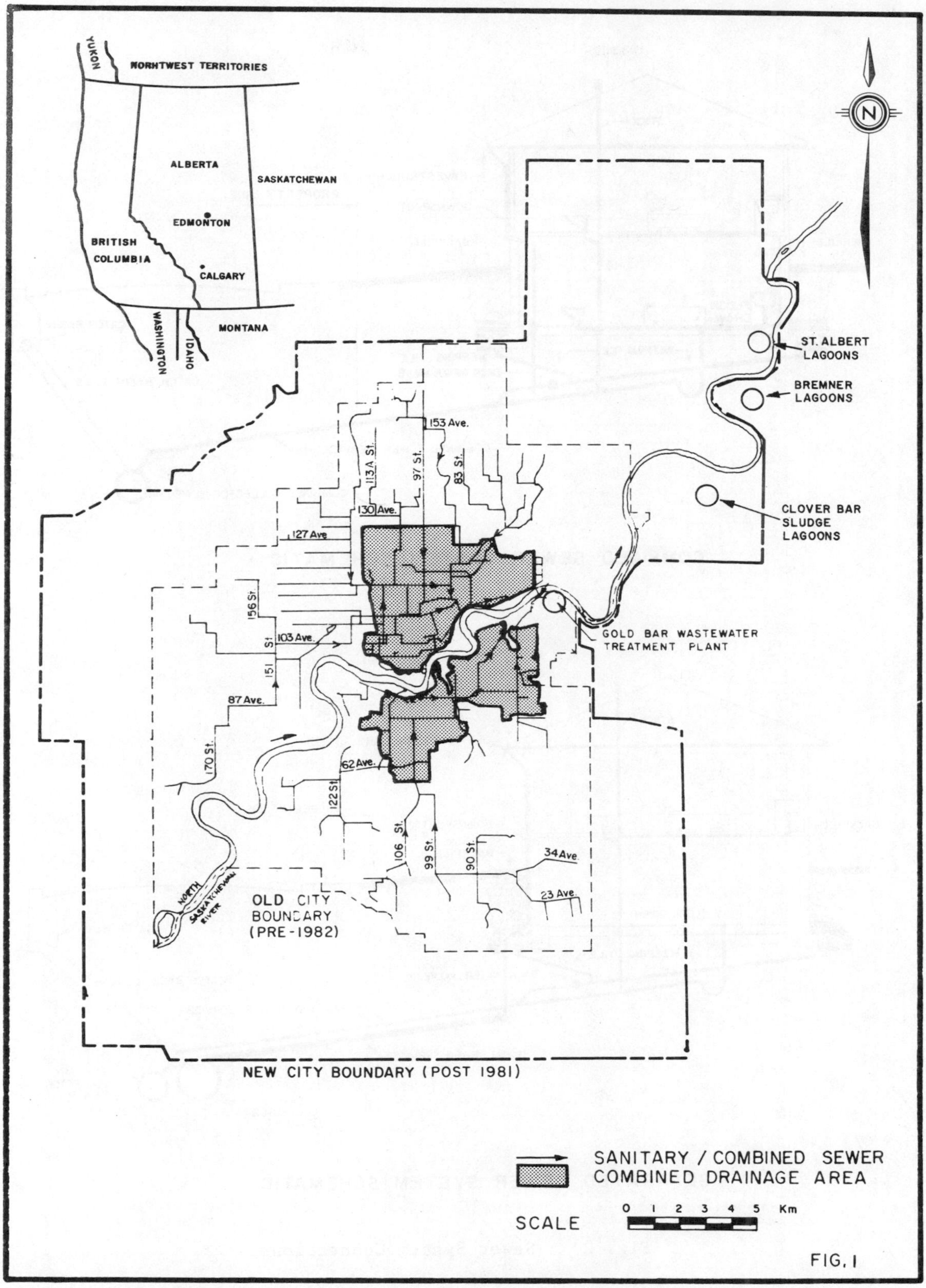

Fig. 1 City of Edmonton Boundary Limits

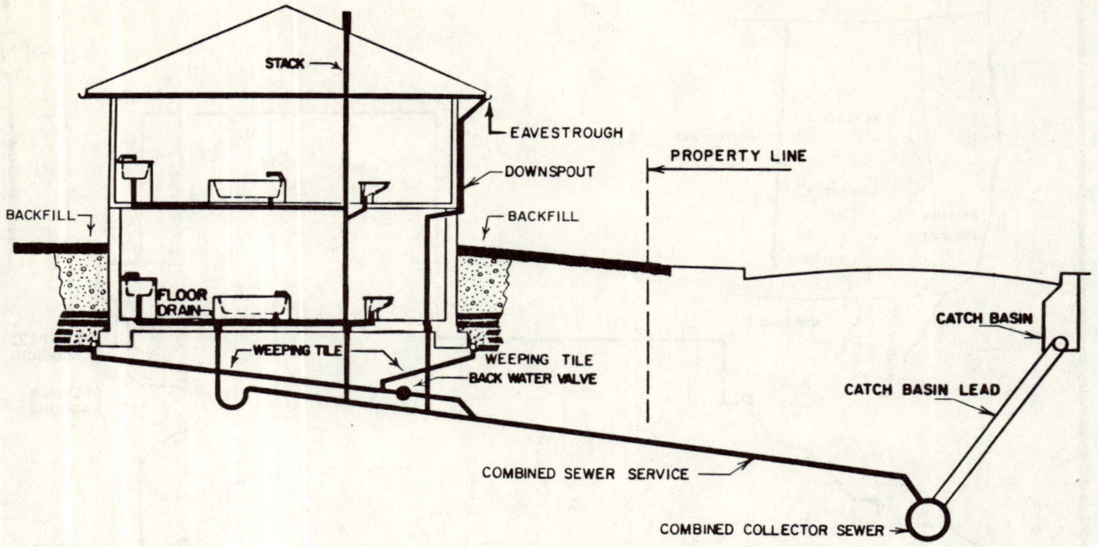

COMBINED SEWER SYSTEM SCHEMATIC

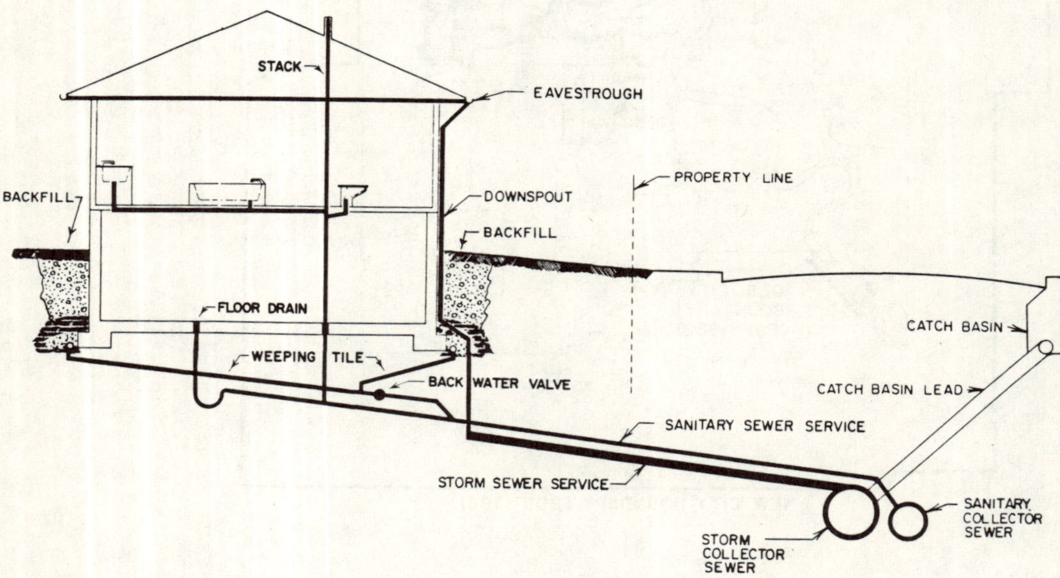

SEPARATED SEWER SYSTEM SCHEMATIC

Fig. 2 Sewer System Connections

Fig. 3 Comparison of Intensity - Duration - Frequency Curves

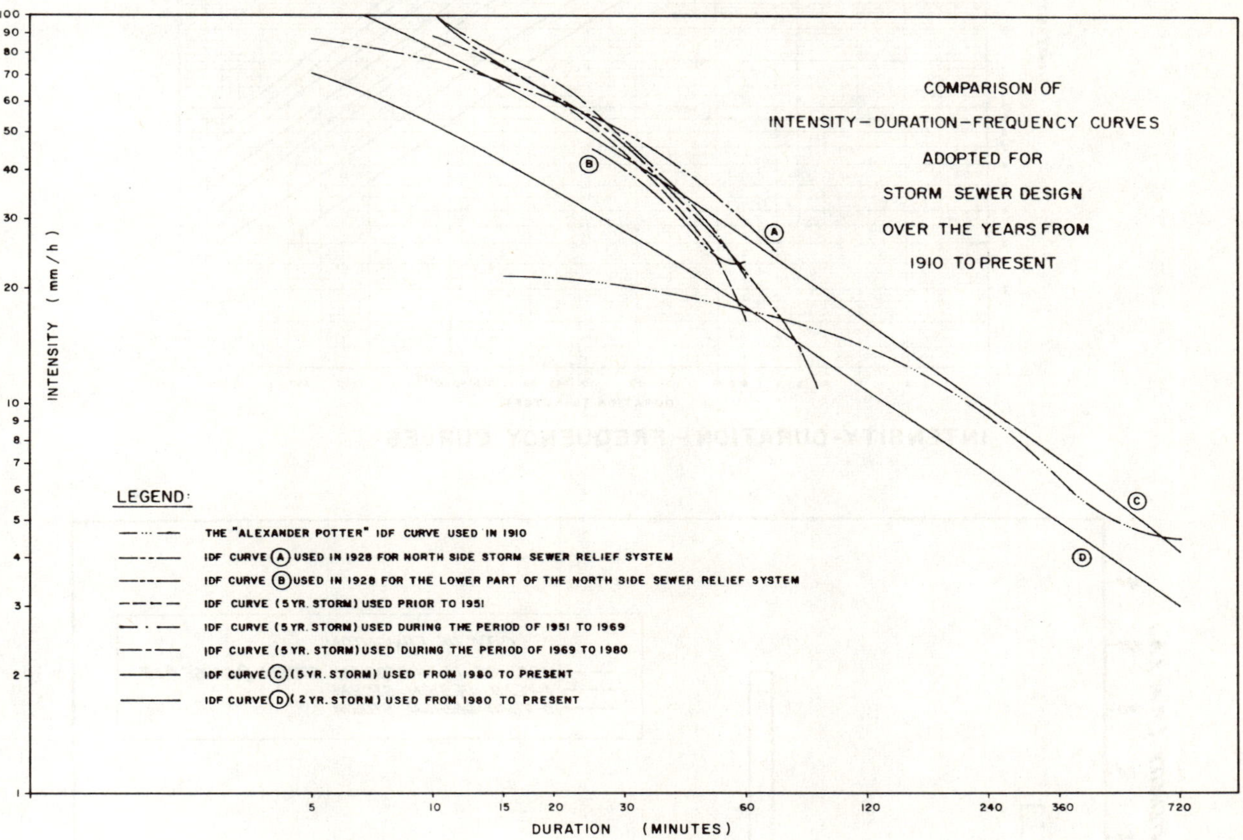

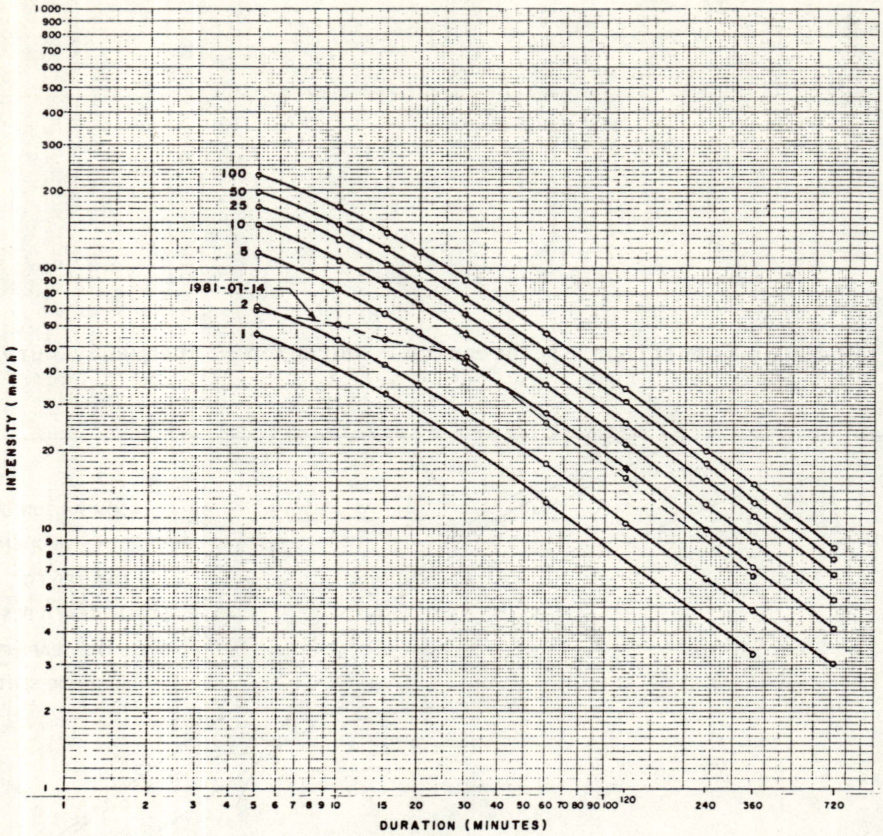

INTENSITY-DURATION - FREQUENCY CURVES

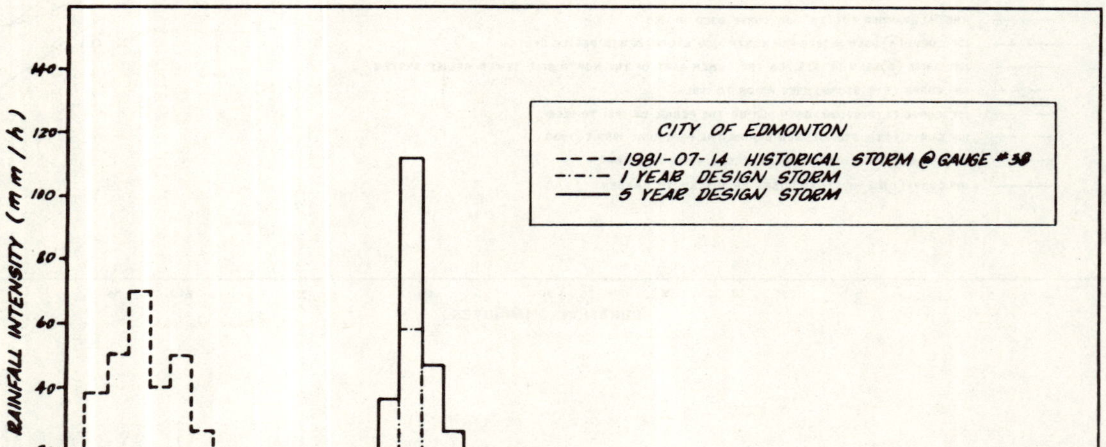

Fig. 4 City of Edmonton Design Curves

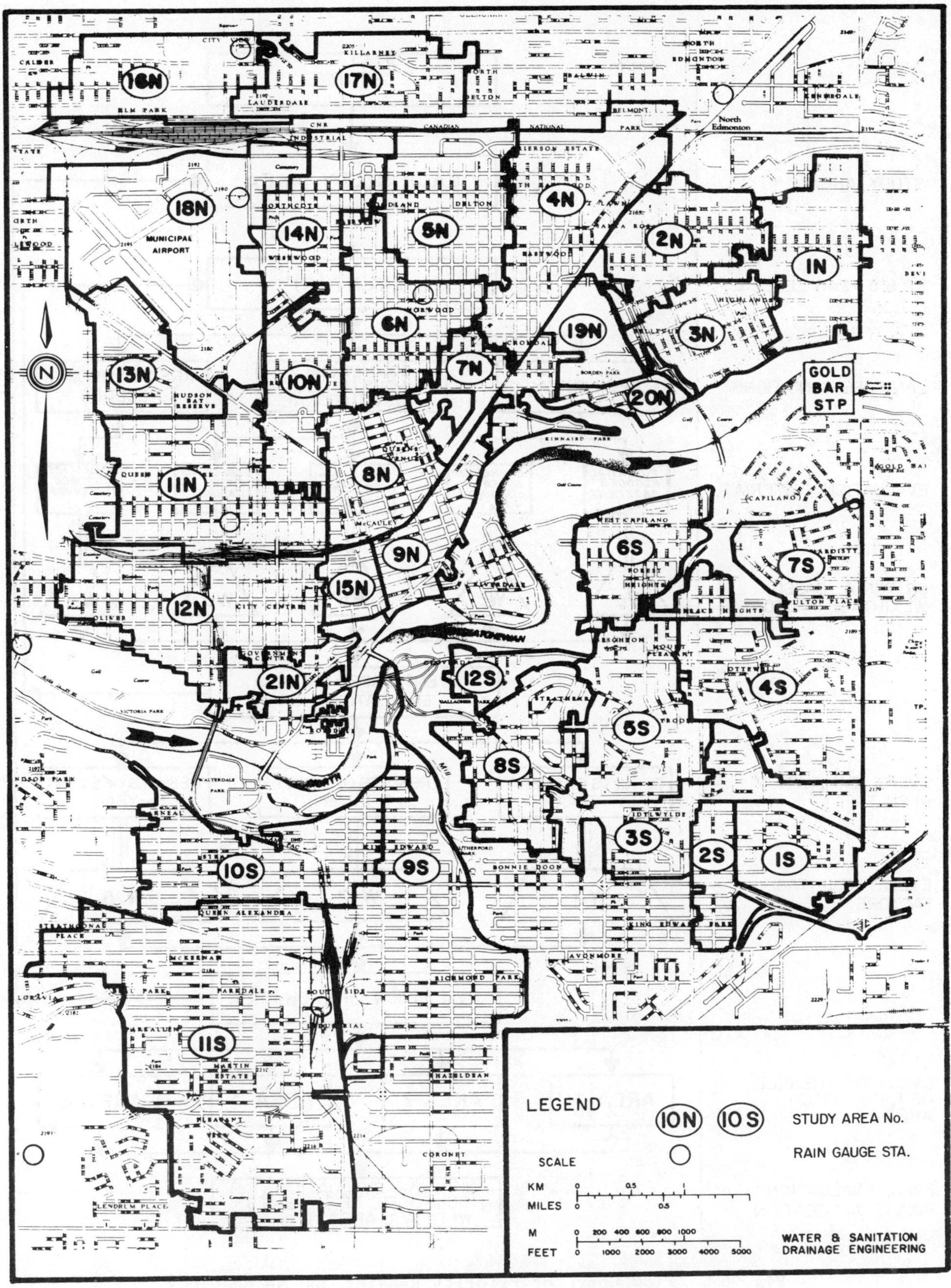

Fig. 5 City of Edmonton Combined Sewer Areas

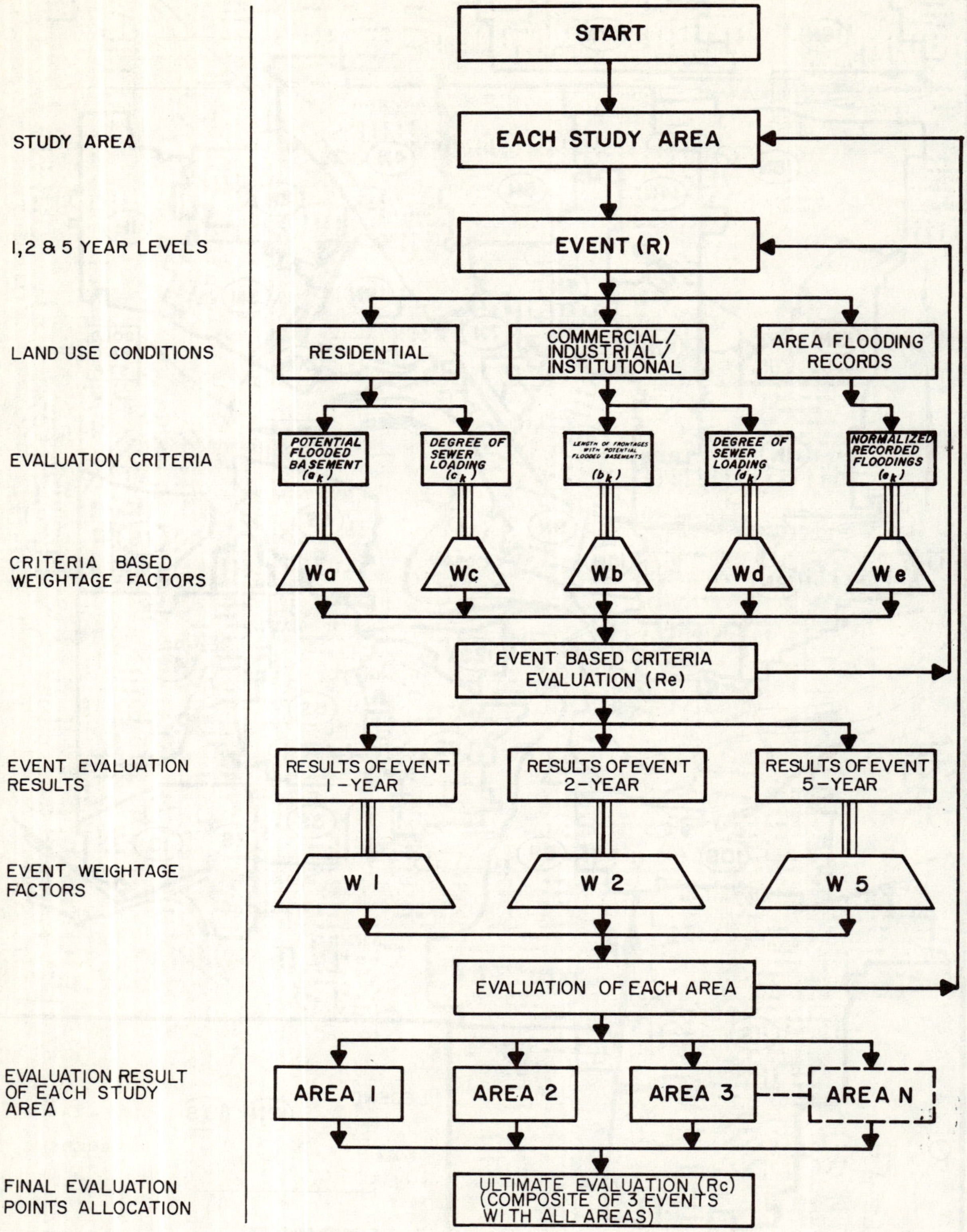

Fig. 6 Flow Chart of Priority Ranking and Methodology

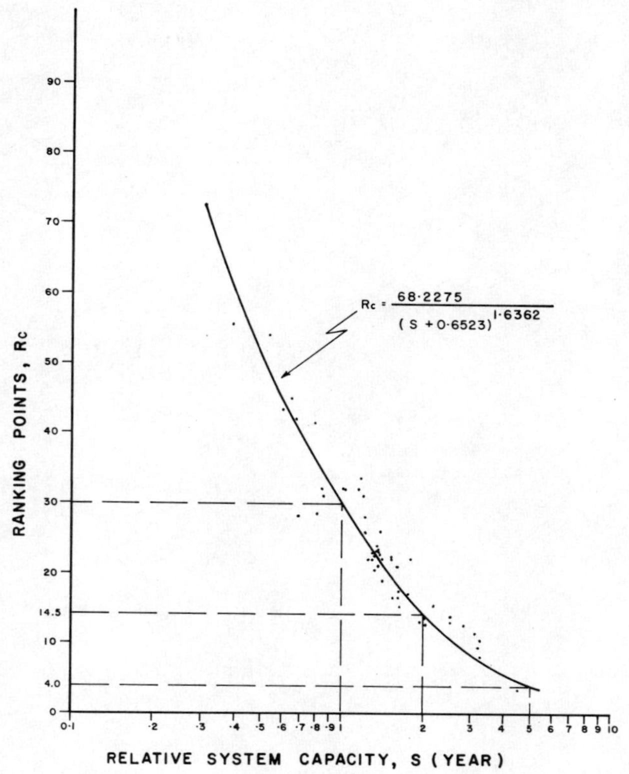

Fig. 7 Sub-Area plot, Ranking vs. Capacity

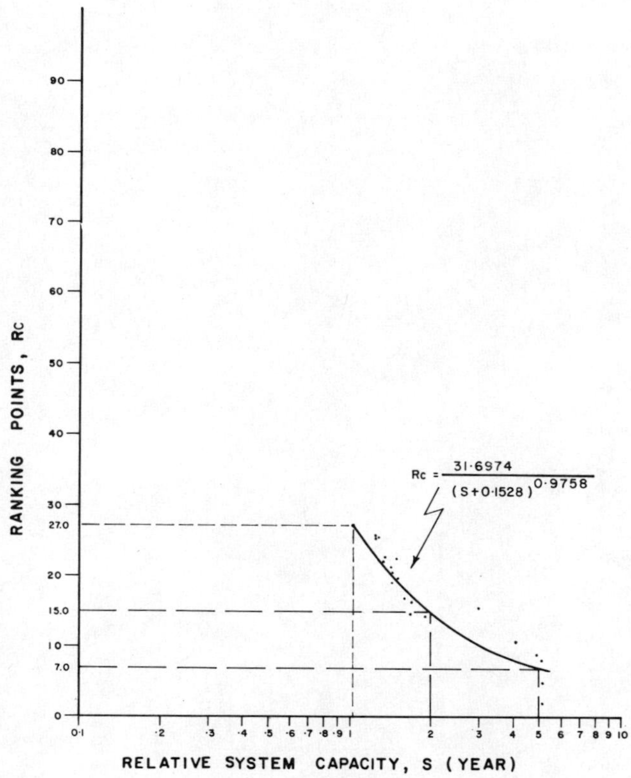

Fig. 8 Study Area plot, Ranking vs. Capacity

International Conference on the

PLANNING, CONSTRUCTION, MAINTENANCE & OPERATION OF SEWERAGE SYSTEMS

12-14 September, 1984

PAPER B3

THE SOCIAL COSTS OF SEWER REHABILITATION

by

D C PETERS, BSc., Ph.D., C.Eng., M.I.C.E

WRC ENGINEERING
PO BOX 85
FRANKLAND ROAD
BLAGROVE
SWINDON
WILTSHIRE SN5 8YR
ENGLAND

SUMMARY

Many sewerage systems in the UK are in poor structural condition and their hydraulic performance is often inadequate. Consequently substantial works are undertaken each year to deal with sewers which have failed structurally or where lack of capacity is causing pollution or flooding. The immediate cost of these works is high, currently around £150 million per annum, but the impact on the communities in which the works take place can be much greater. These adverse effects on society include traffic disruption, loss of trade to business, damage to buildings and underground services and general nuisance caused by the works. Collectively they are loosely called the **social costs.**

The paper sets out examples of how each of these problems can arise and attempts to quantify the extent and economic implications.

The strategy for sewerage rehabilitation now being adopted has, as one of its objectives, the minimisation of these social costs and the paper examines the ways in which this can be achieved and the implications for future works.

Examples are given of how differing levels of social and direct costs can arise from alternative sewer rehabilitation works and the advantages attainable through the new approach.

Held at The University of Reading, England. Organised and sponsored by BHRA, The Fluid Engineering Centre, Cranfield, Bedford in conjunction with the Water Research Centre.

1. INTRODUCTION

The Water Industry is a service industry and has a long and proud tradition of benefit to the community. Over recent years the emphasis in sewerage has increasingly switched from the provision of new services to the maintenance of existing services at acceptable levels. As a consequence the need is now to **rehabilitate** existing systems rather than construct new ones. This rehabilitation is aimed at 3 principal problems - flooding, structural deterioration and pollution from overflows.

In 1980 WRC began work on the development of a strategy to allow rehabilitation works to be undertaken more cost-effectively. Although financial savings to the Authorities were seen as the principal objective, it was recognised that savings in social costs could also be important. A study was set up to assess the scale and nature of the social disruption associated with rehabilitation works. It became evident that the economic cost of the disruption was significant and that it should feature in the strategy set out in the **WRC Sewerage Rehabilitation Manual** (Ref 1).

2. WHAT ARE SOCIAL COSTS?

The word cost is often synonymous with cash. It can however have a broader meaning implying all the adverse results of some decision. That is what economists mean when they refer to cost-benefit analysis and that is what it means here. The term **social** suggests a particular part of this overall cost, and so it is.

Basically, the **social costs** arising out of sewerage works are those consequences which are not borne by the Authority, but which fall directly on members of the community. They may be actual cash losses (either directly or through material damage) or less tangible losses such as environmental damage or loss of personal amenity. However they arise, the key factor is that they do not appear in the Authority accounts. This is important, and warrants consideration of what actually influences Authorities' actions.

3. WATER AUTHORITY POLICY

Water Authority Policy is dominated by service objectives (which define what type of service it should provide),

and what it can spend on achieving them.

The principal service objectives of sewerage are to:
- drain foul sewage from premises
- drain roofs and paved areas

and do so without unacceptable risks of
- flooding,
- structural collapse
- pollution of water courses

The term 'unacceptable risk' is necessary because there can be no absolute guarantee against any of these. What level of risk is acceptable is a matter for individual Authorities.

Ideally, the appropriate level of service would be defined by setting the optimal balance between the financial and social benefits obtained and the financial and social cost of achieving them, i.e. cost-benefit analysis. However, most of the benefits of sewerage rehabilitation are social rather than hard cash.

In 1981/82 the ten regional Water Authorities spent £191 million on sewerage capital works of which £153m was for rehabilitation works. Of this, £84m was spent on improving hydraulic performance, avoidance of flooding and pollution; £45m on works to improve both hydraulic performance and structural condition and a further £25m to improve structural condition alone (Ref 2).

The financial advantages to the Authorities of improved hydraulic performance are minimal, the benefits being almost entirely social. Indeed, the cash and material element of the social benefits are only a small part of the whole, the main benefits being the removal of health hazards, anxiety and loss of amenity - all intangibles.

Since these works absorb the bulk of the rehabilitation spending, and since the benefits of structural improvement are also partly social; the principal justification for sewerage capital spending is social benefit (i.e. the avoidance of social costs).

Unfortunately because there is currently no adequate basis for measuring these, the general use of **cost-benefit analysis** is impracticable. Instead the normal approach is to use **cost-effectiveness analysis.** A level of service is set subjectively and consideration thereafter centres on the most **cost-effective** way of achieving it.

This approach suffers from several drawbacks. In particular:
- it fails to justify the level of service proposed and
- it makes it difficult to examine the potential economic advantages of different levels of service

Because of this it is likely that the industry will come under increasing pressure to move towards cost-benefit analysis, covering both financial and social costs. Such project appraisal techniques are in use in other parts of the public sector, e.g. transportation. The problems in doing so are substantial as the methodology necessary to predict social costs has not yet been developed but the prospect is sufficiently real for the Industry to have set up a working party to examine it.

In the meantime, the only practical alternative is to use the cost-effectiveness approach. In doing so we do not quantify the social benefits but that is no excuse for ignoring the social costs associated with the works themselves; social costs which would be included in a full cost-benefit analysis.

If these social costs are significant they should influence how rehabilitation is carried out.

The reality is that they can be significant and the remainder of this paper considers the implications of this.

4. THE COST OF REHABILITATION

Sewerage rehabilitation works generate costs (in the broad sense) which divide into two types:

- financial costs to the authority - the **Direct Costs**,
- financial, material, and intangible losses to the community - **the Social Costs.**

4.1 Financial Costs to the Authority

These include:

- Planning, design and supervision,
- Payments to contractors, suppliers and DLO,
- Reinstatement of highways,
- Compensation to owners of buildings, services and land overtly damaged,
- Compensation to business for loss of profits.

The first three items are normally accounted for to their full value. The remaining two items represent compensation to third parties, which is (quite properly) subject to requirements of substantiation and reasonableness. While the folklore of inflated claims is well known, in reality many legitimate claims are not pursued for want of adequate records, organisation or energy on the part of those suffering the loss. Overall it seems likely that the total individual loss exceeds substantially the total compensation paid, the difference becoming a **social cost.**

4.2 Social Costs

The losses sustained by the community may affect individuals, groups or organisations. The main identified elements are:

- Delays and diversions to road traffic,
- Long term damage to underground services and adjacent buildings,
- Damage to roads and services on temporary diversion routes,
- Disruption of local economic activity,
- Loss of amenity caused by noise, dirt and smell.

In all these (bar the last one) there will be elements of straight financial loss as well as intangible losses.

5. SOCIAL COSTS ASSOCIATED WITH REHABILITATION

5.1 Traffic Delays

The vast majority of sewers lie beneath roads. Many of the original trunk sewers were laid under what were then, and remain today, the main thoroughfares. These are now usually busy traffic arteries where the consequence of excavation works is serious disruption. In order to quantify how important traffic disruption could be in sewerage works, WRC has carried out a number of on-site studies in which actual delays and diversions were measured. The cost of these were estimated using the Department of Transport values employed in the cost-benefit analysis of road schemes (Ref 3). The results are summarised below.

UNION STREET SEWER RENEWAL - GLASGOW 1974

The works involved the replacement, mainly by open-cut, of 385 m of sewers in a busy city centre, Union Street carries up to 1200 vehicles per hour at peak times, including 150 buses and it is also a principal shopping street. (see Fig 1). It was impossible to keep the street open to other than buses and service vehicles and closure extended over 22 weeks (Ref 4).

Serious congestion occurred on the other city centre routes and substantial sums were spent on traffic management and control. The estimated costs were: (1982 prices)

Civil Works	-	£410,000
Traffic Control	-	£ 49,000
Police	-	£101,000
PTE	-	£ 8,000
Delays to vehicles	-	£200,000 to £900,000

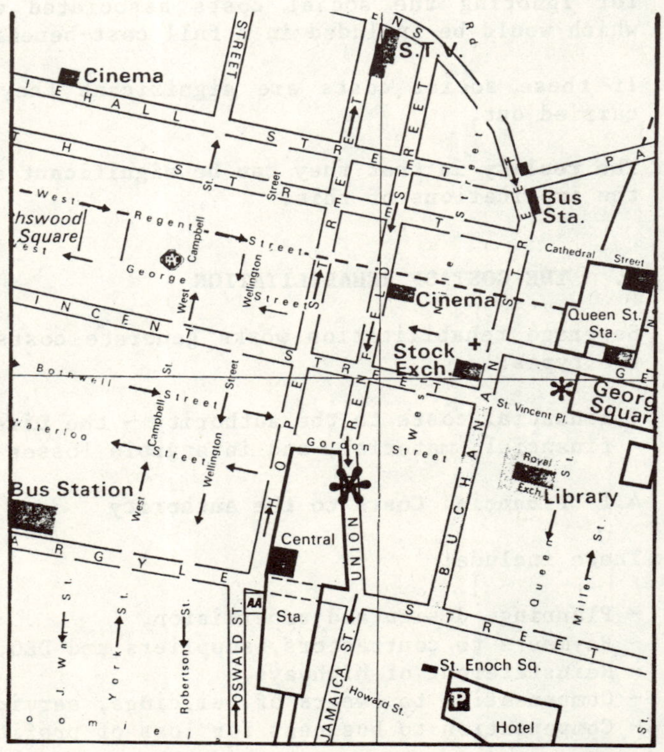

Fig 1

RIVERSWAY, PRESTON, LANCASHIRE - 1981

A 900 mm cast rising main failed beneath the eastbound carriageway of Riversway, Preston. The repair works closed this carriageway for one month. During this period traffic was diverted onto the old Blackpool Road which ran parallel to the dual carriageway. (Fig 2).

Although Riversway carried 10,000 vehicles per day, diversionary routes had adequate capacity so that congestion was limited and the total additional mileage was small. The cost of the repairs was £32,000. The cost of the traffic delays was estimated then at £28,000 (Ref 5).

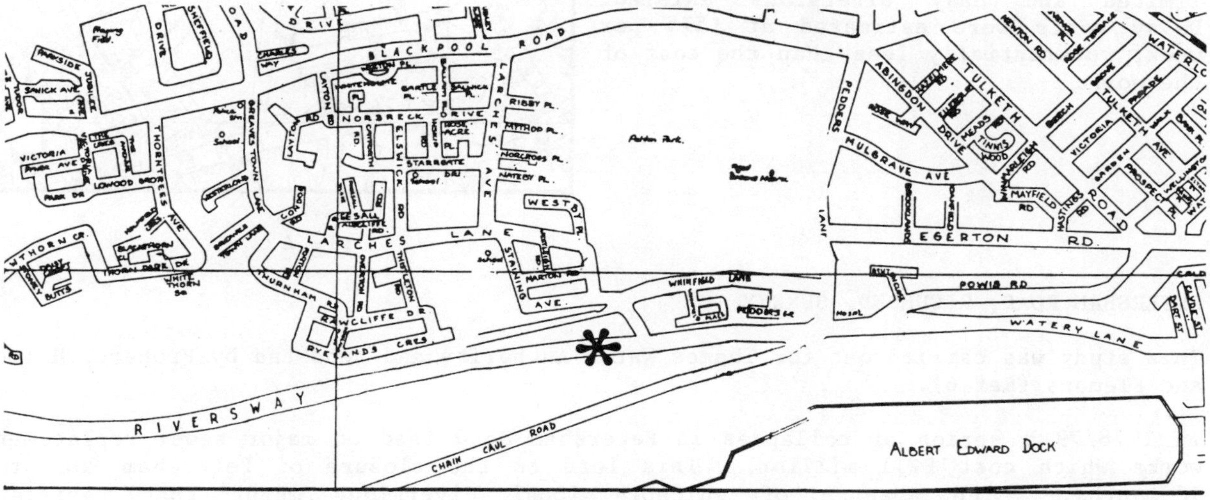

Fig 2

MOTTRAM ROAD, STALYBRIDGE, GREATER MANCHESTER - 1981

A 600 mm brick sewer collapsed beneath the junction of Huddersfield Road and Mottram Road. (Fig 3). A large void opened in the carriageway destroying one set of traffic lights and fracturing a 450 mm water main. Total repair cost was £38,000. As a result, Huddersfield Road was closed in one direction for 4 weeks and two of the five traffic lanes along Mottram Road were closed for 19 weeks. Mottram Road carries 8,000 vehicles a day in each direction and a heavy commuter traffic. Serious congestion occurred particularly during the evening rush hour. Half the vehicles normally passing the site diverted onto other routes. Vehicle delays were estimated at approximately 18,000 vehicle hours and additional journey length at 450,000 km. The cost of the delays and diversions was estimated at £63,000, that is approaching twice the direct cost of the repairs.

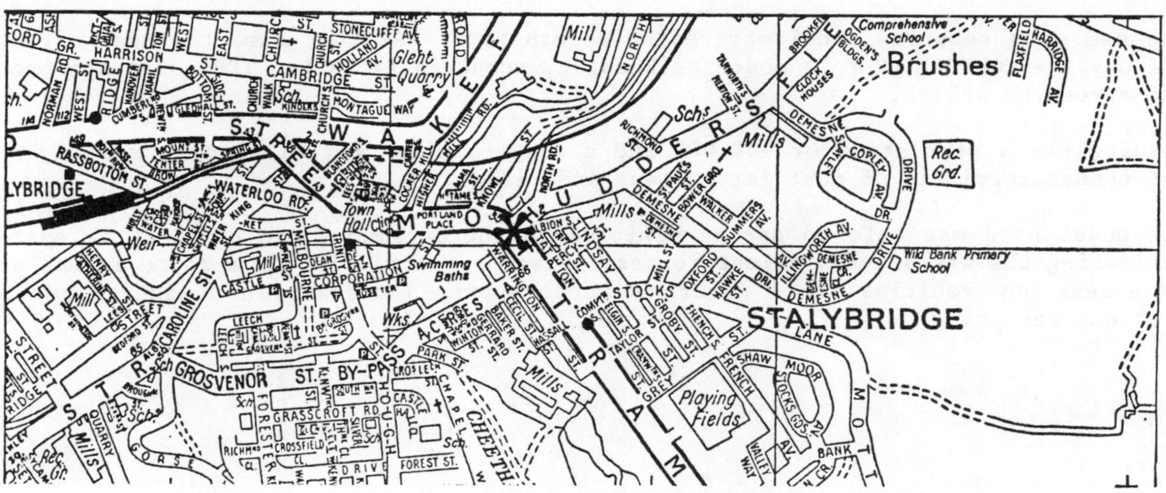

Fig 3

GLEBELAND ROAD, TRAFFORD, GREATER MANCHESTER - 1981

Glebeland Road is a relatively quiet road which carries mainly local traffic and buses. (Fig 4). It is only at all busy during the rush hour. Total daily flow is approximately 6000 vehicles in both directions. In 1981 works were undertaken to repair a manhole necessitating the use of one way working controlled by temporary traffic lights. Delays at the site were limited and easy diversions existed. Delay costs were estimated at £575 per week, substantially less than the cost of the works.

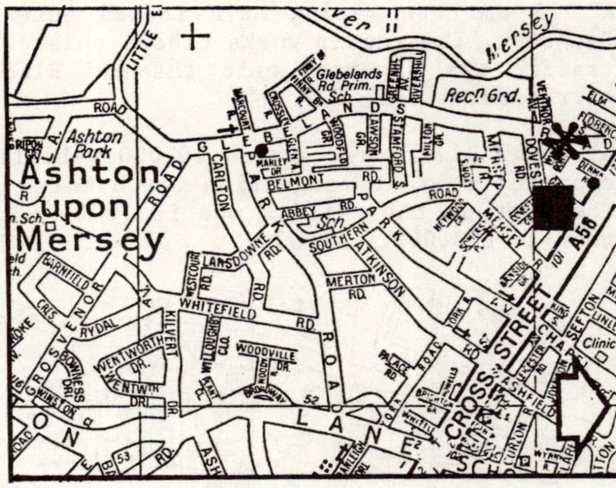

Fig 4

PETERSHAM ROAD, RICHMOND, SURREY

This study was carried out for Thames Water Authority and reported by Probert, Holmes and Flemons (Ref 6).

In 1978/79 a series of collapses in Petersham Road lead to major sewer replacement works which cost £2.1 million. This lead to the closure of Petersham Road for 18 months. The absence of suitable local diversions meant that vehicles, particularly heavy vehicles, had to travel up to 7 km further per journey. Bus services were badly disrupted.

The traffic delay costs were estimated at:

Delays and diversion (private and commercial vehicles)	£2m to £5m
Delays (public transport passengers)	£2m
Additional public transport operating costs	£1m
Total	£5m to £8m

READING TOWN CENTRE

The site studies (above) confirmed that the cost of traffic disruption caused by open cut sewerage works could be comparable to the cost of the works themselves. It was expected that even worse disruption could occur in town centres but the Union Street study had shown the difficulty of site measurement in such cases.

The use of a computer model offered an alternative. At that time the most advanced one was the CONTRAM (Ref 7) model of Reading, produced by TRRL. TRRL were approached and agreed to assist.

Reading has a population of 130,000 and a busy complex central road network. It is not uncharacteristic of most large towns and cities.

The model used was able to assign vehicles to the optimum route through the system, reflecting the way drivers select routes. Given a restriction on a route it was able to assess how vehicles would divert and the overal increases in journey times and distance resulting.

Nine sites were chosen along the line of the trunk sewers where open cut works would be expensive and cause varying degrees of disruption. The location of these is shown on Fig 5.

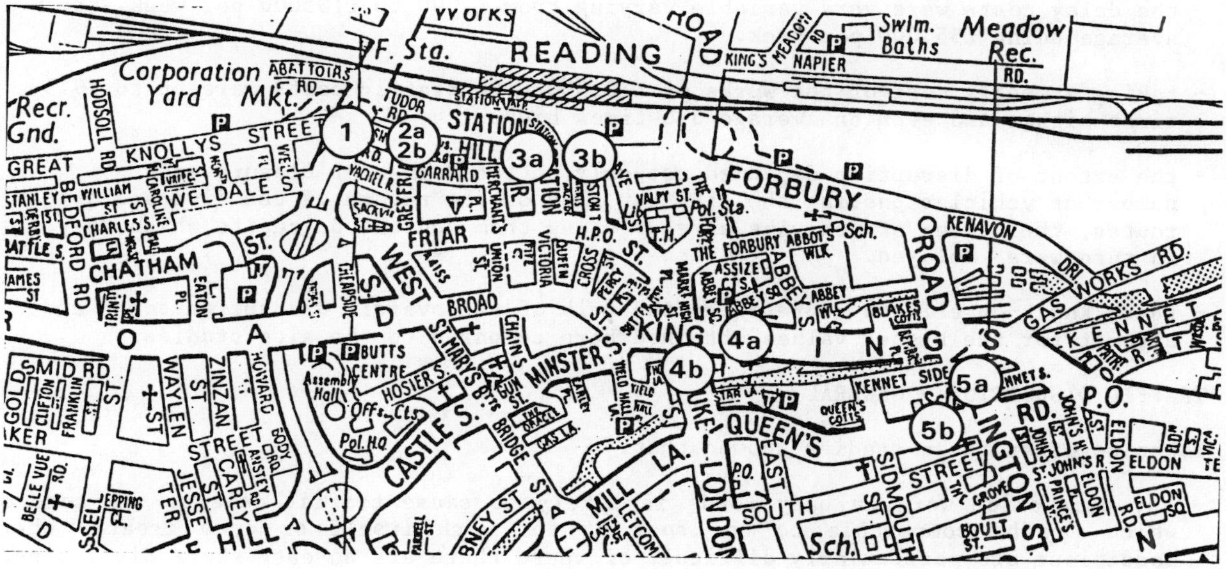

Fig 5

SITE	NATURE OF CLOSURE	VEHICLES PASSING SITE VEH/HR PEAK	NET DELAYS VEH.HRS /WK	INCREASED JOURNEYS KM/WK	WEEKLY COST £	TRAFFIC COSTS / ENG. COSTS (of £10000/WK)
1	Partial closure of Caversham Rd, Full closure of Stanshawe Rd	1650	9528	2172	28000	2.8
2a	Full closure of Stanshawe Rd and Greyfriars Rd,	1980	19512	17472	59000	5.9
2b	Full closure of Stanshawe Rd, Partial closure of Greyfriars Rd	1980	312	672	1000	0.1
3a	Full closure of Station Approach west bound	585	2424	0	7000	0.7
3b	Full closure of Blagrave St	1170	35484	12864	105000	10.5
4a	Full closure of Kings Rd	1326	4920	10188	16000	1.6
4b	Full closure of Duke St	1326	13548	16584	42000	4.2
5a	Full closure of Watlington St	936	11100	24864	36000	3.6
5b	Partial closure of Queens Rd	1443	17316	11604	26000	2.6

Fig 6

The extent and cost of the resulting disruption is shown on Fig 6.

The main points to emerge were:

- the delay costs were very variable varying from £1000 to £105000 per week, the average being £35,000 per week.

- taking a weekly cost of the works of £10,000, the traffic costs were up to 10.5 times higher and even on average 3.5 times higher.

- the extent of disruption depended not only on the degree of closure and number of vehicles passing the site but also on the nature of the diversionary routes, the number of vehicles already using them and the degree to which they in turn were delayed.

- these knock-on effects represented the bulk of the overall disruption costs and account for the higher values obtained here compared to the site studies.

TRAFFIC DISRUPTION - GENERAL

Taken together these studies showed that:

- any excavation work on busy roads is likely to cause traffic delays, the cost of which may be comparable to the cost of the work itself or even greater where conditions are particularly difficult or where there are no easy short diversions.

- in busy town and city centre roads, the delay costs are likely to be many times the costs of the works

- these effects are not limited to very major schemes but can be caused by any substantial excavation works

- although the majority of road users are on private journeys (inc. commuting), the value of time assigned to them is small (approx. 50p/hr). The value of time assigned to business and commercial users is much higher (200 p - 400 p/hr) and despite their smaller numbers these costs represent over half the total cost.

- Fuel wasted by all users is also significant at around 15% of total costs.

5.2 **Damage to Underground Services**

Underground services, including other sewers and water mains, may be damaged by sewerage works in two ways:

(i) Those that are exposed within an excavation are at obvious risk. They are most likely to suffer while being exposed, when damage may range from rupture of external coatings to complete fracture. Providing damage at this stage is noticed it can be repaired, the cost normally being rechargeable to the contractor and hence ultimately the authority. They are not then social costs.

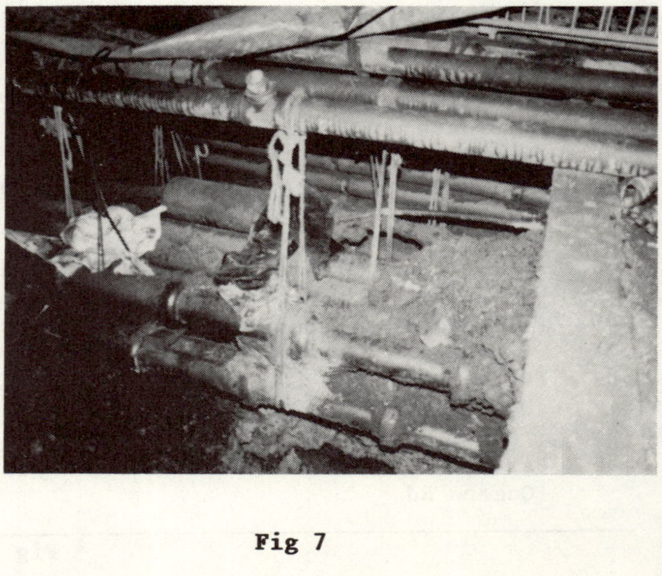

Fig 7

Less obvious damage may occur however. Minor damage to coatings may accelerate deterioration and shorten the service's working life. The disturbance inevitable with excavation and backfilling may also lead to increased stresses in rigid services, increasing the risk of fracture and stress related deterioration. Cast iron pipes are believed to be particularly vulnerable in this way.

(ii) Excavation inevitably leads to subsidence of adjacent ground. Good practice can limit this in planned works but sewer collapses often lead to totally uncontrolled subsidence. The ground movement, which is usually horizontal as well as vertical, induces bending stresses in services beside the excavation.

The problem is not necessarily one of immediate failure, rather that the increased stresses and flexure of joints leads to reduction of life expectancy and hence higher future maintenance costs. Again small cast iron pipes appear most vulnerable.

The scale of this problem is difficult to estimate. One statutory undertaker - British Gas - is, however, sufficiently concerned to have initiated the Consultative Agreement that now applies to all Water Authority and British Gas excavation works. This requires the Water Authorities to bear the cost of replacing or diverting gas mains deemed to be at risk from adjacent sewerage works. The cost of this over the last year was £1M. This may also be indicative of the scale of the damage to water mains.

5.3 Damage and Disruption Caused by Traffic Diversions

When a main traffic route is obstructed, traffic looks for alternative routes. The routes chosen will be those which are the most convenient to the driver, not necessarily those which are designed to take the weight or volume of traffic temporarily using them. The scale of the problem is unknown although there is plenty of hearsay evidence of its occurrence. What is clear is that a small number of heavy lorries can cause serious damage to road surfaces designed (if they ever were) for light vehicles only. There may also be damage to services beneath those roads, mainly the shallow gas and water mains.

The loss of amenity caused to residents living on temporary diversions may also be considerable. Heavy traffic using residential streets not only causes noise and dust, but also poses very real risk of accidents.

5.4 Disruption to Economic Activity

Prior to 1979, it was normally held that loss of trade to business caused by sewerage works did not entitle the business to compensation. In 1979, a court case (Leonides v Thames Water Authority (Ref 8) changed this view and authorities have since paid agreed claims for loss of profit where sewerage works have been the cause. In this respect they have ceased to be social costs and become direct costs. Experience to date suggests that the number and size of claims is generally too small to influence how work is done. It must be recognised that this may change as more businesses realise that they can claim and that some businesses are more vulnerable, and militant - garages being notable.

Whether the payments fully compensate the loss is doubtful; certainly many apparently legitimate claims go unpursued. The unclaimed part of the total losses remains a potential social cost.

It has been suggested by Read (Ref 9) that recurring disruption to the centre of Manchester was not only causing individual loss but was destroying confidence in the whole business community.

There is, however, a complication. Most of the losses represent loss of turnover, i.e. customers who would go elsewhere. But while one firm undoubtedly loses, another usually gains. Depending on how far one casts the net, there may appear to be no overall loss. Are they then social costs?

On a national or even regional basis, probably not. On a local basis they may be. There is also the question as to whether in equity it is reasonable to ignore damage to one party just because someone else gains.

Fig 8

5.5 Loss of Amenity

Loss of amenity covers the effect of noise, dirt, smell and inconvenience caused by open excavations. Traffic disruption is a particular case that has already been dealt with.

It is not just the presence of a hole that causes loss of amenity: it is people's reaction to it. If they don't mind, there is no loss and it has long been evident that certain groups in society mind more than others. Again, Authorities are faced with questions of equity. On the basis of the perceived social costs, the greatest protection should go to upper and middle income housing areas as they are the ones that complain most. Such an approach would not be acceptable on moral or political grounds however, and all groups are usually treated equally.

This aspect of social costs could be very significant, certainly people's reaction to flooding is the key to justifying flood protection.

Fig 9

6. SOCIAL COSTS - HOW TO AVOID THEM

If we accept that sewerage rehabilitation could involve substantial social cost then we are faced with finding a way of minimising those costs within the industry´s overall financial regime.

Most of the elements of social disruption arise directly from open excavations required to repair or replace sewers in highways. Minimise these and the problem recedes.

The simplest advice to those contemplating rehabilitation of a sewerage system would be - don´t, leave well alone. Unfortunately, this is untenable. Many of the sewers suffer from structural defects which if left unattended will cause collapse, subsidence, and far greater problems, both financial and social. Other sewers, often the same ones, lack hydraulic capacity. This causes flooding and pollution, both essentially social problems. Inaction would perpetuate social problems far worse than those caused by rehabilitation. If we must act then, how?

In the past the easiest, cheapest solution was to replace the sewer with a newer larger one. Such an approach is, however, the most disruptive way possible. On small scale works, restricted working hours and space can minimise problems, but this is impractical on the large works which represent the bulk of rehabilitation projects.

One answer, the traditional answer, is tunnelling. By limiting open excavation to a few shafts, often in side streets, the disruption can be greatly reduced. Unfortunately, this approach is usually much more expensive than open-cut work. Authorities have been understandably hesitant to adopt tunnelling merely to avoid disruption unless it was likely to be exceptionally severe or in the face of intense local pressure.

Now alternatives exist

Sewer renovation can restore many structurally defective sewers more cheaply than they can be renewed. At the same time it also causes far less disruption. Most of the techniques involve only limited excavation, a few involve none and the work is completed more quickly. Consequently both the scale and duration of any disruption is cut so that the overall social costs are only a fraction of those caused by open-cut renewal.

The replacement of sewers suffering from hydraulic overload can also often be avoided. Use of new methods of hydraulic analysis permit the design of upgrading measures which allow much of the existing system to be retained. Again not only is this cheaper but it minimises excavation and hence social disruption.

This approach, which is fully described in the **WRC Sewerage Rehabilitation Manual,** concentrates effort on those sewers where open-cut excavation is both most expensive and most disruptive. These **critical sewers** are typically large deep sewers which lie for historic reasons under the busiest streets. The approach proposed in the Manual is justified by savings in direct costs, and social disruption is only taken specifically into account at present in identifying the critical sewers. However, because the approach seeks (on financial grounds) to minimise sewer replacement, it will also cause much smaller social costs than conventional open cut work. What is particularly notable is that **savings in social disruption no longer always have to be bought through higher engineering costs.**

Nonetheless, there will remain circumstances where the least disruptive approach is not the cheapest or where work can only be justified at all when savings in social costs are taken into account. This latter point applies particularly to structural rehabilitation of the less critical sewers where the financial advantages of collapse prevention may be marginal but where collapse would cause serious disruption.

In all such cases Authorities must be uncertain as to their priorities. In extreme cases they will no doubt accept the higher costs as they have done in the past. For the others they will be understandably reluctant to incur higher costs until Government indicates that it accepts this in setting its financial controls.

7. IMPLICATIONS AND CONCLUSIONS

7.1 The principal benefits of sewerage are social. The industry cannot ignore the social costs associated with how it provides and maintains the service.

7.2 Most of the sewerage systems in the UK are in need of rehabilitation. The WRC Sewerage Rehabilitation Manual sets out a strategy which is much less disruptive than conventional sewer replacement techniques. This approach will normally be cheaper financially so that savings in social costs have no longer to be always set against higher engineering costs. The adoption of the strategy will result in substantial savings in both the financial and social cost of rehabilitation.

7.3 There will remain circumstances where social cost savings will justify works which cannot be justified on financial grounds alone. There are indications that the government expects the industry to move towards a cost-benefit approach to project appraisal and that means including social costs and benefits. The industry needs clear guidance as to how it should react when inclusion of social costs leads to higher overall financial costs. This is applicable to **cost effectiveness** analysis as well as **cost-benefit** analysis so the advice is needed **now.**

7.4 The inclusion of social costs in scheme appraisal presumes the existance of methods for their prediction. Such methods must be simple and flexible in use if they are to be applied in everyday use. At present these do not exist.

The **cost-effectiveness** analysis approach in current use requires only a means of predicting the social costs associated with engineering works. Much of the ground work has been done and the remaining need is to work this up into practical guidelines. This work is underway and should be completed relatively quickly.

The general use of **cost-benefit** analysis will require in addition the development of methods of predicting the **social benefits** associated with flood alleviation and pollution control . This is a much greater undertaking.

Some work has already been done on flood alleviation (Ref 10) and WRC is also actively engaged in this area.

Much remains to be done, however, before cost-benefit analysis becomes a practical reality in sewerage. It is likely that the necessary research and the assessment of the implications will take several years.

ACKNOWLEDGEMENTS

The author wishes to thank all those in the industry and elsewhere who co-operated in the research reported here.

In particular thanks are due to:

North-West Water Authority
Yorkshire Water Authority
Thames Water Authority
Preston District Council-Borough Engineers Department
Thameside Metropolitan Council - Borough Engineers Department
Trafford Metropolitan Council - Borough Engineers Department
Manchester City Council - City Engineers Department
Strathclyde Regional Council - Department of Sewerage
Transport and Road Research Laboratory
Department of Transport
Middlesex Polytechnic

REFERENCES

1. WRC/WAA Sewerage Rehabilitation Manual.
   WRC Engineering 1984.

2. **GLENNIE E B and ROBINSON S.** The Sewers of England and Wales. The System and its Structural Problems.
   WRC External Report still to be published.

3. Department of Transport, Highway Economics Note No 2
   July 1980

4. **TOPPING W J and GOW W F.** Sewer Reconstruction Works in Glasgow with Particular Reference to Union Street.
   Paper presented to Institution of Municipal Engineers (Scottish Branch) at Ayr, on 4 October 1974.

5. **SCHOFIELD J.** A Study of Traffic Disruption Caused by a Sewer Failure on Riversway, Preston.
   WRC External Report No 46E, WRC Engineering Centre, October 1981.

6. **PROBERT L A, HOLMES J E V and FLEMONS K J.** Petersham Road, Richmond, Sewer Failure.
   ICE Conference Paper No 9 in Restoration of Sewerage Systems.
   Thomas Telford Ltd., 1982.

7. **LEONARD D R, TOUGH J B and BAGULEY P C.** CONTRAM: A Traffic Assignment Model for Predicting Flows and Queues During Peak Periods.
   Transport and Road Research Laboratory Report 841, 1978.

8. **LEONIDIS V THAMES WATER AUTHORITY**
   Local Government Law Reports 1979
   Volume 77, pp 722 - 729.

9. **READ G.** One of the Priorities of a Major City.
   Water Bulletin, 12 August 1983, pp 12 - 13.

10. Flood Alleviation Benefit-Cost Analysis.
    Cranfield Conference Papers.
    Flood Hazard Research Centre
    Middlesex Polytechnic, July 1983.

International Conference on the
PLANNING, CONSTRUCTION, MAINTENANCE & OPERATION OF SEWERAGE SYSTEMS
12-14 September, 1984

THE STRUCTURAL DETERIORATION OF SEWERS

G.M.A. Jones

M.Sc.,Ph.D (Wales)

WRC Engineering,
P.O. Box 85, Frankland Road,
Blagrove, Swindon,
Wilts, SN5 8YR.

Summary

The nature of typical structural defects found in brick and pipe sewers in the UK is described in broad detail.

In assessing sewer condition internal visual examination based on CCTV or photographic survey plays an important role in identifying defective sewers. However, the fact that a sewer has structural defects does not necessarily mean that it is in imminent danger of collapse.

When assessing the condition of deteriorated sewers consideration should be given to the nature of the surrounding ground, groundwater conditions and hydraulic conditions within the sewer, in association with aspects such as quality of connections and general fabric condition.

Current understanding of potential mechanisms for ground loss and void formation around sewers is summarised. This summary is based on an ongoing programme of field investigation supplemented by laboratory and theoretical studies. The major risk factors associated with surcharging of deteriorated sewers in certain ground conditions are highlighted.

Held at The University of Reading, England. Organised and sponsored by BHRA, The Fluid Engineering Centre, Cranfield, Bedford in conjunction with the Water Research Centre.

1. INTRODUCTION

The ongoing work described in this paper forms part of the background research necessary for the production of the WRC/WAA Sewerage Rehabilitation Manual (Ref. 1). Some of the contents of this paper will be well known to people experienced in sewer surveys and no claim to complete originality is made. The sewerage research effort at WRC Engineering is tailored towards the Rehabilitation Manual and its planned updates. A key plan of the current programme is shown in Fig. 1.

The assessment of deteriorating structures is more a matter of judgement than the design of new ones, and this is particularly true of sewers. Our understanding of why sewers collapse is still imperfect and the information available to those who assess their condition is limited. There are many minor defects in sewers compared to the number of collapses that occur, and even apparently serious defects do not always lead quickly to collapse. It is important to identify the factors that cause deterioration, in particular the type of ground, the ground water level and the load on the sewer. This information should be combined with the internal inspection record in assessing the condition of the sewer.

Collapse investigations carried out by WRC (Ref. 2) suggest that most of the defects that cause collapse would have been visible from a good quality CCTV or photographic survey. Discerning the defects that are liable to lead to collapse, from the much larger number that are not, is a difficult task. The aim is to provide improved guidance to engineers to assist in this task. Three significant causes of collapse would not be visible from a remote visual survey. These are faulty connections (where the connection itself and not the sewer is defective), manhole backdrop detailing and voids in the backfill behind the sewer wall: a reminder that visual survey, while a very useful tool, has its limitations.

Knowledge of the structural performance of deteriorated sewers and the surrounding ground is also of considerable importance to the preparation of appropriate guidance for the design of renovated sewers.

This paper concentrates on those sewers which are of principal importance in the context of the UK network. That is, to clay or concrete pipe sewers and brick sewers with one or two rings of brick. Observations with respect to ground conditions around sewers can be extended to other forms of sewer if required.

2. HOW SEWERS DETERIORATE

The process leading to collapse can be divided into three stages:

STAGE 1 - formation of an initial defect,

STAGE 2 - deterioration arising out of the defect,

STAGE 3 - collapse of the weakened structure.

2.1 Stage 1 - An initial defect

Collapse of a sewer normally originates where an initial, often minor, defect allows further deterioration to occur. Such defects may include:-

- cracking caused by excessive vertical load or bad bedding,
- leaking joints,
- damage caused by badly made connections.
- loss of mortar, missing brick.
- interference damage.

2.2 Stage 2 - Deterioration

Deterioration often involves the surrounding soil. Where water can flow into and out of the sewer, soil can be lost into the sewer. Ground loss can also occur when ground water or effluent from defective connections, flows through the bedding material parallel to the sewer.

The sewer material itself may deteriorate. Joint material and mortar can be eroded, and concrete can be attacked by some chemicals.

2.3 Stage 3 - Collapse

Collapse is often triggered by some random event that may not be related to the cause of deterioration, perhaps a storm or an excavation nearby. This means that it is not possible to predict when a sewer will collapse. However it is possible to judge whether a sewer has deteriorated sufficiently for collapse to be likely.

It is important to appreciate that the concept of progressive deterioration occurring at a regular rate until eventual collapse occurs is normally inappropriate. We are dealing with a situation where there is an underlying trend of progressive deterioration which is accelerated by random events during the service life of the sewer. Hence 'stepped' rates of deterioration, as shown in Fig. 2, are likely. In any time interval t_1 or t_2 during the life of the sewer, deterioration rates may well be significantly different. Any attempts to monitor deterioration and estimate service life by short term field research are therefore likely to be unsuccessful.

3. THE STRUCTURAL BEHAVIOUR OF SEWERS

3.1 Stresses in sewers

The materials used in the past for sewer construction have mainly been clayware, brick and concrete. They generally have more than adequate compressive strength: clayware and concrete also have some tensile strength. Consequently rigid pipes made from clayware and concrete are strong enough to resist some vertical loading on their own. Brick sewers are dependent on soil support for satisfactory performance due to the limited or negligible tensile capacity of brickwork.

A longitudinally cracked sewer can sustain a vertical load well in excess of that necessary to initiate cracking, provided the soil on either side is sufficiently stiff. This has been demonstrated for longitudinal cracking caused by uniform overload (Refs 3, 4). Therefore it is usually not necessary to know the strength of the sewer material in order to assess its stability.

3.2 Ground loss and deteriorated sewers

The loss of ground into deteriorated sewers can result in the formation of voids adjacent to the sewer or under the surface of highways. Alternatively, loss of ground can result in the formation of zones of low density soil, which permit progressive deformation of deteriorated sewers or settlement of roads and adjacent structures.

The processes require the passage of water through gaps in the sewer wall and the transport of soil with it. This is most likely to occur in coarse silts and fine sands, but other soil types are also susceptible. In a deteriorated sewer that surcharges, exfiltration is followed by infiltration, and where this occurs regularly soil is particularly likely to be brought into the sewer.

A detailed examination of the factors contributing to the susceptibility of soils to the processes leading to ground loss is beyond the scope of this paper. However, in general terms the following behaviour can be recognised:-

1) In soils which have little or no cohesion, such as sands and silts, movement of soil particles occurs. This can take the form of migration of fine particles through

a coarse soil matrix, or the washing out of particles of all sizes.

2) In soils which have significant cohesion, such as clay, erosion of soil particles can occur. This takes the form of attack on surfaces of soil exposed to flowing water. Erosion occurs when the cohesive attraction of surface particles is overcome by flow.

3) In mixed soils the processes of particle migration and erosion can occur together.

Fig. 3a and b illustrate the behaviour of silts and sands in the vicinity of a crack type defect. When the sewer is suffering from infiltration, a small initial ground loss can occur but there is a tendency for the soil to segregate adjacent to the defect and form a stable self filtering system over the defect. Depending on the grading of the soil and its density it is possible for defects of up to 10mm width to be spanned in this manner. The defect size can be several times larger than the largest particle size.

Fig. 3c illustrates the behaviour of a similar soil in the vicinity of a defect when hydraulic surcharging occurs through the defect. The main feature to note is that surcharging can destroy the stable self filtering system over the defect. Ground loss can now occur through defects of 2mm width or less. It is no longer possible for the defect size to be greater than the largest particle size. In deteriorated brick sewers fluctuations of water level within the sewer may result in disturbance of the self filtering mechanism.

Fig. 4a and b indicate the behaviour of clays in the vicinity of a crack. Here again surcharging through the defect is a risk factor although the principal mechanism is now one of erosion rather than migration. Poorly compacted clays with low plasticity are likely to be most susceptible to erosion. Well compacted clays with high plasticity are likely to be relatively resistant to erosion. Since the process is a function of the area of the soil exposed to flowing water, longitudinal sewer defects present an added risk factor.

In mixed soils the processes of migration and erosion can act together resulting in loss of ground. The presence of lenses or layers of susceptible soil can have a significant effect on rates of ground loss. In broad terms the processes involving migration or washout of soil particles are more likely to lead to situations of rapid ground loss than those involving erosion of soil particles.

It is clear that there are many factors affecting the susceptibility of a particular sewer to suffer from loss of the surrounding ground. Clearly the major factors emerging are the structural condition of the sewer (in particular its 'leakiness'), the nature of the surrounding ground and the hydraulic 'duty' within the sewer. At a detailed level other effects such as soil grading, soil chemistry, soil density, defect goemetry etc will have an influence.

As previously stated, when ground loss occurs there may be loss of density in the soil around the sewer or actual void formation may occur. The likelihood of a stable void occurring depends on the position of the water table and the soil type and the following comments are relevant :

a) In the long term all unsupported voids below the water table are unstable.

b) Above the water table, voids in clay soil are generally stable. In fine or well graded granular soils voids may exist through capillary suction.

c) When a sewer surcharges and exfiltration occurs in fine granular soils, capillary suction is destroyed and any voids may grow larger or migrate upwards. In clay soils erosion is likely to take place in the vicinity of the sewer and there is no tendancy for the void to migrate.

Voids or areas of low density soil are often found immediately below a road surface. The void will have formed close to the sewer and then migrated upwards. Soil loss of this type is often recognised by road subsidence or by settlement damage to adjacent buildings. Silts and sands are more susceptible to this process than clays.

Since the behaviour of soils is governed by the ground-water environment, events such as bursts or leakage of water mains may have a profound influence on the stability of sewers in the vicinity.

In shallow sewers in clay type soils the effect of seasonal moisture variations in the soil surround may be significant. Moisture variations result in volume changes and movement of the soil surround. This may lead to opening of joints, widening of cracks etc. Subsequent deterioration is then accelerated. Unusual drought conditions may be a major influence.

3.3 Fabric decay

Although the main sewer materials - clayware, concrete or brick - are normally very durable, decay can occur either in them or more commonly in minor elements such as jointing or mortar. There are three common forms:

(a) Loss of jointing in pipe sewers

Many clay pipe sewers built before 1920, were laid with joints sealed by puddle clay. Although effective in the short term it has usually long since been lost leaving leaky joints.

(b) Erosion of mortar in brick sewers

Mortar, particularly the older and softer lime mortar in brick and masonry sewers, can be eroded or attacked chemically leaving bricks loose and the sewer subject to leakage.

(c) Chemical attack of concrete

Chemical attack caused primarily by sulphides in sewage or industrial trade effluent can lead to the collapse of concrete sewers. It is generally rare in the UK but the locations at risk are easy to identify (Ref. 5).

3.4 Loads on deteriorated sewers

A brick sewer, particularly one which has lost mortar, requires a ring compressive stress to maintain its stability. Where this stress is low the sewer can become "sensitive" to small disturbances which may cause bricks to fall out or move within the structure.

A similar effect can occur with multiple cracked pipe sewers where the various pieces can be held in place by the frictional effect of the broken edges provided there is sufficient load on these joints.

Deteriorated sewers in granular soils often experience low stresses from the surrounding soil due to the very efficient arching capability of the circular or near circular shape in frictional soil materials. However stresses on sewers in clays are closer to the full over-burden stress since large deformations are required to mobilise the full soil strength.

3.5 Influence of water pressure

Sewers below the water table will inevitably leak and infiltration will occur.
In the case of a perfectly watertight sewer, water exerts an approximately uniform pressure which causes no problems with circular sewers. With brick, egg or oval sewers, bulging of the sewer walls can occur due to a large external head of water. This is potentially serious and can affect the overall stability of the sewer. Drilling holes in the sewer wall may not relieve the pressure, since local seepage pressures may be increased.

Where the water table is above a sewer with a worn or missing invert, instability due to piping failure is possible, and is more likely in cohesionless soils.

4. IMPLICATIONS OF SPECIFIC DEFECTS

Visual survey plays an important part in the process of assessing sewers. Procedures for the factual description of defects are now established in the Manual of Sewer Condition Classification (Ref. 6). However, it is clear that a broad understanding of defects and their significance is required if sensible assessment decisions are to be made.

The aim of this section is to discuss the implications of specific defects for pipe and brick sewers. No claims for completeness are made since there is a necessity to be concise in a paper.

4.1 Specific structural defects - pipe sewers

a) Longitudinal cracks and fractures

The classical case of longitudinal cracking involves cracks running the length of the pipe at 12, 9, 3 and 6 o'clock. The crack at the crown, 12 o'clock is normally the most obvious. The cracks at 9 and 3 o'clock only become visible as deterioration proceeds as the 'hinge' is at the inner wall of the pipe. The crack at 6 o'clock is normally obscured by flow. If the soil support is inadequate or becomes so, the cracks will develop into fractures.

b) Circumferential cracks and fractures

These are most commonly found near joints. They are generally caused by uneven support, arising from bad laying practice or ground subsidence. In pipelines with joints made 'rigid' with mortar, longitudinal movement will often result in circumferential cracking in the pipe barrel providing stress relief. The defects are not serious in themselves but can allow leakage and hence deterioration.

c) Multiple cracks and fractures

These may arise from simultaneous vertical loading and uneven support, from point loads caused by stones in the backfill, or from the chance coincidence of cracking with inherent lines of weakness in the pipe, or from interference.

Fractures are more serious than cracks, since soil is more likely to be washed into the sewer, and pieces of pipe may fall out or be dislodged.

d) Broken pipes

A pipe is broken when pieces of a cracked or fractured pipe have visibly moved from their original position or are missing. This will normally represent a further stage in the deterioration of a cracked or fractured pipe and is a very serious defect.

NOTE: As defined in the Manual of Sewer Condition Classification (Ref 6) broken pipes may include minor chipping to spigots which is not normally a serious defect. It is necessary to ascertain precisely what is meant by a broken pipe.

e) Deformed pipes

Deformation of pipe sewers occurs when the pipe has cracked longitudinally. The segments of cracked pipe then adjust according to the level of soil support available and deformation occurs. Significant deformation is generally clear evidence that secondary deterioration is taking place.

Final collapse is unlikely to occur until deformation exceeds 10% of vertical diameter. Random 'events' in the service life of the sewer (e.g. surcharging) may influence the soil support providing stability to the cracked pipe. In significantly deformed pipes collapse could occur quickly in response to an external influence.

4.2 Specific structural defects - brick sewers

This section applies to 1, 2 or 3 ring brick sewers only.

a) Mortar loss

Mortar loss is very common but assessing its extent is difficult. Mortar may appear to be missing but actually be present below the surface due to the configuration of the brickwork. Black and white CCTV is inadequate for assessing surface loss; colour photographs generally offer a more effective basis for judgement. In assessing mortar loss it should be noted that in small diameter sewers no internal pointing may have been possible during construction. Similarly, some sewers were built using tapered bricks laid without mortar.

Mortar loss can allow water to pass through the fabric initiating loss of the soil support which is vital to brick sewers. It can also allow bricks to slip or the whole fabric to settle and lose the compression between bricks.

b) Longitudinal cracks and fractures

These are difficult to identify unless the mortar is in good condition. Reliable identification requires a good definition survey. Colour often provides a means to differentiate between a crack and adjacent mortar.

Minor cracks and fractures could have arisen at the construction stage, through inadequate compaction of side backfill leading to initial settlement.

A crack in the crown of a single ring sewer is less serious than one in a multiple ring sewer where it indicates lack of compression in the inner ring, and hence the risk of it falling. Fractures in either case are serious defects.

Good quality visual survey is necessary to identify these defects. Care should be taken to ensure that the important crown area is adequately surveyed.

c) Circumferential and multiple cracks and fractures

Circumferential and multiple cracks/fractures are uncommon in brick sewers and usually indicate a localised problem. They are most commonly found where differential settlement has occurred, at manholes, junctions or changes from open cut to heading work.

d) Displaced and missing bricks

Bricks can move when the mortar that binds them is lost. Displaced bricks in the sewer walls are often an indication that mortar loss in the invert is becoming sufficient to promote instability of the sewer. This type of displacement results in deviations in line and level along the sewer. Once movement occurs displacement or loss of crown bricks often results.

Displaced and missing bricks are a more urgent problem in single ring than in multiple ring sewers but in both cases progressive collapse is likely. Where bricks are totally displaced or missing from the crown of single ring sewers or from both rings in double ring sewers there is clearly a real risk of collapse.

The risk of further disturbing loose bricks should be borne in mind when surveying, cleaning or renovating brick sewers.

A line of displaced bricks at the springing may result from the hand trimmed construction method, the lower half of the sewer being laid on a hand trimmed trench bottom, while the upper half is surrounded by backfill that is less well compacted. This can lead to a shear fracture at/or about springing level. Depending on the degree of subsequent movement, the sewer can then be at risk of collapse since the integrity of the brick ring has been destroyed.

e) Spalling and wear

Spalling of the brick surface can be caused by excessive compressive stress, poor bricks, chemical attack and/or by repointing with very strong mortar. Wear is usually associated with scour around the invert level. In both cases it is necessary to establish the most likely cause and the degree of loss before considering remedial action.

f) Deformation

The shape of a brick sewer is an important indicator of the degree of deterioration which has occurred and in particular the loss of soil support at the sides.

Picture distortion can cause CCTV alone to be an unreliable way of assessing shape unless strict quality control measures are applied to check the geometric accuracy of the monitor screen picture. Colour photographs are better.

It must be remembered that the sewer may not have been built to a precise shape and that some deformation may have occurred on backfilling during or immediately after construction.

In single ring sewers, considerable deformation can often be accommodated before final collapse occurs. In multiple ring sewers even slight deformation (less than 10%) can lead to loss of compression in the inner ring and progressive falling out of the bricks from the crown.

4.3. Specific structural defects - brick and pipe sewers

The following comments apply to defects in both pipe and brick sewers.

a) Defective connections

Connections to pipe sewers are formed either with purpose made branches or, more commonly, by knocking a hole in the main sewer and connecting the branch into it, with or without a saddle piece. Purpose made branches are less vulnerable but suffer from cracking at the junction point.

Many more problems are encountered with ad hoc connections. There are several common forms.

i) Damage to the main sewer, caused when the connection was made, increases the susceptibility of the sewer to deterioration.

ii) Damage to the connecting pipe caused by poor cutting or differential settlement after laying. This allows deterioration around the connections which may then collapse and affect the main sewer. This type of defect is difficult or impossible to detect with conventional survey along the line of a sewer.

iii) Intrusion of the connecting pipe into the main sewer leading to blockages and reduced capacity. This can also cause surcharging and accelerated ground loss through sewer defects.

iv) A gap between the main sewer and the connecting pipe that allows ground loss to occur.

In brick sewers connections are made either by purpose made connecting blocks built into the fabric at the construction stage or by knocking bricks out and pushing a pipe in. The latter cause most trouble as bricks may be displaced and leakage may occur at the joint, in the same way as referred to above. In man entry surveys visual inspection can be made of the connections for one to two metres up the drain. This should be done as the drains may contain defects that affect the stability of the sewer.

b) Line and level

Deviations in level, particularly local low spots, probably indicate subsidence of the sewer caused either by mining subsidence or local loss of ground into the sewer. Inspection of the road surface is useful in both cases. Changes in level can be detected visually or from changes in water level but are difficult to measure using current survey techniques.

4.4 Other observations - brick and pipe sewers

a) Roots

Where a sewer is above the water table there is a risk of roots from nearby trees being attracted by the water in it. Roots can penetrate very fine cracks and it is doubtful whether any sewer is proof against them. They are not serious structural faults in themselves but they may exacerbate existing defects such as cracks; they may cause blockages and surcharging thus accelerating deterioration; while the measures, such as winching, used to clear them can cause further structural damage.

In periods of drought the effects of moisture take up by roots in the ground surrounding sewers could result in deterioration. This would particularly occur in clay soils where drying is associated with significant volume changes. These changes would cause deterioration to the sewer itself by disturbing joints/opening defects etc. Subsequent deterioration would be accelerated.

b) Rats

Rats can form voids when they build tunnels behind the sewer wall.

c) Infiltration

Infiltration can be caused by a high groundwater level, by stormwater that has been forced out of the sewer, or by a leaking water main. Infiltration acts as undisputed evidence that the sewer structure is permeable. In certain soils such as fine sands or silts infiltration can transport soil into the sewer resulting in loss of support and/or surface subsidence. Since the structure is permeable regular surcharging would be an added risk factor due to the mechanisms described in Section 3.2.

The amount of infiltration visible in the survey record may not be typical, since sewers are generally inspected when the flow is low.

d) General deterioration

Where a sewer is suffering from general deterioration and its service condition is poor, hydraulic roughness and capacity are affected. This increases the likelihood of hydraulic surcharging in a general or localised manner, leading to potential acceleration of the deterioration process.

5. THE ROLE OF SUPPLEMENTARY INFORMATION

The processes leading to sewer collapse are complex and two defects which appear similar can lead to very different consequences. It is therefore important that the internal inspection record is supplemented with whatever other evidence is available. As much of the desirable data will sometimes only be obtainable from an excavation a procedure should be set up to record and archive details from all excavations onto or adjacent to critical sewers. In specific situations for critical sewers a limited site investigation may be appropriate. The following information should be sought:-

Depth of cover and surface loading - Is the sewer deep and therefore more likely to be overstressed? Or if very shallow, is it vulnerable to traffic loads or interference damage?

Construction details - such as the number of rings of brickwork, mortar type, pipe class, jointing materials and age.

General soil type - What is the natural soil at sewer level? Is it likely that the sewer is surrounded with this, or with imported backfill?

Groundwater level - Is it above invert level? - If "yes" was infiltration noted?

Does the sewer surcharge? - If so how frequently or how badly?

Is there surface evidence of deterioration - local subsidence, history of road patching, trench lines crossing the sewer?

Has the sewer collapsed before? - If so what happened?

Have there been problems with similar sewers in the district? - What can be learnt from them?

6. TECHNOLOGY FOR ASSESSMENT?

Assessment 'tools' have been suggested which would enable more refined decisions to be made in a routine manner. The three 'tools' currently being supported and/or developed by WRC are void detection, shape measurement and in-situ competence testing.

Each of these tools is of potential benefit for assessment purposes. None of them will enable sensible decisions to be made in isolation from good visual survey and supplementary information about the sewer. The philosophy at WRC Engineering is that if and when 'tools' become available they will supplement the current guidelines for assessment which are based on visual survey and supplementary information.

In foreseeing the role of technology for the assessment of sewers it is crucial to guard against the gathering of information for its own sake. Information can only be of value if it is used in decision making processes. Decision making can be impaired by having too much as well as too little information.

Using the coded defect descriptions from visual survey it is possible to use a computer to rank sewers in order of 'badness' based on the internal condition. The system is based on the principle that the defect types used in reports can be put in a pecking order of severity and scores assigned to each according to their relative severity. Although such programs can be used as a course screen they cannot be used as the sole criterion for sensible decision making. For example, the presence of relatively minor defects can lead to major problems of ground loss if the surrounding soil is of a susceptible nature.

It is possible to foresee that the computer could play a major role in the assessment of sewers. To develop appropriate programs requires a thorough understanding of the problem. Such decision making programs would need to incorporate the knowledge and understanding of informed engineers together with field information. In medicine 'expert systems' are now becoming available for the diagnosis of illness. The same approach may become feasible for assessing 'sick' sewers. Such systems could incorporate features such as cost constraints in the assessment process. However, such systems cannot be justified unless the database of knowledge is sound.

7. CONCLUSIONS

Assessing the structural condition of sewers is at present an art and any means of assessing them must leave room for "engineering judgement".

As our knowledge increases and further research is undertaken into the processes which govern the stability of deteriorating sewers, it will be possible to refine the assessment procedure. However, the element of "engineering judgement" is always likely to remain to some extent.

General remarks have been made regarding why sewers in a severely deteriorated condition do not collapse. The importance of ground conditions and soil type has been emphasised.

Specific structural defects for both brick and pipe sewers have been discussed. An indication of the possible causes and consequences of these defects has been given.

The knowledge regarding structural deterioration currently available clearly suggests that the concept of a uniform rate of deterioration is inappropriate. Deterioration is influenced by random 'events' in the life of a sewer. With brick sewers the processes of deterioration will often accelerate as mortar is completely lost and soil comes in contact with flow inside the sewer. Although the shape of the deterioration against time graph is unknown and is likely to remain so, it is clearly not a linear relationship and this should be bourne in mind when predicting the life of the asset.

In developing appropriate assessment advice for use by drainage engineers, the practical approach must prevail. The end objective of work being undertaken at WRC Engineering is to incorporate the findings of the field and laboratory studies into an improved structural assessment procedure for critical sewers. Such a procedure can never be rigorous and no such claims are made. The intention is to provide a framework for decision making which is consistent and technically justifiable, while still allowing the necessary flexibility for experienced drainage engineers to apply judgement in particular situations. This approach also lends itself to the application of appropriate economic constraints based on considerations such as consequences of collapse, the cost of renovation or replacement, the available level of funding etc. The basis of such a decision making procedure is illustrated in Fig. 5.

For non-critical sewers, say 80% of the sewer network a reactive approach ('crisis management') is likely to be the most cost effective approach. For critical sewers a planned programme of survey, and where necessary rehabilitation is sensible. For the most critical sewers, say 5% of the network the intention must be to try and achieve a failure free situation. For the remaining critical sewers, say 15% of the network, the intention is to significantly reduce the likelihood of failures. It is felt that, with a sensible framework for making decisions, it is reasonable to realise these objectives in a cost effective manner.

Many sewers have defects, some are critical some are not. Visual survey and any site investigation undertaken will reveal defects and anomalies. An approach which reacts to all defects and anomalies is unjustifiable technically and economically, even for critical sewers. In isolation, survey of whatever form, will identify questions without answers or problems without solutions. In conjunction with sensible decision making procedures survey plays a crucial role in the cost effective management of sewer systems.

8. REFERENCES

1. "WRC/WAA Sewerage Rehabilitation Manual", WRC Engineering, 1984.

2. MacPhee, J., and Noriega, N.C.: "Sewer Deterioration Studies Collapse Investigations Report on Findings", WRC Engineering, ER 115E, August 1983.

3. Trott, J.J., Nath, P. and O'Reilly, M.P.: Longitudinally Cracked Pipes and their Structural Capacity" in "Restoration of Sewerage Systems", Thomas Telford Ltd., London 1982.

4. Jones, G.M.A.: "Structural Tests on Cracked and Renovated Pipes", WRC Engineering, ER 51E, March 1982.

5. Stephenson, G. and Britton, R.J.: "The Extent of the Hydrogen Sulphide Problem in the United Kingdom Sewerage Network". WRC Engineering, ER 104E, May 1983.

6. "Manual of Sewer Condition Classification", DOE/NWC Standing Technical Committee, Report No. 24, May 1980.

STRUCTURAL ASSESSMENT – SEWERS

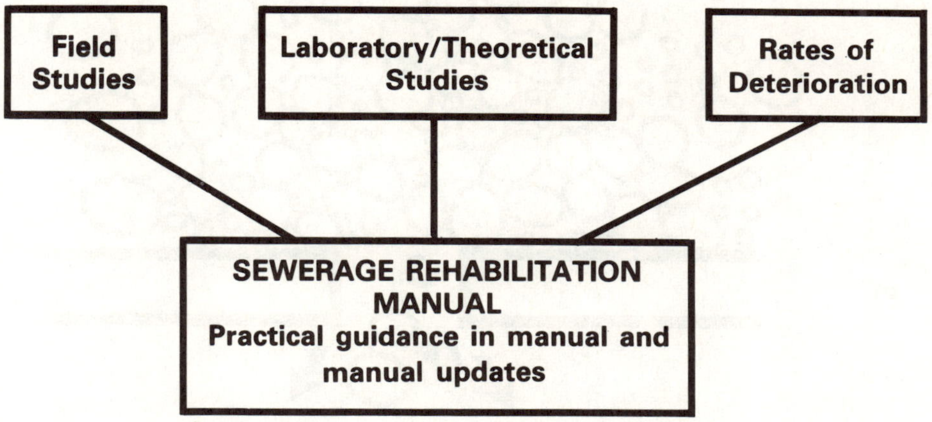

FIG. 1

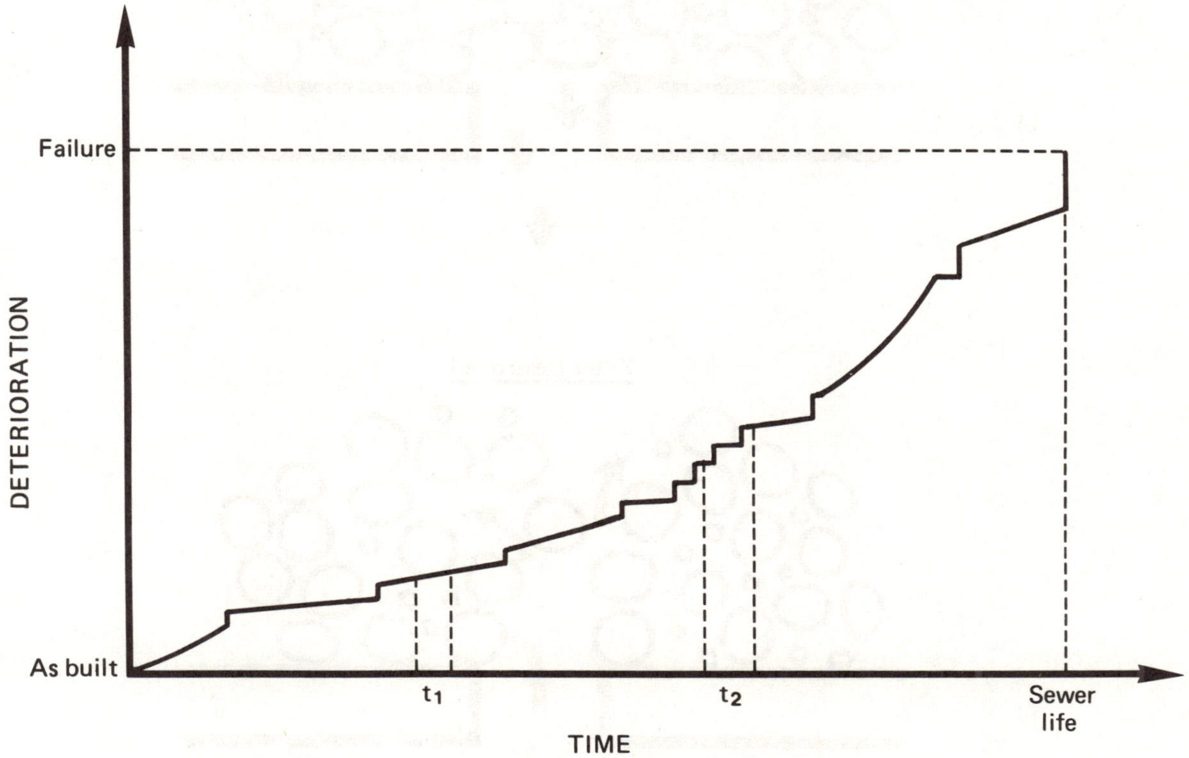

FIG. 2. DIAGRAMMATIC REPRESENTATION OF DETERIORATION HISTORY

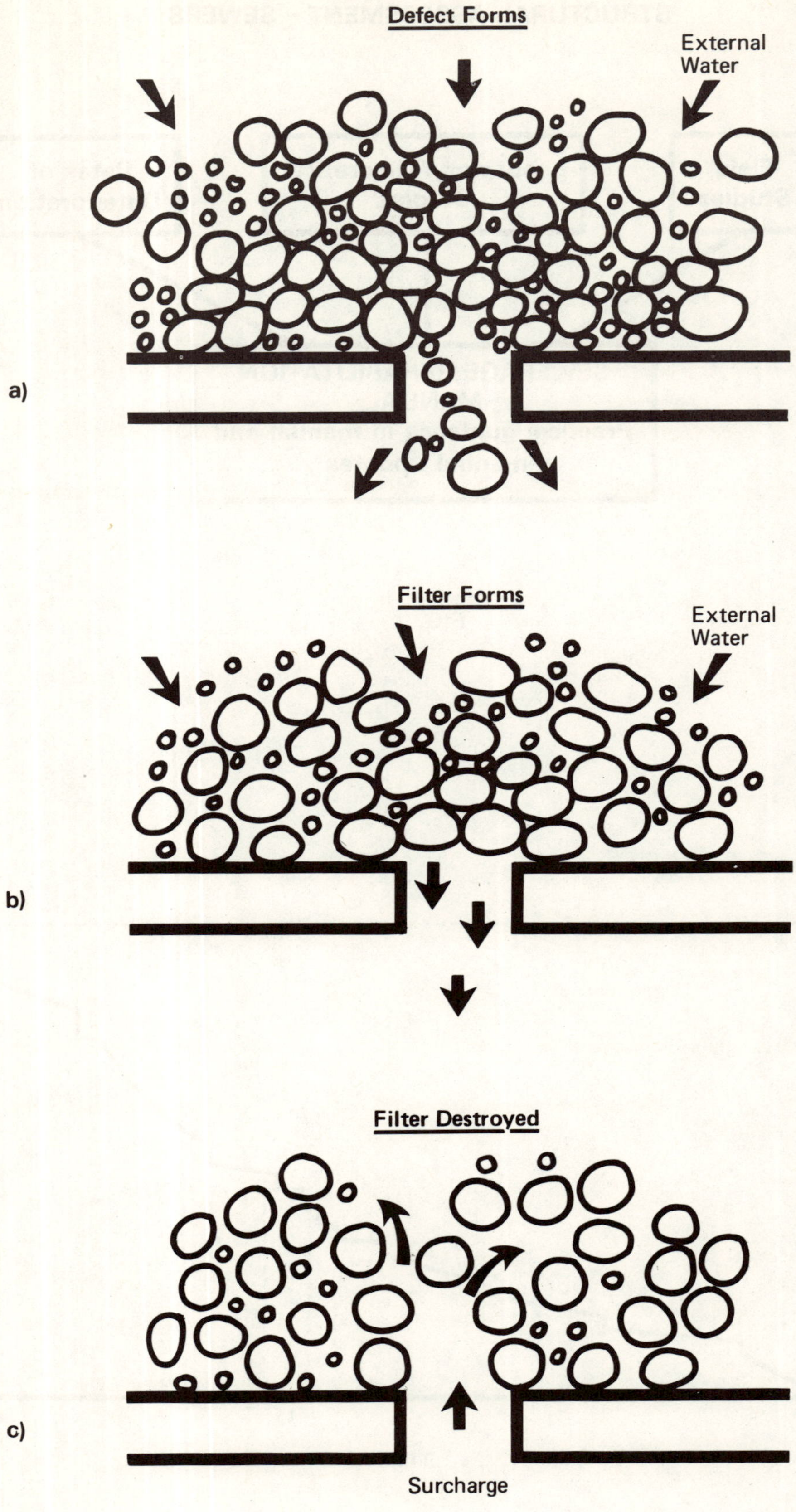

FIG. 3. BEHAVIOUR OF SILTS AND SANDS

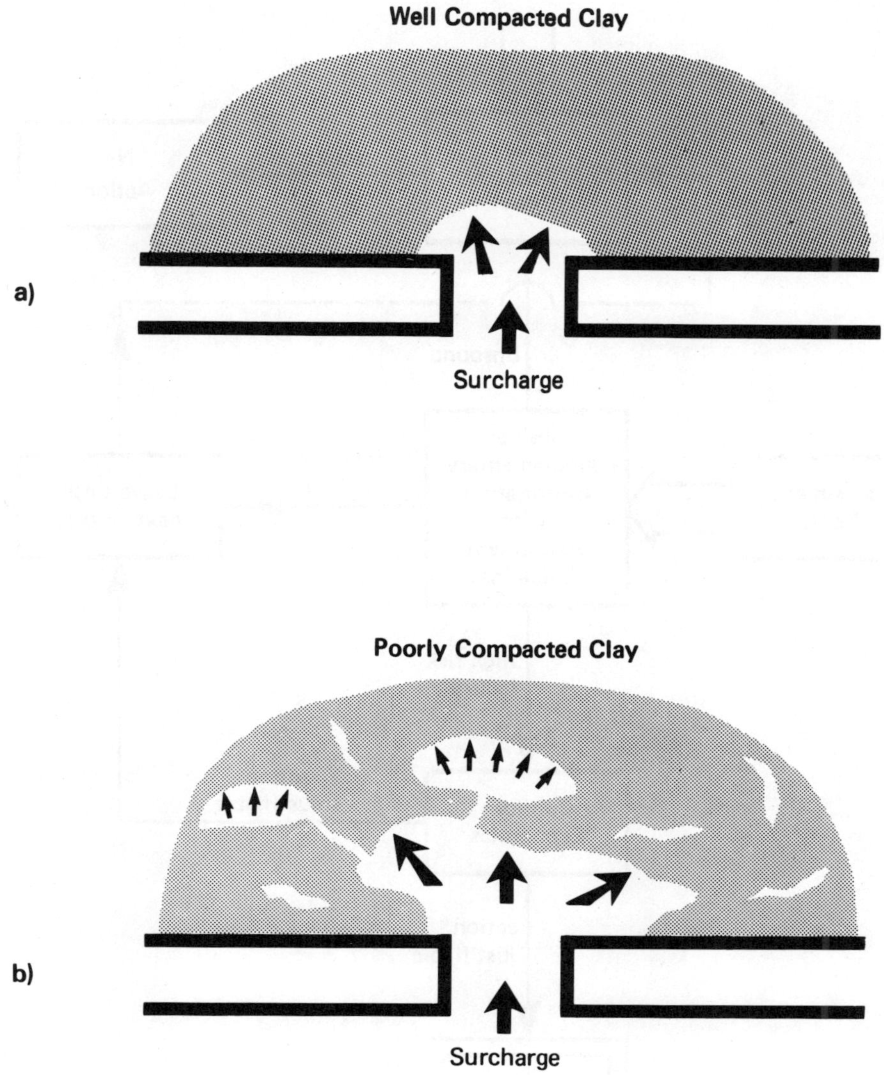

FIG. 4. BEHAVIOUR OF CLAYS

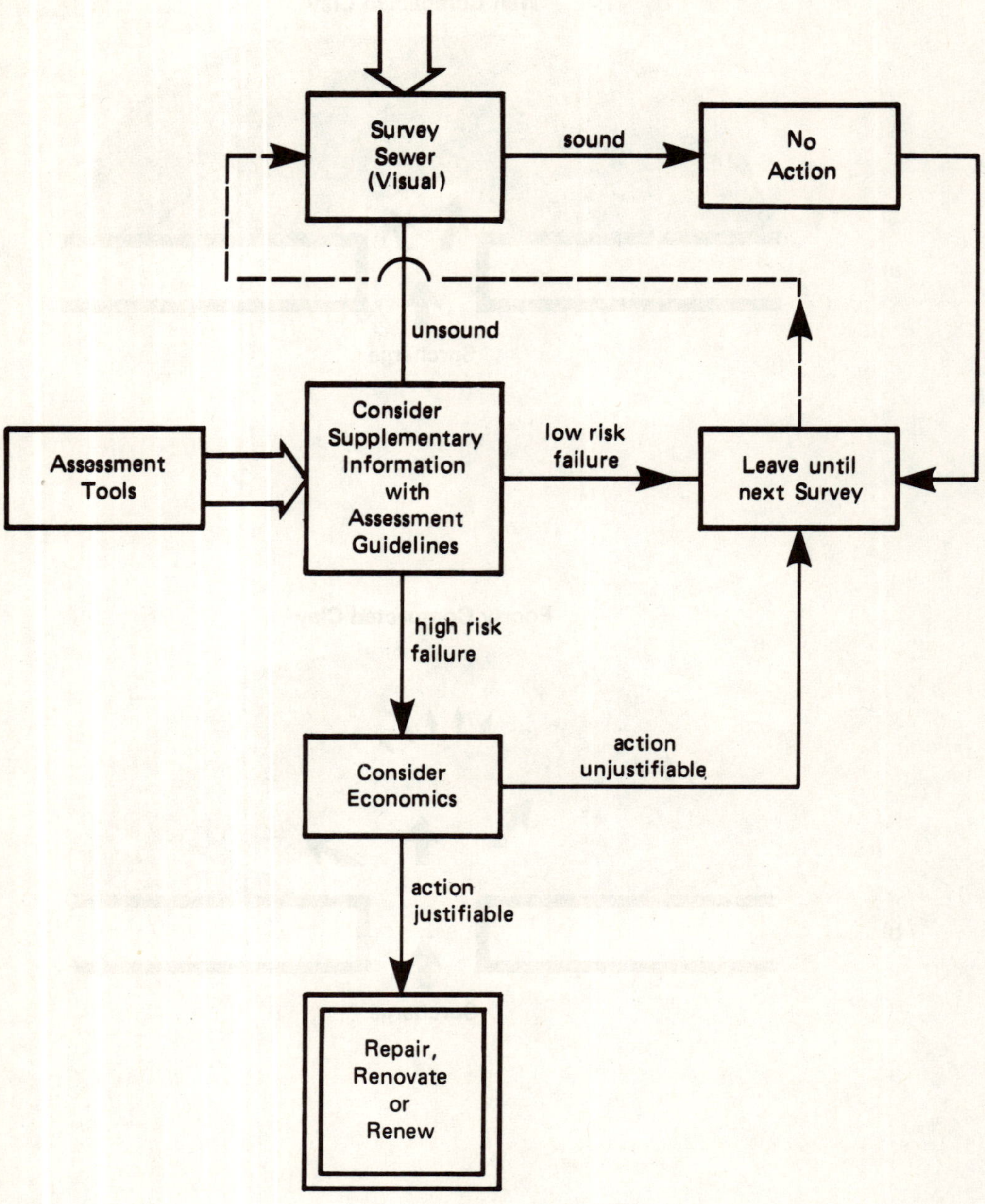
Survey Sewer (Visual)
sound
No Action
unsound
Assessment Tools
Consider Supplementary Information with Assessment Guidelines
low risk failure
Leave until next Survey
high risk failure
Consider Economics
action unjustifiable
action justifiable
Repair, Renovate or Renew

FIG. 5

International Conference on the
PLANNING, CONSTRUCTION, MAINTENANCE & OPERATION OF SEWERAGE SYSTEMS
12-14 September, 1984

PAPER C2

INSPECTION AND RENOVATION OF SEWERS: STATE OF THE ART IN SYDNEY

W.J.P. Goodman, B.E. (Civil), B.Sc., M.I.E.(Aust)

and

P.S. Hope, B.E. (Civil), M.A.W.W.A.

Sydney Water Board
Box A53 P.O. Sydney South 2001
New South Wales
Australia

Summary

Sydney, like many other cities around the world, is facing a future where the sewers are growing old and the need for inspection and renovation is becoming increasingly important. This paper summarises the problems experienced by the Sydney Water Board with its sewers and describes the development of a programmed inspection of the sewers together with some of the renovation techniques utilised by the Board to date.

Held at The University of Reading, England. Organised and sponsored by BHRA, The Fluid Engineering Centre, Cranfield, Bedford in conjunction with the Water Research Centre.

1. INTRODUCTION

Sydney, the capital of the state of New South Wales in Australia, is a city of more than 3 million people. Sydney was founded in 1788, but, until the middle of the 19th Century, it was not sewered. The inner city area and nearby suburbs were progressively sewered by 1900 and, with the growth of the city, the outer areas have been gradually reticulated. The Board was originally set up in 1888 with responsibility for water supply and sewerage collection being transferred from the city and nearby suburban councils. Prior to 1925, construction work was carried out by the Public Works Department. Since that date the Board has constructed its own sewers. The Board has also taken over a number of local sewerage schemes which were originally under the control of various municipal councils. Over 3 million people are now connected to the sewerage system which consists of approximately 17,000 km of conduits of various sizes and types.

The Sydney Board's area of operations extends 40 km north to Palm Beach, 120 km south to Kiama, and west to Mt Victoria which is 130 km from the centre of Sydney. Approximately 200,000 people are still not serviced by the sewerage systems, although an estimated 50,000 live in isolated pockets which cannot be economically sewered.

Most of the early sewers in the inner areas were constructed of either S.G.W. pipes or brickwork and were laid, in most cases, in sandstone trench or tunnel. These sewers have generally given very little trouble over an average life of about 100 years. More recent construction has been in concrete for large sewers and over the last 20 years, various plastic, asbestos cement and fibreglass pipes have been introduced in small quantities in the reticulation sewers, together with V.C.P. which is still the preferred material.

Since 1950, this Board has, where possible in the sandstone areas, constructed the main sewers in tunnel with the roof and upper walls remaining as exposed rock. More recently, tunnel boring machines have been used with resultant smooth profiles, where concrete lining of the invert only was required.

Because of the steep topography, sewers have occasionally crossed creeks and gorges in aqueduct. Early examples were of concrete box sections supported on piers while more recently self-supporting cast iron and steel pipe have been used.

Around 1920, some 225 mm unreinforced concrete pipes were used. Unfortunately, these pipes proved susceptible to groundwater attack and also erosion of the invert. These concrete pipes have now all been replaced.

2. MAIN SEWERS

In the late 1950s and early 1960s, a number of collapses involving three of the Board's main sewers occurred. All of these conduits were constructed of concrete and were attacked by aggressive groundwaters. Major replacement was involved in each case costing several million dollars. As a result of these collapses, a major inspection programme was initiated of the traversible sewers which totalled about 500 km. All sewers down to 900 mm diameter or equivalent were traversed over the 5-year period from 1962 to 1967. These inspections were followed up by a repair programme which involved work varying from minor to extensive and in one case replacement of about 1½ km of defective sewer was needed.

On completion of this 5-year inspection programme, a further schedule was prepared for reinspection. This programme established initial estimates of the rate of deterioration. Further inspections refined the rate of attack and enabled the inspection programme to be set up in such a way that sewers were inspected at intervals depending on both condition and rate of deterioration. Periods between planned inspections now vary from one year to ten years, depending on conditions noted at previous inspections and the type of construction.

Until 25 years ago, methods of inspection of large sewers involved walking on foot or in the case of the smaller ones, traversing in a crouching position using a supporting board. Recent developments include the use of 4 and 6-wheeled trolleys, inflatable canoes and floats.

Deterioration is primarily due to three factors in the main sewers in Sydney; internal attack on concrete due to hydrogen sulphide, external attack on concrete due to sulphated groundwaters and deterioration of joints in brick sewers.

Siltation in many of the larger sewers is a recurring problem. The use of power bucket machines and hydraulic sewer cleaning machines has been introduced over recent years to economically deal with this problem.

3. RETICULATION SEWERS

In the smaller, non-traversible sewers, the bulk of which are constructed of V.C. pipes, the problems are generally cracked pipes, bad joints, poor connections, infiltration of groundwater and tree root infestation.

Recent developments in inspection techniques include the use of colour closed circuit T.V. and the use of video in conjunction with it. About 15 years ago, black and white C.C.T.V. was first used in Sydney and after investigation, colour was introduced about 5 years ago. To date, approximately half of the C.C.T.V. inspections have been on a routine basis with areas known to be troublesome being inspected and other less troublesome areas sampled. The other T.V. work has been in conjunction with specific troubles in the sewers. C.C.T.V. inspection is used as routine prior to submitting any small sewer for renewal or renovation. Location of connections, both live and unused, is facilitated by the use of C.C.T.V. This is particularly valuable in the inner city areas where the old records are unreliable.

4. RECORDING PROCEDURES

Computer storage of field service reports of problems in the sewers has recently been introduced in Sydney. It is expected that, given a reasonable period, computer records will highlight troublesome areas and lead to both more realistic inspection programmes and subsequent renovation work, where necessary. In conjunction with the actual recording, the Sydney Board is developing parameters for assessing the condition of the sewers.

5. RENOVATION TECHNIQUES

5.1 Insituform Lining.

The insituform process was introduced into Australia in 1976, the first trials being carried out in Sydney. The initial work in a live sewer was two lengths of a 525 mm V.C. sewer in Camperdown, a suburb south of Sydney, which was lined in 1978, for a cost which was calculated to represent 17.6% of a conventional relay. Since this initial job, the Board has lined numerous circular V.C. sewers ranging from 150 mm up to 600 mm and oviform sewers ranging from 380 mm x 260 mm up to 1050 mm x 750 mm have also been lined, totalling approximately 20 km.

Generally, lining with insituform has been both successful and economical with the cost being typically less than 25% to 30% of a conventional relay. A substantial part of the cost has been associated with reconnection of the house services. However, the recent introduction of an internal cutter, which is controlled from the surface using C.C.T.V. equipment, should result in further substantial savings. The big advantage of insituform lining is seen as the avoidance of major disruption of traffic, the speed with which the work can be carried out, elimination of costly road openings, the ability to insert linings without dewatering the ground and its relatively low total cost.

The Board's experience is limited to polyester resins, as suitable epoxies for use in insituform are not yet readily available on the Australian market at a competitive price. During the early use of insituform, problems were experienced with the initial setting of the resin, occasionally premature setting at the factory being experienced. There were some problems with shrinkage in the larger sewers. Addition of a retarder led to difficulties with cold groundwater apparently affecting the exothermic reaction, with subsequent incomplete curing of the liner.

Some problems have arisen during the lining of the oviform sewers, primarily due to the shape. Buckling of the liner under external water pressure has occurred with the failure starting in the almost flat section on the sides below springing level. Extensive investigations and tests have been carried out to develop an economical design thickness of liner to withstand such pressures. In particular, the 1000 mm x 660 mm brick Kensington Submain, which is typically 5 m to 6 m deep with normal groundwater levels just below the surface, was subject to failure in the first length treated due to a combination of groundwater pressure and incomplete curing. It was found that controlled curing is critical with thick liners particularly where cold groundwater is in contact externally with the liner during the curing period. Part of the problem was apparently due to boiling of the resin in the centre of the liner. The change in resin composition has overcome this problem and tests carried out in a full sized mock-up have been used to develop liners with sufficient strength to withstand external pressures to which the Kensington Submain and other sewers lined in the future will be subject. Insituform thickness of the order of 22 mm was required in the Kensington Submain under the above conditions. Pinning back of the liner was considered but the brickwork was, in places, not sufficiently competent. Following on successful testing in the Kensington Submain, lining of this sewer has now recommenced with the estimated final cost of the work being of the order of Aus$2 million for about 2 km of sewer. Renewal by excavation and replacement of the conduit in this case is considered impracticable, as the sewer passes beneath buildings. Dewatering of the ground, no matter which route was selected for a relay, would lead to extensive property damage. No estimate has been taken out for comparison purposes, but it is expected that the cost for a conventional relay would be between Aus$10 million and Aus$20 million.

5.2 Fibreglass Liners.

About 10 years ago, the Board repaired the Rocky Creek Submain using fibreglass. This sewer, which is 1000 mm x 660 mm concrete oviform, had been constructed on a steep grade, mainly through parkland beside an adjacent creek. It had ample capacity for the area it served but extensive erosion of the invert had occurred. It was decided to line it with fibreglass up to springing level. The fibreglass liners were made in approximately 2 metre lengths and were approximately 6 mm thick with the wearing surface having zircon in the gelcoat. Because of the environmental damage that would occur in bypassing the flow, the liners were inserted under live sewer conditions, fastened in place with stainless steel ramsets and then grouted to fill up the cavities where erosion of the invert had occurred. A recent inspection shows that the lining is in first-class condition, showing no detectable wear after more than 10 years.

The second fibreglass lining job involved the Abattoirs Submain, constructed of precast concrete oviform sections each about 1 metre long, which had been attacked by various chemicals and gases resulting in exposure of the reinforcement in the walls and obvert. In this instance a full fibreglass liner was inserted and grouted in place, and again there has been no detectable wear since the work was carried out about 8 years ago.

Fibreglass liners have recently been inserted in a circular 900 mm concrete pipe which was subject to gas attack after only 6 years' service. The sewage which caused the trouble drained from a partially sewered catchment, in some instances through 3 pumping stations in series, and consequently was fairly septic when it reached the 900 mm sewer. The upstream end of the length attacked was at the discharge of the rising mains from a large pumping station. Ventilation of the sewer was inadequate in the initial sections resulting in corrosion of the concrete by sulphuric acid generated by hydrogen sulphide gas. The attack had penetrated to the reinforcement in places. Fibreglass liners were selected after calling tenders in which a number of alternatives were also submitted. Concrete liners with P.V.C. protection, insituform, H.D.P.E. and various fibreglass alternatives were considered. The selected method involved a 5 mm thick fibreglass bell jointed liner, grouted in place with a polyurethane material. The job included a curve in the sewer which had been formed by straight pipes approximately 1.2 metres long forming a series of chords. Upstream and downstream of the curve were straight sections. Openings were excavated on the straight sections and fibreglass liners approximately 6 metres long inserted by removing the top of the concrete pipe. It was possible to store the sewage in the pumping station and carriers upstream of it for 6 to 8 hours each day while the work was carried out.

5.3 High Density Polyethylene Lining.

A number of jobs involving sewers up to 400 mm diameter have been carried out using H.D.P.E. The technique involves opening up a sufficient length of trench to permit the insertion of a long continuous liner which is winched through. It was necessary to excavate where connections were involved and in some instances this was a substantial part of the cost. This method of renovation was used both in shallow V.C. sewers subject to cracking and tree root infestation, and deeper C.I. lines which had been subject to salt water attack. In one case, involving short (600 mm long) 225 mm diameter V.C. pipes with lips at the joints, problems with winching were experienced as a result of the inflexibility of the liner over the short lengths between joints and subsequent build up of frictional resistance.

5.4 Special Epoxies.

Repairs to the Western Main Sewer, which is typically brick oviform up to 2.5 m high x 2.0 m wide, are in progress using special epoxies for rejointing the brickwork. This sewer, which is some 80 years old, was constructed with mortar joints which have deteriorated over the years both as a result of gas attack in the roof and erosion on the side walls. Over the years periodic rejointing using cement mortar had been carried out which typically has lasted about 20 years. With increasing labour costs and the development of epoxies, the Sydney Board has looked to the use of special epoxies which can be applied in the humid, damp conditions to rejoint the brickwork, using a pressure gun. This technique is proposed in other old brick sewers in the near future.

5.5 Polyurethane Grouting.

There are a number of sewers where groundwater infiltration through the brickwork is the principal problem. The recent improvement in techniques, which enable grouting from inside a small brick or concrete sewer using polyurethane grout, has enabled renovation work to be carried out economically. A section of the Clovelly Submain, which is a small concrete oviform, had most of the joints leaking under groundwater pressure. This has been successfully sealed using polyurethane grout. This technique has been used also on a number of small sewers; in some instances with the grout being applied externally by pipes inserted from the surface.

A rig is being developed for pressure grouting of the joints of oviform sewers from inside the conduit.

5.6 Structural Repairs.

Repairs have recently been carried out to two major sewers where structural failure appeared imminent. In one case, a main concrete sewer 2.8 m wide x 2.4 m high with about 6 metres of cover developed a crack in the roof extending for about 35 metres. This crack appeared to be due to problems encountered during construction many years before caused by a fall in the roof when the sewer was being constructed in shallow tunnel through clay. Repairs consisted of removing the overburden, constructing piers down to rock which was near invert level, casting longitudinal reinforced concrete beams on top of each row of piers on each side of the sewer and then a new reinforced concrete roof on top of the existing cracked one. The old roof was fastened to the new one with stainless steel bolts. During the repair work the existing cracked roof was propped from inside the sewer.

The second repair involved a reinforced concrete box sewer 3.8 m wide x 2.6 m high in aqueduct which had been severely attacked by sewer gas. The aqueduct was immediately downstream of twin 1800 mm diameter inverted syphons some 300 m long. Earlier repairs consisted of treatment with special epoxies on the underside of the roof and upper walls. This protection had failed after approximately 20 years. It was decided to cast a new roof on top of the existing one. The new roof was dowelled to the top of the walls and the old roof fastened to it, again using stainless steel bolts.

6. THE FUTURE

The Sydney Board, like many other authorities around the world, is facing a future where the sewers are growing old and the need for renovation is becoming increasingly important. Rising costs, particularly wages, together with increasing traffic problems, have pointed the way towards renovation rather than renewal.

New techniques will need to be developed to deal with the old sewers which are steadily deteriorating, despite regular inspections and minor repairs. These techniques will need to be able to cope with problems both in brick oviform as well as concrete conduits of various shapes.

The need for adequate funds to enable the work to be carried out at an early stage before the sewer reaches the point where renovation is no longer practicable, must be appreciated if major disasters, involving extensive disruption to traffic, damage to property, health hazards and other problems, are to be avoided.

ACKNOWLEDGEMENTS

The approval of the Sydney Board to present this paper and the assistance of various officers is gratefully acknowledged.

International Conference on the
PLANNING, CONSTRUCTION, MAINTENANCE & OPERATION OF SEWERAGE SYSTEMS
12-14 September, 1984

RENOVATION DEVELOPMENT TRIALS

K Reed, BSc., MICE and P.B. Rumsey, MSc(Eng), MICE

Applied Technology Group
Water Research Centre, Frankland Road,
PO Box 85, Blagrove,
Swindon, Wiltshire, England

1. SUMMARY

Various new renovation systems, and items of equipment have been developed to augment the range of renovation techniques currently available. The developing systems and equipment require the availability of sites such that their performance under live sewer conditions, both during installation and in service may be assessed. The aim of the performance assessment is that developing systems may, ultimately achieve the ´established´ status already held by some of the currently available systems.

The paper sets out the aims and framework of a formal programme for conducting site trials to be monitored by WRC and gives brief descriptions of systems and items of equipment which will be assessed during the course of the trial programme.

Held at The University of Reading, England. Organised and sponsored by BHRA, The Fluid Engineering Centre, Cranfield, Bedford in conjunction with the Water Research Centre.

2. INTRODUCTION

2.1 Replacement and renovation of sewers has become of major importance to drainage engineers in recent years as a result of concern about the structural condition of our older sewers. A limited number of renovation techniques have been applied to combat the problems of sewer dereliction and during the past decade some systems have achieved an 'established' status.

2.2 These established systems do not necessarily meet the total requirements of the renovation market, or provide solutions which are either economic or fully suitable for a particular situation. Accordingly research and development has continued in the refinement and extension of their use. At the same time new systems and equipment are being developed to cater for the multiplicity of situations and conditions prevalent in the country's sewer systems.

3. DOE RENOVATION DEVELOPMENT TRIALS

3.1 Following initial development work it is important that new systems and equipment be given the opportunity to operate under site conditions; an important stage in any development programme. It allows experience to be gained in installation and operational procedures, as well as in the development of specifications and methods of measurement, and is an essential step towards gaining general acceptance for use in industry. It is important to provide a framework within which an established status can be achieved as quickly as possible. At the same time it is essential to ensure that developing systems and equipment are not accepted generally (nor viewed in the same light as established techniques) until there is reasonable confidence that they can be used/constructed safely and effectively.

3.2 This premise was recognised by Government at an early stage and in November 1982, Tom King, then Minister for Local Government and Environmental Services, wrote to Chairmen of Regional Water Authorities. Mr King requested that possible renovation sites within authority areas be made available to companies offering new renovation systems such that full-scale contract trials could be carried out. Mr King's intent in making such a request was to progress new or significantly improved products and processes on to the market more quickly and more effectively. At the same time it was hoped to achieve the aim of improving the technological base of UK industry in the field of sewer renovation.

3.3 WRC has been commissioned by the DoE to encourage and coordinate the development of new renovation techniques under full-scale contract conditions. An outline framework of operation has been established whereby new systems and equipment can undergo initial assessment trials (pilot trials) and full scale contract trials. This programme commenced formally on 1st February 1984 and will continue for a period of two years.

3.4 In November 1983 WRC wrote to Regional Water Authorities making reference to the approach initiated by Mr King and requesting assistance in the provision of possible renovation sites. At the same time various companies associated with developing renovation systems or equipment were contacted to ascertain their willingness to participate in the Renovation Development Trials. Descriptions of those systems offered by companies who indicated their interest in participation in the Trials are given in Section 5 and 6.

3.5 A discussion document has been issued which includes a basic framework within which the Trials will be operated. The document also details requirements which are to be fulfilled by a relevant system or item of equipment during the course of the Trials. The framework of the Trials includes various defined stages of operation and development each of which will be monitored by WRC. A brief description of these various stages is given below.

3.6 The Pilot Trial involves evaluation of the system/equipment and incorporated materials on both a qualitative and quantitative basis. The work can be carried out in laboratory test rigs, above ground full-scale pipelines, abandoned sewers or, in some instances, in a live sewer situation. Parameters investigated may include material properties; strength; hydraulic characteristics; infiltration control and installation procedure. The economics of new systems may also be examined although at the early stage of development it is difficult to accurately forecast the cost of a developing system on the open construction market.

3.7 Following successful completion of pilot trials each manufacturer/contractor will generally be required to prove his system on a full scale contract basis at four sites. These sites will be of increasing difficulty to allow problems encountered in simple situations to be corrected or resolved before a more difficult site is tackled. The variation of parameters which may be considered to increase the degree of difficulty of a site include: length of sewer; line and level; deformation; site access; cleaning requirements; number and position of laterals; flow condition; and sewer size. The intention of aiming to include such variations is to provide a means of proving the system capable of dealing with any situation likely to be encountered during renovation and considered to be normal to such work. However, it is essential that the contractor/manufacturer be given the security of being able to incorporate new developments as the series of contracts evolves. This requires clients to accept a greater element of risk than might normally be the case in a conventional contract. Proper understanding of the client's and contractor's objectives by both parties is particularly important, as is an effective and close working relationship. This may hinge around special contractural procedures developed and agreed between the parties on a one-off basis before commencement of the contract.

3.8 Day to day supervision of these renovation contracts is to be carried out by RWA/Agency engineers with frequent monitoring by WRC. The purpose of the monitoring procedure is to identify installation problems associated with a specific system and determine how they might be overcome; monitor progress rates; detail cost data; and assess a system's capability for forming or allowing for lateral connections. It is intended that monitoring of the renovated sewer will continue after completion of the contract to observe the performance of the renovation system with the sewer in service.

3.9 Before any recommendations are made to the Water Industry as to the general acceptance of a renovation system the manufacturer/contractor is required to determine, and prove by suitable testing, the long term properties of materials employed in the system. The proving of such properties should ensure that an acceptable in service life expectancy can be achieved. Specifications have been prepared for suitable test procedures to prove short and long term structural properties, chemical and abrasion resistance, adhesion and shear bond strength, with general recommendations for other properties specifically required for a particular renovation system.

3.10 Finally, before a technique can become established a suitable design procedure should be available. Either it should be used with one of WRC's recommended design procedures or a new and proven design approach should be established.

4. SPECIFICATION AND GUIDANCE TO DESIGN ENGINEERS

4.1 The experience gained by WRC in monitoring development trials will be used in the development of model specifications for each technique. In addition, guidance will be given in updates of the Sewerage Rehabilitation Manual as to the choice, economics and structural design of the available systems.

4.2 The success of this centrally organised programme of formal trials will rely on the effectiveness with which information and experience gained with new systems can be disseminated. At the same time however, commercial pressure and confidentiality must remain an accepted part of the programme.

4.3 It is recognised that development and replacement of both new and existing systems and equipment will continue and due account will be taken of this. This will not preclude the incorporation of systems within the Trials programme at a later date.

5. DEVELOPING SYSTEMS AND EQUIPMENT - MAN ENTRY SEWERS

5.1 Most of the established sewer renovation techniques were developed for use in man entry sewers. Pipe and segmental linings, and linings installed by inversion processes, are used in many of these techniques, employing a wide range of materials including polyolefins, GRP, GRC, resin concrete and precast gunite. It is generally recognised that there are far fewer difficulties to be overcome where lateral connections, placing of lining, grouting, etc can be achieved with the availability of man access to the sewer.

5.2 Alphacrete Construction Linings (UK) Ltd

A reinforced mortar coating is being developed by Alphacrete Construction Linings (UK) Ltd. to provide a structural coating that should prove adaptable to changing sewer dimensions. The geometry of the sewer should be rationalised either before or during installation to allow existing structural theory to be utilised in the coating design.

Galvanised steel wire mesh is fixed to the sewer walls and the mortar is applied through a spray jet before being floated off by hand. Initially the sewer is thoroughly cleaned using a low pressure wet blast technique. Development is concentrating on improving rates of progress and control of quality on site. Testing is to be carried out to show suitable resistance to corrosion and abrasion with minimum thickness of coating, and full scale loading tests are being undertaken by Alphacrete to demonstrate the strength of the installed system.

5.3 Dunlop Ltd

Dunlop Ltd have developed a lining for use in man entry sewers which incorporates the use of pre-formed extruded hollow uPVC planks assembled to suit the size and shape of the sewer. Temporary internal support is normally required both during placing of the planks and grouting. Further development is expected which will make the system more versatile by allowing it to be used as a structural renovation solution although a suitable structural design procedure has yet to be established.

5.4 Sovereign Sewers (Acell) Ltd

A new GRP lining using a phenolic resin is being developed by Sovereign Sewers (Acell) Ltd for use as a one piece or segmental sewer lining. This type of lining is already established on the market using polyester resins and so development is directed to an assessment of the advantages and viability of the phenolic resin. Joints between units incorporate a shark´s tooth rubber insert which will provide a seal against grout ingress.

6. DEVELOPING SYSTEMS AND EQUIPMENT - NON MAN ENTRY SEWERS

Development of a total system for the renovation of non man entry sewers requires incorporation of many techniques including :

cutting intruding obstacles and cleaning sewers
placing of lining
reconnection of laterals
localised repair

Developments in all these areas are outlined below.

6.1 CLEANING AND CUTTING

6.1.1 Prior to the insertion of a lining into a sewer, the sewer must be cleaned and intruding lateral connnections removed. A wide range of water jet cleaning equipment is now readily available together with a varied selection of nozzles to suit specific requirements. These systems operate at varying pressures ranging from 140 bar (2,000 psi) to 420 bar (6,000 psi) with high and low volume outputs available throughout the pressure range.

6.1.2 The removal of intruding connections using high pressure water jet cutters has been widely used and continues to be developed. However pumps and ancilliary equipment required to achieve pressures of 700 bar (10,000 psi) are very costly and the quality of cut at high pressures requires to be carefully monitored.

6.1.3 To minimise costs and enable renovation contractors to make use of existing sewer cleaning pumping equipment (which generally operates at pressures of about 140 bar,) WRC instigated development work to entrain abrasive material into low pressure water jets. Such low pressure jetting produces smooth clean cuts on clayware and plastic materials and is less likely to cause secondary damage to the sewer and surrounding soil than high pressure jets which retain high energy levels even after cutting.

6.1.4 **Avent PRS**
Unipump Parts Ltd

Two firms offering cleaning and descaling facilities using high pressure water with a variety of suitable nozzles are Unipump Parts Ltd and Avent PRS.

6.2 SEWER LININGS & COATINGS

6.2.1 **John Kennedy (Civil Engineering) Ltd**

Conventional sliplining of circular sewers is not considered a developing technique as it is widely used and largely understood. A new development by John Kennedy (Civil Engineering) Ltd is the sliplining of egg shaped sewers with site welded polyethylene pipe. A polyethylene egg shaped lining has been developed which is manufactured to a high tolerance and does not deform during welding. Welding of the 2m lengths is carried out within a working shaft and is automatically controlled. Good rates of progress of lining insertion have been achieved. Development has concentrated on the method of lining manufacture as well as the quality of the welding and the lightweight grout which is incorporated as part of the system. Manufacture of the pipe is carried out using accepted principles of plastic moulding technology, but both short and long term properties of the moulded PE have yet to be established.

6.2.2 **Avent PRS**

Avent PRS are developing the Nu Line system of lining using short lengths of polyethylene pipe which are induction fusion welded and chemically grouted individually in the sewer. Each pipe is transported along the sewer and jointed to the previously laid section. Installation is carried out from existing manholes which govern the length of individual pipes that can be inserted. Surface disruption is minimised as no excavation is required. Development of site control of the technique, rates of progress and evaluation of the polymeric expanding foam grout is continuing during site trials. All operations can be monitored by CCTV and the pipes can be pre-cut for lateral connections prior to insertion.

6.2.3 **IPD Ltd**

IPD Ltd are at an early stage in the development of a system for sliplining a sewer with polyethylene or polypropylene pipes from within a manhole. Short lengths of the pipe are jointed within the manhole by a ′snap fit′ joint formed within the wall thickness of the pipe and the assembled length is successively winched along the sewer.
Annulus grouting is carried out from within the lining using sacrificial grout injection valves. At the time of preparation of this paper, full details of the system, and its effectiveness, are not available.

6.2.4 **Costain-Streeters Lining**
Nuttall-Fosroc
Ram Services Ltd

High-build spray-applied structural coating systems are under development by a number of contractors and manufacturers. The Costain-Streeters system has been used successfuly in a live sewer situation, and other systems may achieve this stage in the near future. The potential advantages of such systems are minimal reduction of cross sectional area, the ability to work from existing manholes, the economy of eliminating the need for grouting and the lack of specialised work required for lateral reconnection. All the developing spray-on systems use cold-cured polyurethane or resin based coatings and incorporate pump batching and delivery of the coating material to a rotating spray head that is pulled through the sewer at a constant rate. The formulation of the coating material is important to give the right spraying characteristics as well as the desired structural properties, water tolerance, and crack covering capability. A suitable structural design procedure has yet to be established and it is anticipated that such a procedure will only be applicable to sewers which are deformed to a limited extent.

6.3 ON-LINE REPLACEMENT

6.3.1 **ALH (Contracting) Ltd**
Avent PRS
D J Ryan & Sons Ltd

Joint displacement and a high degree of deformation of the existing sewer can severely limit the size of lining that can be installed. In some sewers the resulting reduction in hydraulic capacity is undesirable and an increase in capacity is preferred. On-line replacement of the sewer using percussive moles is being developed to meet the requirements of such situations. The existing sewer is broken and expanded and a lining drawn in simultaneously behind the mole. Trials are being carried out on three different systems developed by contractors ALH (Contracting) Ltd, Avent PRS and D.J. Ryan and Sons Ltd in association with percussive mole manufacturers or distributors.

WRC is implementing work to assess the effect of moling on adjacent services or structures and to examine the performance of pipes installed by moling (in particular, the effects of possible surface damage, stress concentration and ground support). The potential of the technique in replacing old sewers with polyethylene or polypropylene pipes of the same or larger diameter has been demonstrated. Improvements in reliability are desirable, and site trials are continually exploring this area. Further development will centre on the length of run possible between working pits, the capacity of each system to break out concrete surround to pipes and the capability to increase the size of the sewer.

6.4 RECONNECTION OF LATERALS

6.4.1 The cost of reconnecting laterals to relined sewers can make up a high proportion of the total cost of a contract. In addition, excavation to give man access to such connections can negate the important benefits associated with renovation of non man entry sewers (i.e. reduction of disruption to services, traffic, commerce, and the general public). Two separate approaches have been made to this problem: to make the connections with remotely controlled equipment operating within the sewer linings, or from within the lateral drain; and to develop minimum excavation techniques.

6.4.2 **WRC Engineering**

WRC is developing remote controlled equipment to reconnect lateral drains to lined sewers from within the sewer. A system for lateral reconnection is undergoing development using the Europa cutter (described below). It involves the detection of the lateral remotely from within the lining, bridging of the annulus between the new lining and the old sewer to allow grouting, and the application of an abrasion resistant spray-coat to the exposed annulus grout face at the lateral. The system uses a robotic manipulator which has been designed to operate remotely at distances up to 100m from the control unit in pipes down to 225mm diameter. Methods of remote lateral detection currently available use a magnetic or electromagnetic emission source which is implanted in the lateral drain and later detected by a device mounted on the cutting tool. It is hoped that present development work using microwave techniques will remove the necessity for implanting emission sources.

6.4.3 **Europa Engineering**

Europa Engineering have developed a remote cutter for cutting lateral connections and this equipment incorporates a lateral drain detection unit operating on the electromagnetic principle which can be interchanged with a magnetic unit.

6.4.4 **Avent PRS**
IPD Ltd

IPD Ltd has developed a remote reamer cutter for cutting lateral connections and Avent PRS are able to cut such connections using a high pressure water jet cutter.

6.4.5 **John Kennedy (Civil Engineering) Ltd**

John Kennedy (Civil Engineering) Ltd has developed a cutter which operates from within the lateral drain. This cutter has an articulated body and can to a limited extent negotiate bends in the drain.

6.4.6 **ALH (Contracting) Ltd**

Several methods of minimum excavation or achieving reduced excavation costs have been successfully attempted by Avon Lippiatt Hobbs (Contracting) Ltd. These include excavation by auger and by vacuum excavator. At the same time as developing these techniques, ALH have developed a method of remaking vertical drain connections remotely from ground level.

6.5 LOCALISED REPAIRS

Two systems for localised repair of non man entry sewers are being developed. These will be useful in sewers that have isolated damage.

6.5.1 **Renoline**

The Renoline system uses resin impregnated fibreglass matting moulded round an inflatable packer. When the packer has been placed at the repair location it is inflated to form the matting against the sewer wall.

6.5.2 **AMK Ltd**

AMK Ltd are developing a system using a cylindrical former with inflatable packers. This is positioned at the defect to be repaired and, after the packers are inflated, an epoxy resin is pumped to the annulus round the former. The technique can be used to seal joints and cracks, to fill voids behind the sewer wall and potentially to provide a localised structural coating to the sewer. Development is continuing to improve the resin material, and the finish of the cured resin.

7. **CONCLUSION**

7.1 The DOE sponsored programme of Renovation Development Trials, as presently conceived, will last until early 1986 and it is hoped that before that time many of the systems and equipment described in this paper will complete their development and testing stage, and achieve accepted (and justified) status within the industry. The specific advantages of each system are expected to give each its own place in the sewer renovation market. Together they are expected to make a highly significant contribution to the market and greatly increase the potential for reducing the cost and social disruption of sewer renewal, particularly for non man-entry sewers.

Considerable efforts are being made by many companies involved in sewer renovation to further progress the development and improvement of systems and equipment. This effort, together with the current series of development trials will hopefully assist in further establishing the present UK lead in sewer renovation which can be exploited worldwide.

International Conference on the
PLANNING, CONSTRUCTION, MAINTENANCE & OPERATION OF SEWERAGE SYSTEMS
12-14 September, 1984

NEW TECHNIQUES IN NON-MAN ENTRY SEWER RENOVATION

J. A . EMERY
C.Eng. M.I.C.E., M.I.H.T., M.A.S.C.E.
Luton Borough Council, Luton, Beds.

SUMMARY

This paper details experimental innovative techniques used in lining a section of an old combined sewer in Luton, Bedfordshire. Some of the practical difficulties, which are considered to be applicable to all non-excavation, non-man entry sewers renovations, are described.

Held at The University of Reading, England. Organised and sponsored by BHRA, The Fluid Engineering Centre, Cranfield, Bedford in conjunction with the Water Research Centre.

"The objective should be not to dig up any busy road when there is a reasonable alternative available." (Ref. 1.)

1. INTRODUCTION

Luton, with a population of 165,000 lies some 50 kms north of London. The Borough Council, under an agency agreement, is responsible to the Thames Water Authority for the operation and maintenance of its sewerage system.

As a result of a survey of sewers in Luton it was found that some of the older sewers were in need of repairs or replacement and a programme was commenced using conventional methods.

Luton Borough Council and Thames Water Authority, with the support of the Water Research Centre, decided to treat 467 metres of 300mm diameter combined sewer in High Town Road as experimental, and to allow three innovative techniques to be tried with the minimum of restriction and the maximum of encouragement.

2. HIGH TOWN ROAD

High Town Road was once the main shopping area for the old town and is today a narrow, busy road lined with shops and wholesale warehouses. An adjoining dairy and engineering works generate heavy goods traffic. It forms one of the main bus routes into the modern town centre and in the narrowest section funnels into a one way system with a contra-flow bus lane.

The 300mm (12") diameter combined sewer was laid around 1890 at an average depth of 2.5 metres, generally following the centreline of the road. The sewer was clayware, using the Jennings' collar, patented in 1854 (Fig. 1). Ground movement has caused the line and level of the sewer to vary, opening joints and allowing infiltration of the surrounding chalk soil. The chalk precipitated out where the gradient of the sewer lessened, causing blockage and surcharging. Total collapse had also occurred along one length.

A gas main had previously been relaid in the road and the resulting disruption had caused considerable loss to shopkeepers and resentment amongst the local residents. The Borough Council had to adopt a very diplomatic approach, and by meeting and discussing with the local people managed, to some degree, to replace resentment with resignation.

3. SURVEY AND PREPARATION

The initial survey by C.C.T.V. revealed displaced joints, cracked pipes, missing collars, ponding and laterals protruding up to 100mm into the sewer. The 94 lateral connections were of 100mm or 150mm diameter and many were suspected dead. The investigation of connections was not made easier by the presence of traps below the pavement and the absence of rodding eyes.

An extract from the survey reads "A metal object was detected, estimated to be the size of a man hole cover, buried beneath the shop floor", and gives a good indication of the type of difficulty encountered.

Twenty seven protruding laterals needed to be removed before any form of renovation could take place. This was done with a machine drawn through the sewer and the laterals cut with rotating jets of water at a pressure of 700 bar. (Fig. 2), which was the first time this method had successfully been used in a live sewer in the UK.

The jets were directed at the protruding lateral and their movement controlled by an electric motor with the process monitored by C.C.T.V. This same machine, with modifications, was later used to cut the lateral openings in the Nuttall's Insituform' liner. No problems were experienced during renovation in dealing with lateral protrusion. Apart from slow progress on the first two or three cuts, no problems were experienced.

4. SYSTEMS EMPLOYED

After narrowing down possible options, three systems of non-excavation, internal reconnection renovation were selected. The lengths and locations of these systems are shown in (Fig. 3.)

(i) Edmund Nuttall Ltd - 'Permaline'
Resin impregnated felt "Insituform" Inversion liner. Position of laterals determined by concaving of the liner over the hole, high pressure water used to cut.

(ii) Avent P.R.S. 'Nu-line'
Induction welded Polyethylene Liner, pre-cutting of lateral hole, chemical foam grout.

(iii) Internal Pipe Drilling - I.P.D. Ltd.)
Snap fit polypropylene liner, lateral detector, internal hole cutter, cementitious grout.

A more comprehensive description of these three systems is given in Appendix 'A'.

5. PRELIMINARY TRIALS

In conjunction with the Water Research Centre an attempt was made to determine the existing flow rates in the unrenovated length of sewer with a view to measuring the "before and after" effect of renovation. After the installation of their prototype flow measuring equipment it became apparent that the flows in the sewers were insufficient, in the time available, to provide meaningful readings and the attempt was later abandoned. The equipment which measures velocity and depth has since been modified and is now commercially available. A full description of the equipment is given in a paper by Dr. M. J. Green of W. R. C. (Ref. 2.) The crude but useful result is that the flooding that took place before renovation hasn't taken place after, which is, after all, the object of the exercise.

A further problem discussed with W.R.C. was that of location of connections. A Gamma ray device was proposed and actively investigated. This consisted of a battery powered detector/transmitter which could be placed in each lateral connection prior to lining and held in place with an expanding "umbrella" clip. The Gamma ray source was to be protected by a lead shield which had a hole 0.5-1.0 mm dia drilled into it toensure that Gamma rays were as near parallel as possible. It was eventually decided that the method was perhaps too sophisticated and might prove unacceptable to the Environmental Health Department and the Health and Safety Executive for use in the sewers of High Town Road.

5.1 Nuttal's'Permaline'

This system was already commercially available and consequently it was not necessary to carry out preliminary trials. The cutting of the connection was, however, innovatory. The high pressure water jet cutter used for cutting the protruding connections was used for cutting the line at connections. This work was carried out by M.T.C. Drains and Sewers Ltd.

5.2 Avent P.R.S. 'Nu-line'

Preliminary system trials were conducted at Water Research Centres' outstation at Wroughton where a simulated section of the Luton sewer had been constructed above ground. The more severe misalignments and certain positions of laterals, anticipated as possible problem areas, were reproduced.

Meanwhile various diameter pipes were drawn through the sewer at Luton to ascertain the maximum diameter liner that could be installed. This was found to be 280mm diameter and a polethylene pipe of Pn 3.2 rating was selected.

It was through this work in a flowing sewer that the differences between a simulated and live situation began to emerge.

The Wroughton rig had been constructed from modern spigot and socket pipe and this could not reproduce the steps caused by slippage and misalignment of the Jennings' joint. Equipment had been designed to ride over a step of 22mm and this had to be increased to 40mm before the smooth passage of equipment through the sewer could be obtained.

The original system concept allowed for radial alignment of the lateral opening in the liner to be achieved by guiding the pipe along steel tapes laid along the sewer. The tendency of paper and rag in a flowing sewer to cling to any object laid there made this method impossible and a new alignment system was devised.

There are several colloquial expressions that refer to the adhesive qualities of the solids present in sewage effluent. This was confirmed by experience and the results made more tolerable by the adoption of a ribald sense of humour.

The Wroughton trials proved useful, but it was apparent that further knowledge would need to be gained empirically in a real sewer. For this approach the system had the advantage of causing no disruption to the service or inconvenience to the householder and so a policy of try, observe, analyse, modify and try again was adopted.

6. ROAD AND PEDESTRIAN TRAFFIC

The narrow road, position of manholes and lamp posts meant that the rear wheels of buses and heavy goods vehicles passed over theedge of the open manholes.

If the recommendations of Chapter 8 (Ref. 3.) had been faithfully followed then cones and signs would have obstructed or been obstructed by, vehicles making collections and deliveries to the shops. The use of high intensity xenon flashing beacons were found to be the most effective means of alerting motorists of the hazards.

Where cables and hoses needed to cross a pedestrian walkway they were covered with heavy rubber mats. The regular appearance of a blind lady walking past the working area acted as an excellent incentive to keep equipment within defined and protected zones.

7. CLEANING

Prior to operations each section of sewer was cleaned with high volume, low pressure water jets by Luton Borough Council's Cleansing Department. This was supplemented with a sponge rubber swab winched through to remove the odd stone lodged within joints.

Difficulty was experienced in one 50 metre length with the drawing through of equipment and the water jetter returned. More persistent efforts brought forth four bucketfulls of silt, two butcher's meat hooks, a screwdriver and a kitchen spatula.

In one section that had dropped and was subject to ponding, a mixture of wire, washers, clips and silt had formed into a lump in the invert some 50mm thick. Unsuccessful efforts were made to remove it mechanically and finally a water jet at 250 bar was played upon it. Some of this 'crud' was removed, but so also was the crown of the sewer, which allowed the obstruction to be by-passed, though not in the manner planned.

8. OPERATIONS

As the I.P.D. Ltd. work is still in progress at the time of writing this paper, the following sections refer mainly to the Avent 'Nu-line' system. Nonetheless many of the problems described are common to I.P.D. Ltd. which they are also successfully solving.

The empirical approach was based around problem analysis and observation, unfortunately diagnostic in the majority of cases. Results can be classified under three headings.

(i) System concepts
(ii) Equipment, design and reliability
(iii) Manpower

8.1 System concepts

With the exception of the method of radial alignment, the original concepts worked well. The combined approach of aligning, welding, sealing and grouting the liner pipes containing the lateral opening was simplified, with the alignment checked by C.C.T.V. before welding and the sealing and grouting carried out as a separate operation.

8.2 Equipment

Manufacturers were approached with a description of the working environment. The ability of in-sewer equipment to withstand being dragged over the sharp edges of clayware and concrete, subjected to periods of total immersion in various aggresive effluents, the possible presence of inflammable gases and the likelihood of becoming jammed several metres underground, are typical of the requirements.

Reliability of the equipment is slowly being achieved by persistence and experience in what is best described as a hostile environment.

8.3 Manpower

Advertising for persons experienced in sewer renovation is likely to result in a scant response. Skills and expertise are gained the hard way, mistakes becoming the stepping stones forward.

Both male and female labour were employed on site; despite womens' liberation the local residents were often surprised by the emergence of a girl from a sewer manhole.

The factor of fatigue was observed as the cause of several mistakes and also on the quality of decision making on site. This was reduced by the use of check lists, daily pre-briefing and regular meal breaks.

9. CONCLUSIONS

If one takes the analogy of the development of civil aviation to sewer renovation then a few pioneers have taken to the air in their flimsy, crude machines of different designs.

Luton can be compared to the first crossing of the English Channel and this has given the incentive for further engineering development, with some interesting and sophisticated devices on the way.

Nevertheless, far more than this is required if the necessary structure is to be evolved to allow commercial airline operation. The equivalent of the Civil Aviation Authority as a regulatory body would appear to be forming within the Department of the Environment and the Water Research Centre.

Whether the airline is a national or private enterprise operation, fare paying passengers are required. Local authorities are the U.K. equivalent and their active involvement is needed if these early aviators are to stay aloft, and practice their trade before venturing onto international routes. It is to be hoped that the local authorities will not require the impetus of legislation to disallow opening up the road, as already exists in certain cities abroad [(1)] but will act on their own volition.

The work carried out and still in progress in High Town Road has been the result of skilled enginering and creative, imaginative speculation. It is essential that further inquiry into sewer renovation techniques continue, to which end the hypothetico-deductive method of Popper (Ref. 4.) is to be recommended.

> "So long as a theory withstands detailed and severe tests and is not superceded by another theory in the course of scientific progress, we may say that it has 'Proved its mettle'".

10. ACKNOWLEDGEMENTS

(i) The author wish to thank J. G. Vickers B.Sc. (Eng)., C. Eng., F.I.C.E., Borough Engineer of Luton Borough Council for permission to present this paper.

(ii) The technique using remotely fused joints and integral lateral re-connection was devised by Mr. N. L. Rice. It is currently marketed under the trade name Nu-Line (patent applied for) by Avent PRS of Calne, Wiltshire.

11. REFERENCES

1. Select Committee on Science & Technology - "Report on Water Industry" Dec. 1982 - H.M.S.O.

2. Green M. J. "Hydraulic Analysis of Sewerage Systems". In Proc. 1st International Seminar on Urban Drainage Systems Southampton 1982. Editors R. E. Featherstone and R. Jones. Published by Pitman Advanced Publishing. Paper 2/11.

3. "Traffic Signs Manual" - Chapter 8 Traffic Safety Measures for Road Works - Department of the Environment Scottish Development Department Welsh Office 1973 H.M.S.O.

4. Popper K. L. "The Logic of Scientific Discovery", revised English Edition, London, Routhedge & Kegan Paul 1968.

APPENDIX 'A'

SEWER RENOVATION SYSTEMS USED AT HIGH TOWN ROAD, LUTON

(i) Edmund Nuttall Ltd - "Permaline"

The material used for lining was polyester needle felt impregnated with isophthalic resin, forming a flexible tube. This was then fed through the sewer under a 6m head of water. Water, heated to $75^{o} \pm 5^{o}C$ was circulated through the liner to cure the resin. This resulted in a hard, smooth surfaced, structural lining.

Location of connections was relatively straightforward as the uncured resin felt liner took-up a concave shape where it passed over the connection. This was easily observed using CCTV Cameras. Connections were re-formed by cutting a circular hole using a modified version of the high pressure water jet cutter used for cutting the protruding connections.

(ii) Internal Pipe Drilling Ltd. (I.P.D. Ltd.)

Lining was carried out using short lengths of polypropylene pipe having a wall thickness of approximately 8mm. These were inserted, as a 'train", from the existing manholes. Individual lengths of pipe were joined by means of a simple 'snap-fit' joint formed within the wall thichness of the pipe. A device known as 'MUSCLES' was lowered into the manhole and attached to two chains, each having a breaking strain of 12 tonnes. The 'MUSCLES' device and attached pipes were then hauled through the sewer from an adjacent manhole by means of a heavy duty linear winch maintaining the pipes in compression. (N.B. The 'MUSCLES' device is capable of re-rounding deformed pipe and re-aligning displaced joints. It is also technically possible to circumferentially burst the existing sewer pipe and insert a liner of greater diameter.)

Connections were re-formed by cutting through the newly installed lining using a small pneumatic motor, driving a hardened steel edge cutting mill. (This piece of equipment is known as 'NIGEL'.) Location of connections was made by means of accurate pre-measurement. A feature of the cutting mill is that once the lining has been penetrated it can be made to cut precisely the shape of each individual connection. Monitoring the cutting is a unique CCTV Camera having a virtually omnidirectional head and capable of entry into the connection. See (Figs. 4, 5 & 6.) for equipment used by I.P.D. Ltd. For the lining work so far carried out by I.P.D. Ltd., it has not been necessary to use grout between the new liner and the old sewer due to the contiguity of the two pipes. However, it is possible to fill any annualar space with cementitious grout from within the lining. This can be achieved by pumping the grout through special purpose non-retrievable valves, which act in a one-way mode, thus preventing grout re-entering the sewer.

(iii) Avent P.R.S. 'Nu-line'

Lining was achieved by means of short lengths of polythylene pipe passed throught the sewer from existing manholes. A male and female clip form was machined at either end and an induction ring fitted. These sections were winched through the sewer and clipped together, a coil was then drawn through the sewer, the control box sensing the presence of the induction ring and the joint fused to form a smooth, pressure tight seal. A section requiring connection to an existing drain was preceded by a graduated cage containing a CCTV Camera. This enabled the size and location of the drain to be referenced to the previously laid section which enabled a hole to be cut in the next length of pipe coincidental with the sewer connection. A connection unit was then fused inductively into place and this unit winched into position. Pressurising this section inflated a bag into the connection and sealed it, the wall of the bag acting as a shuttering in the annulus between sewer and liner and through the wall a primary foam grout was injected to locally seal the connection to the liner. The bag was then deflated and withdrawn. The final operation was to seal the remaining annular space with a secondary grouting operation.

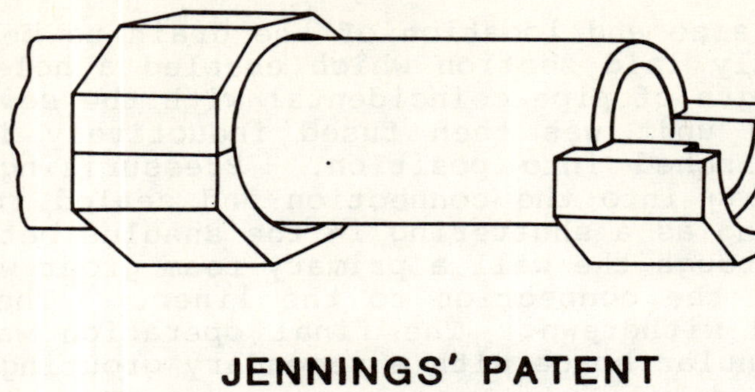

FIGURE 1. JENNINGS' PATENT PIPE (CIRCA 1860)

FIGURE 2. HIGH PRESSURE WATER JET CUTTER

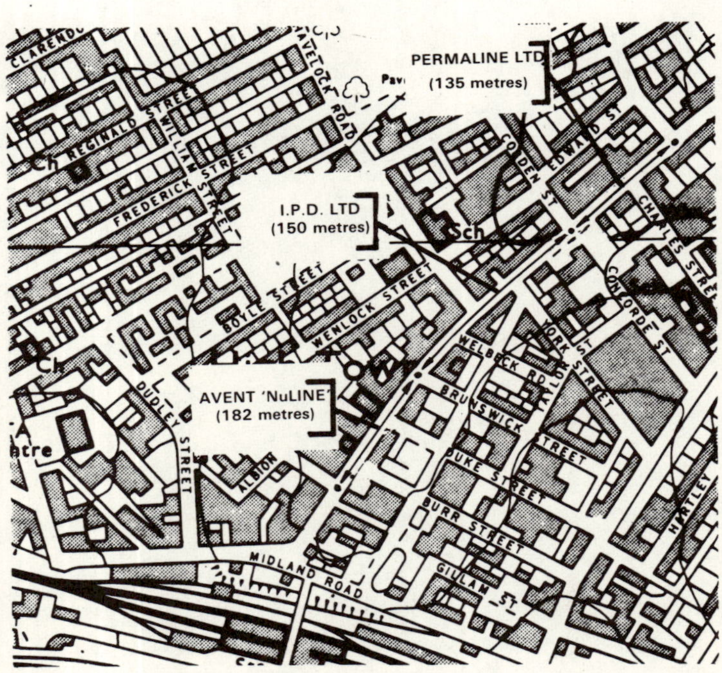

FIGURE 3. HIGH TOWN ROAD - LOCATIONS OF LINING SYSTEMS USED

FIGURE 4. 'MUSCLES'

FIGURE 5. 'NIGEL'

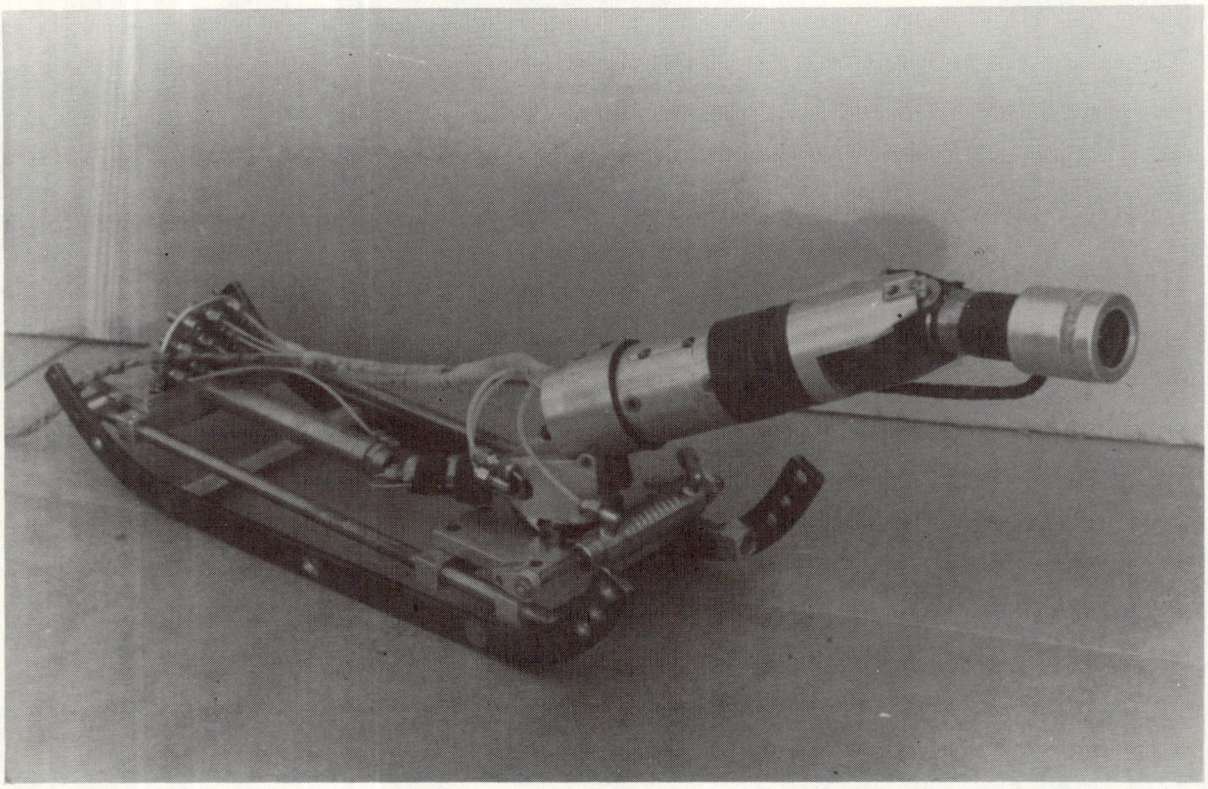

FIGURE 6. 'OMNIDIRECTIONAL CCTV CAMERA'

International Conference on the
PLANNING, CONSTRUCTION, MAINTENANCE & OPERATION OF SEWERAGE SYSTEMS
12-14 September, 1984

PAPER D2

SMALL DIAMETER PIPE MAINTENANCE AND RENOVATION

M. A. JONES M.A., M.I.C.E.
AVENT P.R.S.
Porte Marsh Road,
Calne,
Wiltshire,
SN11 9BW,
England

SUMMARY

A number of new techniques and case studies in pipe renovation and clearing are discussed. These include Jet-Cut clearing of pipes using very high pressure water, Nu-Line renovation using H.D.P.E. pipes installed from manholes and remote lateral reconnection and size for size on-line pipe replacement.

Held at The University of Reading, England. Organised and sponsored by BHRA, The Fluid Engineering Centre, Cranfield, Bedford in conjunction with the Water Research Centre.

1. INTRODUCTION.

Great strides have been made in the technology of small diameter pipe clearing and renovation in the past few years. Until recently the engineers job was relatively simple when assessing pipework problems in that he only had to decide at what point there was economic justification for excavation. There are now an increasing number of options available which are best used at different states of deterioration and in different circumstances.

2. SEWER CLEARING AND CLEANING TECHNIQUES.

This increasing number of options will create problems for engineers in choosing acceptable processes. Table 'A' gives the author's view of the type of choices available at present for clearing and preparing pipework for renovation. This is a very necessary first stage of most projects after the general closed circuit television (CCTV) survey has identified the problem area as, until a survey is made of a clean sewer, it can be very difficult assessing the true scale of the problem. It is sometimes also the case that, having thoroughly cleared and cleaned a sewer, problems with surcharging can be cured with little or no additional work.

The table shows a number of different techniques now available to suit the size of pipe and type of problem. While the major faults have been included and "generally suitable" and "occasionally suitable" categories allocated, particular circumstances may favour one method more than another. For this reason the information should be used for guidance only.

Experience has proved that a guarantee of thorough clearing can only be given when CCTV is used during, or immediately after, the operation. Even with simple silt clearing operations using tanker jetters, an inexperienced operator can leave the majority of material in the sewer should monitoring not be adequate. This can cause expensive delays where cleaning takes place immediately before a renovation operation. In general, the existing techniques of water jetting and power winching and boring will still be the most economic for the types of work where they are already used successfully. New processes now in operation and under development will extend the range of jobs that can be undertaken.

The use of very high pressure water is a tool that has become available to pipeline operators fairly recently and various units have been developed by a number of contractors. The author's firm have developed the "Jet-Cut" service and the key to this is a single highly mobile unit which is adaptable to site conditions and can be used on a wide variety of clearing works. Pure very high pressure water (800 Bar-12000psi) was chosen as the best available cutting medium. The advantages are the lack of any type of pollution or waste material, highly controllable and powerful cutting and the fast dissipation of cutting energy as it leaves the cut material. The very fine jet of water is directed accurately using the new "Clearway Cutter". This allows for rotation in either direction at any speed from 0 to 10 RPM and carries a CCTV camera for monitoring. Having this equipment which can cut through concrete, scale and tree roots the major object of the crew is to position the cutter accurately and efficiently. In order to achieve this a low pressure jetting unit is fitted to remove any silt that may be present and a variety of other equipment such as manhole lifters and rods are carried. To show the versatility of this type of service a number of case studies follow.

2.1. Case Studies.

Wokingham District Council:- Approximately 1,000m of 325mm (14 inch) cast iron trunk sewer running under gardens and across parkland, roads and a stream was restricted by up to a 50mm (2 inch) layer of hard scale. A special unit for rough terraine was made up to remove this scale using very high pressure water with the contract being based upon a fixed rate of approximately £30 per metre. The contract was carried out between December 1983 and February 1984 with no overpumping or disruption to the sewer and with minimal damage to gardens and grassland. Progress rates of approximately 150m per week were achieved despite the nearest water supply being 500m away at one point. A major problem experienced was removing silt before descaling because of the rough surface and the large diameter. A tanker jetter had eventually to be employed for this purpose with the consequent problems of access. Following this the descaling presented no major problems with the cleared pipe being returned substantially to its original condition. Debris was jetted away from the descaler and removed at manholes.

Edmund Nuttall, Permaline Division, Bournemouth District Council:- The contract in January 1984 was to remove eight number intruding connections from approximately 300m of 300mm (12 inch) and 375mm (15 inch) pipework before lining operations. Despite the fact that the line was in poor condition the connections were removed for a fixed price of £210 each.

Malvern District Council: Approximately 80m of 450mm (18 inch) clayware pipe 7m deep had one bad blockage and general root infestation. This was cleared for a fixed price of £1,000.

Oxford City Council: Approximately 100m of 450mm uPVC sewer was up to 75% block with cement grout from a building site. This was a very hard material and care had to be taken in order to avoid damage to the plastic pipe. The main problem was that the level of grout in the pipe was such that it was difficult to do more than flake material from the face. The work was completed at a cost of £5,000.

These projects illustrate the varied types of work which can be successfully completed using only high pressure water but even this range will be greatly increased when further developments reach the commercial stage. One interesting area of work in which both B.H.R.A. and W.R.C. are involved is the use of abrasives with the water jet. This should allow for cutting intruding iron connections and reinforcing steel from sewers as well as speeding removal of other very hard materials or more normal obstructions at a lower working pressure. There is also work being carried out on bulk removal of silt and debris from sewers using a combination of vacuum, jetting and mechanical pulverisation techniques.

3. SEWER RENOVATION TECHNIQUES.

Following pipe clearing and cleaning a realistic assessment can be made of the faults on any particular section. These can range from good condition but under capacity right through to badly deformed with some partial collapses. Once the pipe has reached this latter stage, however, there are not many options open apart from relay. Tables B1 and B2 give a range of probable and possible options for various types of problem on sites where excavation is either very difficult or relatively easy. The attempt here was to cover the two extremes and allow for engineering judgement between these limits as to which techniques to pursue further. The other significant factor which can be seen from the tables is the presence and number of connections. A small number would again fall between the two sections in each table and due allowance would have to be made for this.

New systems are rapidly becoming available and their introduction will be dealt with through the formal channels of the W.R.C. Renovation Strategy so the tables have been compiled to cover existing services only. These range from the long established techniques such as Insituform to the brand new ones such as Nu-Line. They also vary extensively as to price and specification thus requiring a detailed appreciation by the engineer of the relative pros and cons of each system.

TWO NEW RENOVATION SYSTEMS

3.1. Nu-Line

This has been developed and is operated by Avent P.R.S. for the Brickhouse Dudley Group. It enables a structural H.D.P.E. lining including reconnection of laterals to be installed from manholes. One of the key features of the system is the patented induction fusion joint which allows pipes to be drawn individually into position within the sewer and then fused together. This is achieved by machining pipe to a spigot and socket finish and encapsulating a nickle alloy band into one end. When the pipe is pulled together using an inflatable bung a sensing device ensures the correct position and then current is passed through a coil adjacent to the metal band. This excites a current within the band which heats and forms a molten pool of plastic around itself. When the current is turned off the plastic sets leaving the two pipes fused together.

A purpose built vehicle carries all of the equipment needed including a C.C.T.V. system, compressor winch and generator. The power winch is usually set at the upstream manhole and the vehicle at the lower manhole. Depending upon the precise layout of the site this set up can usually be achieved without undue influence on traffic patterns and, for cases where traffic lights are unavoidable, the road can be easily cleared during peak periods. Pumps are generally not required unless flow is heavy. Another advantage is the relatively large H.D.P.E. pipe that can be used since only a single pipe length (usually 1m long) is pulled through the sewer at a time. This can obviously pass slight obstructions, dropped joints and changes in direction where a complete line would suffer from cumulative build up of friction. The result is that the normal H.D.P.E. pipe outside diameter is approximately 20 to 25mm less than the original pipe diameter. The pipe wall is generally slightly thinner than that used for slip lining but it can easily be adjusted to suit specific design criteria. A reduction of approximately 40mm on inside diameter can be expected.

Once the pipe has been thoroughly cleaned and the vehicle and equipment are in position pipes are winched up the sewer one at a time joined and grouted in. Grouting is carried out on each pipe during the jointing process by means of an in-line grout pump. This sits inside the new pipe and includes tanks for the constituent parts. These are pumped, mixed and injected into the annular space near the joint in a liquid form. The Polyurethene grout then foams covering the joint area up to the previously injected material and foams back to just before the open end of the pipe. Repetition of the process gives a fully fused and grouted liner.

Connections are dealt with by a patented process. A pre-survey identifies the approximate position of each lateral and pipe lengths are adjusted so that a connection should fall approximately in the middle of the next pipe to be inserted. A special measuring cage is pulled up to the last H.D.P.E. pipe and a moveable right angle viewing CCTV camera within the cage enables the exact size and position of the lateral to be plotted. A hole is then cut in the next pipe length before it is pulled in. The position is checked again using the CCTV camera before the pipe is welded to the previous one. There then only remains the problem of bridging the gap between the old and new pipe. This is achieved by pulling a pig into position and inflating a rubber bag through the hole in the H.D.P.E. pipe into the lateral. Grout is injected into the annular gap around the connection and when this has solidified, in 3 to 4 minutes, the bag can be withdrawn and the connection is complete. In this way the blocking or diverting of laterals during the relining operation is rendered unnecessary.

The main advantages of this system over existing techniques are structural relining, a minimal decrease in cross sectional area, no excavation, no overpumping, no disruption outside working hours, minimal disruption during working hours and low mobilisation cost.

3.2. Size-for-Size.

This type of operation where a pipe can be replaced with the same size or larger H.D.P.E. pipe is now becoming very popular and a number of companies are involved. Generally a pneumatic hammer or mole provides the power to an expander which breaks the old pipe and compacts the pieces into the surrounding soil. An H.D.P.E. pipe is pulled behind and a winch at the forward manhole keeps the device on track and assists forward movement. Equipment is being made smaller such that insertion and removal can be undertaken from existing manholes. At present projects have taken place up to 300mm diameter and have involved both manhole to manhole and lead in trench to reception pit formats. The economics and suitability of the various arrangements has to be judged on an individual basis. For example, the acceptability of multiple mechanical joints has to be weighed against butt fused joints and the cost and disruption of excavating lead in and reception pits against driving short headings from the launch and reception manholes required to increase the working area.

The obvious disadvantage of this type of system is that it does necessitate some excavation at laterals as this pipework is damaged as the mole passes. There are, however, other options to digging a man sized excavation at these points. One is to divert flow from connections to the manhole and the other is to adopt a semi-remote reconnection method. This involves drilling a small hole down to the new pipe, remotely fusing a riser pipe to the main and then connecting onto the original lateral pipework nearer its source at shallow level.

4. SOME RENOVATION CASE STUDIES.

4.1. Nu-Line.

Client: Gosport Borough Council acting as Agents for Southern Water Authority.

Contract: To reline 480m of 300mm foul sewer with a 225mm and some 150mm connections 4m deep under the main A32 Fareham Road into Gosport with 280mm O.D. H.D.P.E. pipe.

Value: The pipe was installed at a rate of £100.00 per m.

General Description: The existing concrete pipe was subject to chemical attack, following discharges from factories, which had caused extensive aggregate exposure. A check on possible chemicals and their concentrations showed that H.D.P.E. was an acceptable new pipe medium.

The physical situation was such that alternative methods to traditional re-laying had to be examined. These had to allow for no excavation since the pipe was approximately 4m deep under the main access road to Gosport which has very heavy traffic in both directions during rush hours and maintains a moderate traffic flow through the rest of the day. It was also politically difficult to accept a method of work which would cause significant disruption as a series of road works and pipe laying projects on the same road had caused significant disruption to the local business community and population.

Southern Water Authority were aware of this project and its requirements and after studying the available options decided that the Nu-Line system fitted the requirements and it was decided to place the project on Southern Water's Development Contract list. This enabled a fixed price contract to be negotiated with Avent P.R.S.

The contract included cleaning, the removal of a number of intruding lateral connections and relining and did not allow any equipment on or obstruction of the road between 7 am and 8.30 am and between 4.30 pm and 7 pm. The cleaning by tanker jetter caused some problems in that the exposed aggregate in the invert tended to hold back the silt and debris and further stones were continually washed out of the pipe wall. These were overcome by using additional care on jetting and by close monitoring of cleared sections.

Lengths of pipe between manholes varied between 50m and 130m and the manholes were generally a minimum of 1.2m along the line of the sewer. On the sections longer than 100m a double socket pipe was set in the middle of the run and lining carried out from this point first to one manhole and then the other. On sections of less than 100m the first pipe was placed at the upstream manhole and lining carried out to the downstream one. A 225mm (9 inch) connection and two 150mm (6 inch) connections were made in the last section lined.

Once some initial equipment problems had been resolved

progress rates of between 30m and 40m per week were achieved despite the restricted working. Some more serious difficulties occurred because of grout spillage. The very rough surface texture of the original pipe meant that any grout spilt could not be removed mechanically and on two occasions the Jet-Cut service had to be used to remove material. It was also found on one occasion that when grout was inadvertantly pumped inside the H.D.P.E. pipe instead of into the annular gap it adhered very strongly to the pipe wall and the equipment. This was eventually removed mechanically but did demonstrate that the grout used does have the required properties for its annulus filling application. The contract was completed in April 1984.

4.2. Size for Size.

Client: Oxford City Council acting as Agent for Thames Water Authority.

Contract: To replace approximately 180m of 225mm I.D. surface water sewer, including 3 connections, with 225mm High Density Polythene pipe.

Value: £13,500 fixed price.

General Description: The sewer was in poor structural condition and ran through heavy clay under the footpath in London Road, Oxford. Open cut would have necessitated removing lamp posts, signs and a bus shelter and have caused disruption to pedestrians and property owners.

The contract was completed in February 1984 and was carried out using a standard Williams and James pneumatic mole in an expander designed and constructed by Avent P.R.S. The 250mm O.D. 6 Bar H.D.P.E. pipe was butt fused together on the surface, initially for a continuous 80m run. It was then connected to the expander and offered up to the original pipe via a short lead in trench. Directional stability and forward movement was then assisted by a 4 ton Thompson winch at the reception pit. The lateral connections were remade in excavation using a new and simple resin encapsulation kit which leaves a plastic spigot end suitable for connection to a clayware pipe using a Hepsleeve Connector.

This pipe was at one point within a metre of a gas pipe and G.P.O. cable ducts and the contract therefore included for some investigation of the effect of the operation on these services. After moling for approximately 5m, trial excavation showed no evidence of any movement or damage to the services.

Problems were encountered particularly at a road crossing and under some house drives with heavy concrete surrounds to the pipe. Some of these did stop the expander and excavation was required to remove the concrete but in at least two locations the expander did break through the concrete surround. Improvements in design and power will obviously increase the ability of the expander to crack these surrounds.

Speeds of between 0.3 and 0.5m per minute through normal ground and 0.05m per minute with the concrete surround were achieved.

4.3. Slip Lining from Manholes.

Client: Wrekin District Council as Agent for Severn Trent Water.

Contract: To reline from existing manholes approximately 95m of 225mm foul sewer with 180mm O.D. 6 Bar H.D.P.E. pipe and grout the annular gap.

Value: £4,000.

General Desciption: The sewer ran under a 10m high embankment and was below the water table thus rendering excavation completely impractical without difficult dewatering. A trial pull through showed that lining with 180mm O.D. pipe was practical and the contract was then put out to tender. The basis of this was that lining had to be carried out from within the existing manholes. One of these was approximately 1.2m along the line of the pipe and the other relatively inaccessible. Avent P.R.S. proposed using the new Stuart and Lloyds Snaplok pipe and were awarded the contract.

The pipe was jointed in the manhole using the special clamps and pushed into the pipe. Assistance was also given by a tirfor winch at the other manhole connected to a specially fabricated nose cone on the front of the line.

Jointing and insertion were completed within three days in April 1984. The annular space was then grouted using Pozament Aquablend.

This technique appears quick and economic where no connections are present and where a suitable reduction in capacity can be allowed.

5. CONCLUSION.

A number of new techniques have been discussed in this paper and others are becoming available. The message that must be repeated again and again is that the design engineer should be made aware of the variety of techniques available. Given this awareness he can then use his skills to decide upon the most technically and economically acceptable solutions to his particular problems. Realistic tenders can then be sought.

Companies developing new techniques have found that there is no substitute for the experience gained by use in the "live" situation. Very often significant improvements in design and technique can be brought about at an early stage as a result of field experience. The availability of suitable sites and an appropriate degree of flexibility, understanding and involvement on the part of client are all to mutual advantage in the longer term. In this way a proper balance can be achieved between renovation and replacement which will allow economic techniques to prosper and investment decisions for new systems to be made on a sound factual basis rather than speculation.

ACKNOWLEDGEMENTS

The author wishes to thank the Directors of Avent P.R.S. (Pipe Renovation Systems) and the Brickhouse Dudley Group for their permission to publish this paper.

Thanks are also due to Thames Water, Southern Water, Severn Trent Water and their agents for letting contracts which utilise newly developed renovation techniques.

The views expressed in this paper are those of the author and do not necessarily represent those of Avent P.R.S.

TABLE A
SEWER CLEARING AND CLEANING

	Water Jetting	Water Jet with root cutter	Very High Pressure Jet-Cut Systems	Chemical Systems	Power Winching	Power Boring	Under Development		
							Abrasive Water Cutter	Vacuum technique inc. pul-verisers	Small dia. Jet-Cut systems
Pipes 100 mm (4") and smaller Clearing blockages	★★					★★			★
Removing Silt & Fats	★★				★	★			
Removing Scale	★★			★	★	★			★★
Removing Tree Roots		★★		★	★				★★
Removing Int. Connection									★★
150 mm to 300 mm (6" to 12") Clearing blockages	★★		★			★			
Removing Silt & Fats	★★				★★			★	
Removing Scale			★★	★	★★	★	★		
Removing Tree Roots (Fine)		★★	★★	★	★		★		
Removing Tree Roots (Dense)		★	★★	★			★		
Removing Int. Connection			★★				★★		
Removing Int. Steel							★★		
375 mm to 600 mm (15" to 24") Clearing blockages	★★		★			★		★	
Removing Silt & Fats	★★				★★			★	
Removing Scale			★★	★	★		★★		
Removing Tree Roots (Fine)		★	★★	★			★★		
Removing Tree Roots (Dense)			★★	★			★★		
Removing Int. Connection			★★				★★		
Removing Int. Steel							★★		

★★Generally suitable ★Occasionally suitable. Note: The above represents the Author's opinion only.

TABLE B1
SEWER RENOVATION
Excavation impossible or very difficult

Without Connections

	Low Capacity (mm)		Structural shape good (mm)		Structural badly deformed (mm)		Infiltration or exfiltration (mm)		Tree Roots (mm)	
	≤ 300	>300	≤150	>150	≤ 300	>300	≤ 150	>150	≤150	>150
Epoxy Spray			★★	★★			★[2]	★[2]	★★	★★
Insituform			★★	★★	★	★	★★	★★	★★	★★
Joint Sealing							★★	★★	★★	★★
Micro Tunnel	★	★★		★	★	★★				
Nu-Line				★★	★	★		★★		★★
Patch Repair			★★	★★			★★	★★	★★	★★
Size for Size (moling)	★★	★	★	★	★★	★	★	★	★	★
Slip Lining			★★	★★	★	★	★★	★★	★★	★★

With Connections

	Low Capacity (mm)		Structural shape good (mm)		Structural badly deformed (mm)		Infiltration or exfiltration (mm)		Tree Roots (mm)	
	≤300	>300	≤150	>150	≤300	>300	≤150	>150	≤ 150	>150
Epoxy Spray			★★	★★			★[2]	★[2]	★★	★★
Insituform			★	★★	★	★	★	★★	★	★★
Joint Sealing							★★	★★	★★	★★
Micro Tunnel		★[1]			★[1]	★★[1]				
Nu-Line				★★	★	★		★★		★★
Patch Repair			★★	★★			★★	★★	★★	★★
Size for Size (moling)	★[1]		★[1]		★★[1]	★[1]	★[1]	★[1]	★[1]	★[1]
Slip Lining			★[1]	★[1]	★[1]	★[1]	★[1]	★[1]	★[1]	★[1]

1 In conjunction with semi remote reconnection of laterals or diversion of laterals;
★★usually suitable; ★occasionally suitable.
2 After Joint Sealing

Note: The above represents the Author's opinion only.

TABLE B2
SEWER RENOVATION
Excavation reasonably easy

Without Connections

	Low Capacity (mm)		Structural shape good (mm)		Structural badly deformed (mm)		Infiltration or exfiltration (mm)		Tree Roots (mm)	
	≤300	>300	≤150	>150	≤300	>300	≤150	>150	≤150	>150
Epoxy Spray										
Insituform										
Joint Sealing							★	★★	★	★★
Micro Tunnel										
Nu-Line										
Patch Repair			★	★			★	★★	★	★★
Size for Size (moling)	★★		★	★	★★	★	★		★	
Slip Lining			★★	★★	★	★	★★	★★	★★	★★

With Connections

	Low Capacity (mm)		Structural shape good (mm)		Structural badly deformed (mm)		Infiltration or exfiltration (mm)		Tree Roots (mm)	
	≤300	>300	≤150	>150	≤300	>300	≤150	>150	≤150	>150
Epoxy Spray			★	★			★[1]	★[1]	★	★
Insituform				★				★		★
Joint Sealing							★★	★★	★★	★★
Micro Tunnel										
Nu-Line				★				★		★
Patch Repair			★★	★★			★★	★★	★★	★★
Size for Size (moling)	★★		★	★	★★	★	★	★	★	★
Slip Lining			★★	★★			★★	★★	★★	★★

★★Usually suitable; ★Occasionally suitable.
1 After Joint Sealing.

Note: The above represents the Author's opinion only.

International Conference on the
PLANNING, CONSTRUCTION, MAINTENANCE & OPERATION OF SEWERAGE SYSTEMS
12-14 September, 1984

PAPER D3

GUIDELINES ON THE USE OF POLYETHYLENE PIPE FOR USE IN SEWER APPLICATIONS.

BY

G. P. MARSHALL

PIPELINE CONSULTANTS LIMITED

SUMMARY

The paper seeks to provide Civil Engineers with a general guide to factors governing the choice of PE pipe materials for gravity and pumped rising mains. The design basis for PE and the criteria for choosing high or medium density grades are outlined.

A critique of standards and specifications is given together with an outline of a new approach towards designing tests to ensure high toughness is obtained in finished products.

Held at The University of Reading, England. Organised and sponsored by BHRA, The Fluid Engineering Centre, Cranfield, Bedford in conjunction with the Water Research Centre.

1. INTRODUCTION

During the past twenty years, plastics have come to be accepted into widespread use worldwide as suitable materials for the manufacture of pipes both for non-pressure and pressurised applications. Although there are many types of thermoplastic used for specialist applications, the two most common materials used in Europe have been uPVC and High Density Polyethylene (HDPE),

Although HDPE and uPVC are used in broadly similar quantities on the Continent for sewerage and water applications, in the U.K., uPVC has, until recent years, been the predominant material for pipe construction. Although there have been very few problems with uPVC in low pressure and small diameter systems (less than 200mm), the U.K. Water & Sewerage Authorities have experienced very serious service failures on large diameter pressure mains which has reflected badly on the reputation and sales of plastics pipes in general.

At the present time there is much interest in the use of different types of large diameter HDPE pipes in sewerage applications largely generated by two factors:

a) In non-pressure systems the possibility of using very large diameter PE Pipes (up to 1m) in continuously welded 'strings' to slip line and refurbish old, decaying sewers has been found to be attractive. The technique is well established and proven and initial trials by Bolton Metropolitan Borough on behalf of NWWA have been extremely successful. The technique is now being utilised throughout the U.K.

b) The highly successful use of Modified HDPE (of lower density and hence referred to hereafter as Medium Density Polyethylene [MDPE]) on a massive scale by British Gas has led to the development of high quality materials being promoted as suitable for both pressurised water and pumped sewer rising mains.

The case for using PE by the sewerage industry is now being made with great force by many people and it is undoubtedly true that if systems are designed and specified correctly there are many advantages which can give engineers great scope for designing technically sound and economically attractive pipe systems in PE both in new construction and renovation applications.

The present paper seeks to present a general overview of relevant design and specification criteria to be used for High and Medium Density PE pipes to assist engineers in choosing the right materials and hopefully avoiding the problems which have beset other plastics in the past.

Within the UK there is an enormous fund of expertise on PE Pipes both within academy, the Plastics Industry and the Civil Engineering fraternity. If this expertise is harnessed and used then there is every reason to believe that the Civil Engineering Industry can repeat the completely satisfactory experience of British Gas and other users on the Continent of Europe.

1.1 The advertised advantage of Plastics Pipes.

Plastics pipes are frequently promoted by the Plastics Industry as having many advantages over more conventional materials such as ductile iron, asbestos cement, etc. Promotional literature available from plastic pipe manufacturers usually stress the following points in favour of any type of plastic pipe:-

i) They are corrosion resistant - they will withstand chemical attack from soils which are agressive to other pipe materials, and will not corrode in sewage, etc. This is true for PE for which there are no known solvents at temperatures below 50°C. They are resistant to all acids.

ii) They are not bio-degradable and the materials do not contaminate potable water (ref. 1).

iii) Their hydraulic characteristics are advertised as being excellent without build-up of slime and silt with time.

iv) Light weight: All plastics compare very favourably with competitive materials in that they are extremely light weight and can therefore be manipulated and handled on site with great ease. (PE is however, quite soft and mishandling will readily cause external scoring which is not normally a problem but should be avoided).

v) Plastics pipe can notionally sustain far greater imposed strains before failure. Providing the materials are produced so that they only fail in a ductile manner, their inherent flexibility means that traditional problems associated with longitudinal bending imposed by settlement of backfill, traffic loading, etc., should not be significant.

vi) Although the raw materials are extremely expensive on a weight basis, the low density and the lower costs of producing pipe from thermoplastics materials has in general meant that basic pipe costs are often attractive c.f. other materials.
Taking into account the lower laying costs arising from speedier and easier installation has meant that installed costs for plastic pipe are in general competitive.

Disadvantages of thermoplastics are not, in general, stressed by the plastics industry and it is therefore appropriate to review past service experience.

1.2 Previous service performance of Thermoplastics Pipes.

uPVC has been used on a substantial scale for large diameter pipes and although this material is not discussed in detail in this paper , it is pertinent to assess whether the actual service performance of PVC has matched the advertising claims and to try to identify situations where specification and design of plastics pipes in general must be considered carefully.

With uPVC pipes up to 200mm dia. for sewage, drainage or on pressurised potable water mains applications, there have been no significant problems. Although occasional failures do occur, it would appear that these are, in general, ascribable to some form of mal-treatment of the pipe itself.

On PVC pressure mains and a number of pumped sewer installations, where diameters greater than 250mm have been used in substantial quantities (up to 600mm) there have been a large number of multiple pipe failures. Because most of these systems use bell and socket joints there have not been failures longer than one pipe length (typically 6-10m) but the failures have been in a low-strain, brittle catastrophic mode unlike the ductile mode which occurs during standards testing.

It would seem that the primary causes are as follows:

i) Over 95% of those pipes examined by the authors have failed because of localised point loading where pipes have been laid onto extremely poor bedding which has contained boulders, bricks, pieces of concrete, etc. With large diameter pipes, the weight of water in the pipe itself combined with the weight of backfill has been sufficient to generate high local stresses on the bores of the pipe. Small flaws existing in the bores of the pipe directly opposite the sources of the point load have then started to propagate and eventually these cracks have accelerated to give rise to catastrophic failure.

ii) That point loading has frequently been the source of service failures has been confirmed by direct 'in-trench' evidence where pipes have been examined before they have been removed for replacement.

iii) Stress analysis has confirmed that under such circumstances sufficient stress could be generated by the water and the soil loading. It is significant that these failures have occurred only on large diameter pipes where such loading would naturally be much greater.

iv) In some cases, the constant load for soil and water has been exacerbated by traffic loading where pipes have been buried at a shallow depth beneath roads, giving rise to a "fatigue" effect.

v) Pipe Standards in operation when the 'old' uPVC pipes were produced did not contain any tests for notch sensitivity and hence did not simulate service conditions. A new version of BS3505 now in draft form rectifies this situation for UPVC and the WRC Specification for PE pressure pipes does contain a notched impact test. (Not necessarily relevant for long term loading). Although PE is known to be far less notch sensitive than uPVC it is still true that standards tests for PE require testing of pristine pipe and results may not be representative of performance of scored pipe which actually enters service.

vi) It has been proposed that repeated, transient pressure surges can cause failure in plastics pipes (Ref.2). As to whether these surges are detrimental because of fast overpressures or a fatigue effect is unclear - there is no direct experimental evidence. The majority of service failures of large pipe seem to have occurred under static pressure conditions where surge could not be the prime reason for failure.

Although fatigue due to cyclic pressure has caused some uPVC failures the proof is not conclusive that it is the major factor causing failure. Direct, on-site examinations suggest poor laying to be a more prevalent cause of problems.

1.3 Lessons from Past Experience.

It would appear that the major factors governing plastics pipe design which have led to probelms are two-fold:

i) Materials have been produced which have low resistance to crack propagation.

ii) Pipes have been subjected to overloading largely caused by poor control of installation. The 'lightweight factor' which enables easy handling on site can often lead to careless treatment of plastics pipes.

1.4 The Case for High Density Polyethylene.

Because plastics materials are often very similar in appearance and do have general characteristic of time/temperature sensitivities, etc. which are common, many engineers seem to believe that their performance characteristics as pipes will be the same. This is not true.

Polyethylene pipes have very different performance properties to other thermoplastics. As a family of materials they are structurally different in that they are partially crystalline and although of low tensile strength, polyethylenes are far more ductile than other amorphous materials. Indeed, it is primarily because of their low strength that they are so ductile - in common with copper and mild steel in the metals field.

The family of materials broadly referred to as high density polyethylenes has been used for potable water distribution on the Continent and in Scandinavia for over 20 years. In the more northern countries in particular, they have found great favour because of their high impact resistance and extremely good low temperature failure characteristics. Although the modulus and yield stress are low, they retain their flexibility at low temperatures and can also be fusion welded to form one continuous length of pipe which can easily be used to transport potable water in difficult terrain where conventional pipe laying techniques would be extremely arduous and expensive.

2. POLYETHYLENE FOR PIPELINES

2.1 Types of Polyethylene.

There are two very distinct families of polyethylenes, which are broadly classified according to density, viz:

i) Low Density Polyethylene

Ethylene gas together with oxygen is polymerised under very high pressure (approx. 1500 bar at 80 - 300°C) to produce a polymer which has a density in the range 0.915 - 0.925 g/cm^3. This material (initially called polythene - an ICI trade name in the early days) has rather low tensile strength and modulus but it is extremely ductile and can sustain strains up to 800% before failure. The tensile'strength' and modulus are both very sensitive to time and temperature. At 20°C the 'strength' is predicted to decrease from 12 to 5 MPa over 50 years continuous loading and the modulus drops from 160 MPa to 53 MPa when data are extrapolated to 50 years.

Because of the very low strength and stiffness and the high sensitivity to time of these properties, the material is not used for the production of pipe on diameters greater than 50mm. Larger diameter pipes would have to be so thick as to make them economically unviable.

ii) High Density Polyethylene

Using a low pressure process ethylene can be polymerised using catalysts to form a higher density polyethylene commonly referred to as HDPE with densities in the range 0.925 to 0.965 kg/cm^3. (For densities between 0.925 and 0.945, the terminology MDPE is used). Whilst the increase in density produced with these materials does not seem (to engineers unfamiliar with plastics) to be very significant, the materials which result have very different properties.

The HDPE materials have higher tensile strength and higher initial modulli than either their medium or low density counterparts, a short-term tensile yield stress being of the order of 25 MPA for HDPE and 19 MPa for MDPE. Values of modulus (E) for HDPE are in general 10%

higher than those for MDPE, being of the order of 0.8 GPa in the short-term. Again, as for the LDPE Materials, the tensile strength and modulus vary as a function of time and this time-dependence is different for both HDPE and MDPE. The medium density grades are rather less sensitive to time.

It should be fully appreciated that the polyethylene family of materials can be modified by the raw material manufacturers to produce co-polymers (equivalent to alloys) which have markedly different properties particularly with respect to their failure behaviour. Control over chemistry can be utilised by the large chemical companies to produce polyethylene modifications whose properties can be tailored to suit specialised requirements.

2.2 Supply of Polyethylene and Pipes.

When polyethylene is produced by raw materials suppliers it is in the form of a powder which is then subsequently combined with various additives such as stabilisers, anti-oxidants, pigments, etc., by various compounding techniques to make granules. These are supplied to the companies who manufacture pipe by extrusion. Polymer granules are fed into an extruder where they are melted and forced in a molten condition through a circular die to form the shape of the pipe. The pipe is then cooled by water, drawn off continuously and sawn to appropriate lengths or is coiled.

The plastics industry, in broad terms, is structured such that the raw materials suppliers are in competition with each other to supply different grades of materials to various extrusion companies and it is the large chemical companies who carry out much of the basic research into the properties of their individual compounds.

In the United Kingdom, this was generally true until 1970 when British Gas opted to use medium density types of polyethylene for their gas distribution system. Although B.P., Phillips and DuPont, as major chemical suppliers of gas grade materials have had to do an enormous amount of background research and development on their materials, the influence of the Engineering Research Station at British Gas has been to enforce the most stringent series of performance and quality control tests on the companies who subsequently extrude the raw materials. This had led to the development in the UK of an extremely expert pipe extrusion industry who carry out their own research and development on extruded pipe and injection moulded fittings to a far greater degree than anywhere else in Europe or indeed in the world. The BGC specifications requiring suppliers to demonstrate their technical competence are quite unique and have thereby led to the development of a different type of PE Pipe Industry in the UK than elsewhere.

One of the limitations of the UK manufacturers who have supplied BGC is that they have, in general, concentrated on the smaller range of diameters of pipe (these were the range demanded by BGC). In general, most manufacturers supply a complete system in diameters up to 400mm (only Stewarts & Lloyds supply pipe to 800mm diameter). There is no manufacturing reason why such companies should not supply larger diameter pipe for the UK water industry - it is a question of demand.

(It should be noted that the total stored energy in the pipe wall under pressure increases as diameter increases. Hence, the choice of materials whose high toughness and crack growth resistance can be demonstrated is far more critical for large diameter pipes [see 5.2].)

Because foreign manufacturers tend to produce the larger sizes (up to 1.6m in Finland) they have been the major suppliers of PE to the UK market for the sewerage industry (much for slip lining). Almost universally, the materials used for such supply have been higher density PE's. Although HDPE is known to be less ductile than the more sophisticated MDPE co-polymers, there is the compensation of a higher short-term modulus which can result in increased resistance to buckling during grouting or a high water table. For such non-pressure duties, the flexibility of the PE and the advantage of welding long pipe 'strings' are deemed to be the most important factors by specifying engineers. There is a strong case for this view, since, providing the welds are of sufficient quality as to resist the strains imposed during installation, the function of the PE pipe after slip lining such sewers is to act as a seamless, chemically resistant, lining - all imposed loadings are taken by the grout in the annulus. There is perceived to be no need for a material with assured long term strength retention properties. Deformation properties are most important - the deformation criteria being given by the WRC Rehabilitation Programme (ref. 7).

The major danger for the user industry is to assume that there is a direct parallel for use of similar HDPE pipe in pressurised situations where long term strength properties are far more vital and it is necessary to exercise greater care in material selection.

3. LONG TERM MECHANICAL PROPERTIES OF PE PIPE MATERIALS.

3.1 Creep - Constant Stress.

To account for the decrease in strength and stiffness with time, resort is made to strength 'regression' curves and 'creep' curves (for E).

Such data abound for PE Pipe materials and are obtained:

i) For stiffness - by leaving a sample under a constant stress and measurement of the increase in deflection with time. Thus via conventional equations a graph of Modulus (E) vs time can be plotted - as shown in figure 1.

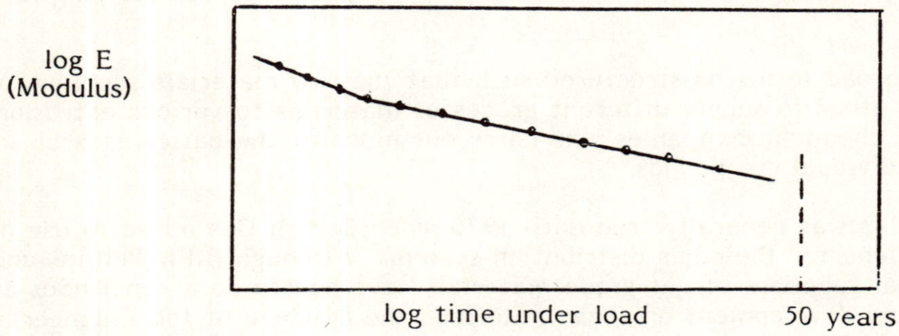

Figure 1

On log-log scales data are usually linear allowing long term extrapolations to be made.

In design, if it is required to know how much deformation an article will sustain after x years, the appropriate linear formula is used with the valuje of E for x years substituted.

ii) For 'strength' - by leaving pipes under different pressures and measurement of time to failure, e.g. figure 2.

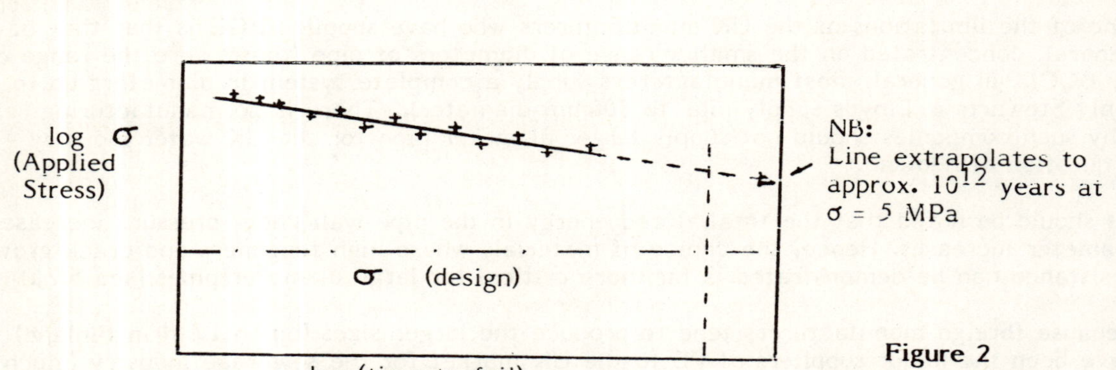

Figure 2

When data are plotted on log-log scales the decrease in strength with time is linear allowing accurate extrapolation - given enough data points are available.

To accept predictions that any material with a 20 year pedigree can be expected to last for millions/billions of years stretches the imagination a little and it is therefore prudent to examine whether there are any known factors which could be used to provide a rationale for a 'worst case' design basis.

The most obvious factor to examine is the assumption that the nature of failure will always be in a ductile manner. After all, uPVC pipes designed on a massive life predicted by ductile regression data have failed in very short time scales by the different process of brittle crack growth.

3.2 Acceleration of Brittle Failure in PE Pipes

Because long times and high temperatures generally produce equivalent changes in the mechanical properties of thermoplastics such as PE, it has become common practice to accelerate testing of PE by carrying out tests at 80°C.

i) Historical

In the early history of PE pipe devleopment (mid '50's) with the original high density materials, it was found that testing pipes at 80°C did encourage ductile/brittle transitions. Further, it was also found that at outset of such transitions, there was a 'knee' in the regression lines and that the development of failure where ductility was restricted on the micro scale caused a decrease in the 'expected' lifetime - as shown schematically in figure 3.

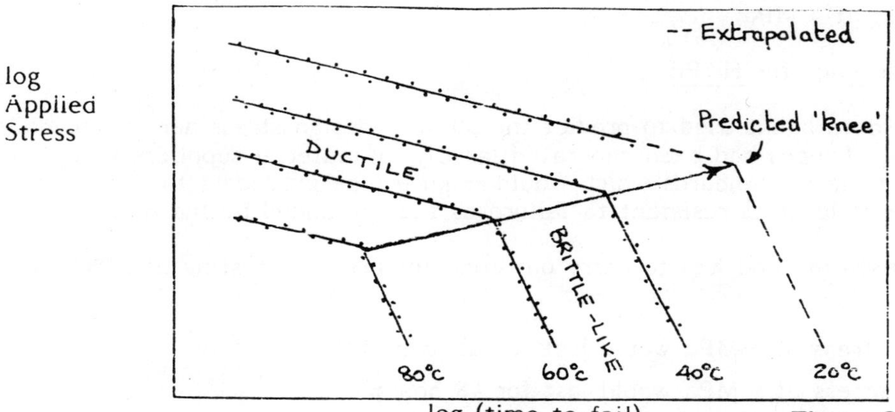

Figure 3

Subsequent exhaustive testing of pipes at 60°C and 40°C showed that the same transition could occur at lower temperatures - albeit at higher stresses and longer times.

With the availability of large data banks (on small bore pipe - 32mm dia.) it became possible to draw a line joining 'knees' to allow the position of the 'knee' at 20°C to be predicted. With HDPE the 'knee' was predicted as occurring at 20 years - obviously beyond the test times then currently available (indeed, it is only in recent times that failures of test pipes held under pressure for 20+ years have confirmed the onset of a transition at 20°C).

To accommodate the concept of a change in failure mode altering the simple linear regression analysis, the German Plastics Industry adopted the following procedure:

a) They accepted that ductile/brittle transitions caused the linear ductile line (A) to switch to a second line of higher slope (B) (see figure 4).

b) They assumed that the slope of the 80°C line was likely to be the most severe case and should therefore apply to all brittle cracking. (Conceptionally correct in that we would expect a reduction in time dependance with reducing temperature).

c) They accepted that the position of the 20°C 'knee' could be predicted from higher temperature data.

d) They assumed that it was reasonable to draw the 80°C 'brittle' line on the 20°C data to allow for prediction of long term strengths/lifetimes at 20°C.

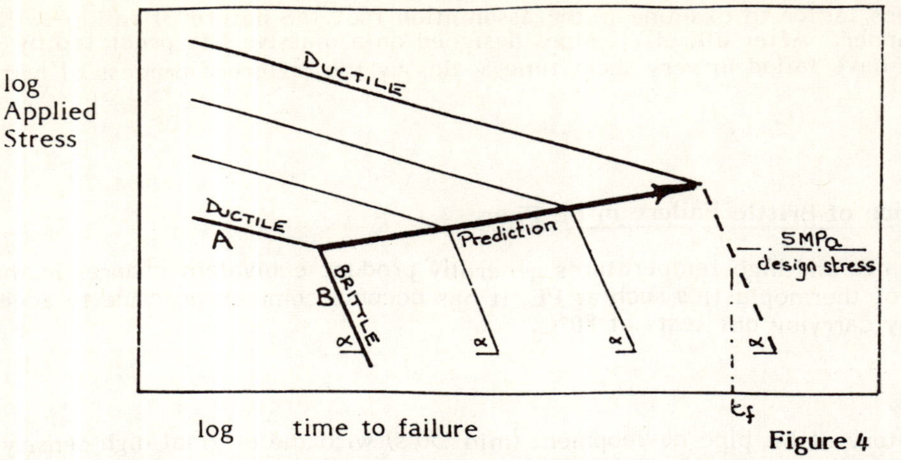

Figure 4

Using this procedure, with all lines drawn through the lower bounds of the test data available gave a prediction that for HDPE the failure time (t_f) (see figure 4) at a stress of 5 MPa would be approx. 100 years.

Alternatively there was a safety factor of 1.3 on stress carrying capacity for a 50 year life with a design stress of 5 MPa.

ii) Predicted Lifetimes for HDPE

Because of all the data banks used to predict the 50 year design stress needed for dimensioning and pressure rating of pipes had been generated largely by material suppliers it was considered necessary to have tests in standards which would require pipe producers to demonstrate that they could produce pipe at least as resistant to failure as that produced by the material suppliers.

For HDPE, the chosen method was to carry out constant pressure testing at 80°C and demonstrate either that:

a) Pipe under a stress of 3 MPa would last for at least 170 hours - or

b) Pipe under a stress of 4 MPa would last for 48 hours.

Pipe which passed such tests was then deemed to be superior to that used for the design stress production and hence was considered satisfactory.

3.3 Predictions of Lifetime for MDPE

With MDPE and modified HDPE, quality assurance is a difficult problem because the materials are so difficult to crack. Most specifications have copied the German (DIN) route and call for demonstration of enhanced lifetimes at higher stresses at 80°C viz:

1) A lifetime of 170 hours at = 4 MPa or

2) A lifetime of 60 hours at = 5 MPa.

Considering the 80°C dataavailable, these tests are almost farcical since the BP data predict lifetimes over 5000 hours at these levels.

With BP Rigidex PE, it would seem that at 80°C the onset of the ductile/brittle knee has been increased from less than 10 hours to approx. 3000 hours.

If the 80°C data are extrapolated as before (figure 4) to predict a 'knee' at 20°C then the ductile/brittle transition is not expected to occur until 500 years. Given this extremely large increase in lifetime or stress carrying capacity offers users of the new polyethylenes two advantages:

a) Users can design with PE at higher stresses (6.3 MPa c.f. 5 MPa - retaining the 1.3 safety factor).

b) There should be greater assurance of a longer life.

3.4 Design for Time Dependance

The concept of using data generated over long time periods for predicting stiffness and strength is often confusing to Engineers familiar with materials whose properties do not vary with time. Short term data taken from Tables in materials suppliers literature are not to be used indiscriminately as they are normally only given for guidance and comparison.

With plastics pipes, two sets of data should be requested by designers:

i) The creep curve giving values of modulus vs time is essential for obtaining values of E to be substituted into conventional elastic design formulae. The value of E to be used should be read from the graph at the appropriate time. e.g. If a calculation is to be made for pipe deformation in the long term due to soil loading, then E at 50 years is appropriate. If resistance to buckling due to grouting is required then the value of E at 2 hours will be more appropriate.

ii) The strength 'regression' curve should be provided. From such a curve, the actual burst resistance at 50 years can be gleaned or alternatively values of strength at 1-2 days obtained, to allow the engineer to establish a suitable test pressure for a main during comissioning.

These data should be available from all reputable pipe suppliers.

4. JOINTING OF POLYETHYLENE PIPES AND FITTINGS

One of the most oft promoted advantages of using PE pipes and fittings systems is that the pipes can be guaranteed 'leak-proof' because jointing is carried out using fusion welding techniques. Basically, the process is relatively simple in that the material is heated against a hot metal plate/former of some kind with a small amount of pressure applied to ensure that a considerable zone on either side of the sections to be welded is heated and softened. After removal of the heating plate the two hot faces of pipe or pipe and fitting are brought into contact under pressure. The pressure is maintained for a set period of time and is then released. The result should be a fusion joint which has a strength greater than the strength of the parent material (because of the increase in section thickness caused by the weld bead which is formed under pressure).

If welding is carried out in a satisfactory manner by trained personnel who have been approved after a set period of training then there is no doubt that practical experience shows that pipe system put together in this way is indeed leak-proof and welded sections are as strong and ductile as the pipe itself.

Butt fusion techniques are normally employed for pipe and fitting sizes greater than 90 mm in diameter. Machines for welding pipe to pipe are available in the UK for sizes up to 900mm and larger sizes are available abroad (up to 1200mm).

4.1 Controlling Welding.

Whilst it is perfectly possible to obtain 'perfect' joints on a regular basis, there are a number of factors which must be considered, as follows:

a) Compatibility.

All the various systems of polyethylene which are currently available for use in the water industry are not mutually compatible with each other with respect to the performance of a weld.

Some types of PE 'soften' at different temperatures and in their softened state have different viscosities which affects both the magnitude of the force needed to cause fusion and the way in which the force is applied. A table summarising the degree of compatibility of the common types of PE's is given in Table 1. When different PE's are fused to each other, trial tests should be made to ensure the weld is good (eg Clause 6, ref.8)

	BP	PHILLIPS	Du PONT	HOECHST
BP		yes	wac-if**	wac*
PHILLIPS	yes		wac-if**	wac
Du PONT	wac-if	wac-if		no
HOECHST	wac	wac	no	

Table 1

*** with added care** **** with care - in factory**

b) Efficiency of Welding Techniques

Although fusion welding is a relatively straight forward procedure, it is necessary to employ welders who have been trained to use the appropriate equipment and are experienced at not only preparing and jointing pipes but who can also discern whether a good joint has been made. Training courses for welders are available within the construction industry, through British Gas and their pipe suppliers and via the Water Industries Training Association.

All fusion weld techniques with PE should give 100% efficient welds - indeed it has been found from testing that in general pipes fail before welded sections. (It is a BGC requirement that welded sections pass the same test requirement on virgin pipe).

c) On-Site Welding

When welds are being made in trenches in inclement weather, a suitable shelter should be used to ensure that constant temperature conditions are approached and that the weld itself is protected from ingress of foreign matter, etc. This is particularly important during the time when the heater plate is removed. This time should be minimised and in every case kept below 12 secs. This is essential to avoid cooling the molten surface.

Poor control over temperature on heating plates, etc. can lead to cold spots in welds where the material does not properly fuse together. Similarly, allowing sand, or dirt to contaminate a weld generates sources of stress concentrations. It is not difficult to weld carefully - merely a matter of 'good housekeeping'.

A German 'Standard' does exist which gives an excellent code of practice for defining on-site welding conditions, details of adequate welding records, etc. (DVS Codes 2207/8).

d) Testing & Inpsecting Welds

Before any contract involving welding of a main designed to sustain pressure it is advisable to carry out approval trials using the actual equipment/welder proposed under simulated site conditions.

By testing tensile samples cut from welded sections, the efficiency of welding can be assessed. In all such tests the unwelded PE should fail in a ductile manner. The weld itself should remain intact if the 'beads' are retained.

Such mechanical testing of welded sections on an occasional random basis is probably the most effective technique of assessing weld 'quality' on site.

Although X-Ray examination can be very effective in discerning flaws in welds (see Ref. 3) the technique suffers from being too sensitive. The latest X-Ray equipment is capable of discerning flaws of micron level and there is then posed the problem of establishing a suitable criterion for defining a limit of an 'acceptable' flaw size.

Although Ultra Sonic techniques have been proposed for assessing welds in PE, this method suffers from insensitivity. Traces are extremely difficult to interpret and scattered noise causes confusion in interpretation using angled probes.

e) Monitoring Welding

In the last two years, a system using a conventional micro computer has been developed which may be used to collect information on temperatures and pressures during welding. The system has been successfully used on large Gas and Water insertion projects to collect information on over 500 site welds. Use of such a technique gives confidence to both Contractor and Client that all welding operations are kept within prescribed limits to ensure uniformity of welding conditions.

With such monitoring, which is an extension of the good housekeeping philosophy, welding may proceed on a 24 hour basis - making the joining process far more cost effective whilst maintaining the highest quality standards (e.g. ref. 4).

5. SPECIFICATIONS FOR PE PIPES

5.1 Coventional Specification

For sewerage applications it is pertinent both in stress analysis and specification of pipes, to correctly identify the degree to which pipe needs to be 'designed' for a given application.

i) Gravity Mains

With non-pressure sewers being renovated by slip lining with PE; survival of the insertion and subsequent design against buckling during service are the primary factors. As stated earlier, the choice of pipe material may be based on stiffness considerations rather than a detailed appraisal of imposed stresses which may affect long term performance. Provided pipe can be purchased which meets very general quality requirements and control is exercised over welding, the first pipe choice will be largely determined by costing under competitive tender, with the lowest price pipe being that chosen. Such a procedure operating over the last two years has proved totally satisfactory. There is perceived to be little need for sophistication. Pipes are generally specified to the dimensional requirement of DIN 8074 with general quality assurance as given in DIN 8075. This latter standard gives methods and recommendations for testing which may be carried out. The cornerstone test is proof of resistance to failure under pressure at 80°C (see Section 3). It should be noted that such testing will only be completed when requested and agreed between customer and supplier.

ii) Pumped Rising Mains

(a) For Relatively Stable Pressure Regimes

Where sewer or effluent pipes are to be used in applications where the pipe is expected to sustain pressure over very long periods, then the situation is different. Because the cost of PE pipe (above 125mm) is directly proportional to wall thickness, cost-effective design is best achieved by using pipes at the correct specified performance ratings and evidence is needed to provide assurance that the pipe will perform satisfactorily.

In such cases, it is more appropriate to choose pipe from suppliers who carry out regular testing of their products. The recently issued specification for MDPE is the WRC Specification 4-32-03 which has been designed to satisfy minimum levels of assurance of quality for potable water applications.

Although 4-32-03 does provide customers with assurance that the pipe supplier is concerned to supply a consistent product, the requirements demanded are heavily biased towards testing of pipe in perfect condition. It is a Specification largely designed with the producers, to show that pipe which leaves the factory is consistent. Experience in the past has shown that because of misuse and poor laying control, pipe which is actually installed may be required to sustain overload and may be in less than perfect condition following installation.

MDPE grades have enormous reserves of toughness and for many routine duties, supply to a 4-32-03 Specification will be quite adequate - particularly for diameters to 250mm where there is much background test data and service experience. For larger diameter pipe where data bases are more scarce and where the case for using PE may be based on the use of pipe at its design limit, it has been considered appropriate (by NWWA and the present author) to draw up a new specification for demonstration of high toughness (secn. 5.2).

b) For Dynamic Fluctuating Pressures

In sewer rising mains which are pumped, it is well known that pressures will regularly vary in the main and surges may be expected. WRC believe that the fatigue due to such fluctuations have caused many failures with uPVC (Ref. 2). The present author is not at all convinced by the data base given as evidence in support of this view but does agree that dynamic fatigue would exacerbate a static loading situation which could itself cause failure of a uPVC pipe with low fracture toughness.

With the higher density polyethylenes, service experience on the Continent on sewer rising mains has been excellent. There are no reported cases of pipe failures due to fatigue. Indeed, although laboratory test data are scarce, such evidence as does exist suggests that at ambient temperature, the effect of dynamic loading is negligible.

It would seem that the main source of potential problem is with the use of injection moulded PE fittings and with stress concentrations introduced around poorly formed welds. The fatigue behaviour of HDPE in larger diameter PE fittings has been extensively studied by Brunel University and an overall view would seem to be that dynamic pressures will cause a reduction in pipe lifetimes only where brittle failure may be expected. e.g. At 80°C and with welded pipe/fittings where insufficient care is taken to match diameters, leaving some areas of mismatch, which can be taken as stress concentrations. (See Ref. 5)

If a raw material with exceptional slow crack growth resistance is chosen where only ductile failure is predicted then providing the regular dynamic pressure range is within the overall pipe pressure rating, fatigue should not represent a severe constraint on design.

Because MDPE has far more capacity to absorb energy than other, more rigid thermoplastics, the material may be expected to resist occasional surge pressure far more readily.

In special cases where pumping is expected to cause damage to pipe bores (- due to stones etc.) then it is advisable to carry out preliminary screen tests using accelerated testing of notched samples to examine whether extensive dynamic pressure cycling will be a design consideration.

For any demanding duty therefore, it appears apparent, from past experience, that the most vital property which needs to be demonstrated by a plastics pipe material is resistance to slow crack propagation over a long time scale. Whether the loading be static or dynamic it is the resistance of the material to defect propagation which is the most essential design property.

This lesson has been learned with uPVC and the new British Standard 3505 will be including tests specifically designed to demonstrate that the new high toughness uPVC pipes which have been developed in recent years have far greater resources of toughness than hitherto. There is no reason why polyethylene pressure pipes should not be required to demonstrate similar properties.

5.2 Specification for High Toughness in PE Pipes

At the request of NWWA who have had some sorry experience with brittle failure of uPVC, a new specification has been drawn up for PE. Based on past experience analysing poor PVC performance, the specification requires manufacturers to demonstrate that PE pipe has satisfactory resistance to crack propagation both in long and short term loading situations.

Using test methods based on a fracture mechanics analysis (e.g. Ref. 6) of failure generation from defects and stress concentrations, the specification seeks to provide reassurance that the finished pipe has tolerance to overload situations and imperfections which have always been found to give problems with other materials. By requiring pipe as produced to be tested under pressure at 80°C with artificially induced, dead sharp cracks in external walls to 10% of wall thickness, the design basis for PE pipe (secn. 3) is being assessed with pipe more closely related to that which may be installed. Further clauses require samples to be tested with large defects in the bore whilst a bending moment is applied and the notched impact energy level of 4-32-03 has been increased. Other tests require measurement of residual stresses and density profiles to ensure forced cooling does not cause problems.

The tests show clear differences exist in the performance of MDPE as compared with older HDPE grades some of which fail the stringent requirements. The MDPE materials produced in the UK by BP which have been derived from the polyethylene so successfully used by British Gas have outstanding performance in such testing. Simulated service loading either by pressure or bending/point loading will not cause embrittlement. As welding gives joints with the same toughness as virgin pipe, such materials will undoubtedly provide engineers with pipe systems which can be used with assurance in demanding sewer applications.

By the application of intensive research and development in the UK and a willingness to learn from past mistakes, it is believed that plastics pipes in the future will perform predictably. The predictions are excellent.

References

1. Water Research Centre Specification 4-32-03. Spec. for PE Pressure Pipe for Cold Potable Water (Sizes Greater than 63mm). May 1983.

2. Kirby, P. C: Plastics & Rubber Mats. & Applns. 5, (1). p.78, May 1980.

3. Marshall, G.P: Sliplining a 42" Gas Main Using PE, (to be published: Inst. of Gas Engineers Jrl.)

4. New Civil Engineer, p.16, March 31st, 1983.

5. Barker, M.B; Bowman. J; Bevis, M: Journal of Mtls. Sci., 18, p. 1095-1118 (1983).

6. Marshall, G.P; 'Design for Toughness in Polymers', Plastics & Rubbers Processing and Applications, 2, p. 169-182, 1982.

7. WRC Sewerage Rehabilitation Manual - Vol. 3.

8. WRC Specification 4-32-04. Specn. for PE Fusion Joints & Fittings for use with Cold Potable Water Pressure Pipes.

International Conference on the
PLANNING, CONSTRUCTION, MAINTENANCE & OPERATION OF SEWERAGE SYSTEMS
12-14 September, 1984

PAPER D4

DEVELOPMENTS IN GROUTING TECHNOLOGY FOR SEWER RENOVATION

S M GILL B.Sc

WRC Engineering Centre
Swindon
Wiltshire
SN5 8YR

Summary

The importance of grouting to many sewer renovation techniques cannot be overemphasised. Work carried out at WRC in recent years has undertaken to make good any deficiencies in grouting technology and to ensure that the best advice regarding materials, equipment and site practice is available to those engaged in the practicalities of sewer renovation.

This paper describes methods of test for the grout properties required for design, developments in lightweight grouts and gives guidance for site practice.

Careful monitoring of grout properties and pressure at the injection point is essential since such measurements provide the only reliable indication of grout quality before it is lost underground.

It is intended that standard specification clauses will be written which will enable more precise control of the important aspects of grouting technique in sewer renovation contracts.

Held at The University of Reading, England. Organised and sponsored by BHRA, The Fluid Engineering Centre, Cranfield, Bedford in conjunction with the Water Research Centre.

1. INTRODUCTION

Grouting is recommended as an essential stage in those sewer renovation techniques which create an annular gap between a lining and the existing sewer wall. The grout may be required to fulfill some or all of the following functions:

(i) Transmit stresses between the lining and original sewer.

(ii) Provide restraint against deformation of the lining by vertical loads.

(iii) Prevent formation of a drain parallel to the existing sewer and hence further deterioration.

(iv) Prevent the possibility of failure of the lining caused by point loading due to uneven bedding.

High confidence in the grouting operation is required for the design procedures described in the WRC Sewerage Rehabilitation Manual.

During the last 2 years a major programme of research into grouting materials and practice has been undertaken at the Water Research Centre with the objective of providing positive advice to the Water Industry.

This paper concentrates on recent developments in materials, particularly lightweight grouts, the importance of grouting to structural design and on site practice. Much information already exists on the "conventional" grouting materials available at present.

2. DESIGN REQUIREMENTS

A good understanding of the design procedures described in the WRC Sewerage Rehabilitation Manual is essential to appreciate the role played by the grout. Three design approaches are noted which accommodate all of the "established" techniques of sewer rehabilitation:

Type I The lining, grout and existing sewer act as a composite section. The grout must therefore be capable of transmitting ground loadings from the existing structure to the lining. The shear bond strength between grout and both existing sewer and lining should exceed 1 N/mm^2, whilst the in-situ compressive strength of the grout should exceed 10 N/mm^2.

Type II The lining acts as a flexible pipe requiring no bond between lining and grout or existing structure. The function of the grout is to support the lining, for which a minimum in-situ compressive strength of $2N/mm^2$ is recommended.

Type III The lining acts as permanent formwork for the grout which provides any structural improvement to the sewer. The compressive strength required will depend on the designed strength of the relined sewer though should typically be 20-30 N/mm^2.

The structural design procedures are based on the assumption that grout fills at least 80% of the annulus.

It is anticipated that Type I and II linings will account for almost all of the renovation market in the UK.

2.1 Measurement of Grout Properties Required for Design

2.1.1 Compressive Strength - required to resist ground loading

There is no British Standard relating directly to the measurement of grout compressive strength. It is recommended that cubes are cast, cured and tested in accordance with BS 4551 : Methods of testing mortars and mortar testing sand. Where the predicted 28 day strength is less than 2 N/mm^2 such as grout for exterior void filling it is considered that cube testing is not necessary.

2.1.2 **Shear Bond Strength** - required to transmit load from ground to lining

Again, no British Standard exists for measurement of shear strength. A test has been developed at the WRC Engineering Centre which can be used to determine the shear strength of any grout/lining combination. The test, described in detail in Appendix 1, takes 28 days to complete and is useful for factory quality control. A simple on site method of test, of vital importance for site control, is proposed which requires only 2 days. This on-site test consists of casting grout or mortar cubes onto the lining panel to be checked and curing for 1 day before pushing off the cubes using a hydraulic jack fitted with a pressure gauge.

A limited number of tests have been carried out at WRC which show a direct relationship between shear bond strength and both lining backing roughness and grout strength.

3. LIGHTWEIGHT GROUTS

The reduction of lining flotation is often highly desirable, particularly for relatively lightweight linings such as Polyethylene. Theoretically the use of a grout of density less than that of Polyethylene (approximately 950 kg/m^3) would ensure that flotation is completely prevented if the lining is filled with water. Unfortunately since grout of this density will simply float on water in the annulus it is unlikely to prove satisfactory. The target density for a lightweight grout should therefore exceed that of water.

3.1 Materials

At present there are 3 types of lightweight grout:

Foamed
Lightweight aggregate
Extremely High Water Content

3.1.1 Foamed

Foamed grouts are formed by incorporating a stable preformed foam into a conventional grout mixture of either sand/cement or PFA/cement. The foam may have either a synthetic or animal protein base. Typically foam is produced at a density of approximately 75 g/l, It must be sufficiently stable to retain its cellular structure during mixing and pumping of the grout.

For annulus grouting it is recommended that the density should be 1100 kg/m^3 $\pm$ 50 kg/m^3.

Sulphate Resisting Portland cement can be used in foamed grouts, although the rate of gain of strength is less than for OPC foamed grouts.

In both field and laboratory trials carried out by WRC 2 major problems have been encountered:

(i) Difficulty in maintaining consistent density.

(ii) Segregation of foam from the mix causing the formation of a very weak upper layer of grout.

Foamed grouts may be produced on either a batch or continuous basis.

For equipment using continuous production the first problem, whilst minimised, is not completely solved since variations in sand grading, moisture content or cleanliness will affect the grout properties.

The second problem is highlighted when grouting in the presence of water. Admixtures can be used to produce a more cohesive mix which is more stable in water although the workability is reduced causing increased pumping pressure and difficulty in placing.

Unless testing is carried out which confirms that the grout workability is acceptable, the addition of such admixtures is not recommended.

Typically, 28 day cube compressive strengths of 3 - 6 N/mm^2 are attainable with sand/cement ratios up to 2:1 and PFA/cement ratios up to 2:1.

3.1.2 Lightweight Aggregate

Lightweight aggregates or fillers may be used to reduce grout density to less than 1000 kg/m^3. When using prebagged materials the control of density is no more onerous than when using conventional, fully dense grouts. Segregation of the lightweight particles may however occur, this must be minimised by correct mix design.

No special equipment is required to mix or pump such grouts.

3.1.3 High Water Content

By the careful addition of admixtures it is possible to produce a grout which although relatively lightweight due to its very high water content is satisfactory in its performance for annulus grouting. The density of grouts produced in this way is unlikely to be less than 1300 kg/m^3.

4. INSTALLATION

Without good grouting practice it is impossible to have confidence in any renovated sewer. It is difficult to confidently assess the position of grout after injection in a man entry sewer without resorting to the removal of cores or the drilling of holes in the relined job whilst for non man entry sewers there is no equipment currently available which can detect the presence of voids. Even if the grout can be accurately located there is no guarantee as to its quality.

Sensible planning of a grouting operation should ensure its smooth running. Control is required both before and during grouting.

4.1 Before Grouting

4.1.1 Grout Selection

General

The selection of an appropriate grout for the job is crucial. Values of in-situ compressive and shear bond strength required for the different types of lining were given earlier in the paper. It should be noted that these properties must be achieved when the grout has sufficient workability to be placed without presenting problems in pumping. The effect of water content on the workability and strength of a PFA/OPC grout is shown on Figures 1,2. The tendency of grout to bleed is markedly increased by increasing grout water content. If bleeding is excessive (>2%) then the formation of a void along the annulus crown will be detrimental to the renovated sewer.

Relationship between in-situ strength and cube strength

Several factors, including curing regime and specimen shape cause differences between the strength of cores removed from the annulus and cubes cast during grouting. Experimental work at WRC, described in detail in Appendix 2 has shown that the in-situ 28 day compressive strengths are approximately 65% of the measured 28 day cube strength. For a type I lining it is therefore recommended that grout should be selected with a 28 day cube compressive strength of at least 15 N/mm^2 and for a type II lining, 3 N/mm^2.

Use of Admixtures

For most annulus grouting the use of admixtures is felt to be unnecessary. In the presence of infiltration, flooded annuli or very restricted annular gaps, admixtures may be beneficial.

It should be noted that no cementitious grout should be expected to perform adequately if fast flowing water is present in the annulus. In such cases, preparatory work should be considered.

To reduce problems of washout associated with flowing water gelling aids (e.g. bentonite) or various proprietory admixtures based on either polyethylene oxide or cellulose derivatives may be used.

In very narrow annular gaps (less than 25 mm) the use of plasticising or super plasticising admixtures may be advisable.

Durability

In the presence of aggressive groundwater of low pH or high sulphate content the durability of the grout is important. Generally, use of high cement content and low water/cement ratio will improve durability. Reference should be made to BRE digest 250 when assessing the suitability of grout to resist sulphate attack. If the pH of the groundwater is less than 5.5, the use of non-cementitious grouts should be considered.

Providing that lateral connections are correctly remade and that joints between lining segments are satisfactory it may be assumed that the grout does not come into contact with the effluent carried by the sewer.

4.1.2 Equipment Selection

The use of appropriate equipment for grouting is vital. The use of free fall concrete mixers or hand mixing should not be permitted. To clarify both this section of the paper and future discussions the following terms will be used for different types of mixer available.

A colloidal mixer (whilst not producing a true colloid) is one in which a very efficient mixing action is achieved by the very rapid rotation of a disc at the base of the mixing chamber. This action imparts a great amount of energy to the grout and ensures a very rapid, thorough wetting of the cement/solid particles.

A high shear mixer is one in which the mixing action is achieved by the threshing action of rotating blades between fixed vanes.

A paddle mixer comprises a simple paddle system rotating in the base of the mixing chamber with baffles fixed on the sides of the chamber to direct the mix into the central area.

The terminology used to describe grout pumps is somewhat clearer. Rotary screw and double diaphragm pumps are the most frequently used pumps for grouting in sewer renovation. The ram or piston type pumps tend to be more often used for other grouting applications such as ground stablisation due to their higher operating pressure.

A programme of experimental work devised by WRC to determine the effect of equipment on grout has shown that factors such as temperature and operator control are likely to prove the major cause of changes in grout properties. The colloidal mixer, having the lowest recommended mixing time, should be used with particular care in hot weather.

4.2 During Grouting

4.2.1 Injection Procedure

Length to be grouted

The position of manholes or other access points will play a large part in determining the length of sewer that can safely be grouted in one "pass". Other factors must, however, be considered. The pumping operation should be completed in less than the time taken for the grout to lose its pumpability. Pump output must therefore be matched with the annular volume to be filled. The amount of grout actually used may be well in excess of the calculated annulus volume. For a brick sewer a figure of 25% excess may be considered typical. The use of very fluid grouts will of course tend to increase this value.

Generally it is recommended that a grouted length should not exceed 50 m.

Position of injection points

The expulsion of groundwater or effluent from the annulus is fundamental to the success of the operation. The grout should displace water without mixing occurring. Unless the sewer gradient is steeper than approximately 1:50 it is recommended that grout is injected from the downstream end. If it is necessary to grout from the upstream end the provision of adequate vent pipes is vital.

For man entry sewers the control of the injection procedure is simplified by the ability to observe grout emerging at successive grout holes and by gently tapping the lining with a hammer to confirm its position. The provision of grout injection points at approximately 10 m intervals is usually adequate. For non man entry sewers the provision of 2 injection points, at crown and springing, is sensible in the event of blockage.

The grout fluidity and the size of the annular gap will determine which of the 2 possible flow patterns shown in Figures 3, 4 occur.

It is considered that in general the pattern of flow corresponds to that in Figure 3 although the use of a thixotropic or low workability grout in a narrow or non-uniform annular gap containing obstructions can result in totally unpredictable flow patterns. In a trial conducted at the WRC Engineering Centre using a thixotropic grout to fill a large (90 mm maximum) annular gap in a 6 m length of simulated sewer, side injection resulted in the filling pattern shown in Figure 5.

Some thought should be given to anticipating the flow behaviour since the formation of voids by trapping air or water may prove detrimental to the long term stability of the renovated sewer.

Grouting Pressure

The careful control of grouting pressure is essential to avoid the danger of deforming or even buckling the lining. In the WRC Sewerage Rehabilitation Manual it is recommended that the grouting pressure should not exceed 35 kN/m^2. The pressure at the pump outlet cannot be used to give a reliable estimate of pressure at the injection point since hose diameter, length, grout type and temperature will greatly influence the pressure loss between pump and annulus. WRC have monitored injection pressures on several annulus grouting operations finding that in most cases, grouting pressures do not exceed 35 kN/m^2. It is recommended that a pressure gauge with suitable scale (e.g. 0-200 kN/m^2, 0-30 psi) should be fitted as close as possible to the injection point.

Staged Grouting

The practice of staged grouting is often used to reduce the flotation and buckling forces on a lining or, when installing segmental linings, to improve invert stability before fixing of crown units. A disadvantage of staged grouting is that the formation of a longitudinal joint becomes almost inevitable. Such joints or other discontinuities in the grout may allow external hydrostatic pressure to be applied to the lining. Whilst allowances for grout inhomogeneity are made in the structural design procedure it is not desirable to permit the possible formation of an alternative drain parallel to the new lining by erosion of the grout along such a joint.

Quality Control

During grouting it is important to follow the recommendations of the manufacturers of the equipment used and to monitor the properties of the grout before it is injected. The control of grouting is more difficult when batching on site rather than using prebagged material.

The following tests should be carried out at frequent intervals during grouting and checked against figures obtained in trial mixes prior to commencing the contract:

(i) **Grout density** - either by accurately weighing a container of known volume or by using a mud balance.

(ii) **Workability** - using either a flow trough or Flow Cone. Care should be taken to use a Cone with approximately 10 mm orifice. The size of orifice must be noted.

MEASUREMENTS OF GROUT DENSITY AND WORKABILITY PROVIDE THE ONLY RELIABLE INDICATION OF THE GROUT QUALITY BEFORE IT DISAPPEARS UNDERGROUND. The frequency of testing recommended reflects the need for confidence in material properties.

Cubes should be cast for testing at 28 days to confirm that the grout strength is satisfactory. If possible a sample of grout emerging from a breather pipe should be tested to provide additional reassurance. It is recommended that a minimum of 3 cubes be taken for each 50 m of sewer grouted or for every 5 m^3 of grout injected.

5. CONCLUSIONS

For sewer renovation techniques dependant on the successful placing of grout between lining and sewer it is essential that considerable thought and care is taken, both before and during grouting.

If the recommendations detailed in this paper are followed it is considered that the engineer responsible for a renovation contract can have confidence in the stability and durability of the annulus grouting.

APPENDIX 1

Shear Bond Test Procedure

A sample of lining unit is placed in the bottom of a 150 mm cube mould with roughened back uppermost. Grout is poured into the cube mould to a depth of 100 mm. The specimens are cured for 5-7 days in damp conditions before demoulding and placing in a curing tank. 28 days after casting the bond test is carried out by clamping the grout portion of the cube and applying a vertical load, to failure, to the edge of the square of lining bonded to it. (See Figure below).

The minimum allowable radius of curvature for a lining sample is 400 mm.

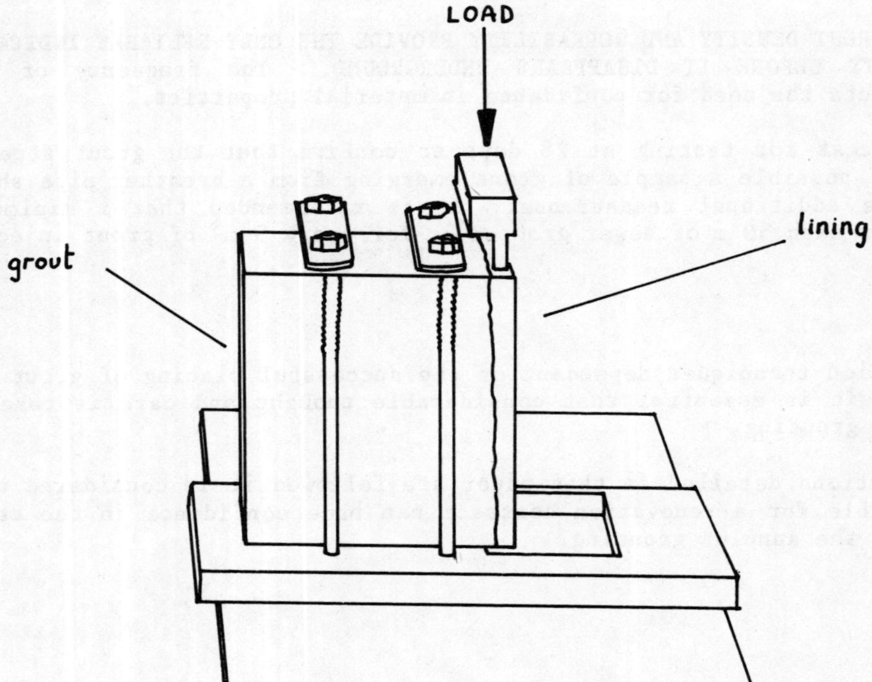

APPENDIX 2

Comparison of Cube/In Situ Grout Strength

WRC Trials

Several trials have been carried out to quantify the differences between cube compressive and in-situ grout strength. The objective is to allow confident prediction of in-situ grout strength from measurement of 28 day cube strengths.

Two concentric glass reinforced plastic (GRP) pipes, 6 m long and approximately 300 mm and 400 mm diameter were used to simulate a relined sewer. Grout was placed in the annulus using conventional grouting equipment. During placing, cubes were taken for testing at 7 and 28 days. After several weeks cores were removed from the annulus grout and tested in accordance with BS 1881 Pt 4.

Interpretation of Results

A direct comparison of cube and core strengths is not possible for 2 reasons. Firstly the difference in curing regime, secondly the difference in specimen shape.

The effect of the curing regime can be estimated by considering maturity. As curing temperatures are lowered, hydration becomes slower until it ceases completely at $-10^{o}C$. The maturity of concrete is the sum of the products of the number of degrees Celcius above $-10^{o}C$ multiplied by the number of hours curing at that temperature,

$$\text{i.e } M = (t_1 + 10)hrs_1 + (t_2 + 10)hrs_2 \ldots\ldots + (t_m + 10)hrs_m$$

The strength of a cementitious material varies logarithmically with its maturity. Thus the strength of a mix at any maturity can be calculated once the maturity at a particular age and strength is known. One complication is caused by the effect of the fineness and composition of the cement which necessitates the use of a series of graphs representing different cements.

The correct gradient is determined by plotting the known 7 and 28 day cube strengths against maturity.

Core maturity can be then plotted against the average measured core strength to enable estimates to be made of the measured core strength after any curing regime.

The effect of specimen shape must now be accounted for. The following two approaches can be used to obtain an estimate of cube strength from core strength.

(i) BS 1881 states that the strength of a cylindrical specimen with length/diameter ratio 2.0 is equal to four-fifths the strength of a cube. A graph in BS 1881 part 4 can be used to estimate the correction factor needed to take into account the actual length/diameter ratio of the core. This value is then multiplied by 1.25 to give an equivalent cube strength.

(ii) In the Concrete Society's Technical Report Number 11, "Concrete core testing for strength" the equation given below is used to obtain an estimate of actual cube strength. This equation takes into account the shape, size and orientation of the drilled core.

$$\text{Estimated actual strength} = \frac{2.3}{1.5+\frac{1}{\lambda}} \times \text{core strength}$$

λ = length/diameter ratio

Since there is no sound reason for rejecting either of these techniques it seems reasonable to use the average value obtained from both calculations.

In 2 separate trials carried out at WRC, the ratio of

$$\frac{\text{in-situ equivalent cube strength after 28 days @ rig ambient temperature} \times 100}{\text{measured cube strength after 28 days @ } 20^{o}\text{C}}$$

has been found to be approximately 65%.

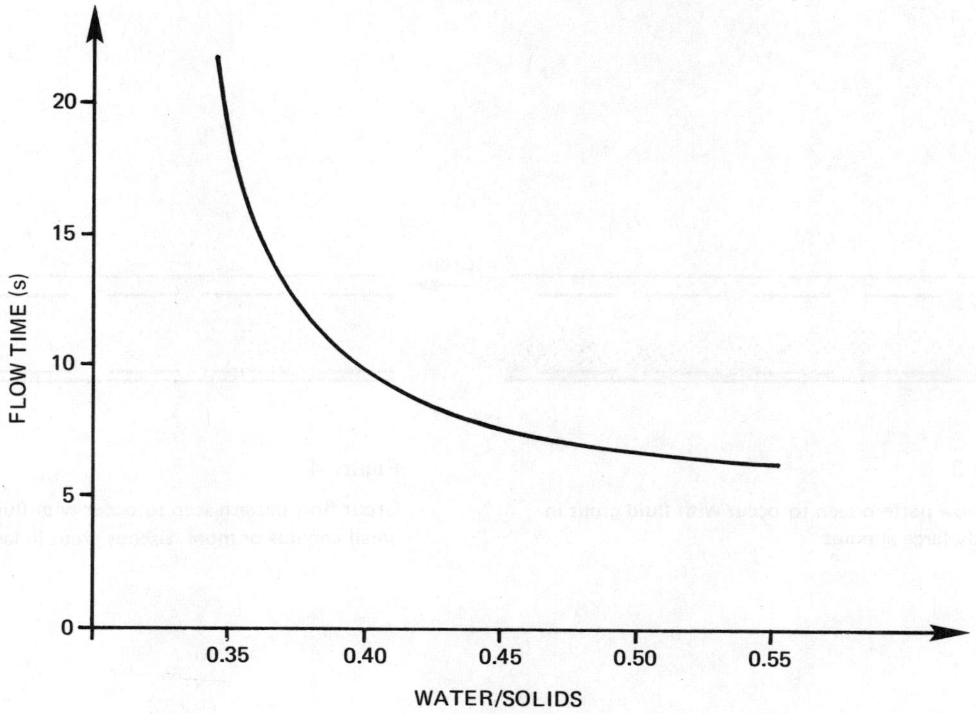

Figure 1
Effect of water content on workability for PFA/OPC 1 : 1
(Measured by 10 mm Marsh Cone)

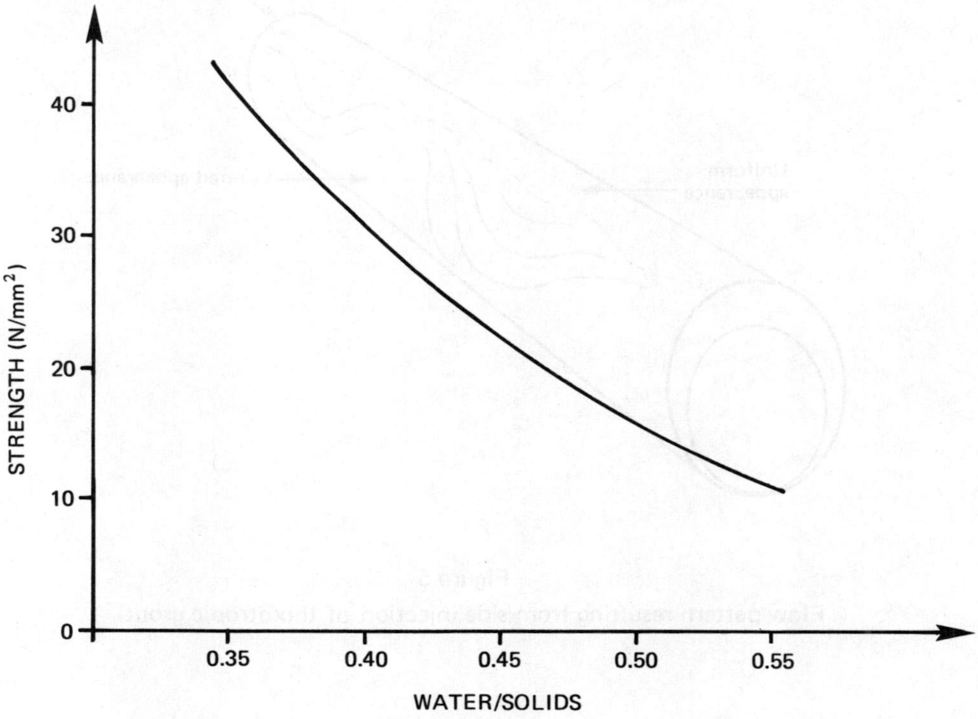

Figure 2
Effect of water content on 28 day cube
compressive strength for PFA/OPC 1 : 1

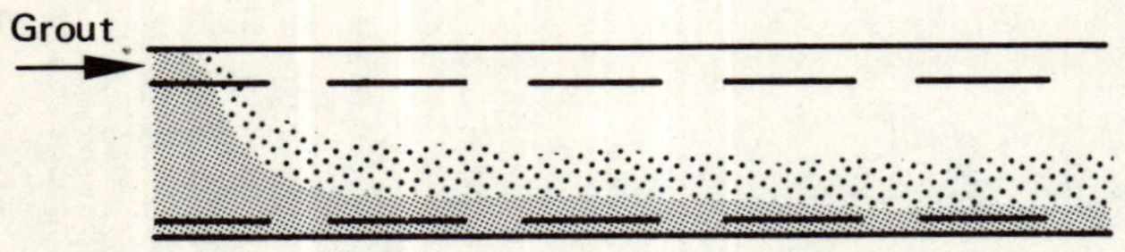

Figure 3

Grout flow pattern seen to occur with fluid grout in relatively large annulus

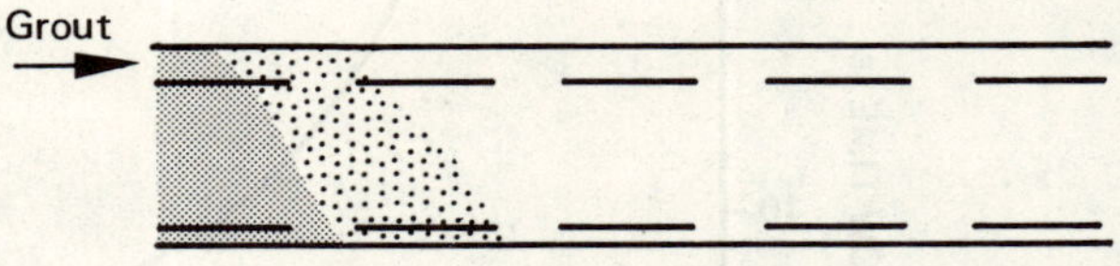

Figure 4

Grout flow pattern seen to occur with fluid grout in small annulus or more viscous grout in large annulus

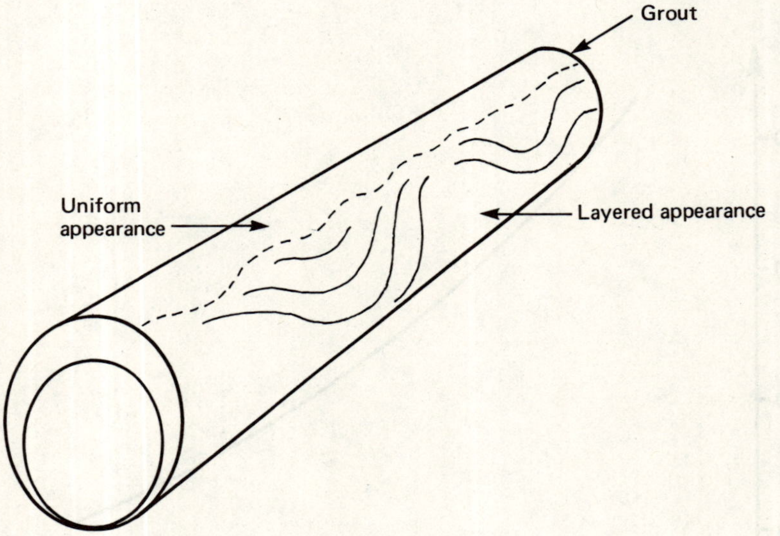

Figure 5

Flow pattern resulting from side injection of thixotropic grout

International Conference on the

PLANNING, CONSTRUCTION, MAINTENANCE & OPERATION OF SEWERAGE SYSTEMS

12-14 September, 1984

PAPER E1

A REVIEW OF METHODOLOGIES AND TECHNIQUES IN SEWERAGE NETWORK PLANNING AND ANALYSIS

Alan T.K. Fok MICE and Hugh G. Fraser MASc.

Department of Civil Engineering
University of Ottawa
Ottawa, Canada

SUMMARY

As a result of increasing innovation and complexity of design, function and maintenance of urban drainage systems, techniques for analysis and evaluation of sewer networks are changing. Current trends are to use physically based simulation techniques rather the purely empirical methods applied in the past. This paper reviews typical methodologies and techniques adopted in current Canadian engineering practice. Selected reprensentative hydrologic, hydraulic and computer methodologies for sewer planning and analysis are discussed. Also present are various runoff control and stormwater management techniques. Two case studies, one for upgrading an existing system and the other for a new development, are used to illustrate applications of the improved procedures.

Held at The University of Reading, England. Organised and sponsored by BHRA, The Fluid Engineering Centre, Cranfield, Bedford in conjunction with the Water Research Centre.

1. INTRODUCTION

1.1 GENERAL

One of the largest areas of municipal expenditure is in the capital and operating costs associated with sewer servicing. This occurs both in the provision of sewer servicing for new developments and in the maintenance and improvement of existing systems.

In North America construction of enclosed conduits for the transportation of sewerage dates back to the middle of the last century. As cities have grown and technology has changed so has engineering practice in providing drainage. Early efforts included enclosure of small streams which provided conveyance for either or both human waste and stormwater runoff. Some of these still exist. Modern techniques include separated systems, "major/minor" storm drainage, the "three-pipe" system as well as a multitude of pumping and control devices. A corresponding change in analytical techniques is also evident.

Design methodology has advanced from the middle 19th century when no formal designs methods were used, followed by the adoption of the Rational Method (also known as the Lloyd-Davies method) in the early part of this century, to the present day use of computer simulation models of varying degrees of sophistication.

1.2 GOALS AND OBJECTIVES

The level of sophistication at which design is conducted depends on the present level of technology and the goals set by political, environmental and economic needs. Technological advances lead to improved design methodologies as well as an expansion and refinement of desired goals.

Urban drainage goals can be broadly categorized in three levels of increasing complexity.

1. Provision of efficient sewer systems for frequent storm events (typically those with return periods less than 5 years).

2. Provision of adequate major conveyance systems for infrequent storm events (often events with return periods up to 100 years).

3. Incorporation of runoff control and stormwater management schemes to minimize the hydrologic/hydraulic and environmental impacts to tributary areas and downstream developments.

One significant result of improved management procedures, which is often overlooked but represents a fundamental improvement over historical design approaches is that design goals are defined with greater clarity. These will vary depending on the external constraints as well as the type of development. An existing system has many built-in constraints that limit sewer upgrading or redevelopment options. With a new system a wider variety of alternatives can be explored subject to the downstream capacity constraints. Definition of goals has therefore become more sensitive to the broader needs and impacts inherent in drainage design that were previously ignored.

To achieve the improved goals, new methodologies and techniques have been developed. These have been summarized in this paper and are presented as follows:

1. Review typical methodologies currently used in hydrologic and hydraulic evaluation of Canadian sewer works.

2. Review presently available techniques of analysis with special emphasis on the use of mathematical computer models.

3. Present some runoff control and stormwater management measures that be incorporated into an innovative urban drainage system design.

4. Present two examples of successful practical applications where the above concepts and analytical techniques have been used by the authors.

2. RAINFALL CONSIDERATIONS

Hydrology in urban drainage design is concerned with determining the maximum runoff flow rates at specified points in the system. In existing developments flow measurements could be used but these are usually unavailable. Rainfall measurements in the vicinity of an existing or new development are more readily available. These are transformed using a mathematical model to obtain estimates of the runoff.

Rainfall is naturally highly variable both temporally and spatially. Generally the spatial variability is ignored unless a real time analysis of the system operation is being conducted (Ref. 1) or a trunk system is being designed. The rainfall time distribution is considered important in Canadian simulation analysis, although some methods are applied which assume a uniform time distribution.

2.1 RAINFALL ANALYSIS AND DESIGN STORMS

Most rainfall analysis is still conducted to determine intensity-duration-frequency (IDF) curves. These curves are developed from a set of maximum intensities for a range of fixed durations obtained from many storm events. In reality, few real storm events ever have a constant frequency for each intensity-duration (ID) throughout the whole rainfall period. A comparison of a set of IDF curves with an ID curve obtained from a real storm is shown in Fig. 1A.

The rainfall input for a hydrologic model can be from an IDF curve, a synthetic storm or a historical storm. Design with the rational method uses the intensity obtained from an IDF curve, for a selected duration and frequency, to compute a peak flow. Synthetic storms such as the Chicago or various SCS (Soil Conservation Service) storms use IDF curves to obtain the storm volume for a particular duration and frequency. The volume is distributed using that particular distribution to obtain the synthetic storm hyetograph. The Chicago storm is normally used when peak flows are the main concern and the SCS storm when the runoff volume may be critical.

The frequency of a design storm is assumed to be the same as the peak flow frequency. Theoretically selection of a design frequency should be based on an economic analysis of the costs and benefits for a particular project. In reality though, this type of analysis is not conducted very frequently. A high level of protection is usually required to minimize public complaints.

Historical storms may also be used in conjunction with or in place of synthetic storms. In many Canadian municipalities recorded historical storms are used in conjunction with synthetic design storm events.

2.1.1 Storm Duration

The selected duration of a storm affects both the peak flow rate and runoff volume. The drainage engineer should analyze events with longer durations to ensure that the drainage system is designed for the most critical duration of the design return period. Summer thunderstorm type events may be of a short duration but they are usually critical for determining pipe conveyance capacity in urban areas. Longer events give more infiltration, thus increasing extraneous flow in the sanitary system. Spring snowmelt/rainfall events can be of a long duration and should be evaluated where storage measures are being used. An analysis using a continuous record should also be conducted when the operation of storage measures is being considered (Fig. 1C).

3. FLOW CONSIDERATIONS

3.1 SOURCES OF INFLOW

The watershed characteristics are important in determining the amount of flow that enters the drainage system. The land use, soil type, antecedent soil moisture, area, slopes, shape and existing drainage patterns all affect the flows in the sewer system. In urbanized sewer areas flow may enter the sewerage system from a direct connection to the surface. This may be from roofs or paved surfaces entering the system through downspouts or catchbasins (Figure 2B & 2C). Flow from pervious areas will be detained for a longer period of time it may seep into the ground and enter the system through weeping tiles or drain toward a surface inlet. This surface water runoff may be transported in a combined or a storm sewer to a downstream receiving body. The extraneous flow may enter the sanitary or combined sewer depending on the system. The combined sewer must operate satisfactorily for both DWF (Dry Weather Flow) and WWF (Wet Weather Flow) whereas storm sewers are only designed for WWF conditions.

Sanitary sewers are designed to operate for a DWF condition. Since the treatment plant and sewer capacity are limited, extraneous flow into the system should be kept to a minimum. Extraneous flow may enter from weeping tiles (Fig. 2A), manhole leakage or groundwater infiltration. If excessive extraneous flow enters the system the sanitary sewer will be acting as a combined sewer. Capacity problems may occur both in the sewer and at the treatment plant if this additional flow is neglected.

3.2 RUNOFF PREDICTION

Runoff predictions may be made from measurements or by using a modelling procedure. Analysis and improvements in existing developments may be based on either monitored or modelled flows, whereas new sewer systems must be designed using a modelling procedure.

3.2.1 Monitored Flows

Monitoring is conducted to establish peak flow levels in existing sewer systems. This requires installment of rainfall and flow guages. Flow into the system in urban areas depends on the size, age and connection from the houses or buildings. The runoff also varies with the soil type on the lot as well as backfilling around the house. Monitoring programs can also be used to verify model simulations.

The determination of extraneous flows in sanitary sewers requires a monitoring program as the flows vary from one area to another. In a study conducted in Winnipeg (Ref. 2) flows from individual lots were first measured (Fig 3.A1). They were then accumulated using a hydraulic model to obtain the extraneous flow from blocks and districts. The study developed a set of design hydrographs (Fig 3.A2) for estimating extraneous flow in the system.

3.2.2 Model Simulation

In most analyses flows have to be simulated with a hydrologic model. Today a wide variety of hydrologic simulation programs are available to design engineers. The hydrologic and hydraulic simulation models are reviewed in chapter 5.

4. HYDRAULIC ANALYSIS

4.1 MINOR SYSTEM

Most sewer systems are designed according to the size of the drainage area and natural flow direction. The hydraulic behaviour under operating conditions may be much more difficult to predict. This is especially true for surcharged pipe conditions. Estimates for a sewer system operation during an observed storm are shown in the plan and profile contained in Figs. 4.A and 4.B. A sewer pipe may surcharge if there are downstream tailwater constraints or the pipe has inadequate capacity to convey runoff from the local and upstream tributary areas (Fig. 4.C). Surcharged levels that reach the ground surface become free surface flow.

4.2 MAJOR SYSTEM

Major system hydraulics are concerned with the surface flow patterns, and how they impact on the minor system. The nature of properties abutting to the street, the street width, slope and gutter dimensions influence the runoff patterns (Fig. 4C). Inlet of surface flows to the minor system are controlled by the catchbasins' location and capacity. Hydraulic computations can establish the flow depth, velocity and percentage of water that will inflow or be routed past the minor system inlets.

5. COMPUTER MODELS AND MODELLING TECHNIQUES

5.1 MODEL SELECTION

With modern techniques in hydrologic/hydraulics depending so heavily on computer modelling, the selection of an appropriate computer model becomes critical.

A multitude of computer models have been developed for the analysis of urban drainage systems, and successfully applied in Canadian cities. Several models typical of those used by Canadian engineers are discussed herein. These are: (i) STORM (ii) OTTHYMO (iii) OTTSWMM (iv) SWAN.

The models are separate but complimentary. The primary objectives of selecting several distinct but compatible models as opposed to one comprehensive model are: (a) to reduce the cost of model simulation by performing the analysis in progressive stages. (b) to simplify the modelling by subclassifying and categorizing the sewer system; (c) to assess the concept of inlet controls by modelling the catchbasin inlets and the street configuration; (d) to increase the efficiency of analysis by improving the input data inventory. The IMPSWM program was established at the University of Ottawa so that a package of complementary computer models would be available to designers from the initial planning to final design stages (Ref. 3).

The STORM and OTTHYMO models are planning models that can be used to evaluate a system with relatively small amounts of data. The OTTHYMO model is a single event model. It is used at a preliminary stage to investigate the predevelopment and postdevelopment flows as well as to evaluate different SWM alternatives. The STORM model is a continuous simulation program. With a continuous rainfall record the overall performance of a proposed design can be evaluated. For more detailed design the OTTSWM model can be used. This model simulates surface flow, catchbasin inlet performance, pipe flow as well as various runoff control devices. When an overall system is being classified a program such as SWAN is very useful.

5.2 MODEL DESCRIPTION

5.2.1 STORM Computer Model

The STORM computer program, developed by the U.S. Army Corps of Engineers and Water Resources Engineers Inc., can be used to investigate long-term rainfall/runoff behaviour. The program can analyze the storage, treatment and overflow operation at a detention facility or a waste treatment plant. It is useful for examining the system response to numerous consecutive events (Fig 1.B). The model is not capable of detailed systems analysis but is well suited for use during a preliminary analysis and sizing of control and major conveyance structures as it is economical and can be applied rapidly.

5.2.2 OTTHYMO Computer Model

Most computer models currently in use were developed either for urban conditions (Post-Development) or for rural conditions (Pre-Development). The OTTHYMO model, a modified version of the USDA's HYMO model (Ref. 4), was developed at the University of Ottawa to bridge this gap. It is an integrated model that can be used to simulate both urban and rural flows. The urban flow is modelled using two parallel quasi-linear reservoirs one for the pervious area, the other for the impervious. The storage coefficient for the parallel reservoirs varies with the peak excess intensity, thus making the urban routine quasi-linear. The rural area runoff hydrograph is determined using the Nash unit hydrograph (Ref. 5) It is particularly useful in the development of master drainage plans where pre and post development flows are to be analyzed prior to development. Although a single event model, it provides a much greater degree of detail and flexibility than does STORM. OTTHYMO provides a level of detail between preliminary and detailed analysis and produces results which can be reconciled with detailed models such as OTTSWM.

5.2.3 OTTSWMM Computer Model

Traditionally, the storm sewer designer has assumed that 100 percent of the design flow from any area would be intercepted by catchbasins. In reality, catchbasins on a continuous grade are inefficient collectors of water. Water not collected by one catchbasin flows downstream to the next. The process is repeated until the storm subsides or until the water is finaly intercepted at a low point. This process is modelled in OTTSWM.

The computer program OTTSWMM was developed, from its precursor the USEPA's SWMM model (Ref. 6), at the University of Ottawa. The depth and velocity of flow in gutters are first calculated, using a kinematic response to rainfall intensity. The inlet capacity of the catchbasins are determined and the amount of flow entering the minor system is computed. In system routing is conducted using a kinematic routing routine. If surcharging or a complex system is being analyzed a dynamic routing routine is required. This is provided by the Extended Transport subprogram which can simulate pressure flow, in-system storage, wiers, looping effects and submerged outfalls. The development of the OTTSWMM model has proven to be a powerful tool when assessing alternative design configurations or when analyzing an existing system.

5.2.4 SWAN Computer Model

The computer program SWAN (Ref. 7) performs a steady-state analysis of extensive sewer networks and a computerized inventory of all pipe segments. This is particularly useful when an existing system is being analyzed. The segments of a large sewer network can be stored as a data base for analysis. Plan and profile plots for particular sewer segments identifying inverts, manhole depths, pipe lengths, diameters and capacities can be easily be made.

5.3 MODEL VALIDATION

In order to get the best use of the model application, some basic checks are necessary. The checks are essentially of three types: i) Physical - The simulation model should be chosen to ensure that the basic hydrologic and hydraulic routing and continuity relatations are reasonably represented. ii) Calibration - The recorded flow or water level should be used for comparison with simulated results. (iii) Model Performance - The simulation of real events should be compared to recorded flooding complaints.

6. DRAINAGE SCHEMES

Innovative urban drainage planning aims at minimizing adverse effects of urban runoff in external areas (eg. neighbouring developments or receiving waters), and in the internal service area itself. To achieve this, the simplistic evaluation approach adopted in conventional design practice needs to be extended and upgraded. This section presents some runoff control and stormwater management measures that may be used in an urban drainage system.

6.1 ON-SITE MEASURES

6.1.1 Roof Leader Disconnection

Disconnecting roof leaders from direct input to a sewer system and redirecting flows to pervious surfaces can reduce or delay peak flows to the sewer network. A number of Canadian cities have successfully carried out voluntary roof disconnection programs. In implementing this practice, proper measures to direct roof water away from foundation walls and neighbouring properties, as well as ensuring that lots are properly graded, should be carried out. The results of this implementation are encouraging as a significant reduction of sewer backup incidents has been reported.

6.1.2 Backflow Protective Devices

An urban sewer system is usually designed to a specific service level (for example, a 2 or 5 year return period). However, as discussed below, there are other additional protective and preventive measures which can be adopted by individual residents to ensure further protection. These measures are effective for larger storm events in new systems. For an existing system that is under-capacity sewer backups can only be prevented for minor events until the sewers are finally upgraded.

One or more of the following devices can provide special protection from sewer backflow into plumbing fixtures. The measures are for individual basement or house frontal protection. Devices such as a screw cap, sewer backwater valve, gate valve or a sump pump should be installed between the house plumbing fixtures and the urban sewer system.

6.2 MAJOR SYSTEM MEASURES

6.2.1 Catchbasin Inlet Control

Catchbasin inlet restriction (inlet control) on roadways can be practiced to prevent surcharging of existing lateral sewers by inducing temporary ponding on the street surface. The benefits of this are to possibly eliminate the need for immediate replacement of an undersized lateral, in an existing sewer system, or upgrade its service level for a new drainage system. The lateral may be enlarged in the future as downstream collectors or trunk sewers normally have adequate capacity.

Inlet control requires detailed analysis and assessment of a dual drainage system to ensure proper functioning of both systems. In particular, inlet control measures should not create or aggravate adverse surface flooding conditions especially during major storm events. An effective method of reducing the inlet capacity of a catchbasin is to install an office plate at its outlet.

6.2.2 Above-Ground Non-Residential Storage

This control measure is most applicable to flat roofed industrial or commercial buildings and to open car-parking lots. Flood peaks can be reduced by detaining runoff prior to discharge to the sewer system.

6.2.3 Detention Storage

One of the current drainage practices is to adopt the policy of "Zero Runoff Increase". This means that no increase in the peak runoff rate for any design storm event (which may have a return period up to 50 or 100 years) is permitted from new developments or upgraded existing systems. Satisfying this criteria normally requires a sophisticated dual-drainage system as well as a large storage facility such as detention storage to control the runoff (Ref. 8).

Detention storage may be provided by a wet pond or a dry pond. Wet ponds or lakes are usually quite large and act both as runoff control devices and sites for recreational facilities. Substantial cost savings are available because of the smaller pipe sizes that may be installed. For smaller developments dry detention ponds are usually preferred. They can be located in parks or in an easement.

6.3 MINOR SYSTEM MEASURES

6.3.1 Conventional Sewer Design

Based on current Canadian urban drainage design practice design criteria for sewer works are outlined below:

Hydraulic Considerations

1. The same protection level should be provided through the whole service area. It should be noted however that inflow into downstream trunk sewers may be signficantly reduced due to spatial variation of the rainfall;

2. No surface frontage flooding is allowed in industrial/commercial areas;

3. No significant surcharge in the residential areas is acceptable. Limited surcharging is allowed if the surcharge is relatively small as compared with the pipe cover, and if the surcharge does not adversely affect hydraulic conditions in upstream areas;

4. Maximum flow velocities in the sewer system should not exceed the maximum allowed for pipe strength and anchorage protection;

5. Pipe slopes should be steep enough to ensure that the minimum flow velocities in the sewer system are not below the minimum required for pollution removal purposes.

Construction Considerations

1. The sewer cover (from ground elevation to pipe crown) should be greater than that for weeping tiles or sanitary facilities;

2. New sewer replacements should have the same slope as the existing sewer, except where the flow velocity is beyond the design range as discussed in points (4) and (5) above.

Location Considerations

1. Wherever possible, new sewers should not be constructed on heavily travelled roads or publicly sensitive areas;

2. Wherever possible, new sewer construction should avoid interference with existing utilities. These include: water, gas, electric and telephone ductlines.

Trunk System Consideration

In many North American cities, a separate trunk system may be constructed to convey runoff from large sized urban areas. The trunk sewer (lower system) is below the lateral/collector system (upper system), the two being connected by drop manholes.

1. The hydraulic performance of drop manhole structures normally behaves as orifice or weir flow. Vortex formation from the upper level sewer outlets may occur. The impact of increased runoff from new developments or from existing drainage reliefs should be assessed in comparison with the drop manhole capacity;

2. Recent publications (Ref. 9) indicated that adverse hydraulic transients in the trunk system may be induced by the rapid variation of flow rates from the upper level system. The fluctuation of water levels in drop manholes due to transients may significantly affect the hydraulic performance of the manholes and the upper sewer networks. The author of the publication stressed that a fully dynamic model should be used to analyze a trunk system which undergoes pressurization.

6.3.2 Underground Storage

Sewers could be greatly oversized to provide underground storage in order to reduce the peak flow to the downstream system thereby significantly reducing the sizes of the downstream sewers.

In order to prevent potential sewer backup into the connected house service system, due to the increase of water level inside an oversized sewer, deep sewer cover is essential. A hydro-brake installation is normally provided to control the outflow to downstream areas after the peak runoff occurrence.

6.3.3 Temporary Emergency Overflow

Sewer Overflow Connection

In some situations, an overflow can be placed from the combined sewer system to a neighbouring storm system or another conveyance system. It should be noted, however, some adverse impacts may result due to the increased pollutant loading to receiving water bodies.

Booster Pumping

In cases where lack of grade limits outflows from an area thereby causing flooding, it may be possible to install permanent or portable pump facilities as a control measure. Such pumps can speedily remove flows to a suitable discharge point and therefore reduce or eliminate flooding. However this option implies a high capital, maintenance and operation cost. Furthermore, potential pollution problems may result by permitting combined sewer flows to be discharged to the surface. These two measures should only be implemented in emergencies to prevent sewers from backing up into surrounding basements.

6.4 SUMMARY

The techniques discussed above have been used across Canada. Applications have been made in both new developments or where the existing sewer facilities were being upgraded. Where redevelopment is occurring flows may increase beyond the capacity of an existing system. Implementation of some of the runoff control measures can bring flows to within existing pipe capacities. Two examples of typical urban drainage planning and analysis are summarized in the next chapter. They illustrate that current state of the art runoff control and stormwater management techniques can be adapted in existing or new urban developments.

7. PRACTICAL APPLICATIONS

7.1 CASE STUDY 1: FLOOD RELIEF OF AN EXISTING DRAINAGE SYSTEM

Techniques for hydrologic/hydraulic analysis were applied to a sanitary sewer area of 2000 ha in the City of Winnipeg, Canada (Ref. 2). Frequent sewer backups and basement flooding has occurred due to heavy extraneous inflow.

In the overall study, the response and amount of extraneous inflow were obtained as described in section 2.3.1. With the inflows an hydraulic modelling analysis was conducted to determine the existing sewer capacity (Fig 5A). Several relief schemes were then evaluated and compared including roof leader disconnection, lot grade and settlement improvement, reconnection of weeping tiles to storm sewers, sewer replacement and trunk interceptor, storage in both in-line and off-line, increased pumping facilities and temporary overflow connections.

The factors affecting selection of alternatives included: the degree of protection level required; cost effectiveness; flexibility for future improvement; records of complaints; sewer loading and surcharging conditions; sewer physical conditions based on its cover depth, chronology and TV inspection.

The recommended relief implementation plan (Fig. 5B) consists of: (i) overflow pipes with flat check-gates placed to connect the existing sanitary sewers to the storm system or to the receiving streams; (ii) in-line storage (2700 mm diameter) equipped with hydrobrakes and located in areas where overflow connections are infeasible; (iii) sewer replacement at locations where the pipe cover was too shallow to install in-line storage; and (iv) new pumping facilities to divert the runoff in order to reduce its downstream loading.

This relief plan upgraded the sewer system, the frequency of occurence being reduced from a one year return period to a ten year return period. The cost of a conventional solution would have been eight million dollars this was reduced to one million dollars by using stormwater management measures.

7.2 CASE STUDY 2: DEVELOPMENT OF A FUTURE SUBDIVISION

The present case study reviews the process of defining an urban drainage scheme for a new development area.

In 1978, a Preliminary Drainage Review report (Ref. 10) was prepared. The report reviewed the existing rural conditions and constraints of the 2000 ha area (of which the proposed subdivision is a part) in North Edmonton, Canada (Fig 6.A). Several alternative drainage schemes were defined and assessed. These included: (i) a conventional gravity sewer system (ii) pumping facilities and (iii) a detention facility with a controlled release to the existing downstream sewer system. Based on functional, environmental and economic criteria, the alternative with a controlled storm runoff release was selected for further investigation.

A master drainage plan report was prepared in 1979 (Ref. 11). It assessed the drainage impacts on the 1000 ha future development area as well as on the downstream developed area. This was done for both pre and post development conditions. The preliminary layouts of detention lakes, main streets, parks, and development density were also addressed. Computer modelling was used in the hydrologic simulation it included a continuous rainfall/snowmelt runoff simulation for a period of sixteen years (1960-1977). The hydraulic response of the interrelated lake system was also assessed.

The final report on Stormwater Management Design (Ref. 12) was prepared in 1980. It detailed the analysis of the selected stormwater management scheme previously described. The report was submitted along with detailed engineering drawings for the entire subdivision (250 ha) as shown in Fig. 6B. It contained detailed hydraulic designs and a sophisticated modelling analysis related to the minor and major systems. A careful analysis of the overland flow paths, street gutter and catchbasin capacities was conducted. Also, the hydraulic response between the detention lakes and their interconnected sewer systems were analyzed in detail. Spillways, sediment control facilities and shoreline protection were provided in this final report.

With this design there was no peak increase above predevelopment 1/5 year flow for the post development 1/100 year flow. The design improved the downstream flow conditions for frequent storm events. In general, a high level of service was provided for both the major and minor systems. Each of the detention lakes provides the same level of service throughout the construction phases and final operation.

ACKNOWLEDGEMENTS

The writers would like to thank Dr. P.E. Wisner IMPSWM Director, University of Ottawa, for his encouragement during the preparation of this paper. In addition the writers wish to acknowledge the assistance and cooperation of Mr. S. Vatagoda, Mr. M. Ahmad and the staff of the City of Edmonton, Water and Sanitation Department during the course of this study.

REFERENCES

1. James W., Henry D. and Haro H.: "Integrity of Rainfall Time Series for Models of Existing Combined Sewer Systems" In Proceedings of Stormwater and Water Quality Management Modelling Users Group Meeting, 18-19 October 1982 Ed. by P. Wisner, Univ. of Ottawa, pg. 146.

2. Fok A.T.K., Wolfe W.R. and Perks A.R.: "Analysis of Heavily Infiltrated Sewer Systems", a paper for the Second International Conference on Urban Storm Drainage, June 1981, Illinois, U.S.A.

3. Wisner P.: "IMPSWM Applications for Urban Drainage Studies" "Implementation of Stormwater Management in Canada", ASCE Foundation Engineering Conference on Stormwater Detention Facilities, held in New Hampshire USA., August 1982.

4. Williams J.R. and Hann R.W.: "Hymo, A Problem-Oriented Computer Language for Building Hydrologic Models", Water Resources Research, Vol. 8, No. 1, February 1972.

5. Nash J.E.: "The Form of the Instantaneous Unit Hydrograph", International Association of Scientific Hydrology, Pub. 45, 3(1957).

6. Huber C.W., Heany J.P. and Nix S.T.: "Storm Water Management Model, Version III - Documentation of New Procedures", Department of Environmental Sciences, University of Florida, Gainsville, Florida, January 1980.

7. Erdman and Anthony Consulting Engineers: "Applications Manual for Sewer Analysis System-SWAN", January 1983, Rochester, New York, USA.

8. Fok A., Vatagoda S. and Rowney C.: "Some Considerations of Detention Lake Design", Proceedings of eighth international Symposium on Urban Hydrologic, Hydraulic and Sediment Control, July 1981, Kentucky, USA.

9. Song C.C., Cardle J.A. and Leung K.S.: "Transient Mixed Flow Models for Storm Sewers". Journal of the Hydraulics Division Am. Soc. of Civil Engineers, vol. 109,No. 11, Nov. 1983, pp.1487-1503.

10. James F. MacLaren Ltd. "Report on Stormwater Management in North and Northeast Edmonton", May 1978, Edmonton, Alberta.

11. Stanley Assoc. Ltd.: "Master Drainage Plan, Lake District, Edmonton", November 1979, Edmonton, Alberta.

12. Proctor Redfern Butler and Krebes Ltd.: "Lake Optimization Study. Neighbourhood 6 & 7, Lake District", August 1980, Edmonton, Alberta.

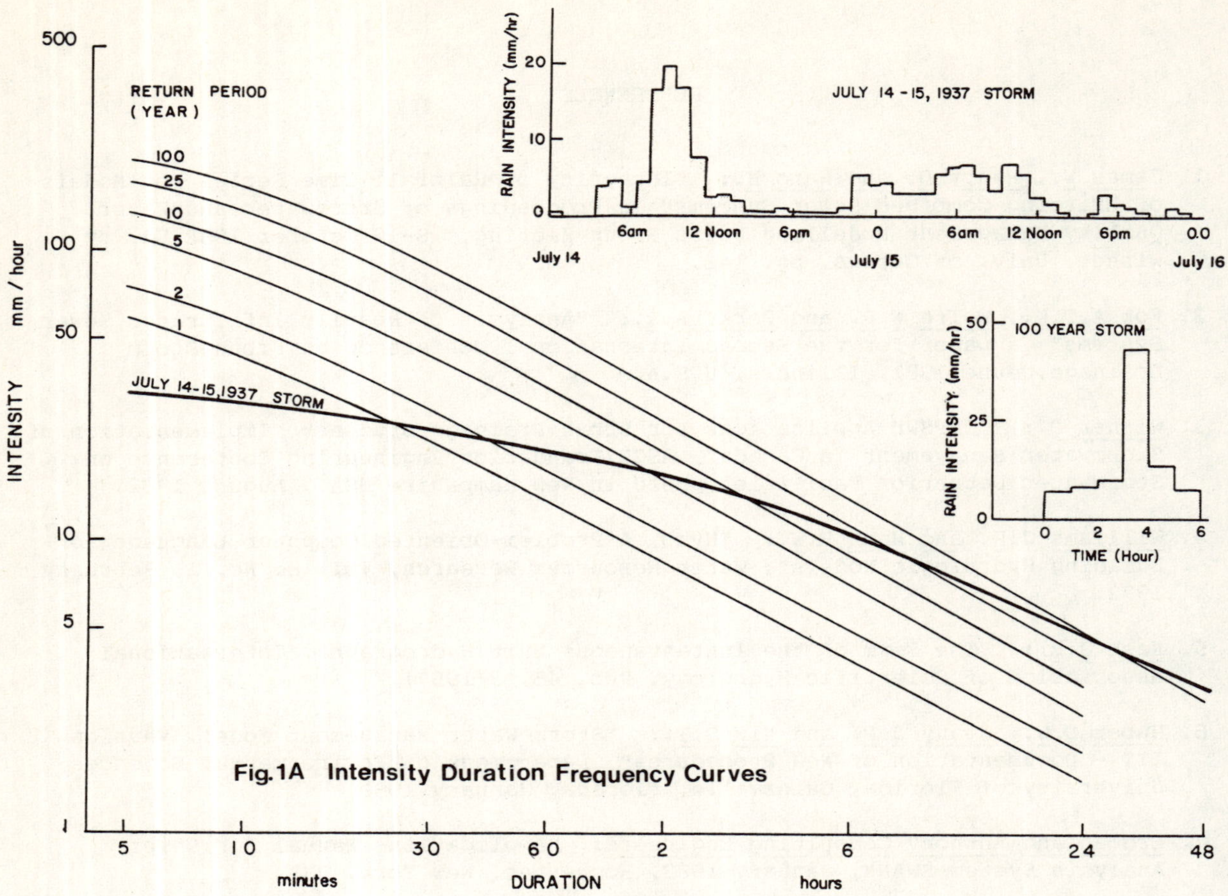

Fig.1A Intensity Duration Frequency Curves

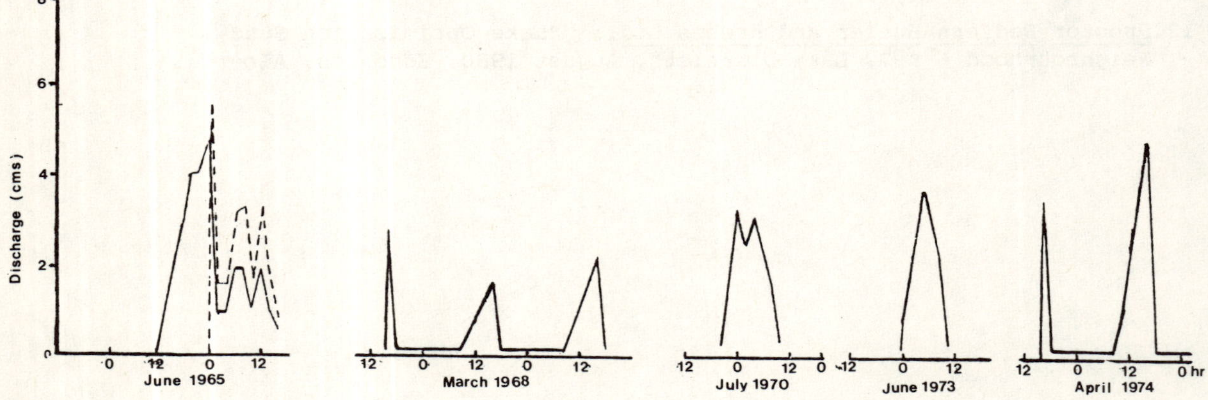

Fig.1B Long Term Simulation
(Overland Flow Conditions for Castle Downs II
Events between 1961 – 1977)

Figure1 Design Events

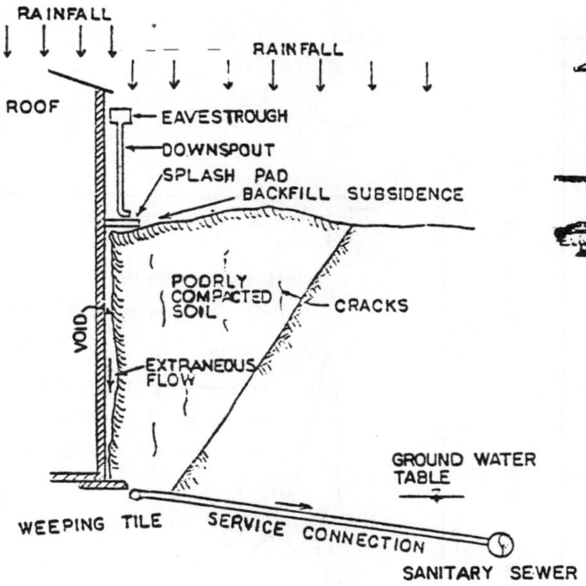

Fig. 2A Extranéous Inflow

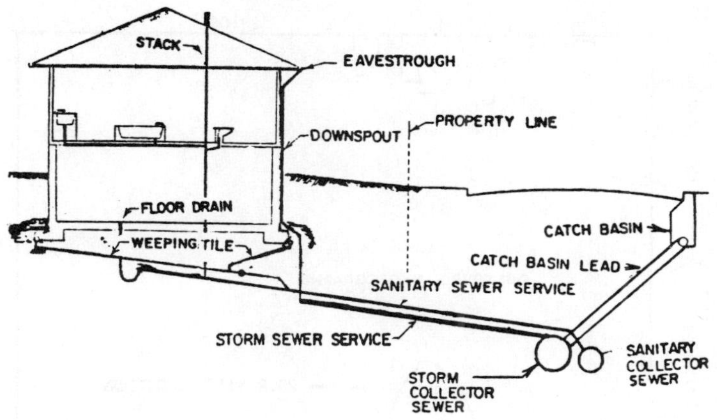

Fig. 2B Schematic of a Separate Sewer System

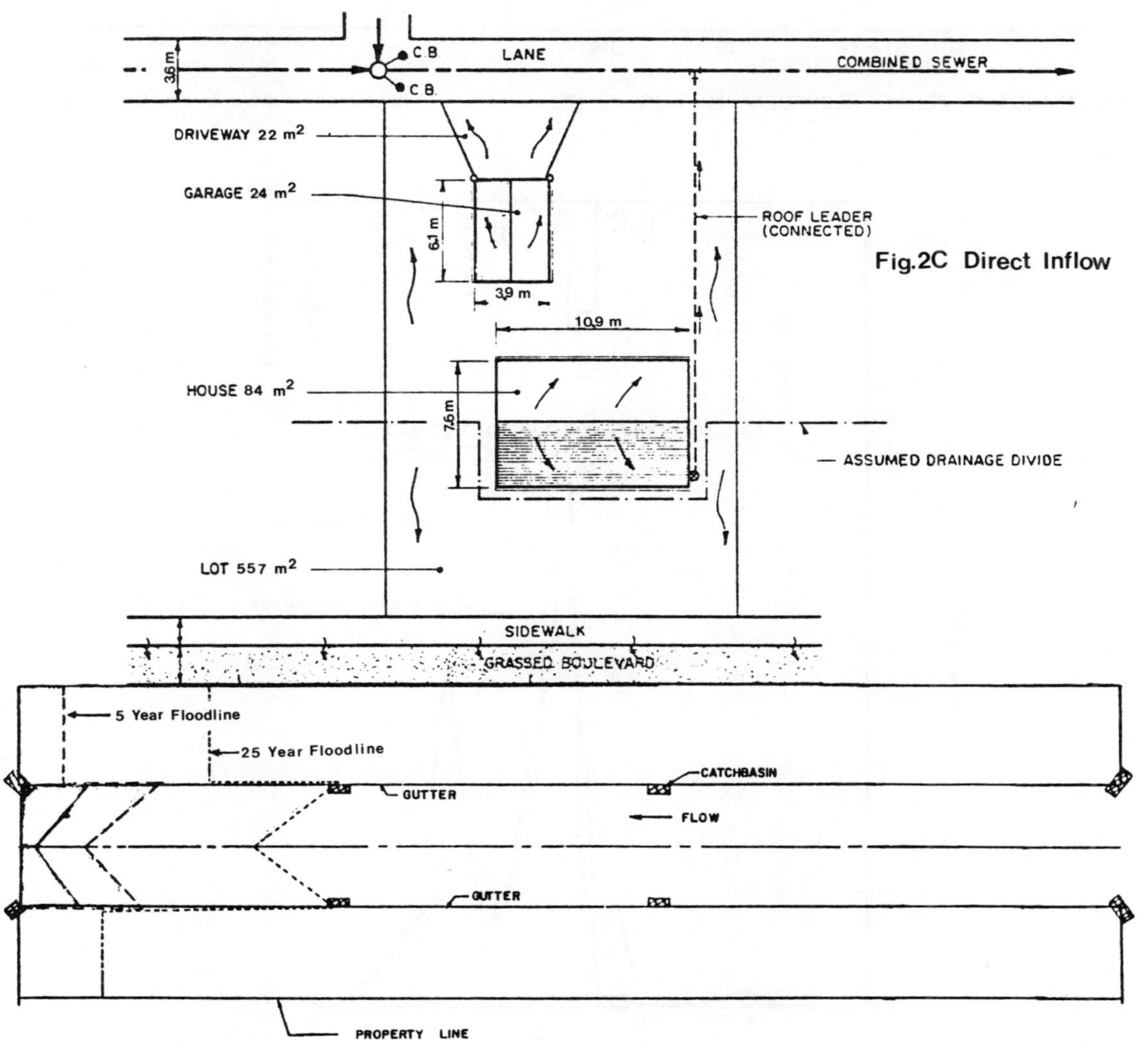

Fig.2C Direct Inflow

Figure 2 Sources of Inflow

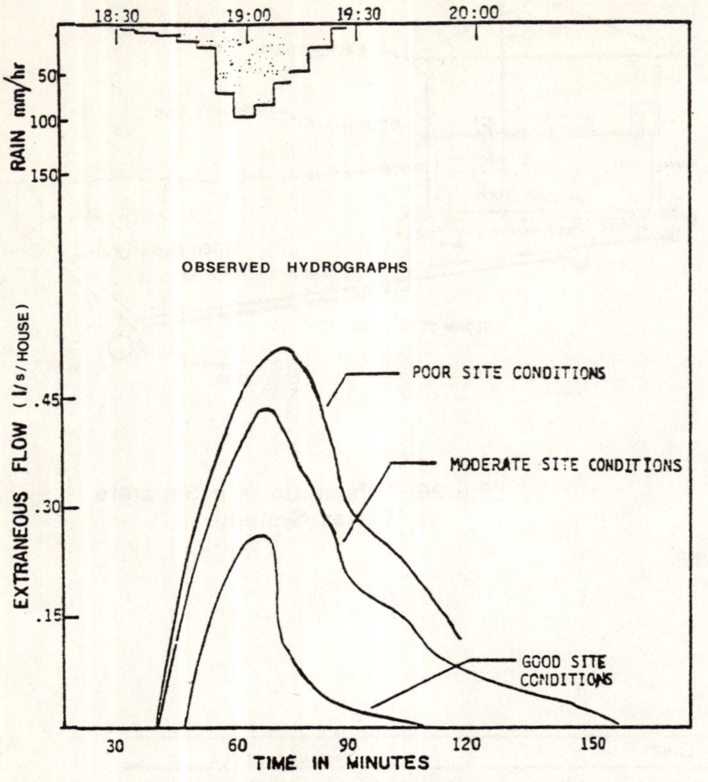

Fig.3A1 EXTRANEOUS FLOWS RECORDED IN AUG. 15, 1972

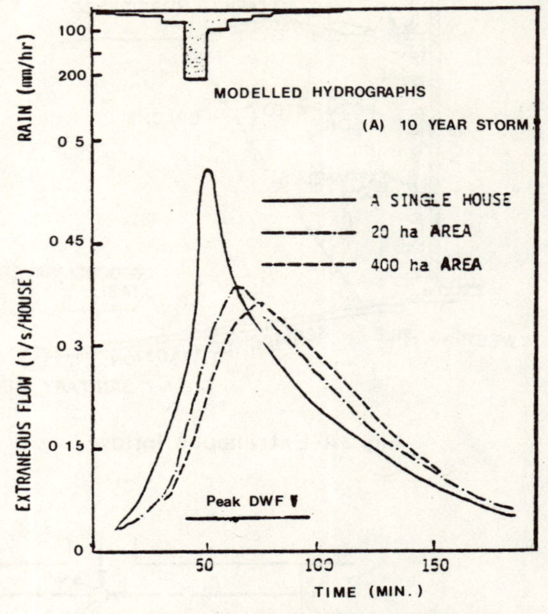

Fig.3A2 EXTRANEOUS FLOW HYDROGRAPHS

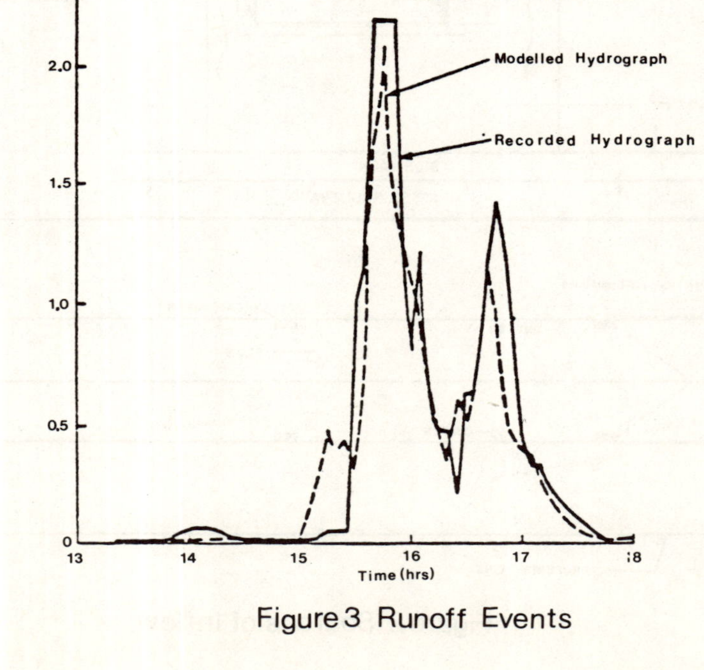

Fig. 3B OBSERVED AND MODELLED DIRECT RUNOFF

Figure 3 Runoff Events

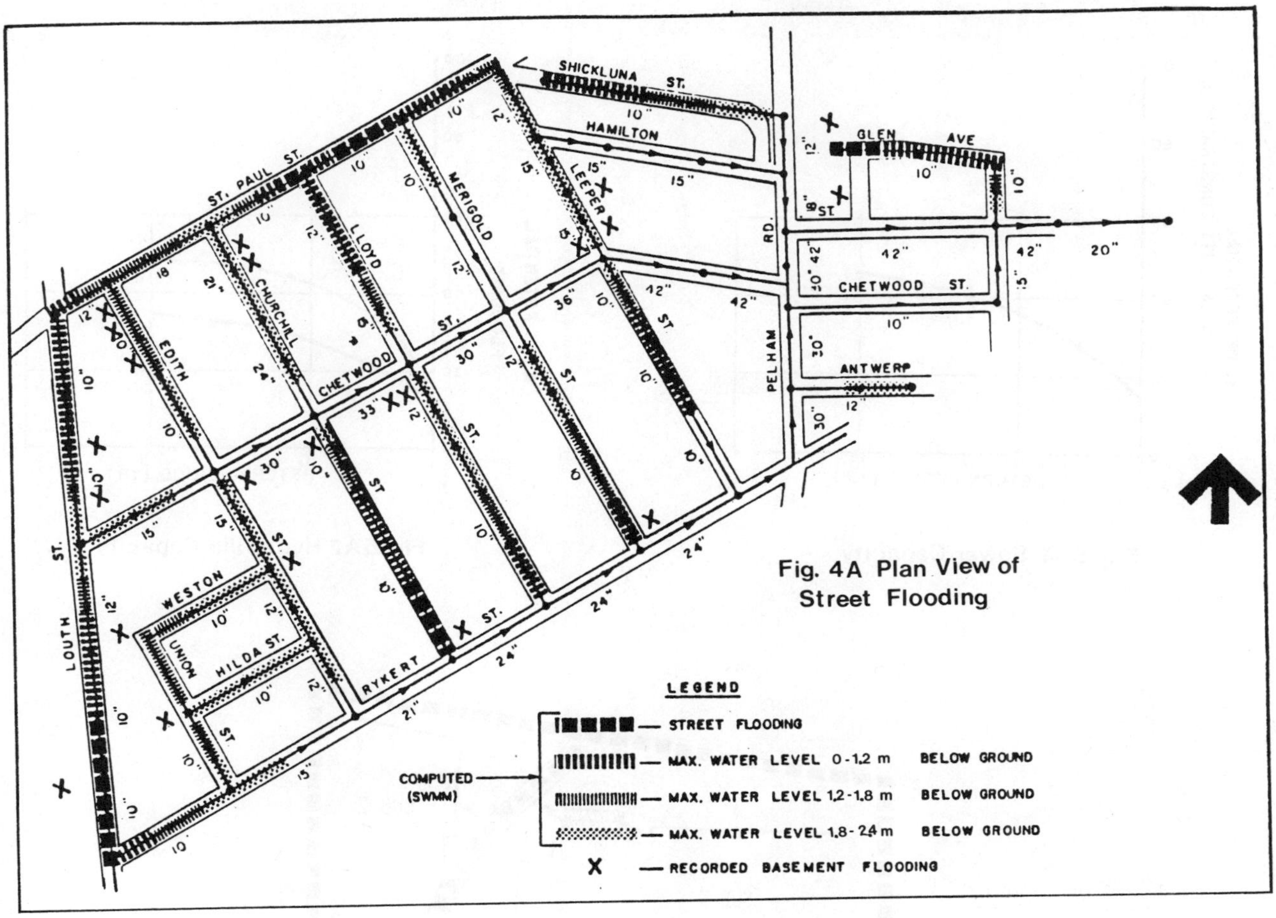

Fig. 4A Plan View of Street Flooding

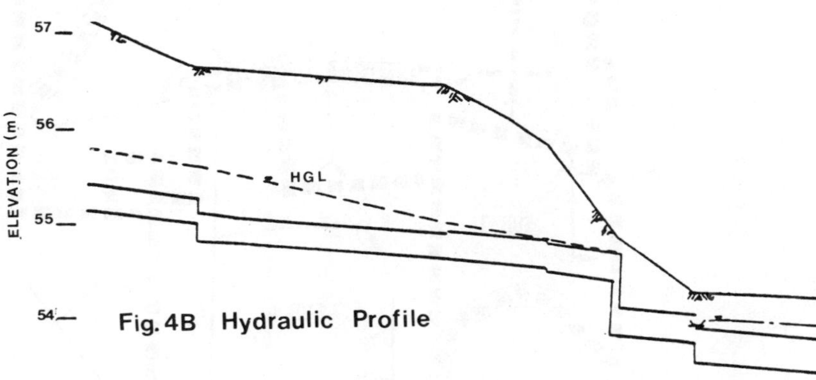

Fig. 4B Hydraulic Profile

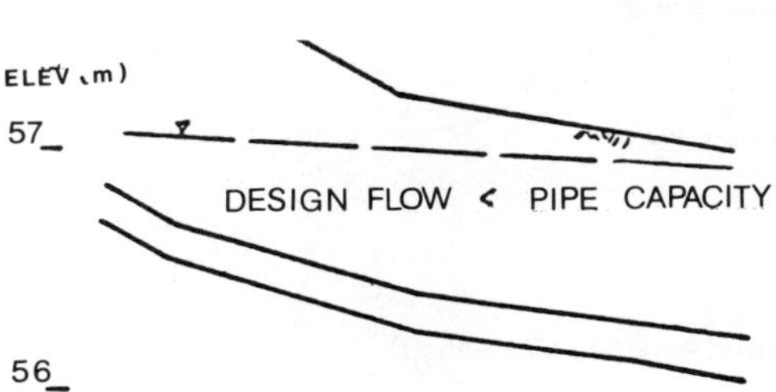

Fig. 4.C1 Tailwater Constraint

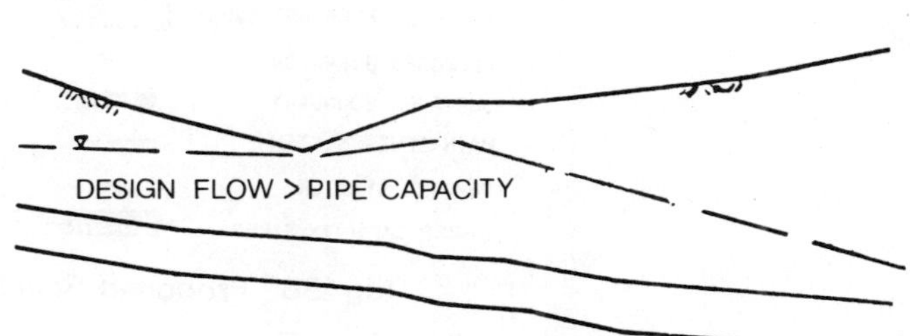

Fig. 4.C2 Capacity Constraint

Figure 4 Hydraulic Simulation

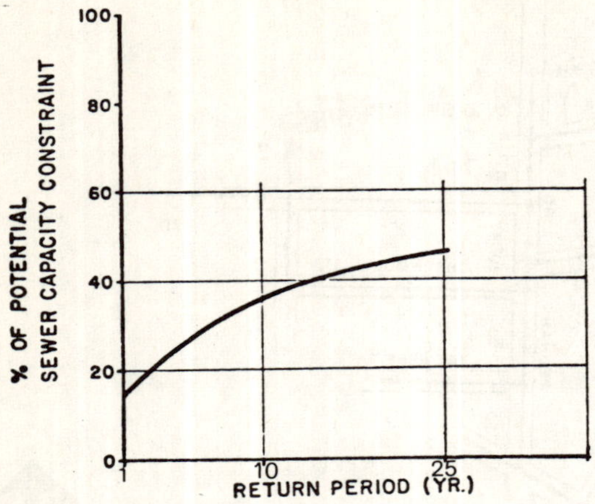

Fig. 5.A1 Sewer Capacity

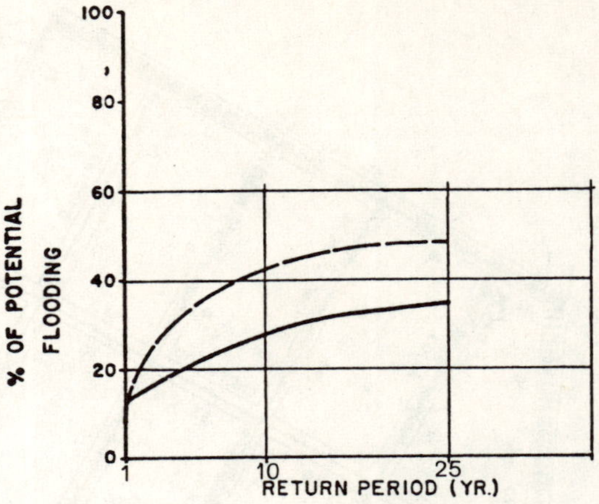

Fig. 5.A2 Hydraulic Capacity

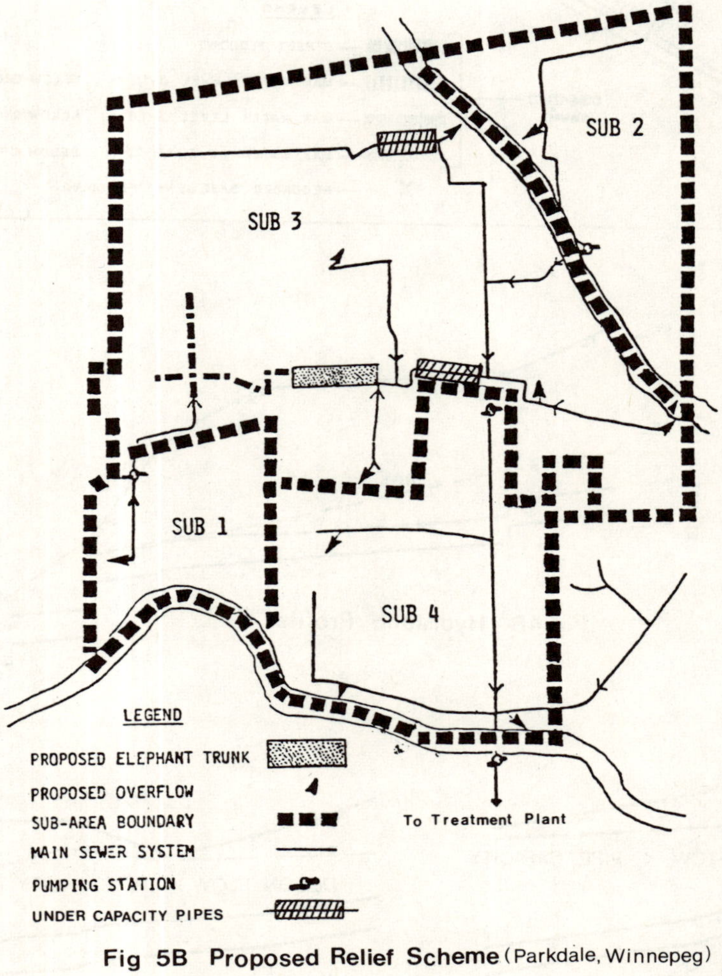

Fig 5B Proposed Relief Scheme (Parkdale, Winnepeg)

Figure 5 Existing Drainage System

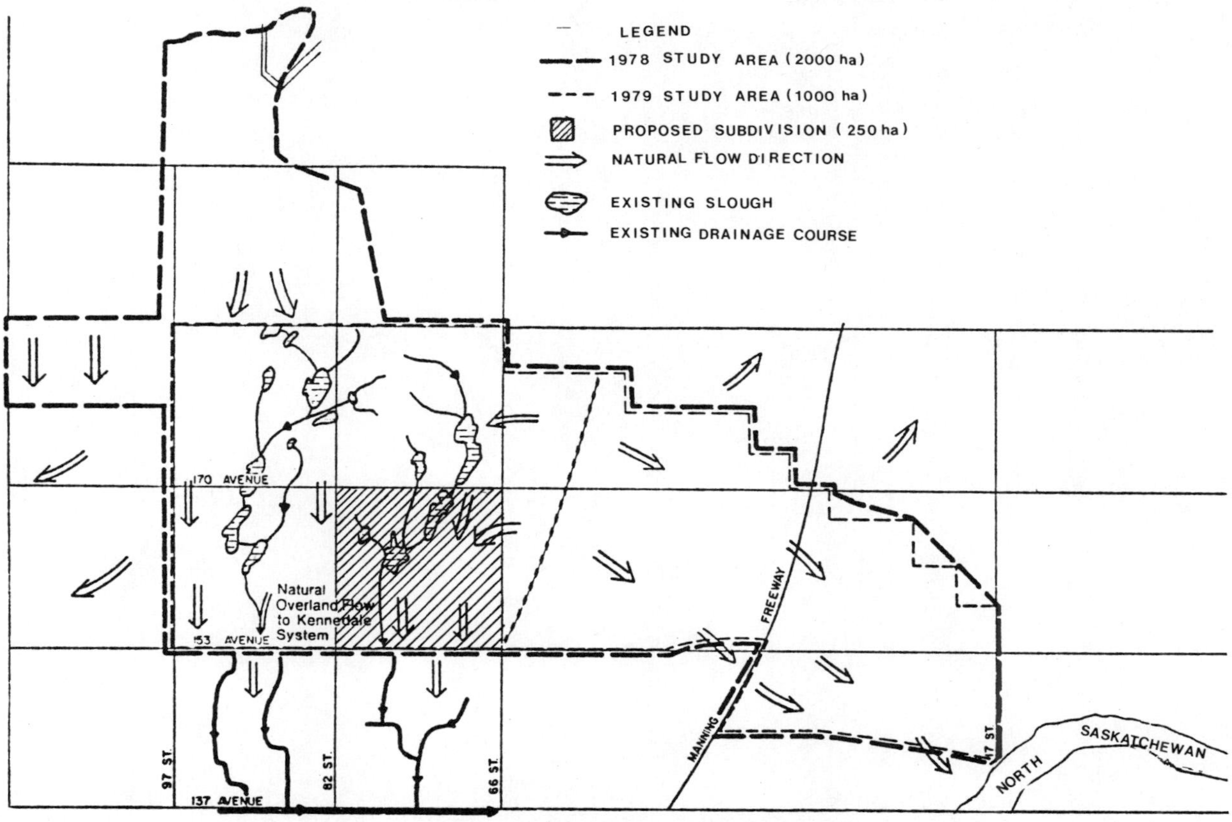

Fig. 6A Predevelopment Flow Pattern

Fig. 6B Proposed Drainage System

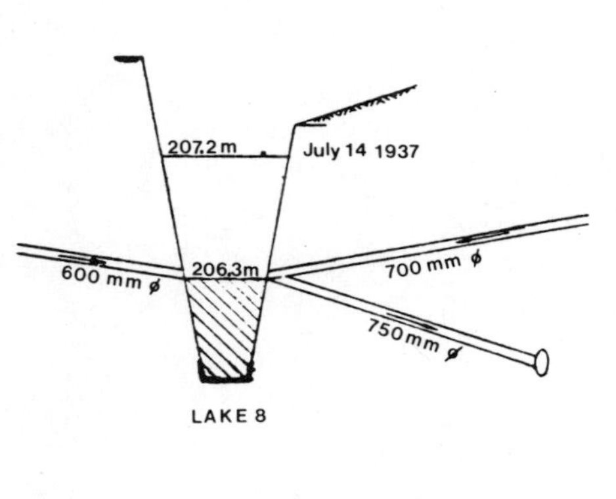

Fig. 6C Lake Profiles

Figure 6 New Drainage System

International Conference on the
PLANNING, CONSTRUCTION, MAINTENANCE & OPERATION OF SEWERAGE SYSTEMS
12-14 September, 1984

PAPER E2

THE ASSESSMENT OF SEWER SYSTEM HYDRAULIC PERFORMANCE

by

I T CLIFFORDE, BSc; C.Eng; MICE; MIPHE
and
M J GREEN BSc; PhD

WRC Engineering
PO Box 85
Frankland Road
Blagrove
Swindon
Wiltshire SN5 8YR
England

SUMMARY

The strategy and procedure of the WRC/WAA Sewerage Rehabilitation Manual has major implications for hydraulic modelling and upgrading practice.

This paper outlines the hydraulic requirements of the strategy and considers the difficulties which may arise in application of the recommended procedure. WRC's experience in the construction, verification and use of simulation models is used to justify the proposals of the Manual; to illustrate ways in which difficulties can be overcome and to demonstrate how the objectives of the strategy can be fulfilled.

Future enhancements in hydraulic modelling and the Rehabilitation Manual are also considered.

Held at The University of Reading, England. Organised and sponsored by BHRA, The Fluid Engineering Centre, Cranfield, Bedford in conjunction with the Water Research Centre.

1. INTRODUCTION

A strategy for the upgrading of sub-standard sewer systems has been proposed in the WRC/WAA Sewerage Rehabilitation Manual(1). Hydraulic simulation models are essential in the use of this strategy. Present models are a major advance on the previous design orientated techniques, but are far from being the perfect tool. At present the limitations of the programs require the user to employ many simplifications in constructing a mathematical model of any substantial sewer system. Many of these constraints will remain into the foreseeable future but others can be anticipated to be resolved in the next generation of models.

This paper examines the capabilities of the most widely used UK simulation model (WASSP-SIM, the Simulation Program of the Wallingford Procedure) in the context of the Rehabilitation Manual and draws on experience gained by WRC to illustrate how difficulties in modelling extensive sewer systems can be overcome. Enhancements for the future, arising from this experience are proposed and discussed.

Verification of sewer system models becomes increasingly relevant as the ability of the models to truly replicate conditions experienced in the physical system improves. The paper goes on to illustrate the need for verification and to justify the expenditure of substantial monies for this purpose by reference to WRC's experience in carrying out flow surveys over the past three years.

In conclusion, the potential arising from comprehensive and accurate knowledge of sewer system performance to allow more cost-effective solutions to hydraulic problems is briefly discussed.

2. HYDRAULIC ASPECTS OF THE REHABILITATION STRATEGY

The rehabilitation strategy and its underlying philosophy have been adequately described in another paper presented to this conference(2). In reading this paper, the following two points concerning the strategy should be borne in mind:

- hydraulic modelling is only part of an integrated approach covering structural, operational as well as hydraulic problems and,

- the procedure should be applied to a complete drainage area, which may mean a sewer system serving tens of thousands of people.

The parts of the procedure relevant to hydraulic modelling may be summarised in the simplest terms as follows:

- construction of a model

- verification of the model

- use of the model to design upgrading works.

Having a general understanding of the role that hydraulic modelling plays in the rehabilitation of sewer systems it is appropriate to consider the capabilities and limitations of the available modelling techniques relative to the sewer systems to be represented.

3. THE SIMULATION PROGRAM OF THE WALLINGFORD PROCEDURE (WASSP-SIM)

A full description of the capabilities of WASSP-SIM is given in the program manual(4).

WASSP-SIM, like all programs which simulate surcharged flow conditions, works on the basis of building a complete picture of the sewer system at each time step. This process naturally utilises a great deal of computing capacity and it is for this practical reason, and of limiting run-time, that the number of pipes and the simulation period is restricted. In the case of WASSP-SIM these limits may vary but are commonly 400 pipes and 480 time steps. The numbers of ancillaries and other features are likewise limited for similar reasons.

4. CONSTRUCTION OF A DRAINAGE AREA HYDRAULIC MODEL

A drainage area (as defined in the Manual) of 100,000 people is not uncommon. The National Assessment(5) estimates an average population of approximately 250 persons per kilometre of sewer whilst a survey by Balmer & Meers(6) has suggested 50 metres run length between manholes as typical.

On the basis of these figures a drainage area may therefore be expected to contain up to 400 km of sewer or 8000 individual sewer lengths.

Systems containing 50 or more overflows and similar numbers of pumping stations and other ancillaries exist in the UK.

Modelling therefore frequently involves considerable simplification both to meet the numerical limitations of the program capacity and to avoid the need for collection and handling of unmanageable quantities of data. Furthermore there is the question of reducing complex features to a form amenable to mathematical representation i.e. unusual overflow arrangements, flow loops etc.

The Manual proposes a procedure whereby the bulk of the model's capability is concentrated in the areas where major expenditure is to be anticipated (i.e the critical sewer) by allowing gross simplifications in peripheral areas where it is only necessary to gain a reasonable understanding of the runoff patterns at the outfalls. Nonetheless the requirement for a significant degree of simplification in the "core-area" remains and, whilst the Manual provides guidelines and advice on how this may be achieved in general terms, each particular catchment is likely to throw-up individual problems which will have to be solved by the modeller on the basis of his own experience.

The general approach to be employed when simplifying a complex arrangement of sewers or ancillaries is that the model behaviour should approximate to that of the physical system, particularly at the critical stage when the system is operating at the level of relevant performance criteria.

The following section gives brief descriptions of two of the more complex sewer systems which have been successfully modelled by WRC which may be helpful to other modellers in dealing with the problems associated with their systems.

5. EXAMPLES OF DRAINAGE AREA MODELS

5.1 York

The combined sewer system for York city centre serves approx 100,000 people and is thought to comprise about 300 km of sewer and 9,000 manholes, although documentation is poor so these estimates are uncertain. The main features of the system are illustrated in Fig. 1 and can be summarised as follows:

(i) The main sewers are large, laid to slack gradients and are deep due to the flat topography of the area.

(ii) There are approx 50 storm water overflows (SWO's) throughout the system discharging to the Rivers Ouse and Foss and various minor tributaries.

(iii) There are 3 inverted syphons where major sewers cross under rivers.

(iv) The whole system drains to a single, major pumping station.

Operational features of the system relevant to the modelling exercise are:

(v) Many sewers are overloaded, causing periodic flooding.

(vi) The performance of some SWO's is affected by river levels.

(vii) Levels in the wet-well of the pumping station are maintained at a high level to minimise pumping costs, thereby causing backing-up in the incoming sewers.

(viii) Many sewers suffer from significant quantities of sedimentation.

(ix) Much of the system suffers infiltration which can occur in large volumes after heavy storms.

Initial construction of the model for the York system was primarily concerned with reducing the extensive network to less than 400 pipes and in adequately representing the many overflows throughout the system. (Figure 1 illustrates the extent and form of the model).

The number of modelled overflows was somewhat reduced by eliminating those spilling only at very high surcharge levels such that they did not operate during verification events and would not be retained in any upgrading scheme. A few others were found to be derelict or redundant and were thus ignored. In total 38 overflows were included in the model.

Initial runs of the model showed poor standards of fit for verification events and serious instability (eg oscillations) in several parts of the model. Figs. 2 and 3 illustrate typical examples of early hydrographs.

The model was refined progressively (over a series of runs) in the following ways:

(i) 6 overflows having the configuration shown in Fig. 4a where found to produce instability in the model. This was eliminated by substituting the arrangement shown in Fig. 4b which was more acceptable to the model. Fig. 5 illustrates the different model behaviour for the two configurations.

(ii) At other overflows exhibiting instability or poor flow balance, the initial choice of values for weir and orifice coefficients was reviewed and values closer to those suggested in the Wallingford Manual substituted. This, in general, alleviated the problems of instability and flow balance and, where necessary, adjustment was made to weir lengths and orifice sizes to produce accurate hydraulic performance.

(iii) In several instances sub-catchments exhibited significantly incorrect volumes of runoff. In 3 cases, field investigations revealed the presence of previously unknown inter-connecting pipes between sub-catchment systems. Inclusion of these pipes in the model markedly improved flow predictions.

(iv) The inverted syphons are fairly complex structures involving twin cast iron pipes with associated dividing chambers and flow control arrangements. In the model the syphons were represented as a single "equivalent" pipe but with a high head loss index to allow for the entry/exit and bend losses.

(v) The pumping station is only significant to the performance of the gravity sewer system draining to it in that the high operating levels in the wet well impose a control effect on the system discharge. This feature was introduced into the model as a downstream level hydrograph which had noticeable effect throughout much of the trunk sewer system.

(vi) The modelling of an extensive sewer system in relatively few pipes causes a serious underestimate of system storage volume in the model. As the system at York was extensively surcharged during verification events, this detracted from the standard of fit between predicted and observed hydrographs. The inclusion of extra manhole storage equivalent to the unmodelled volume significantly improved the situation.

(vii) Finally the hydraulic roughness of every pipe in the model was reviewed in the light of the latest WRC studies of this topic(7). Due allowance was thereby made for the poor condition of certain sewers and the effects of sediment as well as the basic constructional features of the sewers.

After refining the model in these ways the predictions of the model were much more acceptable as typically illustrated in Figs. 6 and 7 (which are at the same locations as Figs. 2 and 3).

Fig. 8, however, illustrates one phenomenon exhibited by the system which has not been replicated. This is the so-called York "sponge-effect" which is characterised by a tendency throughout the catchment towards higher than anticipated rates of flow during the falling limb of a storm hydrograph and overall volumes of discharge in excess of those predicted. This behaviour is thought to be explained by short period infiltration. Naturally this process is less rapid than direct runoff to the sewers and results in increased flows in the latter stages of the storm hydrograph. This phenomenon cannot be readily modelled by WASSP-SIM. Nonetheless in general it can be said that an acceptable model of the York sewer system has been produced.

5.2 **Torquay**(8)

The sewer system of Torquay is very different to that of York, although still combined in character. The population at 60,000 is rather smaller and the sewers are much steeper and shallower. The primary feature of the area is the lack of any natural water-courses in the bulk of the catchment and hence the absence of SWO's on the sewer system. In fact what are now the sewers were originally the water-courses which have been culverted as the catchment has developed. Further development led to overloading of these original culverts, necessitating the provision of extra flow capacity by adding parallel pipes as appropriate. Generations of this process have led to a system where the main sewers are frequently duplicated and triplicated having multiple cross-connections and manholes where sewers converge and redivide with flows splitting in a somewhat arbitrary manner. Fig 9 gives an indication of the prevalence of these features and the general form of the system.

The system also contains:

- three major and one minor overflow at the bottom of the system which spill excess storm flows to the harbour area.

- a major pumping station collecting flows from a large sub-catchment and discharging into the main system.

- a natural stream taken into the head of the sewer system.

- outfalls which are influenced by tide levels.

A TRRL model comprising, by coincidence, approx 400 pipes already existed for the Torquay network and this was used as the basis for the WASSP-SIM model, hence no difficulties were experienced in simplifying the system to meet the capacity limitations of the program.

The complex arrangements of multiple pipes with numerous cross connections and common manholes proved to be largely beyond the capabilities of the program to model in any literal way, causing major problems of instability and oscillation. After considerable experimentation, it was determined that the majority of double/triple runs could be reasonably modelled as single pipes of equivalent capacity, particularly when frequent interconnections made the transfer of flows readily feasible in the physical system. Where bifurcations remain generally separate for long distances, the need to model the split is greater and, fortunately, so was the capability of WASSP-SIM to model the arrangement. Fig 10 shows the final form of the model.

The other features of the system were built into the model over a series of runs to refine the predictions. The overflows, being complex in character have not always been modelled directly on the basis of their physical characteristics, but in each case an arrangement to represent their hydraulic behaviour has been identified within the capabilities of the program with the aid of the verification data.

The pumping station has been modelled with six pumps having discrete on/off levels, giving accurate representation of the physical system.

The stream input to the head of the sewer system was measured during verification events and hydrographs have been generated by appropriate manual techniques for design storms. Both were incorporated into the model as input discharge hydrographs without difficulty.

Figs. 11-13 show hydrographs for a typical location in the Torquay network and demonstrate the improvements achieved through the process of refinement using verification data.

6. FUTURE MODELLING ENHANCEMENTS

Ways in which simulation models can, and almost certainly will, improve in the future fall into three areas:

(i) Hardware improvements

(ii) Improved ancillary representation

(iii) Specific detailed refinements.

Computer technology is a new field which is developing rapidly. It has been explained that the principal limitations on the size of flow simulation models are those of computing capacity and run-time. It is to be expected that in a comparatively short time engineers will have economical access to machines which have greatly increased capacity and which operate much faster. This being the case, models can be expanded to suit. Nonetheless there will still remain the constraint of data availability. Management information systems may make the handling of large amounts of data more acceptable but the need to go out and get information on poorly documented old systems will still exist. This will be no easier in the future and will probably become the main constraint on model size.

Flows in straightforward pipe systems are fairly well represented in present models. The principal scope for improvement is in the representation of ancillaries, in particular the non-standard structure commonly found in old sewer systems which have developed piecemeal over long periods of time.

Many such structures cannot be analysed mathematically and so provision must remain for the modeller to specify the performance. In other cases there is scope to improve the range and flexibility of existing programs to represent the types of structure found in real sewer systems.

At a detailed level, there are certain specific aspects in which present models are recognised to be deficient. These are obviously specific to individual programs and, as such, the comments here are mainly relevant to WASSP-SIM. Spatially variable rainfall, free-surface backwater effects, on-line tank representation and above ground attenuation effects are all under consideration, as are the possibilities for increasing the range of pipe shapes accepted by the program, the pumping station sub-routine, together with rising main modelling.

Finally at a much broader level, there is the question of modelling water quality. WASSP-SIM is at present exclusively concerned with quantities of flow. Pollution and water quality is an area of concern in this and many other countries and as such the subject of modelling of quality as well as quantity in all types of sewer systems cannot be ignored. Some simulation models, e.g. SWMM[9], already possess this facility and WRC are hoping in the future to extend WASSP-SIM and the Rehabilitation Manual to include this capability and also to consider the impact of sewer discharges on receiving waters.

7. VERIFICATION OF HYDRAULIC MODELS

Being able to construct a hydraulic model of a sewer system which produces predictions of flow which are both detailed and credible is a major step down the road to solving the problems of that system. However, it is not enough just to produce results which are believable, to enable really cost-effective solutions to be proposed, they must also be accurate. This is where verification of the model plays a major role.

Given perfect knowledge of a sewer system and its contributing catchment and the absence of features which defy rigorous mathematical representation, a model could be produced which would give predictions in which an engineer could have high confidence. However, WRC experience to date suggests that even for sewer systems which are considered to be well documented, application of the detailed procedures of the Manual throw-up sufficient queries and reveal enough inaccuracies that no engineer could have any real confidence in an unverified model.
A method by which data to verify a model can be obtained quickly is the form of flow survey described in The Rehabilitation Manual.
The cost of obtaining data in this way is not cheap (nor is the task of constructing a simulation model) £25,000 is typical for a medium/large network, but this sum must be compared with the potential savings to be derived from, say, a 40 - 50% reduction in cost of a multi-million pound upgrading scheme.

The principal purposes for carrying out such a survey can be itemised as follows:

(i) To confirm the assumptions made during construction of the model regarding the connection of contributing areas in the catchment.

(ii) To confirm the layout of the sewer system as derived from records and limited field inspection.

(iii) To observe the performance of ancillary structures where this is not apparent from their basic configuration.

Other areas in which flow surveys can provide useful data are:

(iv) Establishment of levels of base flow due to foul sewage inputs in combined sewers, infiltration, stream inflows etc.

(v) Establishment of a guide to sediment patterns in the system.

WRC have carried out upwards of 40 flow surveys in the UK over the past three years and in many cases have also constructed or been associated with the construction of simulation models of those sewer systems. In the vast majority of cases the verification process has revealed significant errors in the initial model and worthwhile savings have resulted from the improved model.

A recently published WRC Report describes in a comprehensive manner the advantages which have been derived from flow surveys carried out to-date by WRC(10). Some examples follow.

7.1 Assignment of Contributing Areas

Fig 14 illustrates an example of areas in a catchment being attributed to the wrong part of the sewer system at Oswaldtwistle. The wide discrepancy between modelled and measured flows caused field investigations to be undertaken which confirmed the error. Correction in the model produced the hydrographs of Fig 15 which shows much improved fit.

Another instance of incorrect areas was witnessed in the Tang Hall sub-catchment of York. The York catchment has been described in section 5.1. Fig 16, again shows the measured hydrograph but here the sewer records indicated that Tang Hall was a separate storm and foul system and that only foul flow was draining to the monitor. The evidence of a storm water component in the Tang Hall foul sewer is clear from the recorded hydrograph. An assessment of the impermeable area in this sub-catchment was made and incorporated in the model as indicated in Fig 16, giving improved simulation. Agreement between measured and modelled flow at locations lower down the network was also markedly improved.

7.2 Assessment of Contributing Areas

Fig 17 shows the comparison of measured and modelled flow hydrographs for a 1200 mm diameter pipe in Littlehampton. The difference between the two hydrographs, apart from a small rainfall timing error, shows an overall wide discrepancy. The initial approach to obtaining impermeable areas had been to make a rough estimate for contributory areas but, once the hydrograph discrepancy had been flagged up, a more realistic approach was adopted. Fig 18 shows the hydrographs following the re-assessment of contributory areas and much better agreement is achieved. The following points have also assisted in improving the fit:

(1) The DWF for the start of storm was initially over-estimated.

(2) Observations of silt in the pipe during the flow survey allowed a more realistic roughness (ks) value to be employed[7]. In Fig 17, a global roughness ks = 1.5 mm, was used whilst in Fig 18 a value for ks of 15 mm was used in the sewers where silt was present. The effect is noticed by the attenuation of the modelled peak flows.

7.3 Performance of Ancillary Structures

The most common type of structure that has to be modelled is an overflow. An example of how flow measurements assist in assessing the performance of an overflow is seen in Fig 19 and Fig 20. The example is taken from a flow survey in Glastonbury where bricks had been removed from the wall of the manhole allowing storm water to pass into an adjacent pipe. Although hydraulic modelling has yet to take place, the flow measurements revealed that the continuation flow, (the monitor was located in the downstream pipe) is 50 to 60 l/s where the hydrograph flattens instead of peaking (Fig 19) and that the monitor in the overflow some 300 m downstream records the peaks (Fig 20). Hence the performance of this non-standard overflow can be assessed.

8. CONCLUSION: USING A VERIFIED FLOW SIMULATION MODEL

The basic reason for constructing a sophisticated model of an existing sewer system is to enable the most cost-effective upgrading solution to be found. Traditionally hydraulic upgrading has been achieved by replacement or reinforcement methods, i.e construction of a new sewer in place of, or supplementary to, the existing conduit. These solutions are effective but costly and have been favoured simply because they have hitherto been the only answers in which engineers have felt truly confident, given the tools at their disposal. The advent of comprehensive simulation models which can be effectively verified gives the engineer a tool with which he can investigate other forms of solution with equal confidence.

A variety of methods which attenuate or reduce flows in a sewer system are described in the Rehabilitation Manual. The potential for use of these methods may vary with the practical circumstances and the nature of a particular network, but one or more will generally offer a viable alternative to the traditional solution of a new pipe. In certain circumstances a replacement or reinforcement solution may still prove to be the preferred solution, but the preference should be on the grounds of hard economic analysis. Renovation techniques ensure that even structurally deteriorated sewers can have acceptably long service lives, whilst the hydraulic model used in conjunction with sensible performance criteria can ensure that the finished job will provide adequate standards of service.

Case studies illustrating the use of alternative forms of upgrading are included in the Rehabilitation Manual and will be supplemented as further examples evolve, through the implementation programme for the Manual.

9. REFERENCES

1. **"WRC/WAA Sewerage Rehabilitation Manual"**
   WRC Engineering 1984.

2. **FIDDES D. "A Sewer Rehabilitation Strategy for the United Kingdom".**
   BHRA Conference Paper 1984.

3. **WILLIAMS D W and BARTLETT J M.** "Application of Hydraulic Analysis Techniques to Investigate an Overloaded Sewerage System".
   3rd International Conference on Urban Storm Drainage. Goteborg, Sweden, June 1984.

4. **"Design and Analysis of Urban Storm Drainage - The Wallingford Procedure".**
   NWC/DoE Standing Technical Committee Report No 28
   NWC, Sept 1981.

5. **"Sewers and Water Mains - A National Assessment".**
   NWC/DoE Standing Technical Committee Report No 4
   NWC, June 1977.

6. **BALMER R and MEERS K. "Surveying Severn-Trent's Sewers"**
   ICE Conference Paper No 14 in Restoration of Sewerage Systems
   Thomas Telford Ltd, 1982.

7. **HENDERSON R J. "A Guide to Hydraulic Roughness in Sewers"**
   WRC External Report No 131E
   WRC Engineering, March 1984.

8. **TAYLOR A J and CLIFFORDE I T. "Hydraulic Analysis of the Sewerage System of Torquay"**
   WRC External Report No 124E.
   WRC Engineering, December 1983.

9. **HUBER W C et al. "Storm Water Management Model User's Manual"**
   US Environmental Protection Agency. 1980.

10. **WILLIAMS D W "Hydraulic Analyses - The Role of Sewer Flow Surveys".**
    WRC External Report to be published shortly
    WRC Engineering

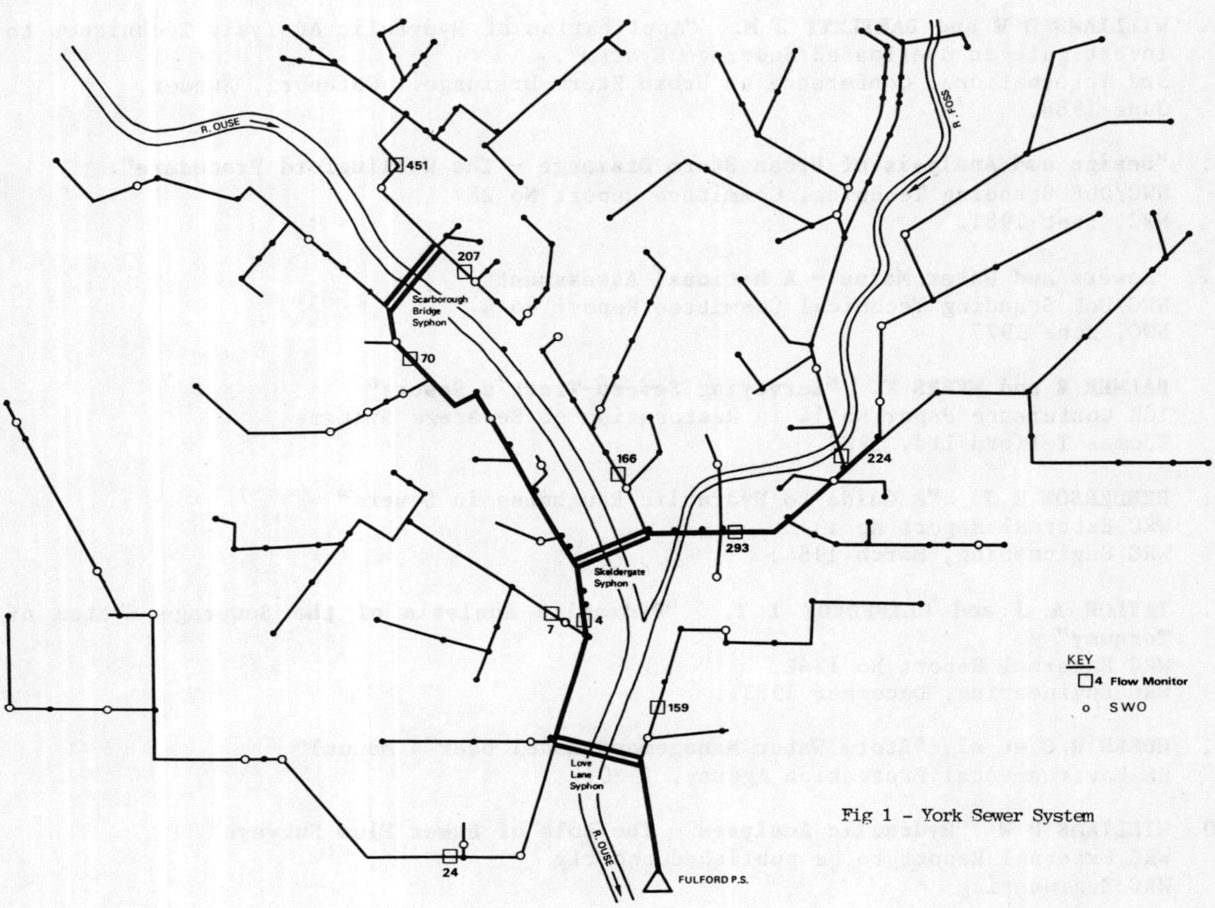

Fig 1 - York Sewer System

Fig 1 - York Sewer System

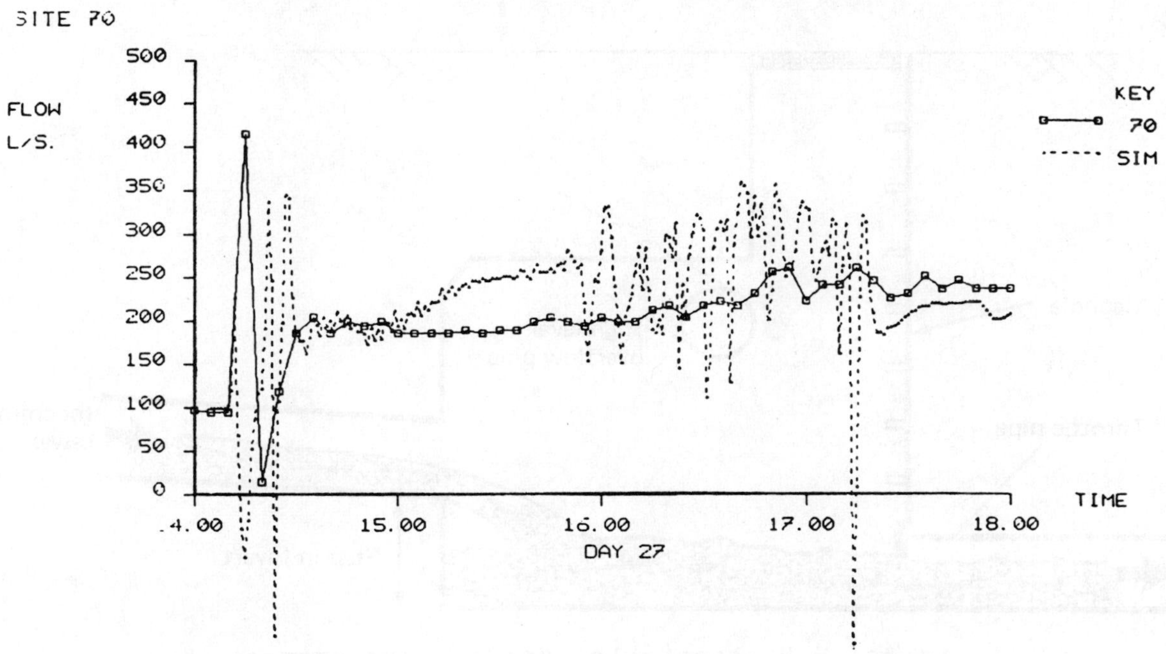

Fig 2 - York Hydrograph - showing oscillations due to model instability

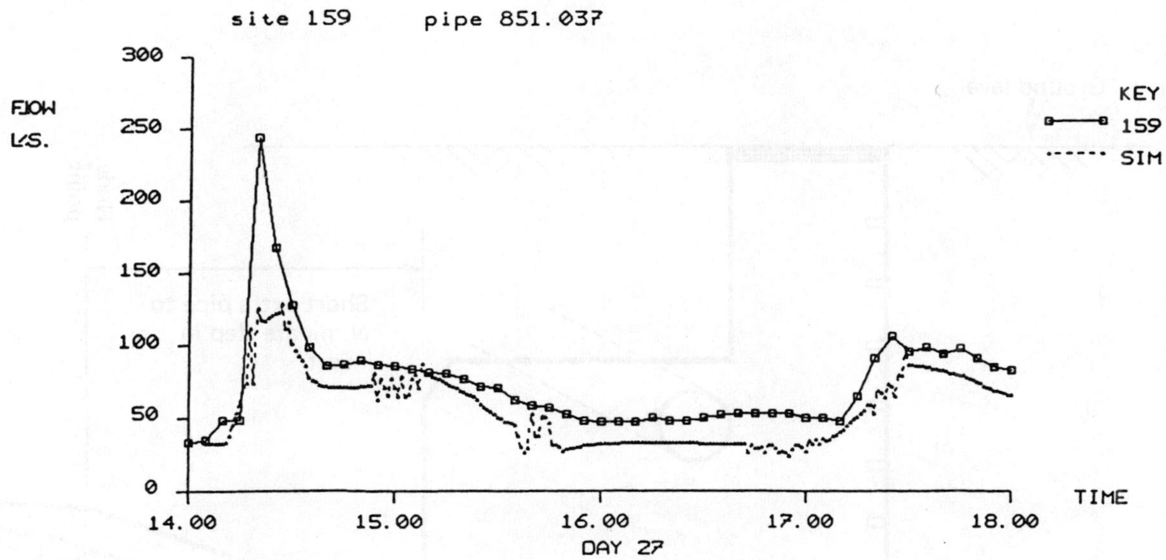

Fig 3 - York Hydrograph - showing "chopped" peak due to poor upstream overflow representation

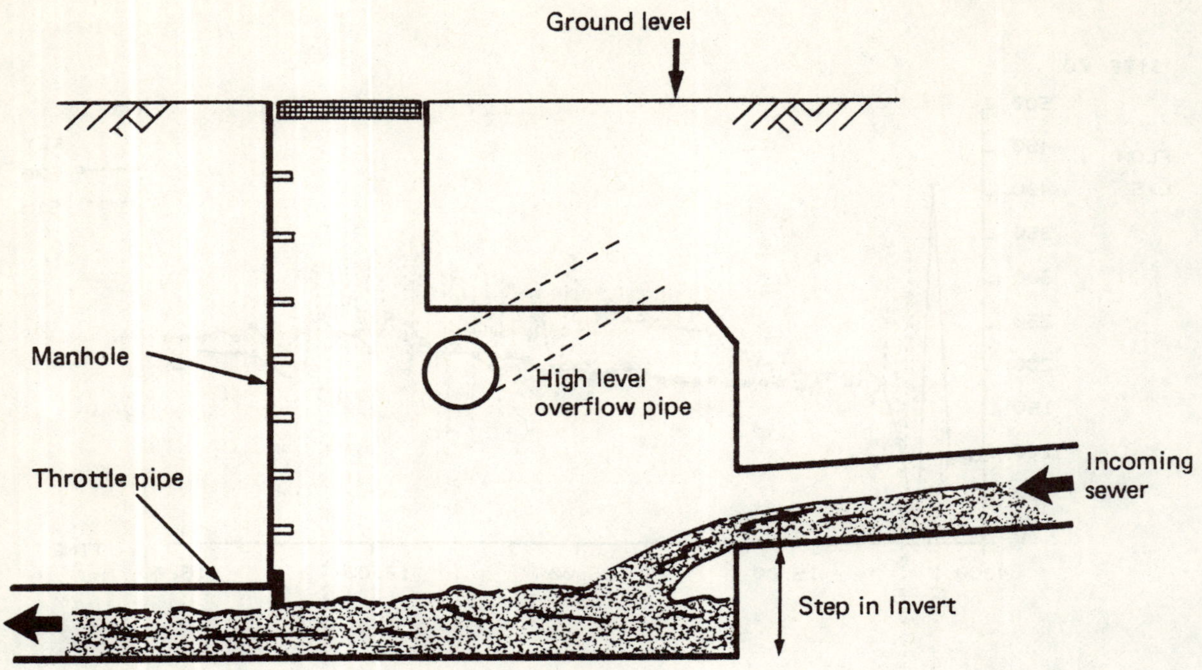

Fig 4a - Unacceptable overflow arrangement

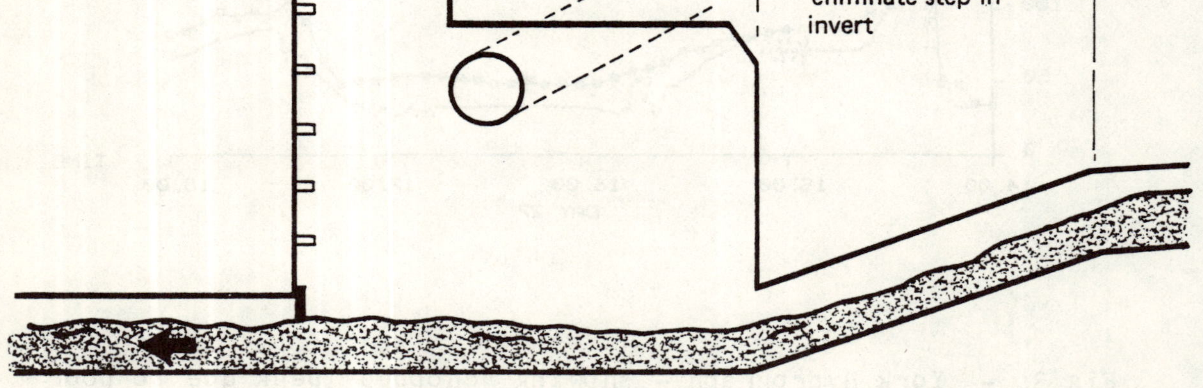

Fig 4b - Modified overflow arrangement

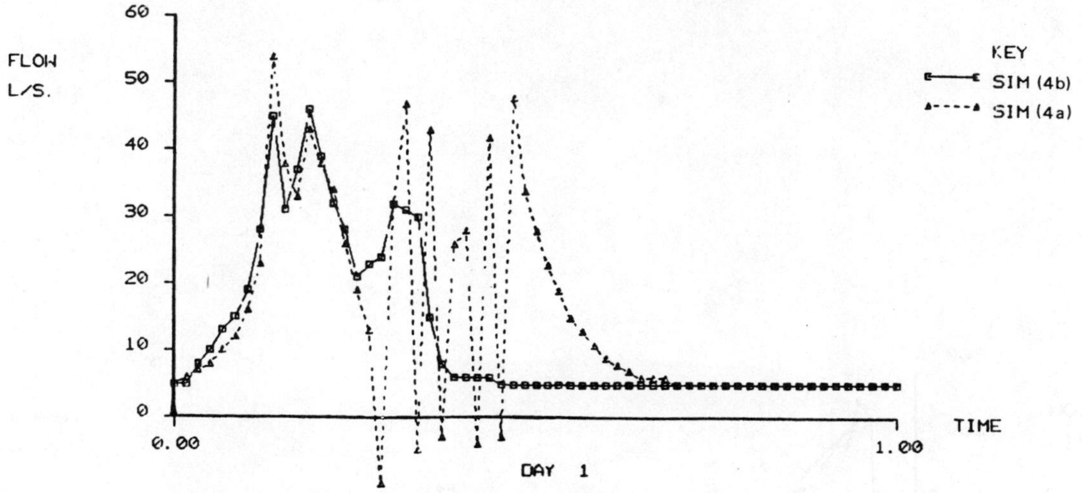

Fig 5 - York Hydrograph - contrasting simulations for overflows of configurations in Fig 4

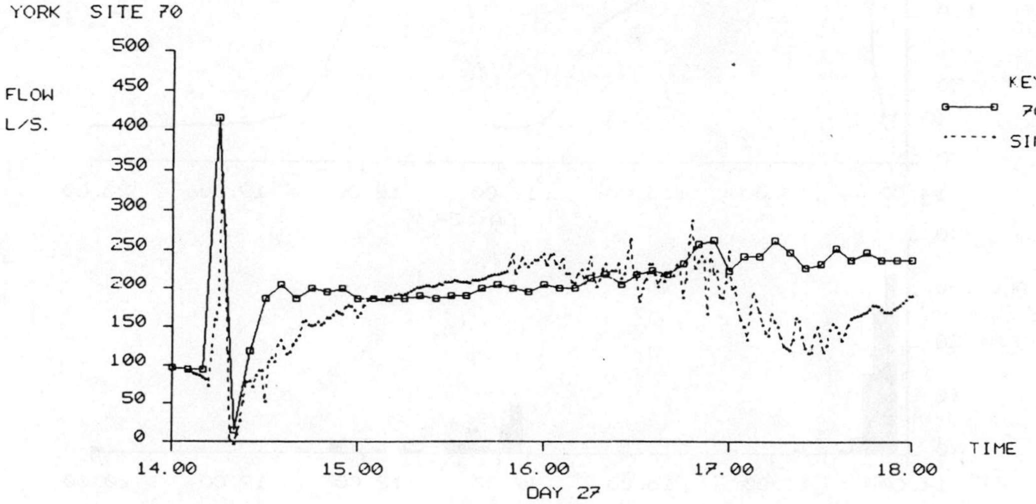

Fig 6 - York Hydrograph - good fit

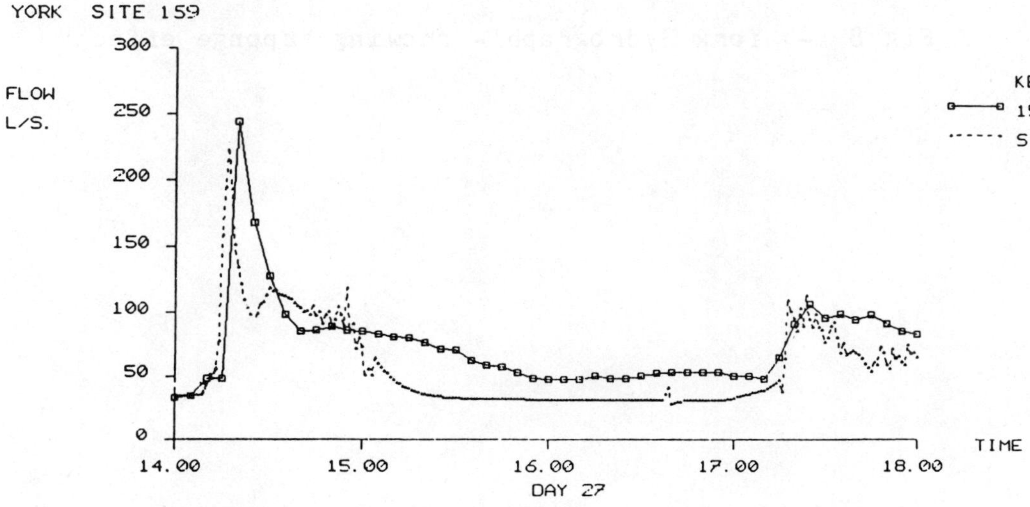

Fig 7 - York Hydrograph - good fit

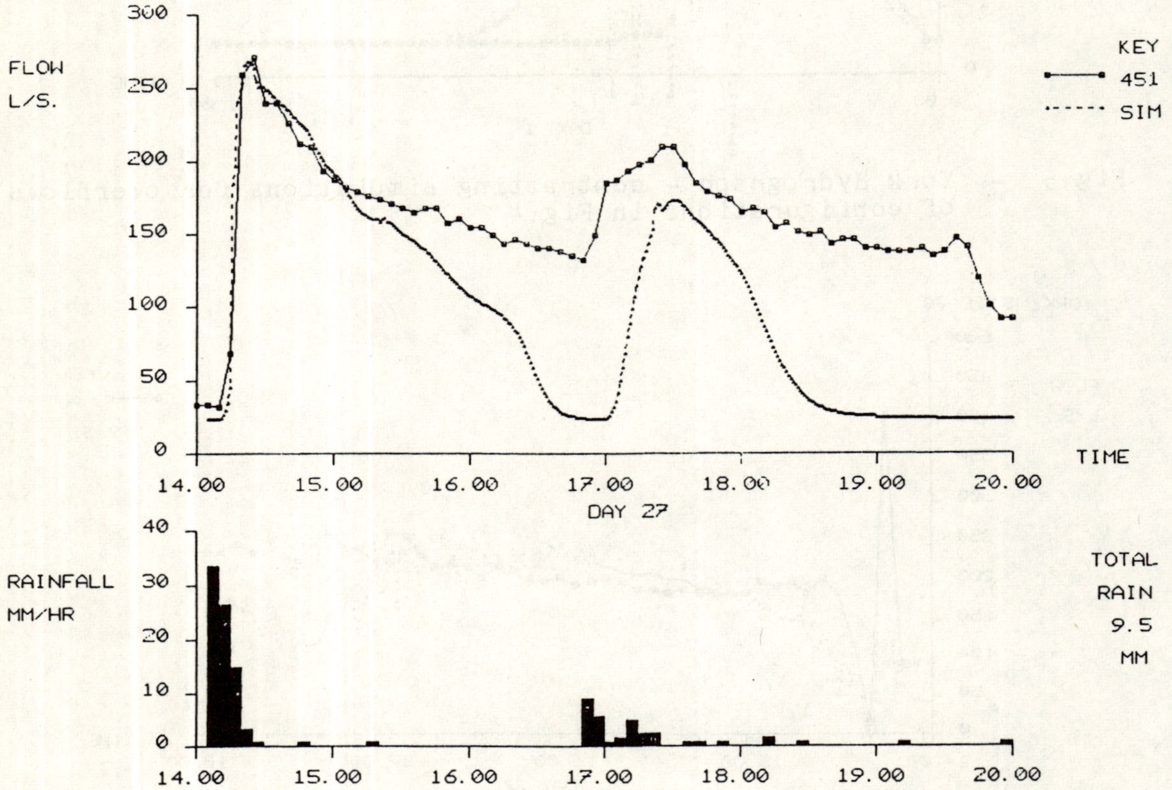

Fig 8 - York Hydrograph - showing "sponge effect"

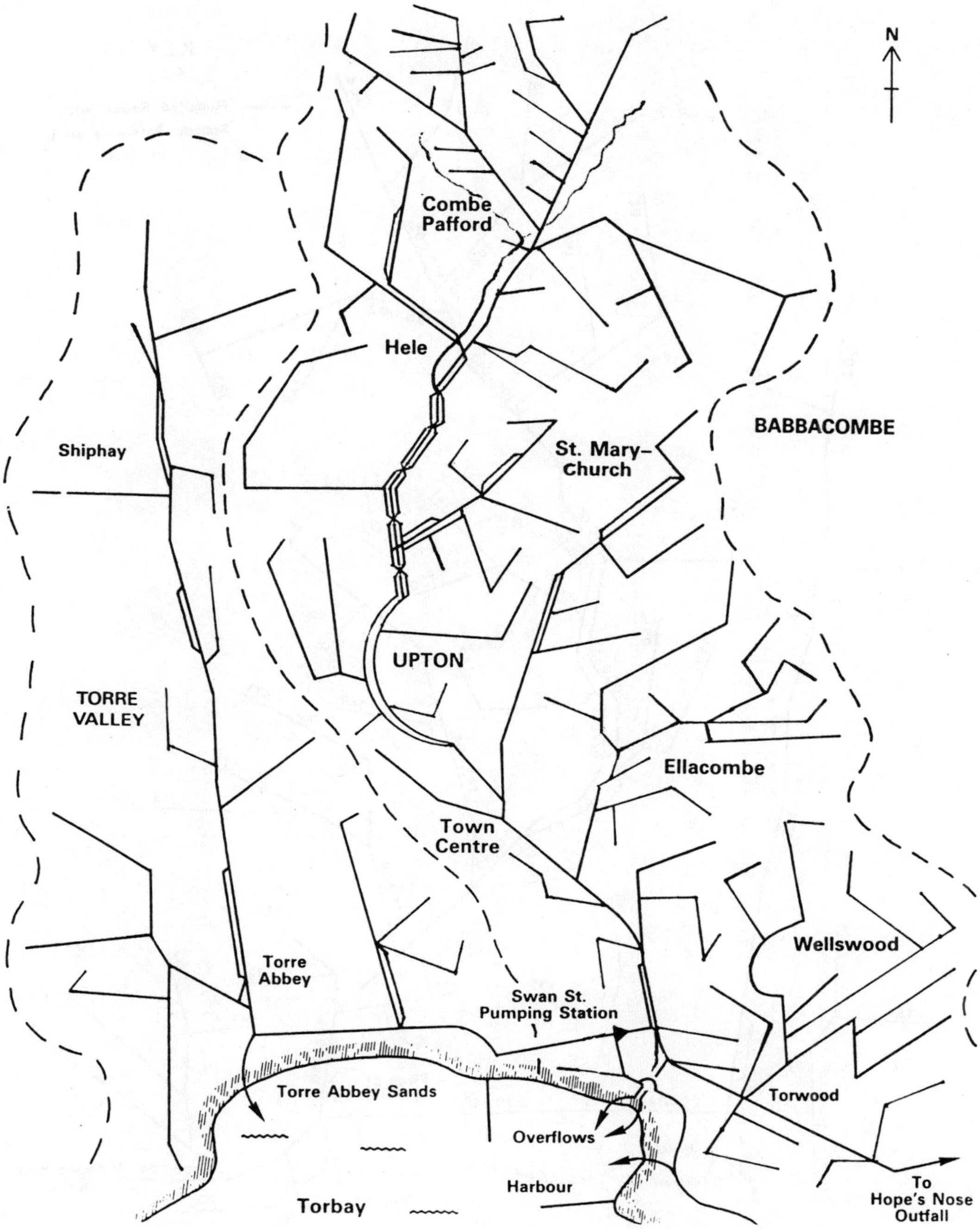

Fig 9 - Torquay Sewer System

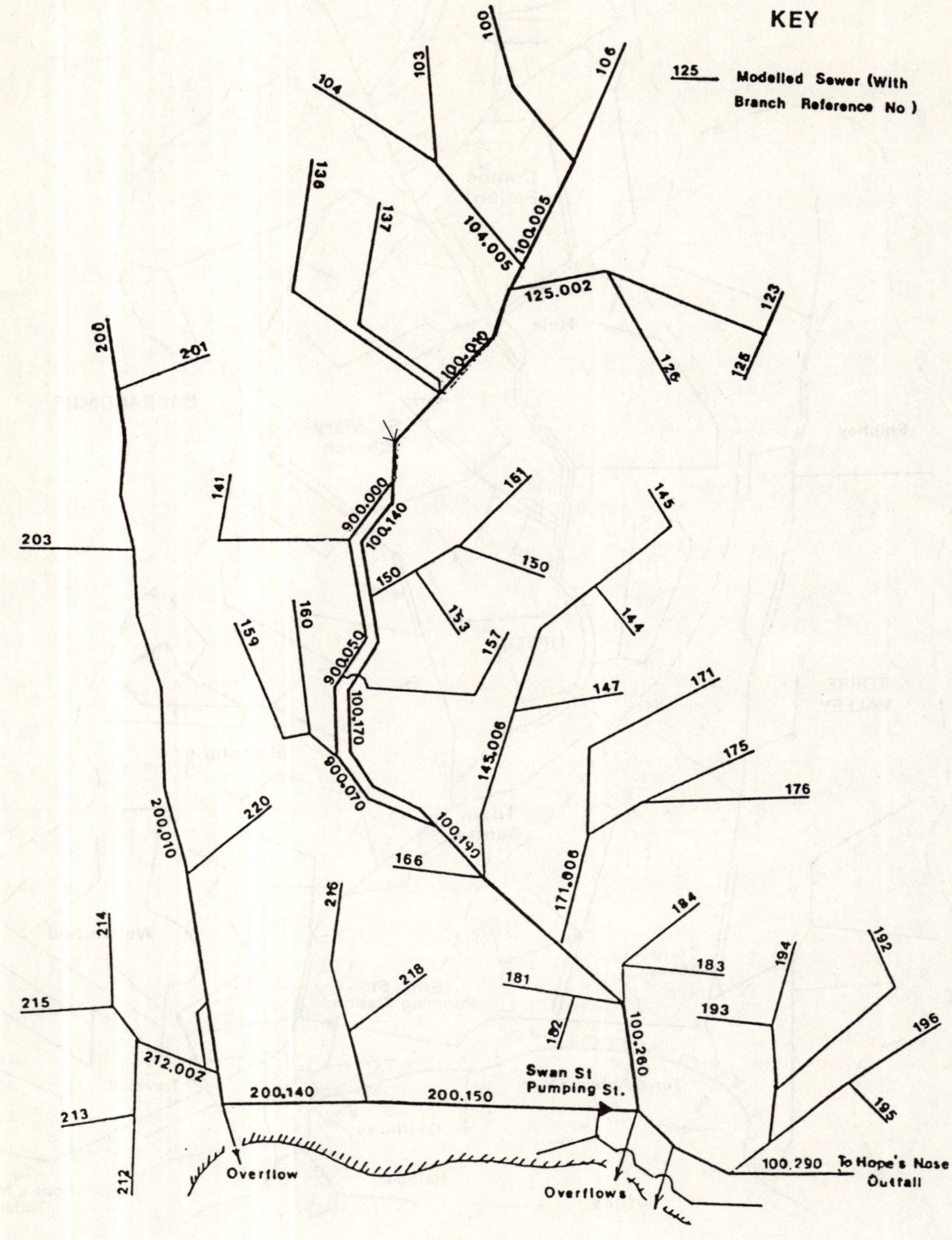

Fig 10 - Torquay Model

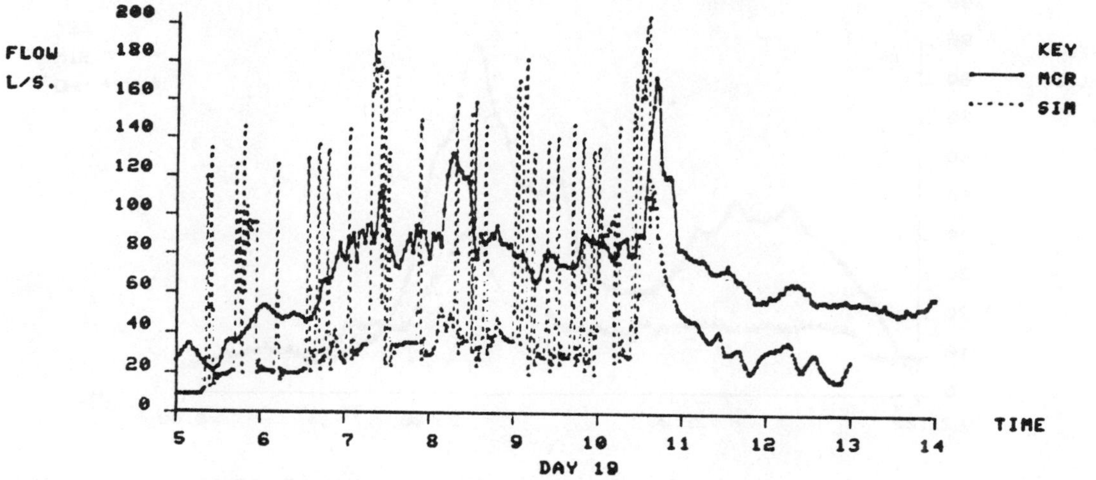

Fig 11 - Torquay Hydrograph - initial run showing oscillations

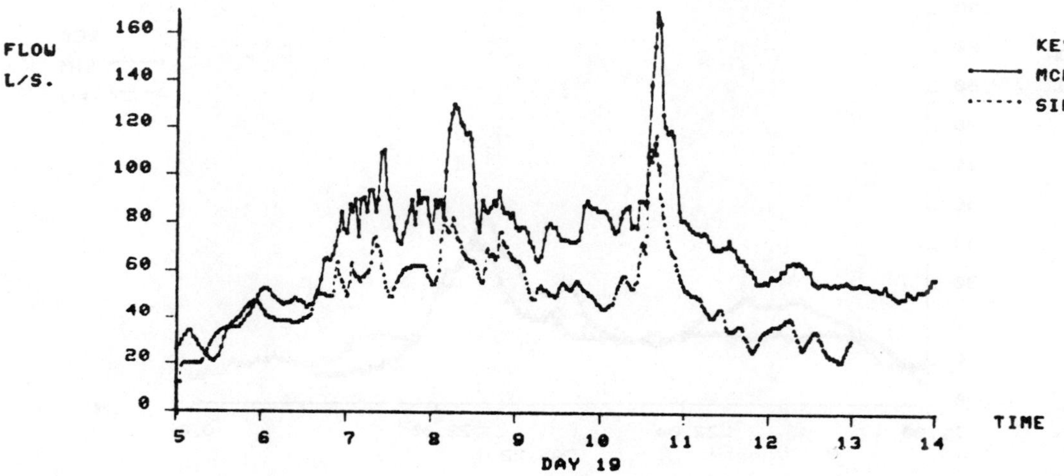

Fig 12 - Torquay Hydrograph - oscillations eliminated but poor fit

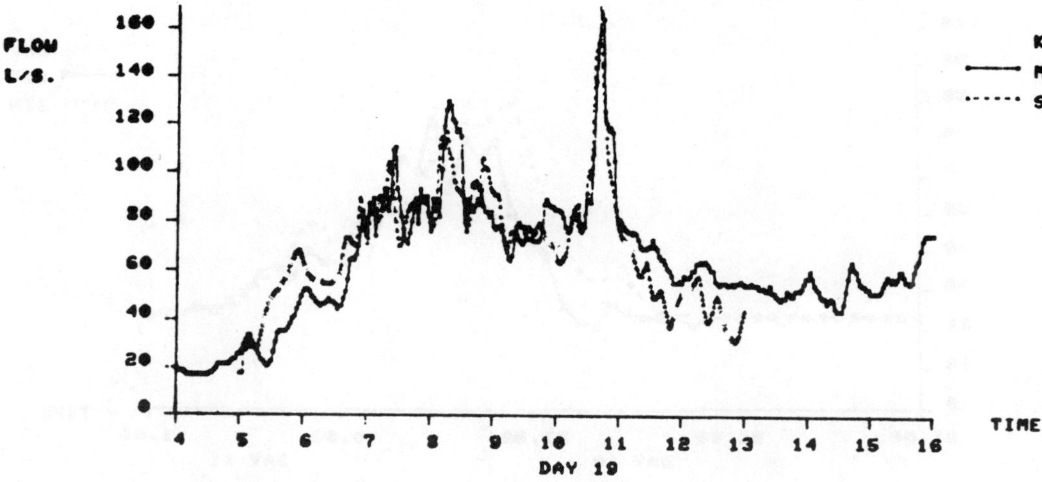

Fig 13 - Torquay Hydrograph - improved fit after re-assignment of contributing areas

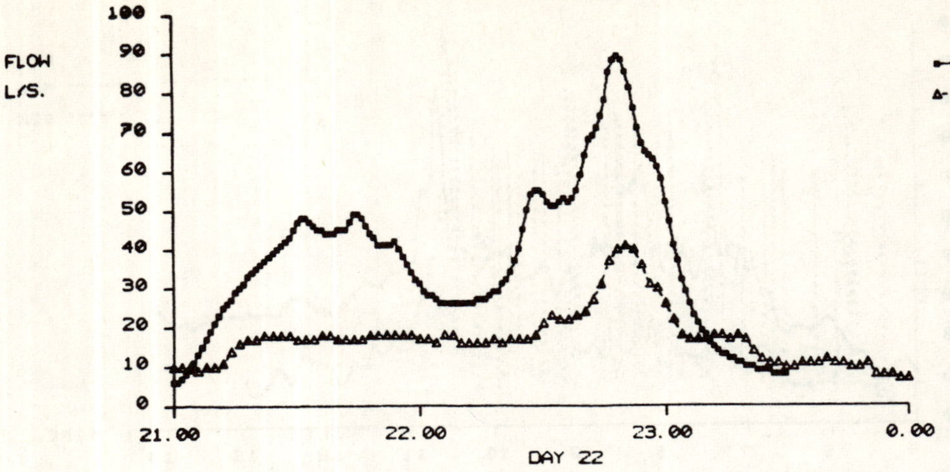

Fig 14 - Oswaldtwistle Hydrograph - showing gross discrepancy due to wrongly assigned areas

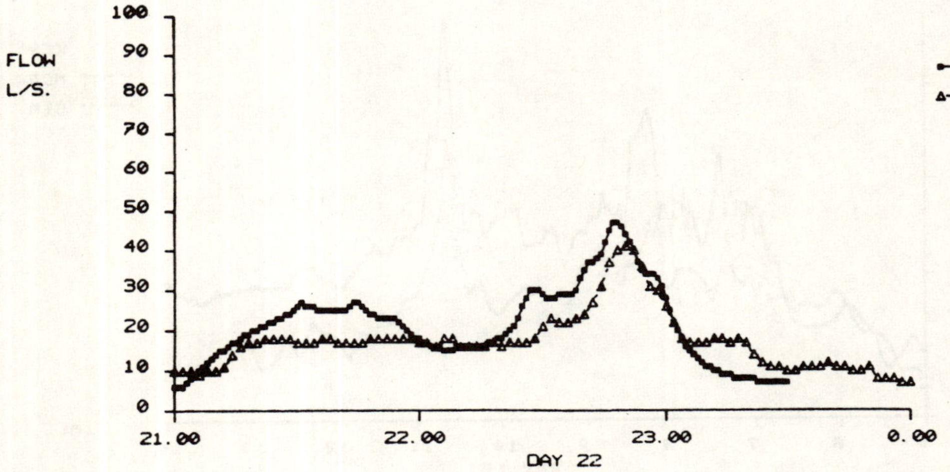

Fig 15 - Oswaldtwistle Hydrograph - improved fit after reassignment of areas

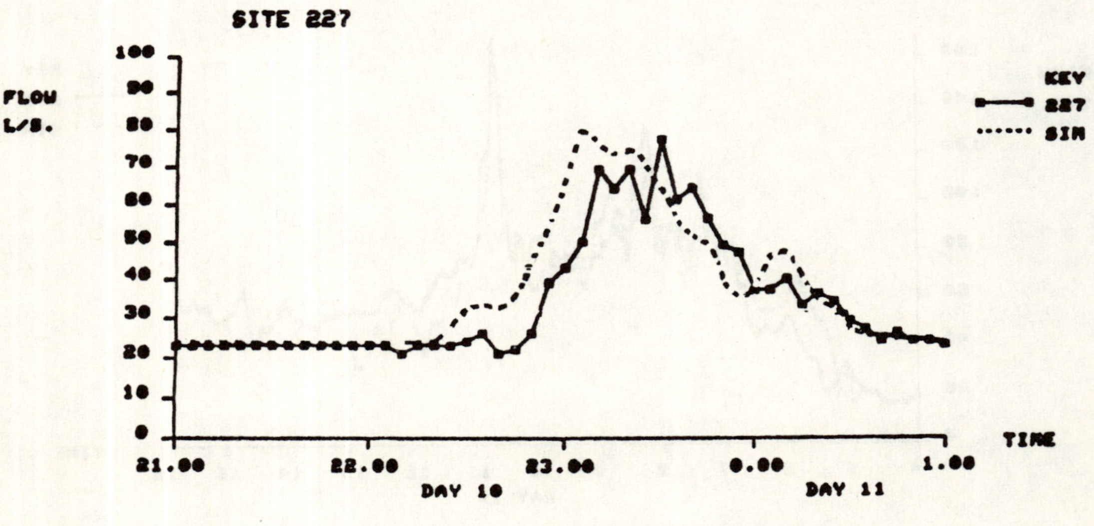

Fig 16 - Tang Hall Hydrograph - showing recorded and final model response to rainfall in a nominally "separate" area

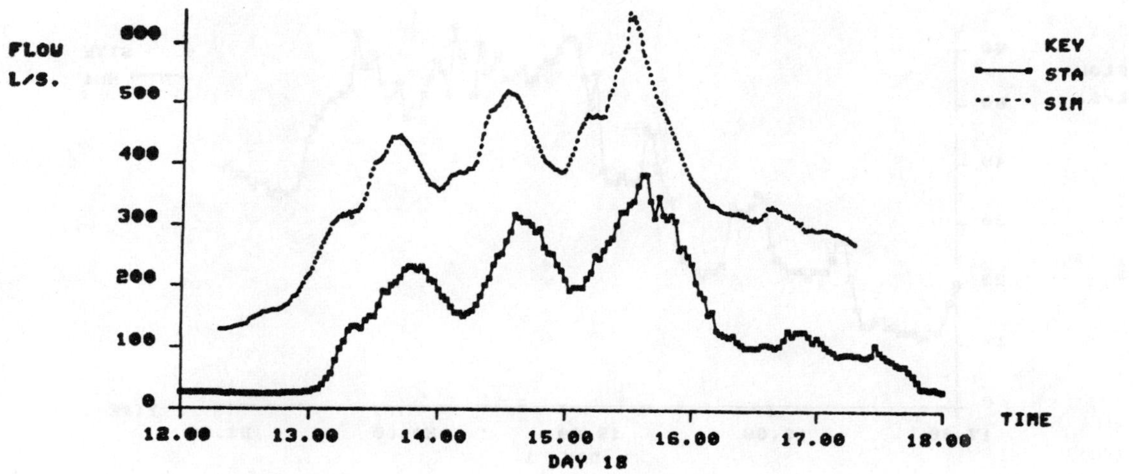

Fig 17 - Littlehampton Hydrograph - showing poor initial fit

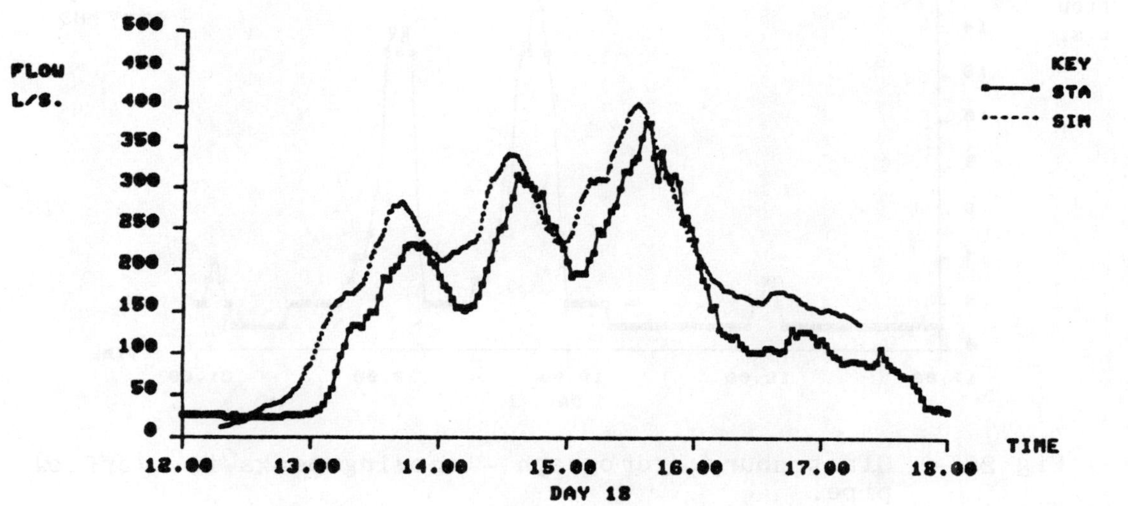

Fig 18 - Littlehampton Hydrograph - showing improved fit after reassessment of areas

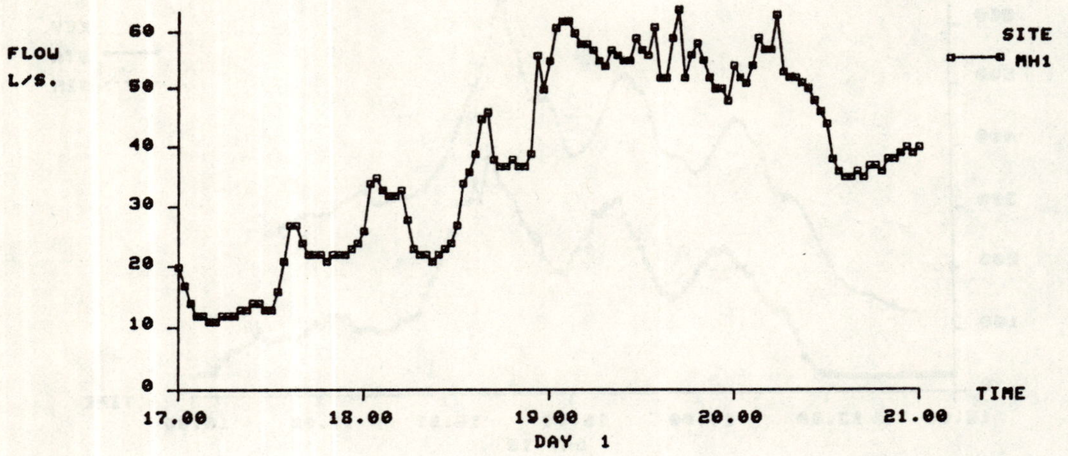

Fig 19 - Glastonbury Hydrograph - showing "chopped" peak

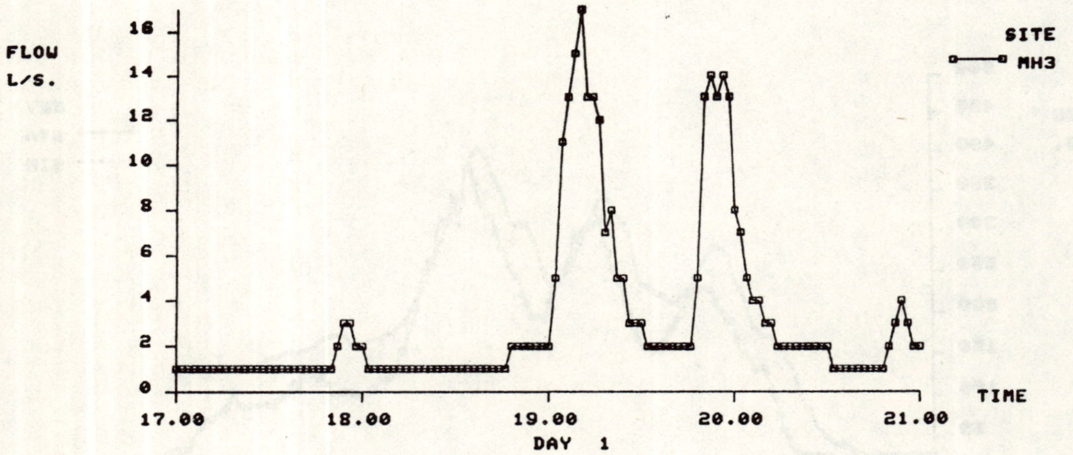

Fig 20 - Glastonbury Hydrograph - showing peaks in overflow pipe.

International Conference on the
PLANNING, CONSTRUCTION, MAINTENANCE & OPERATION OF SEWERAGE SYSTEMS
12-14 September, 1984

SEDIMENT TRANSPORT IN SEWERS AND THE DESIGN IMPLICATIONS

P. Ackers
C. Eng., M. Sc(Eng), FCGI, FICE, MIWES, MASCE

Binnie & Partners
Artillery House, Artillery Row,
London, SW1P 1RX, England

SUMMARY

Self-cleaning criteria for sediment carrying sewers and storm drains are not well established. Some recent research indicates that much higher velocities (and therefore gradients) may be required in large conduits than are implied by conventional approaches.

One of the difficulties is that the extrapolation of laboratory information is uncertain, especially as there is insufficient field scale information to support the theoretical concepts forming the basis of extrapolation.

The present state of knowledge is reviewed within alternative theories to indicate both the band of uncertainty and the problems of establishing a balanced engineering criterion.

Held at The University of Reading, England. Organised and sponsored by BHRA, The Fluid Engineering Centre, Cranfield, Bedford in conjunction with the Water Research Centre.

NOMENCLATURE

A	=	area of flow section
A_c	=	initial motion condition (in A-W equation)
C	=	empirical coefficient (in A-W equation)
C_v	=	volumetric concentration
d	=	sediment diameter
D	=	pipe diameter
F_{gr}	=	mobility parameter in A-W equation
g	=	gravitational acceleration
G_{gr}	=	transport parameter in A-W equation
k_s	=	equivalent sand roughness
m	=	empirical exponent
n	=	fine/coarse transition "parameter"
n_m	=	Manning coefficient
P	=	wetted perimeter of flow section
Q	=	volumetric discharge of fluid
R	=	hydraulic mean depth (A/P)
s	=	relative density of sediment
S	=	hydraulic gradient, or invert gradient for uniform free-surface flow
v_*	=	shear velocity, (gRS)
V	=	average velocity (Q/A)
V_o	=	effective threshold velocity for $C_v = 0$, usually $= V_{tp}$
V_s	=	limiting velocity for deposition
V_{tp}	=	threshold velocity
w_e	=	effective width for sediment transport
X	=	sediment mass flux as ratio of fluid mass flux (equivalent to a concentration by weight)
y	=	flow depth when A-W equation is applied to a rectangular channel
$\emptyset$	=	transport parameter used by Novak and Nalluri
λ	=	friction factor, $8g\,RS/V^2$
λ_o	=	λ for the pipe without sediment transport
ν	=	kinematic viscosity of the fluid
ψ	=	inverse mobility parameter used by Novak and Nalluri

1. INTRODUCTION

Although information on the concentration and grading of sediment in sewers and storm drains is sparse, it is well known that silt, sand and coarser solids gain access to them in significant quantities. There are many instances where drainage systems require expensive maintenance to remove accumulating sediment deposits, as well as many cases where the inadequacy of combined systems in conveying flood flows can be attributed at least in part to loss of capacity through sediment deposition.

Various criteria have been proposed for self-cleaning conditions, most usually expressed as a limiting velocity (Ref.1) but sometimes as a limiting shear stress related to a particle size (Ref.2). The old "Maguire rule" for laying a 12 inch pipe at 1 in 120, a 24 inch at 1 in 240 etc. was a shear stress criterion, as average shear stress is proportional to the product of slope and hydraulic mean depth.

In the early 1950's, an important programme of research was undertaken at the University of Iowa. This was summarised by Laursen (Ref. 3) and included work by Craven and Ambrose. Craven (Ref.4) examined transport under full-bore conditions, using 50 mm and 140 mm dia.pipes and uniform sands in the range 0.25 to 1.62 mm. He considered the maximum transport rate with a given hydraulic gradient with no deposition in the pipe; and also the transport function with a mobile sediment bed.

Ambrose (Ref.5) extended the research to cover free-surface flow with a sediment bed, but discovered that this mode of operation is essentially unstable: he had to generate a uniform sediment deposit under full-bore conditions, the performance then being examined under free-surface conditions.

Laursen's summary (Ref.3) identified several features that are crucial to the problem. He recognised that for full-bore flow the joint application of a friction equation and sediment transport equation defined both the discharge capacity and the transport capacity. One form of sediment function expressed the limiting condition of maximum transport without deposition, and hence if the sediment supply was less than this the pipeline would operate without a sediment bed, whilst if the supply was greater deposition would occur. A different function was derived to describe transport in the latter case. Thus, if a given pipe is supplied with known water and sediment fluxes, then with full-bore flow the hydraulics are defined: only one solution of the resistance and transport functions exists, yielding the necessary depth of deposit (which might in some cases be zero) and hydraulic gradient.

The Iowa researchers also recognised that free-surface flow with constant discharge of water and sediment and with a sediment bed was inherently unstable. If a clean pipe operating at a prescribed gradient and flow rate with a free-surface was supplied with sediment exceeding its no-deposit capacity, sedimentation would begin to occur. This would increase the roughness (especially if dunes occurred) and reduce the velocity and the transport capacity. This would accelerate deposition (consistent with the surplus rate of supply) until finally the pipeline would run full-bore. If it could generate an increased hydraulic gradient through surcharge, the gradient and depth of deposition might then achieve values simultaneously satisfying both the resistance and transport equations.

Important research on particle movement has been undertaken more recently at the University of Newcastle-upon-Tyne by Novak, Nalluri and Ojo (Refs 6, 7). The Newcastle programme has covered the initial motion condition of discrete particles on a smooth bed; of patterns of grouped particles; and of sediment transport at the limit of no-deposit. This programme of work led to a review of sewer design in relation sediment deposition (Ref.8).

May (Ref.9) reviewed the subject in 1975, but found that different authorities had produced a range of criteria that left considerable uncertainty over their application and relevance to practical situations. Meanwhile, the Hydraulics Research Station was conducting a major series of laboratory tests, with reference also to field observations, research which was reported in 1982 (Ref.10). The laboratory tests were carried out in 77 mm and 158 mm dia. pipes, using two coarse sands of about 0.6 mm median diameter and two gravels of 6 and 8 mm diameter.

In 1978, the writer (Ref. 11) combined the Ackers-White sediment transport formula (Ref.12) with the Colebrook-White resistance formula (Ref.13) to study the relationship between velocity, hydraulic gradient, pipe diameter, pipe roughness, sediment size and concentration, the roughness of the sediment bed and depth of flow. The method produced results consistent with Laursen's conclusions and was more conservative for large diameter conduits than some engineering criteria. However, no field data for large pipes were available as confirmation.

The present paper examines that basic approach in comparison with research results published since then, with some tentative conclusions that may be of importance in terms of design criteria.

2. AVAILABLE THEORY

There is a dichotomy between the sediment transport theories and concepts applied to open channels and those applied to pipes. In part, this has arisen because pipeline transport in the process industries usually involves very high concentrations without a bed of sediment whereas channel transport usually involves lower concentrations with a bed formed of the sediment in transport. Conditions in sewers fall between and hence research in neither the main areas is directly applicable: in sewers and storm drains concentrations are relatively low but usually a deposited bed is to be avoided.

2.1 May's method

In Ref.10, May reviewed previous research and theory on pipeline transport and developed several new concepts, as well as retaining some of those well established in the literature. For example, following Novak and Nalluri (Ref. 8) he utilised a threshold velocity equation for zero transport in a smooth pipe, non-dimensionalised by incorporating gd(S-1):

$$V_{tp} = 0.61\sqrt{g\,d\,(s-1)}\left(\frac{d}{R}\right)^{-0.27} \qquad (1)$$

May's theory of pipe-line transport with little or no deposition is based on the hydromechanics of particle movement and dimensional analysis. Certain assumptions and empirical coefficients were inevitable but were evaluated from his own detailed research. The final equation relating the limiting velocity for deposition, V_s to the transport rate expressed as a volumetric concentrating, C_v,

$$C_v = 0.0205\left(\frac{D}{A}\right)^2\left(\frac{d}{R}\right)^{0.6}\left\{\frac{V_s^2}{g\,(s-1)\,D}\right\}^{3/2}\left\{1-\frac{V_o}{V_s}\right\}^{-4} \qquad (2)$$

In this, the threshold velocity for zero transport, V_o, is evaluated from equ. 1.

Functions were also given (differing for the sand sizes and gravels) for the increase of resistance over the smooth-pipe friction law (which applied to the laboratory pipes), when C_v was non-zero.

Given the sediment characteristics and volumetric concentration, together with pipe size and flow depth, equations 1 and 2 are solved (setting $V_o = V_{tp}$) with V_s as the unknown. Having evaluated V_s, the discharge, Q, is obtained for the given cross-section geometry and a friction equation is applied (e.g. Manning on Colebrook-White) to determine the requisite gradient. (It was not made clear in Ref. 10 whether the self-cleaning velocity and transport function were expected to be independent of pipe roughness, but in private discussion this was confirmed, based on the similarly of the velocity profile close to the pipe wall for smooth and rough turbulent conditions).

2.2 Novak and Nalluri method

In addition to equation 1, the "Newcastle" method of Ref.8 uses their experimentally determined transport function for no deposition:

$$\phi = 11.6\,\psi^{-2} \tag{3}$$

where

$$\phi = \frac{C_v V_s R}{\sqrt{(s-1)\,gd^3}} \tag{4}$$

and

$$\psi = \frac{(s-1)d}{SR} \tag{5}$$

The method of solution is similar to that indicated for May's theory but as the equations are rather simpler in form, Novak and Nalluri explained how, combining with the Manning equation, a direct solution was obtainable for the no-deposit case:

$$S = (s-1)\left(\frac{d}{R}\right)^{1/3}\left\{\frac{C_v R^{1/6}}{11.6 n_m \sqrt{g}}\right\}^{2/3} \tag{6}$$

It can also be shown from the Novak and Nalluri transport function that

$$V_s = \left\{\frac{(s-1)^{3/2} C_v R^{5/3} d^{1/2}}{11.6\, n_m^4\, g^{1/2}}\right\}^{1/3} \tag{7}$$

Hence in this method the limiting velocity is not independent of roughness: for given sediment and flow geometry, it decreases with increasing roughness ($V_s \propto n_m^{-4/3}$) as does S ($S \propto n_m^{-2/3}$). This is different from May's conclusion and perhaps contrary to expectation, although it is a plausible result of the relationship of turbulence to roughness.

Fig. 1 is a plot of Q, S and D based on n_m = 0.010 and 0.015 for pipes which are half-full showing:

- the 'code of Practice' self-cleansing condition, V = 0.76 m/s
- May's velocity criterion
- Novak and Nalluri's formulation.

The sediment is assumed to be a sand of 0.4 mm dia, s = 2.65 at $C_v = 100 \times 10^{-6}$. The range of pipe size is chosen to extend from the laboratory scale (as well as small sewers) through to the largest size of interest in order to illustrate how problematic the requirements for large pipes are.

2.3 Author's 1978 method

As indicated earlier, the Author proposed an approach which utilised the Ackers-White sediment transport formula (devised for wide open channels with a sediment bed) and the Colebrook-White friction equation (covering smooth, transitional and rough-turbulent conditions with roughness coefficient, k_s). These equations are complex but were based on sound physical principles, but several crucial (and at that time insecure) assumptions had to be made when applying them to sediment in pipes. The most significant of these was that the full width of the flow cross-section, i.e. the pipe diameter for pipes half-full or over, was effective in generating particle movement, as would be the case in a rectangular channel with a sediment bed over its full width.

For the present study, the Ackers-White formula has been generalised to cover channels with any shape of cross-section. The basic function relates sediment transport (G_{gr} in non-dimensional form) to the excess of a mobility parameter, F_{gr}, over its value at initial motion, A_c. The mobility number takes account of the velocity (most relevant to fine material) in relation to sediment size and density.

The basic formulation for a rectangular channel is as follows:-

$$G_{gr} = C\left\{\frac{F_{gr}}{A_c} - 1\right\}^m \qquad (8)$$

where
$$G_{gr} = \frac{Xy}{sd}\left(\frac{v_*}{V}\right)^n \qquad (9)$$

$$F_{gr} = \frac{V^{1-n} \quad v_*^{\,n}}{\sqrt{g(s-1)\,d}\,\left\{\sqrt{32}\,\log\left(\frac{10y}{d}\right)\right\}^{1-n}} \qquad (10)$$

A_c, C, m and n are all functions of particle size, expressed as

$$A_c\ C,\ m,\ n = \text{functions } (D_{gr}) \qquad (11)$$

where

$$D_{gr} = \left\{\frac{g\,(s-1)}{\nu^2}\right\}^{1/3} d \qquad (12)$$

Fuller details are given in Appendix 1, including its modification for non-rectangular sections and pipes.

The Ackers-White sediment transport function was established on the basis of laboratory scale tests in flumes but has been confirmed at field scale when extrapolated to large rivers where the ratio of sediment size to flow depth may be two or three orders of magnitude different from in the laboratory. Its use as part of an overall theory for transport in pipes may reduce the doubts about extrapolation from laboratory to field scale, but if it does not fit results at laboratory scale where sediment is transported in pipes, it should not be applied for any scale of piped system of course. It seemed appropriate therefore to test it against May's experimental data, paying particular attention to the relationship between transport rates in a channel with a full width sediment bed and in a pipe with no deposit or very limited deposit in the invert.

May's research does not include fully detailed results for transport over a sediment bed (the volume or depth of deposition was not recorded) and the tests did not extend to cases with continuous beds of significant depth. The "with deposit" results in Ref. 10 are all with small deposits, usually separated dunes or ripples, rather than a continuous deposit occupying say 1/10 of the pipe diameter. The crests of these separated bed forms may, however, have come as high as 0.1 - 0.15D (private communication).

3. COMPARISON WITH HRS EXPERIMENTS

3.1 Sediment transport

Because data with deposition was incomplete in terms of depth or volume of deposition, comparison was first carried out using results at the limit of deposition.

Using the observed velocity and friction factor data, the Ackers-White transport equation was applied to predict sediment concentrations for a selection of the HRS cases. The comparison is shown of Fig. 2 for full-pipe flow and Fig. 3 for part-full flow. The A-W function shows significantly higher transport rates at the limit of deposition than were observed; in other words appreciably lower velocities for no deposit for a given transport rate and hence lower limiting gradients. It is clear that the A-W function for channel flow is not transferable to the no-deposit pipes with the assumptions originally made.

The principal assumption was that the active bed width for sediment transport is the pipe diameter for flows above half full, or the water surface width for flows below half full. Referring to other aspects of open channel sediment transport, implicit in this is the assumption that at the limit of deposit there is 100 percent availability of sediment across the active bed width as if there was already a deposit. This is unlikely to be correct and was the most likely source of discrepancy, as there is no theoretical reason for a physically based channel transport theory to fail merely because the shape of the flow section is changed.

The assumption about the active bed width was tested by repeating the computations with the active bed-width unspecified. Comparion of observation with calculation permitted evaluation of the effective bed-width and this is shown in Fig.4 and Table 1. The comparison covered both the limit of deposit condition and the with-deposit conditions (where available) for pipe-full flow.

Table 1 shows that at the limit of deposition the effective width is very much less than the pipe diameter, under 10 percent with 0.6 mm dia. sand in the 77 mm and 158 mm pipes, but of the order of 50 percent with 6 and 8 mm gravels in the larger pipe. Expressed as multiples of grain size, the average limit of deposit values are:

Sediment dia., d_{50} mm	Pipe dia. mm	Eff. width / d_{50}
0.63	158	8.3
0.57	77	9.8
5.8	158	15.8
7.9	158	10.0

Thus in order to apply the A-W open channel transport function at the limit of deposition it appears an effective width of the order of 10 times the particle diameter is appropriate.

Fig. 4 extends the results for sand to higher transport rates involving a small deposit. A selection of the test data is used (about every fifth point) covering the full range. As the concentration is increased beyond that for no deposit, the effective width increases until it corresponds to the pipe diameter, i.e. the assumption of Ref. 11.

It is concluded that the method would fit the laboratory results reasonably well provided an effective width for sediment transport of about 10 times the particle diameter is assumed at the limit of deposit. Also, if the effective width is taken to be the pipe diameter, the pipe will not be free from deposition under the calculated conditions, although the depth of deposition may be acceptable for engineering design as the loss of cross-sectional area is limited (the influence on resistance is considered later). A further inference is that if deposition is significant, say 0.1 D or more, then the effective width for transport appears to approach the pipe diameter, as was rather arbitrarily assumed in Ref. 11. (It was probably fortuitous that two of the test series were continued to the point where w_e = D almost exactly, but none went beyond.)

3.2 Resistance to flow

One of the conclusive findings in May's work, consistent with previous research on pipeline transport, is that resistance increases with sediment transport even without any deposit to physically roughen the boundary. The loss of capacity due to increased resistance (or apparent increased roughness) in the no deposit case is usually small and for sand is given by:

$$\lambda = \lambda_o + 6.35\, C_v \left\{ \frac{\sqrt{g\, R\, (s-1)}}{V} \right\}^3 \left\{ 1 - \frac{V_o}{V} \right\}^{-2} \qquad (13)$$

For the coarser sediments tested (gravel) the equation is:

$$\lambda = \lambda_o + 1.655\, C_v \left\{ \frac{g\, R\, (s-1)}{V^2} \right\} \left\{ 1 - \frac{V_o}{V} \right\}^{-1} \left(\frac{d}{R} \right)^{-\frac{1}{4}} \qquad (14)$$

May based λ_o on the Blasius equation (Ref. 14) rather than Colebrook-White, largely for ease of manipulation. Selected experimental data, including cases with deposition, have been re-examined to evaluate k_s, the roughness parameter in the Colebrook-White equation, to assess the additional resistance in terms of the equation which is now in general use for the design of sewers and storm drains.

The Colebrook and White equation (Ref. 13) for resistance is:

$$V = -\sqrt{(32gRS)}\, \log \left\{ \frac{k_s}{14.8R} + \frac{1.255\, \nu}{R\sqrt{(32gRS)}} \right\} \qquad (15)$$

In all cases without a deposit the k_s values calculated from the measured velocity and friction factor is a small negative value, confirming that the laboratory pipes were effectively smooth. One series of tests with sand extended to high transport rates with a small deposit and the k_s value increased progressively from zero to about 0.1 mm for $C_v \sim 10$ ppm; 5 mm for $C_v \sim 250$ ppm, 6 mm for $C \sim 550$ ppm. There was a deposit of sediment in all cases where $k_s > 0.1$ mm. This demonstrates that although (as equations 13 and 14 show) there is an increase in resistance due to the energy needed to transport the sediment when there is no deposit, once there is deposition the effective roughness of the pipe and bed increases rather more significantly. Although there would only be a small error in neglecting the additional resistance with no deposit, the effect of deposition on roughness must be taken into account.

4. DESIGN CRITERIA

4.1 Comparison of alternative methods

The following criteria for 'self cleaning' are available:

1. The 'Code of Practice velocity recommendation.
2. The Novak and Nalluri method (equs 3, 4 and 5)
3. May's limiting velocity formula (equ 2)
4. Author's method for limit of deposition, using an effective transport width of 10 d.
5. Author's method for some deposition, using an effective width of D.

There are illustrated in Fig. 5†, assuming that the material being transported is sand, d = 0.4 mm, at a volumetric concentration, C_v = 100 ppm, in a pipe with k_s = 3 mm. Also, shown by a broken line, is the fifth method with k_s increased to 10 mm to allow for minor deposition.

Somewhat different comparisons would emerge for other sediment grades and concentrations, and other pipe roughness. The example chosen is part of a spectrum of conditions that might occur in practice. Frequently, the engineer is concerned with a mixed distribution of sediment, with a proportion of gravelly material as well

† (excluding 2, N and N, for which see fig 1, n_m= 0.015)

as sand sizes. However, the aim should be to exclude coarse material through effective gulley traps and maintenance, as it requires a high gradient for movement. According to equations 6 and 7 (Novak and Nalluri), gradient depends on $d^{1/3}$ and velocity on $d^{1/6}$.

A rather unexpected finding that the required velocity reduces with increasing diameter applies to May's analysis (Fig. 11 of Ref. 10) where a 10 fold increase of sediment size (0.5 mm to to 5 mm) reduces the self-cleaning velocity in a 2 m dia. pipe from 2.7 m/s to 2.15 m/s (C_v = 50 ppm).

4.2 The problem of large pipes

A range of alternative limiting gradients could be extracted from fig. 5, and the variation between the different criteria becomes very significant for large pipes. For a 4 m dia. pipe or tunnel, such as may be used in the drainage of a large city, the required minimum velocities and gradients for the case considered are:

		V m/s	S 10^{-3}
1.	Code of Practice	0.76	0.14
2.	May	6.7	10
3.	Author (no deposition)	4.5	4.5
4.	Author (limited deposition with allowance for increased roughness)	1.2	0.43
5.	Novak and Nalluri (Fig1, n_m = 0.015)	1.3	0.39

5, CONCLUSIONS

5.1 Despite the uncertainty of extrapolation, it seems to be established that for completely deposit free operation, much higher velocities (and hence gradients) are needed in large pipes than previous engineering recommendations envisaged. However if a limited deposition is accepted, corresponding to some dunes and ripples moving through the pipe-line perhaps with patches of clean invert between them, the velocity requirements are more in line with current practice.

5.2 It is highly desirable for information to be obtained on field performance, for a range of pipe sizes and sediment influxes. Without such data, it is difficult to assess the reliability of methods based on theoretical approaches confirmed only at laboratory scale.

5.3 In the meantime, any criteria for sediment transport in large diameter sewers and storm drains are uncertain. The criterion to be adopted is very sensitive to the interpretation of self-cleansing, a 'no deposit' condition appearing to be unduly severe for the transport of sand. Nevertheless, the method based on the Ackers-White transport function with an active bed width equalling the pipe diameter should be considered as the limit of acceptance, although field information (if available) would form a much better basis of assessing design criteria.

5.4 The Code of Practice recommendation of 0.76 m/s as a generalised rule appears inadequate in pipes of large diameter.

5.5 The alternative methods discussed here are based on steady flow, whereas discharges in sewers and drains vary in time, particular in surface water and storm systems. The variation of flow needs to be taken into account in applying self-cleansing criteria to design. Short term deposition is probably acceptable provided it is not cumulative and allowance is made for it in hydraulic performance.

5.6 The methods considered here refer to non-cohesive particulate matter. Other criteria may govern the choice of gradient e.g. sliming, other sewage solids, septicity.

6. ACKNOWLEDGEMENTS

The writer thanks Prof. Novak and Mr. May for their helpful discussions on this interesting subject, and also Mr. S. J. Hargreaves for his assistance whilst a vacation student with Binnie & Partners.

7. REFERENCES

1. B.S.I., Sewerage, Code of Practice CP2005, British Standards Instn, London, 1968.

2. Vickers J.A. et al, Erosion of sewers and drains, Construction Industry Research and Information Association, London, Res. Report 14, 1968.

3. Laursen, E.M. The hydraulics of a storm-drain system for sediment transporting flow, Bull. No. 5, Iowa Highway Research Board, 1956.

4. Craven, J.P. The transportation of sand in pipes, I., full-pipe flow. Proc. 5th Hydraulics Conf, Bulletin 34, Studies in Engineering, State Univ. Iowa, USA 1953, p 67-76.

5. Ambrose, H.H. The transportation of sand in pipes, II, free-surface flow, Proc. 5th Hydraulics Conf. Bulletin 34, Studies in Engineering, State Univ., Iowa, USA, 1953 p 77-78.

6. Novak, P. and Nalluri, C. Sediment transport in smooth fixed bed channels, Proc. ASCE, Vol. 101, HY 9, Sept 1975.

7. Ojo, S.I.A. Study of incipient motions and sediment transport over fixed beds, Ph.D Thesis, University of Newcastle-upon-Tyne, U.K., Sept. 1978.

8. Novak, P. and Nalluri, C. Sewer design for no sediment deposition, Proc. Instn Civ. Engrs, Part 2, Vol. 65, Sept. 1978, p 669-674.

9. May, R.W.P. Deposition of grit in pipes: literature survey, INT 139, Hydraulics Research Station, Wallingford, 1975.

10. May, R.W.P. Sediment transport in sewers, Report IT 222, Hydraulics Research Station, Wallingford, Feb 1982.

11. Ackers, P. Urban drainage : the effects of sediment on performance and design criteria. Int. Conf on Urban Storm Drainage, Southampton, April 1978, p 535-545.

12. Ackers, P. and White, W.R. Sediment transport : new approach and analysis, Proc. ASCE, Vol 99, HY 11, Nov 1973, pp 2041-3060.

13. Colebrook, C.F. Turbulent flow in pipes, with particular reference to the transition region between the smooth and rough pipe laws, J. Instn. Civ. Engrs. Vol. 11, 1939, p 133.

14. Blasius, H. Das ahnlichkeitsgesetz bei reibungsvorgangen in flussigkeiten, V.D.I. Forschungshaft, no. 131, 1913.

TABLE 1. Calculation of effective width of mobile bed from Ackers-White formula for open channels and May full-pipe transfer data

Pipe dia. (mm)	Sed. dia d_{50} (mm)	Observed C_v (ppm)	Calculated effective width (mm)	$\frac{W_e}{D}$	$\frac{W_e}{d_{50}}$	Average W_e/d_{50}
76.7	0.57	58.5	4.53	0.059	7.9	
		170	5.06	0.066	8.9	
		346	5.83	0.076	10.2	
		509	5.14	0.067	9.0	
		740	5.22	0.068	9.2	
		1150	5.91	0.077	10.4	
		1430	5.83	0.076	10.3	
		2110	6.90	0.090	12.1	9.75
158.3	0.63	6.3	6.02	0.038	9.5	
		8.4	3.17	0.020	5.1	
		143	2.22	0.014	3.5	
		36.6	5.83	0.037	9.2	
		36.1	4.27	0.027	6.7	
		38.3	3.32	0.021	5.3	
		41.7	3.17	0.020	5.1	
		90.7	4.59	0.029	7.2	
		176	6.65	0.042	10.6	
		212	6.17	0.039	9.7	
		402	11.24	0.071	17.7	
		353	6.17	0.039	9.7	8.28
158.3	5.8	144	84.1	0.531	14.5	
		207	64.4	0.407	11.1	
		478	96.4	0.609	16.6	
		687	105.9	0.669	18.2	
		918	107.5	0.679	18.5	15.8
158.3	7.9	132	93.7	0.592	11.9	
		214	70.3	0.444	8.9	
		413	82.5	0.521	10.4	
		552	81.7	0.516	10.3	
		612	67.1	0.424	8.5	10.0

APPENDIX

APPLICATION OF ACKERS-WHITE SEDIMENT TRANSPORT THEORY TO PIPES AND NON-RECTANGULAR SECTIONS.

The theory applied to wide open-channels is given in Appendix 1 of ref. 12. Only the amendments needed to cover flow cross-sections other than wide channels are outlined here. The sediment mobility number, F_{gr}, is compounded of mobility numbers for coarse grains, F_{cg}, and fine grains, F_{fg}, which account for their different modes of transport.

Coarse grains

Coarse grains move as a bed process activated by the shear stress on the grains.

$$F_{cg} = \frac{V}{\sqrt{32g(s-1)d}} \left[\log\left(\frac{10y}{d}\right)\right]^{-1} \qquad \text{...... A1}$$

where the log term is closely related to the rough turbulent equation for a wide channel where $R \simeq y$:

$$v_{*cg} = \sqrt{32}V \left[\log \frac{12.3y}{k_s}\right]^{-1} \qquad \text{...... A2}$$

For a non-rectangular section in which $R \neq y$, the comparable form is

$$v_{*cg} = \sqrt{32}V \left[\log \frac{14.8R}{k_s}\right]^{-1} \qquad \text{...... A3}$$

It is therefore appropriate to replace the wide-channel definition of F_{cg} by;

$$F_{cg} = \frac{V}{\sqrt{32g(s-1)d}} \left[\log\left(\frac{12R}{d}\right)\right]^{-1} \qquad \text{...... A4}$$

Introducing the concept of an effective width, w_e, to represent the limited availability of sediment at the bed in comparison with the fill channel width in the alluvial channel case, the sediment load, w_b, (mass rate of transport), per unit of effective width is given by:

$$w_b = X \rho g Q/w_e \qquad \text{...... A5}$$

In the basic transport theory, the efficiency E_{cg}, of coarse grain movement is the ratio of the above (converted to immersed weight) to the grain shear part of the stream power, $\tau_{cg}V$, per unit area of the perimeter, where:

$$\tau_{cg} = (v_{*cg})^2 \rho \qquad \text{...... A6}$$

Thus

$$E_{cg} = \frac{X \rho g Q}{w_e} \; \frac{s-1}{s} \; \frac{32\left[\log(12R/d)\right]^2}{\rho V^3}$$

$$E_{cg} = 32 \frac{gA}{w_e V^2} \; \frac{s-1}{s} \left[\log\left(\frac{12R}{d}\right)\right]^2 X \qquad \text{...... A7}$$

The engineering variables were separated in the original theory by introducing:

$$G_{cg} = E_{cg} F_{cg}^{2} \qquad \text{...... A8}$$

which now becomes

$$G_{cg} = \frac{XA}{sw_e d} = \frac{X \pi DR}{sw_e d} \qquad \text{...... A9}$$

for a circular section flowing full.

Fine grains

Fine grains are assumed to move in suspension under the influence of the turbulence generated by the overall resistance to flow.

The work done in keeping sediment in suspension is:

$$X\left(\frac{s-1}{s}\right)\rho g A w \quad \text{(per unit length)}$$

where w is the fall velocity of the sediment.

The mean shear stress generating turbulent suspension is:

$$\tau_{fg} = \rho gRS = \rho v_*^{2} \qquad \text{...... A10}$$

and the stream power per unit length is ρgSQ.

The efficiency of the fine-grain (suspension) process is therefore:

$$E_{fg} = X\left(\frac{s-1}{s}\right)\frac{Aw}{SQ} \qquad \text{...... A11}$$

In ref. 12, the processing of separating the main variables led to:

$$G_{fg} = E_{fg}\left[\frac{v_* \nu}{gd^2(s-1)}\right]^{3} D_{gr}^{3} \qquad \text{...... A12}$$

where

$$D_{gr} = \left\{\frac{g(s-1)}{\nu^2}\right\}^{1/3} d \qquad \text{...... A13}$$

Using Stokes law for the fall velocity, w,

$$w = \frac{gd^2(s-1)}{18\nu} \qquad \text{...... A14}$$

in combination with equations, A12, A11, and A10 (and omitting the numerical constant) yields:

$$G_{fg} = \left(\frac{XR}{sd}\right)\frac{v_*}{V} \qquad \text{...... A15}$$

General functions

Following ref 12, the general definition of the non-dimensional transport become:

$$G_{gr} = \left(\frac{XR}{sd}\right)\left(\frac{A}{w_e R}\right)^{1-n}\left(\frac{v_*}{V}\right)^{n} \qquad \text{...... A16}$$

where n = 1 for fine material, n = 0 for coarse material. Following the earlier theory for wide channels:

$$n = \text{function}\ (D_{gr}) \qquad \text{...... A17}$$

for intermediate sizes of material.

The general definition of particle mobility utilises equ. A4 for the coarse limit with the fine limit from ref. 12.

$$F_{fg} = \frac{v_* V}{gd^2(s-1)} \qquad \text{...... A18}$$

to give

$$F_{gr} = \frac{V^{1-n}\, v_*^{\,n}}{\sqrt{g(s-1)d}\ \left\{\sqrt{32}\,\log(12R/d)\right\}^{1-n}} \qquad \text{...... A19}$$

<u>Formulae for transport in a pipe.</u>

With D_{gr}, G_{gr} and F_{gr} defined by equs. A13, A16 and A19 it may be reasonably assumed that the formulae derived from open channel data would apply. From ref. 12, these are:

General transport formula:

$$G_{gr} = C\left\{\frac{F_{gr}}{A_{gr}} - 1\right\}^m \qquad \text{...... A20}$$

Transition parameter between fine and coarse limits; $1.0 < D_{gr} < 60$:

$$n = 1.00 - 0.56 \log D_{gr} \qquad \text{...... A21}$$

Initial motion condition, i.e. when $G_{gr} = 0$:

$$A_{gr} = 0.14 + 0.23/\sqrt{D_{gr}} \qquad \text{...... A22}$$

(A = 0.17 for coarse sediment)

Coefficient C:

$$\log C = 2.86 \log D_{gr} - (\log D_{gr})^2 - 3.53 \qquad \text{...... A22}$$

(C = 0.025 for coarse material)

Exponent m:

$$m = 1.34 + 9.66/D_{gr} \qquad \text{...... A23}$$

(m = 1.50 for coarse material)

The calculation procedure is as follows:

1. Determine D_{gr} from known values of d, g, s and ν.
2. Determine values of n, A_{gr}, C and m for this D_{gr} value from equs. A21-A23.
3. Compute F_{gr} from equ. A19 for the flow conditions (R, V and v_*) and sediment.
4. Determine G_{gr} from equ. A20.
5. Use equ. A16 to evaluate X, all other variables on the left being known or assumed (e.g. w_e). If X is known, as for laboratory results, w_e may be considered the unknown to be evaluated.

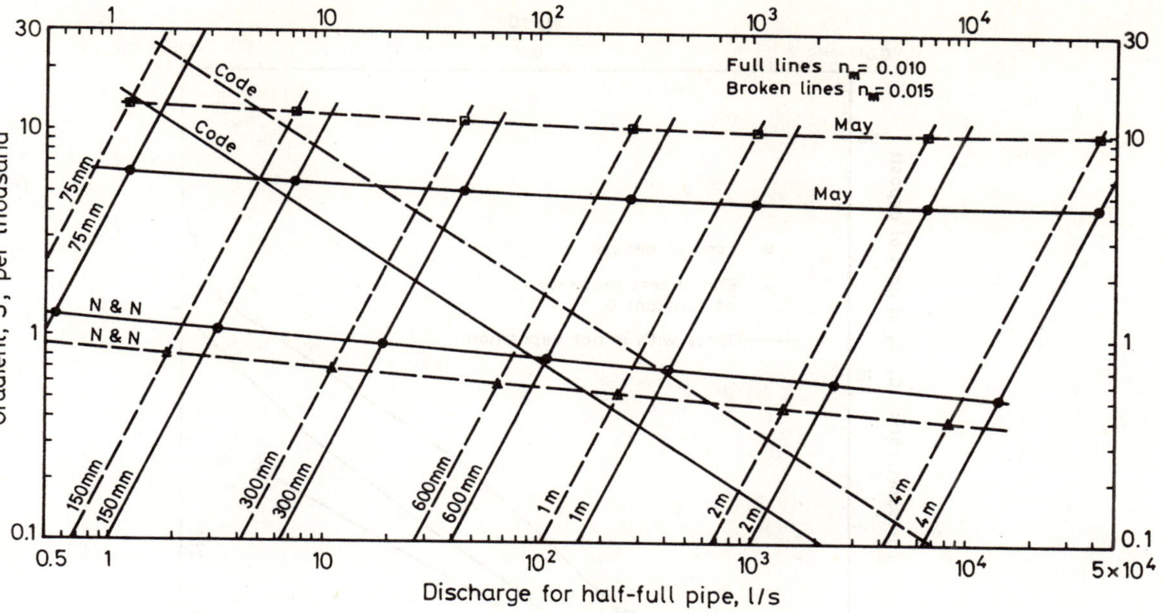

Fig 1. Alternative self-cleansing criteria applied to pipes up to 4 m diameter: Q-S-D plot for half full pipes, 0.4 mm dia. sand, $C_V = 100\times10^{-6}$.

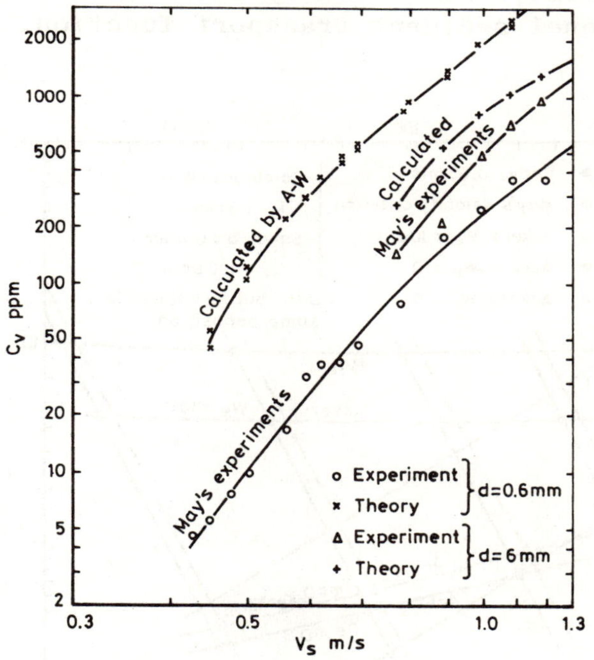

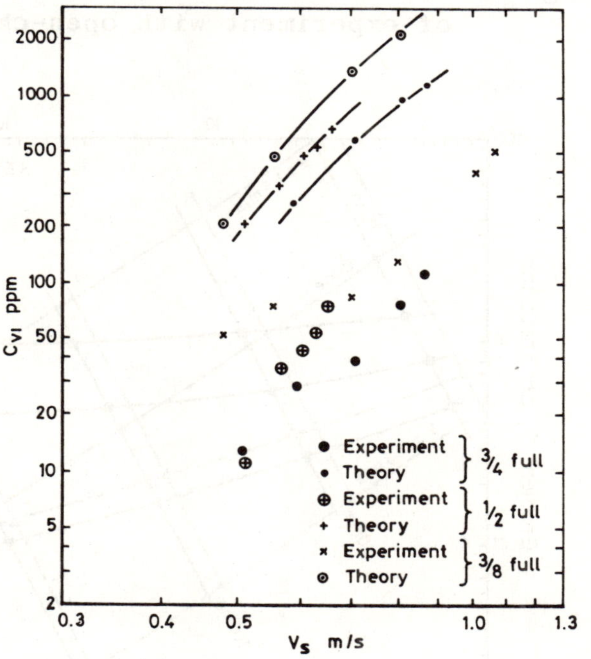

Fig 2. (left) Limit of deposition for 158 mm dia. pipe flowing full: comparison of experimental data (May, ref. 10) with open-channel mobile bed function (Ackers and White, ref. 12).

Fig 3. (right) Limit of deposition for 158 mm dia. pipe flowing part full (0.6 mm dia. sediment).

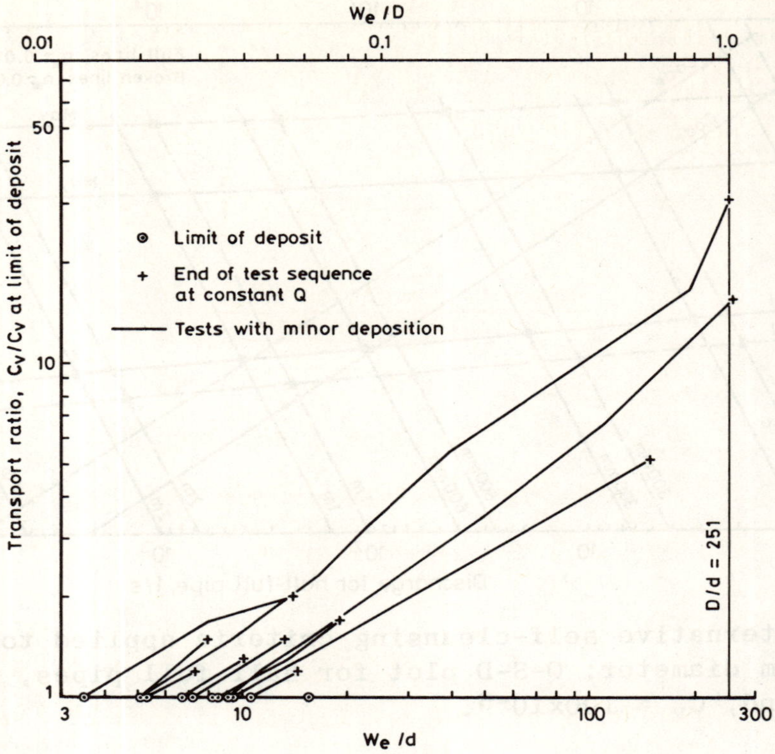

Fig 4. Determination of effective active bed width by comparison of experiment with open-channel sediment transport function.

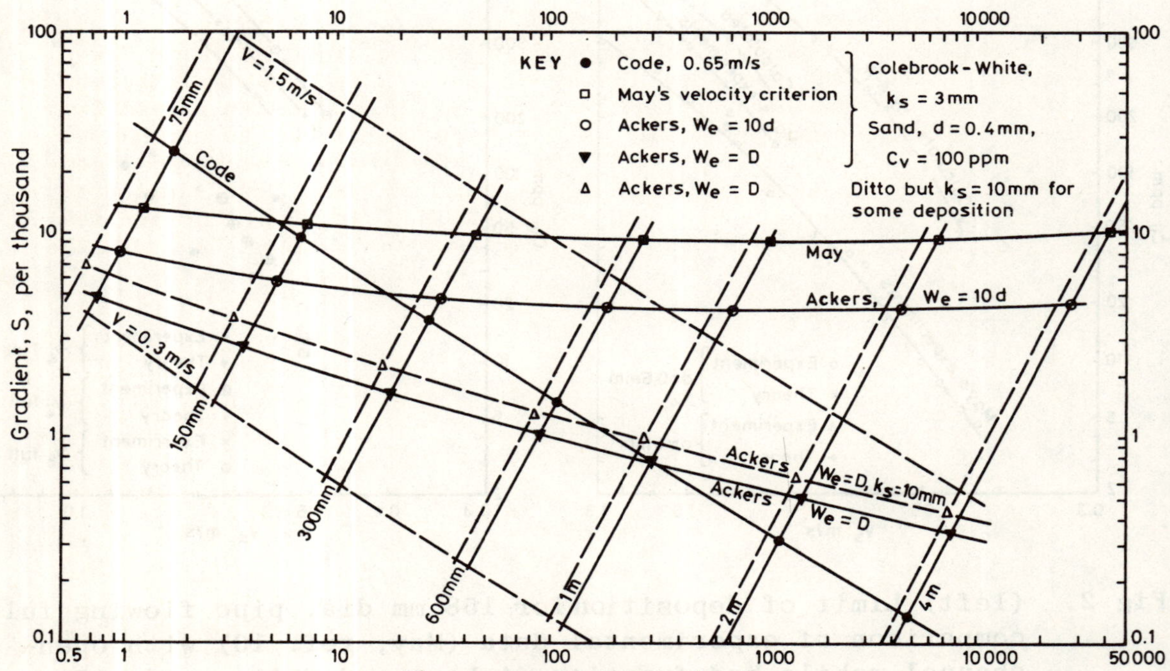

Fig 5. Comparison of self-cleansing criteria: Code of Practice, May and Ackers methods (Ackers method calculated for full pipes but plotted at ½Q for comparison).

International Conference on the
PLANNING, CONSTRUCTION, MAINTENANCE & OPERATION OF SEWERAGE SYSTEMS
12-14 September, 1984

PAPER F1

EFFECTIVE SIMULATION OF EXISTING SEWERAGE SYSTEMS

A R EADON B.Sc. C.Eng. MICE

Principal Regional Engineer
Severn-Trent Water Authority
United Kingdom

SYNOPSIS

Hydraulic simulation of sewerage systems is currently regarded as somewhat of a luxury by drainage engineers and often viewed rather as a research exercise than a worthwhile analysis. Setting aside the reluctance to accept new methods, there are two principal objections which have tended to restrict its application to large complex systems scheduled for extensive improvements. Firstly it is considered that traditional hydraulic calculations are more appropriate, and secondly, the cost of simulation is prohibitive.

The simulation programme incorporated in the Wallingford Storm Sewer Package (WASSP-SIM) has been in use in Severn-Trent Water Authority since 1980. It has been applied to a variety of systems to assess its value and engineers now using the technique hesitate to return to traditional methods. From this work a standard approach has been devised which placates the above objections, has shown that simulation is not only appropriate but desirable, and that it should be applied to all hydraulic calculations on existing sewerage systems.

Held at The University of Reading, England. Organised and sponsored by BHRA, The Fluid Engineering Centre, Cranfield, Bedford in conjunction with the Water Research Centre.

Introduction

Expenditure on urban drainage has always been carefully controlled and hydraulic calculations are rarely carried out until a sewerage system has failed or is likely to fail in the near future. The traditional approach therefore has been to combine analysis and design into a single calculation by design method.

Design methods are wholly appropriate for calculating the hydraulic demand on a new system and this approach has been well proven by the test of time. At the outset therefore it would appear logical that improvements to existing systems should be designed on the same basis and furthermore that an independent analysis is not worth pursuing.

1 Objectives for Hydraulic Analyses

1.1 To examine the situation it is necessary to define the objectives of an analysis and then to be assured that design calculations meet those requirements. Some of the main objectives are suggested under the following headings:-

a) Investigation of Flooding

It must be proven that flooding is due to failure of the sewerage system. It may be for instance that local connections are blocked, that an adjacent system is at fault, or a watercourse has caused backing up. In combined systems the fault may lie with a pumping station or blocked overflow. A full investigation should also examine the frequency of occurrence and relate it to meteorological events. The objective therefore is to locate the fault and estimate the magnitude of the problem.

b) Investigation of Pollution from Storm Water Overflows

Hydraulic calculations associated with overflows are extremely complex and calculations should incorporate dilution effects in both the discharge and the receiving watercourse. It is worth noting that design calculations are not reliable when carry-on flows are affected by surcharge and when outlets to the receiving watercourse are drowned. Overflows are an extremely important feature of some existing systems and cannot readily be eliminated without substantial resewering downstream. The objective of the analysis therefore is to provide an accurate assessment of the level of nuisance and the current performance of the overflow.

(c) Accommodation of Growth

This is probably the one area where most engineers would agree that an accurate analysis is vital. The majority of our systems are subject to growth but fortunately most of them have reserve capacity. It is assumed that this has arisen from a number of factors including:- the adoption of minimum pipe sizes; the principle of sizing up to a commercially available diameter; the use of tunnel sections at nominal sizes; the available storage capacity in manholes; the factors of safety on design and the design horizon used for the existing sewers. An important consideration is that adequacy is measured from existing loading right up to the point of failure (ie flooding or gross pollution). The objective of the analysis therefore is to assess the existing demand and thus to check that any additional loading from new development will not cause failure within a reasonable time.

(d) Sewer Maintenance

The problems of sewer maintenance have been highlighted in recent years by the current emphasis on sewer dereliction. With substantial structural problems, silting or root growth, maintenance costs can be very high and engineers are faced with the problem of assessing hydraulic adequacy. If the hydraulic capacity of the sewer is limited replacement at a larger diameter may be the preferred option, and an objective of analysis is to assess the hydraulic life of those sewers in need of extensive maintenance. Furthermore, there is a relationship between sewer collapses and frequency of surcharging and an analysis must indicate those sewers subject to hydraulic extremes.

(e) Phasing and extent of Improvements

Having designed improvements it is frequently the case that only part of the works is justifiable or affordable in the short term. The effect of constructing part of the works, particularly if the main outfall is excluded, is hydraulically complex. The engineer needs to be sure that improvements will not merely move troublespots downstream and an analysis of the existing system with the proposals incorporated has to be carried out. An objective of analysis therefore is to test the consequencies of phased improvements.

(f) Efficiency or Pumping Stations

The mechanical content of pumping stations has a shorter life then sewerage systems and the efficiency of pumping units is likely to be improved if their design is matched to current demand. Analysis of the existing upstream sewerage is required, to quantify the current range of hydraulic loading.

(g) Design of Sewage Treatment Works

Improvements or extensions to sewage treatment works are often required independently of improvements to the sewerage system. A phased approach is preferred and an analysis of the upstream sewerage will provide the short, medium and long term hydraulic loading essential for efficient design of inlet works and storm tanks. It should be noted that the converse may also be true in that improvements to a foul sewerage system may well affect the performance of a sewage treatment works and the engineer must be aware of these implications.

1.2 It is concluded that the objectives of an analysis cover a wide range of activities. It should be noted that in (f) and (g) above, engineers other than the sewerage specialist require analyses to be carried out and, that in every case it is the analysis of an existing system and its current performance which is at least part of the objective. This being so, an analysis tool must be capable of modelling any current situation and this may include average and peak rates of flows, levels and frequency of surcharging, flooding and pollution.

2 Analysis by Design Methods

2.1 Design methods for storm sewerage fall into two categories 'rational' and 'hydrograph'. Both calculate peak flows by which sewers can be correctly sized and are ideal for the design of new systems on greenfield sites. Hydrograph methods tend to give more economic sizes on large systems by taking pipe storage into account, and since actual storms can be input to most versions they may in theory be used to simulate a real event.

2.2 In practice the range is limited however and design methods can only be expected to give accurate results providing the built-in assumptions are not violated. These assumptions are fundamental to the design process and the three most important ones in this context are:-

a) that peak flows are fully accommodated by the sewer;

b) that every sewer in the system has a free outfall;

c) that flow is always directed towards the outfall.

This means that an analysis by design method will only be accurate providing no surcharging occurs at any time and these are hardly the conditions which would trigger an analysis.

2.3 In the past, however, design methods have been the only means of carrying out hydraulic analyses and despite the above limitations, versions have been devised which run in "analysis mode". The terminology is confusing and the calculations need careful interpretation when the limits are exceeded. For instance, if the calculated peak flows exceed the available capacity, the majority of design methods, having no means of accommodating surcharge, indicate a 'surcharge factor' and pass the calculated flows on downstream. It is sometimes argued that a surcharged sewer will indeed pass on flows in excess of its rated capacity, but in so doing, the upstream sewer will be impeded and the peak flow attenuated. Locally the analysis may not be too far from reality but in a large system under heavy loading, the errors can accumulate and give a very false impression for trunk sewers where an accurate analysis may be required for phasing improvements or carrying out extensive maintenance.

2.4 The majority of engineers are aware of this problem however and usually interpret the results well. They also keep the loading under control by using return frequencies which could be expected to produce flows in the same order as the rated sewer capacities. Where these are 'design' frequencies the method effectively redesigns the entire system and the 'surcharge factors' indicate those sewers which would need to be uprated to produce an ideal system. Analysis by design method therefore, where design frequencies are used, compares the existing system with one that would be provided on a greenfield site and is a plausible basis for an improvement strategy.

2.5 From this point on, analysis by design method has got out of hand however because, in the desire to analyse a system and design improvements at the same time, values for impermeable area and dry weather flow in the calculations tend to be those used for the design of new systems. If allowances for growth are also included this can result in an indication that many of the sewers are several times overloaded. In reality it is not the sewers that are overloaded but the calculations and they are not likely to be of assistance for the type of analysis defined earlier.

2.6 It is concluded therefore that surcharge factors as generated by design methods are no substitute for actual surcharging and a higher form of calculation is required.

3 Analysis by Simulation Method

3.1 Simulation models are now available which at first sight appear to meet the full requirements of an analysis tool. None of the limitations of design methods is present in WASSP-SIM and the calculations not only give rates of flow but also surcharge levels and flooding. In addition, there is more opportunity for the engineer to model the particular physical characteristics of the system under analysis and any type of storm loading can be applied.

3.2 There are still some problems however which were not fully appreciated until WASSP has been used on several catchments and it was possible to check the calculations against known performance. It is a problem which is common to all calculations in that the quality of the results is directly related to the quality of the survey. If information on paved areas particularly had been poorly assessed, the simulation gave unrealistic results.

3.3 The engineer is faced therefore with a very extensive and detailed survey. Above ground the catchment survey would extend on to private property and below ground it would involve a detailed sewer survey. It was also apparent from early tests that simplification of the system, by omitting lateral sewers, reduced the below ground storage in manholes, and exaggerated surcharge and flooding.

3.4 It is concluded therefore that even though a simulation method like WASSP-SIM is an appropriate tool the cost of detailed surveying may be prohibitive.

4 Analysis by Monitoring

4.1 To meet the objectives of an analysis, calculations are not the complete answer and engineers may have to adopt the more fundamental approach of monitoring. Indeed for some of the requirements of an analysis (the verification of flooding complaints for instance or frequency of operation of an overflow), monitoring is easily the most rewarding approach.

4.2 However, sewerage is subject to variable loading by meteorological events and to take account of this an extensive data set would have to be produced, which could take several years for a complete analysis.

4.3 Also monitoring is particular to the specific location and generally will give very little information on conditions elsewhere in the system. An analysis relying entirely on monitoring therefore would require an extensive gauging exercise which again would be prohibitively expensive.

5 Monitoring Plus Modelling

5.1 In considering the two problems;

a) That simulation models require an accurate input.

b) That monitoring is site specific.

a combination of the two is mutually beneficial. The various hydraulic and hydrological relationships which make up WASSP-SIM have been derived from an extensive data set and if those can be accepted, monitoring can be assigned to the specific job of checking input data. This has the advantages of a much shorter gauging period and wider spacing of gauges.

5.2 The process of using gauges to check input data has become known as "verification of the model" and falls sufficiently short of full calibration to be financially attractive.

5.3 Referring back to paragraph 3.2 where it was stated that the simulation performed unrealistically with poor input data, that opinion was based on the known performance of the system. It was achieved by assuming for instance, that a one year design storm of appropriate duration, when loaded into the model, should reproduce the annual flooding records. The validity of using design storms for this purpose needs to be supported by further research work, but the adoption of this technique in Severn-Trent Water Authority as a means of checking data, has been very rewarding. Complex foul sewerage systems serving established urban areas have passed through various stages of development and there may be combined, partially separated or separated sub-catchments within it. Indeed there may be further sub-divisions, with some areas having independent highway drains and some properties with informal or local systems of dealing with storm water. Usually records are not complete, particularly for private property, and the engineer, faced with limited resources for surveying, can only sample check the effective area formally connected to the system. It has been found that by using a combination of simulation model and design storms any gross errors are shown up by the results and indicate those areas where the information may be incorrect and more detailed surveying is necessary. The technique will also indicate gross errors in the sewer records, and this 'crude' verification is becoming an established technique to be used in conjunction with surveying.

6 Monitoring Techniques

6.1 It is recommended that crude verification is carried out before monitoring so that best use can be made of the equipment. Fig.1 shows a probable relationship between the costs of surveying and the costs of monitoring to produce an effective analysis. The curve will not be the same for each system but it shows that in general an effective analysis requires some monitoring in addition to basic survey information and that high surveying costs may be offset by increased monitoring. It is also believed, as the curve shows, that high monitoring costs may be offset by a more detailed survey.

6.2 There is another relationship between the number of gauges and the length of the gauging period which needs to be examined. Rainfall is far from predictable and periods of drought will always be a problem, and when it does rain, the coverage is unlikely to be uniform. A short gauging period will therefore have to take this into account by employing a reasonably dense network of raingauges and sewer flow meters. With a long gauging period there is a greater chance of avoiding 'difficult' storms and fewer gauges may be employed. It is suggested at this stage that the relationship is similar to Fig.2 where the axes are 'gauging period' and 'gauge density'.

6.3 The type of gauge also needs to be defined. Rainfall must be recorded on intensity gauges which should be sited within the catchment. There should never be less than two on a reasonably sized catchment and a density of one gauge for 8,000 population seems to be a workable guideline. Sewer gauges are a little more difficult because an ideal gauge would record rate of flow, direction of flow and surcharge level, and be capable of operating under any hydraulic conditions. This has almost been achieved by the equipment currently used by flow survey contractors and its use is recommended, but it is expensive and high costs have resulted in the development of a short term technique only. This has some disadvantages for example:

a) complete coverage by a storm is less likely;

b) if only minor storms occur, prediction of the extreme event has to rely more heavily on the capabilities of the model;

c) there may be a particular difficulty with establishing antecedent conditions for minor storms or a single major event.

6.4 While the short term technique used by contractors has developed, Severn-Trent Water Authority has been devising a different method. This arose from the idea that monitoring would not necessarily be associated with modelling and would be relatively inexpensive. Monitors for rate of flow were considered too expensive from the outset and the technique developed from the need to monitor surcharging at troublespots to assess the priority for improvement. It was also considered after some experience with WASSP-SIM that the weakest part of a simulation model is the prediction of surcharging and, together with input data, surcharge should be a component for verification.

The Authority's principal application is the analysis of major foul sewerage systems where gauging may already be carried out at the inlet to the sewage treatment works. This is the fundamental gauging point where, from rate of flow, the effective paved area connected to the system in the whole of the upstream catchment can be checked. This uses the principal that:

rainfall x effective paved area = volume of run-off

and may be used providing there are no losses from the system. It is unnecessary therefore to monitor rate of flow elsewhere in the catchment and other gauges may be simple surcharge recorders located in the vicinity of known troublespots. Therefore, with costs reduced, a longer gauging period is affordable which overcomes the difficulties of short term gauging given earlier. The major disadvantage of this method is that if the effective paved area is shown to be in error, then it is difficult to locate the sub-catchment causing it and a more detailed survey may be required at the outset.

In practice however recorded levels within the depth of the sewer are often good enough to indicate changes in rate of flow and errors have been traced from this information.

6.5 A variety of instruments are available for recording surcharge ranging from sophisticated ultrasonic devices with intelligent logging to water sensitive paper and even plastic cups on poles which are inspected after each event. Costs can therefore by very low indeed and gauging periods may be extended to several months so that surcharging predicted by the model can be verified.

6.6 The Severn-Trent Water Authority method of verification has the advantage of in-house economies. It also takes advantage of the independent requirement to monitor sewage treatment works and major terminal pumping stations. In some cases the gauges have been found to be in a poor state of repair or have a limited range, and a combined effort by the relevant departments is arranging to renew or reinforce existing gauges in advance of hydraulic analyses.

6.7 The Severn-Trent Water Authority method of verification is therefore summarised as follows:-

(a) Check existing gauges at major outfalls and renew or renovate as necessary in advance of the analysis.

(b) Install an appropriate number of raingauges within the catchment.

(c) Install surcharge recorders/indicators at troublespots and other critical or strategic points in the sewerage system.

7 Accuracy

7.1 The accuracy of a model may be assessed by several different parameters in accordance with the objectives. A distinction between volume of run-off and surcharging has already been discussed and it is possible that different accuracies will be apparent for each one. A specific accuracy is therefore difficult to define but in all cases engineers are tempted to improve the model by allocating more and more resources to surveying and monitoring.

7.2 Fig.2 is a suggested relationship between "accuracy" and the costs of the investigation which is based on several analyses in Severn-Trent Water Authority. It is shown that, unless there are gross errors in input data, the model will always tend to overpredict because paved areas are assumed to have an ideal connection to the system and surveyors tend to include doubtful areas. This is safe practice for design purposes and the practice should not be unduly suppressed because potential troublespots are likely to be highlighted (and therfore not overlooked) during crude verification.

7.3 The accuracies for analysis by design methods is very much an oversimplification however. Some engineers include for future growth and there is the possibility of gross errors being made on above and below ground surveys. Therefore since simulation models use the same type of input data initially, accuracies can lie anywhere in the shaded area shown.

7.4 WASSP-SIM incorporates some details of the surface run-off sub model and headloss indices at manholes which require investigation, but in general survey costs should not be much higher than for design methods. The better accuracy attained by simulation is mainly due to the substitution of the surface run-off sub model for the traditional assumption of 100% contribution from paved areas.

7.5 Crude verification costs very little indeed although it is suggested that a special effort is made to gather as much information as possible on the recent past performance of the system. The increased accuracy after crude verification is not considered to be an overstatement particularly for inexperienced engineers. But it is conceded that engineers used to the techniques will employ several stages of crude verification during the survey to obtain better data and the model will be more accurate at the outset. The suggested error of + 100% at this stage may be pessimistic.

7.6 The deployment of monitors by whatever chosen method, after crude verification, begins to bring accuracies down to a respectable level. Figures of + 20% are achievable at reasonable cost but then the curve begins to flatten out and as is shown by the 'calibration' phase in Fig.2 better accuracy is not worth pursuing.

8 Programme of Activities

8.1 Previous sections have discussed the development of a new technique for hydraulic analysis of sewerage systems. The traditional approach was less complex and considerably cheaper where emphasis was placed more on designing improvements than analysing the existing sewers. Extra resources for analysis, as stated in the introduction, are difficult to justify at the outset and all of the activities of surveying, modelling and monitoring need to be carefully controlled to avoid unnecessary expenditure.

8.2 The Severn-Trent Water Authority has discovered that an interactive approach to surveying, modelling and monitoring will help to avoid abortive expenditure and the typical programme given in Fig.3 which uses this principal, is recommended.

8.3 The programme is based on an analysis of the foul sewerage system serving a fully developed area of approximately 500 Ha. It is assumed that sewer records are good and for in-house gauging, no civil engineering is required to set up gauging stations. The following notes explain the various elements of the programme.

(i) Progress Meetings

Formal meetings on progress are advisable and some key dates are indicated, (eg immediately after crude verification) where the programme and resources can be sensibly adjusted in response to the reported progress.

(ii) Maps

Both detailed and general maps of the area are required at various stages during the study. It is good practice to have appropriate sizes and scales available to define sub-catchments and record the above ground survey information. Overlays for sewer maps are probably worth the investment and considerable thought should be given to a method of recording the effective paved and pervious areas connected to the sewerage system.

More general maps of a convenient size and scale, are useful for presenting information to non-specialists at progress meetings, and some thought should be given to the type of map which will be required for the final report. A general map should be produced at this stage indicating the known troublespots. Details should include flooding points, collapses, high maintenance areas and the locations of sewer ancillaries. Generally speaking, more detail is required than that used to trigger the analysis, and the engineer should request information from all relevant departments.

(iii) Data Collection

A Below Ground

Even when complete sewer records are available, the engineer should inspect critical parts of the system to make sure that it is being correctly configured. This work should be combined with a search for suitable monitoring stations. The network should not be unnecessarily simplified in the vicinity of troublespots, ancillaries and strategically important sewers, and the use of the sewered sub-area model in WASSP-SIM should be kept to a minimum.

B Above Ground

The subsequent accuracy of the model is almost directly dependent upon the quality of the above ground survey information. It must be clearly understood at the outset that the objective is to produce a model of the <u>existing</u> system, where a significant proportion of the paved area may not be effectively connected to the sewers. The practice of defining sub-areas as 'separate', 'partially separate' or 'combined' and assuming that they will be properly connected at some time in the future is <u>incorrect</u> and likely to cause serious overprediction by the model. Therefore, proportionally more resources should be provided to investigate connections on site, and search deposited plans in local council archives. It is essential that the engineer is involved in this work, since there are significant differences (associated with surface run-off and the way it enters the sewerage system) between WASSP and traditional design methods, which must be taken into account. The definition of 'total area' for instance, needs to be modified for heterogeneous catchments.

This work can easily be overdone however and expensive resources wasted on gathering insignificant data. If data is not readily available, it is worth postponing the effort until after the 'crude verification' stage which will establish those sub-catchments requiring more detailed investigation.

C Foul Flows

The base foul flow or 'flow in dry weather' is required as an independent input to WASSP. It may be input globally, or as an inflow to each sewer, or a combination of both. The actual values will be verified later by monitoring but an estimate is required at an early stage for crude verification. Once again it is important to remember that it is <u>actual</u> flows which are required and not those which are used for designing new sewers on green field sites. Reference to recorded WRW flows and current water supply rates should be made for residential areas, and trade effluent and supply meters for industry.

Some studies may require a more careful assessment of base flows than others and provision should be made for measurement at strategic points in the system. The short-term installation of portable flumes may be worth considering in some instances.

(iv) <u>Data Input</u>

Data input is not difficult providing the rules set out in Volume 2 of the Wallingford Procedure are adhered to. The mainframe version of WASSP is not particularly user friendly and it is advisable to complete the preprinted data sheets before using the keyboard.

(v) Crude Verification

This is a checking procedure which will help to identify gross errors in the input data. The procedure is to load the model with a range of built-in design storms and to compare results with the recent known performance (ie the simulation of a one year design storm should fit the annual flooding records).

When the obvious errors have been corrected, it is recommended that a meeting is arranged with the sewer maintenance foreman to discuss various features of the system. The foreman's knowledge is an important ingredient of the study and better extracted with the aid of the crude model.

(vi) Input Data Review

As a result of the crude verification, detailed survey work can be scheduled to the appropriate areas. The crude model may also indicate doubtful or unforeseen features which require verification by monitoring, so where surcharge recorders are being used, they may be moved at this stage to more useful locations. It is recommended that where contractors are used for flow survey they are not engaged until the crude model is performing reasonably satisfactorily and the most useful gauging locations are known.

(vii) Gauging

At present there are two approaches:- flow survey contractors, and a method developed within the Authority. Both have advantages and a number of disadvantages, which indicate the current state of the art. The choice will depend upon such factors as:

Availability of in-house equipment
Local experience in using gauges
Urgency of the study
Financial restraints
Availability of contractors

In any event it is important that gauging is correctly managed and properly scheduled to avoid abortive use of resources. The modeller should actively supervise this work and be in a position to supply back-up equipment as quickly as possible when gauges fail.

The modeller must also be aware of any external restraints or influences on the sewerage system during the gauging period. Blockages and collapses are particularly important, together with river levels and the operation/ maintenance of ancillaries. It may also be prudent to observe the progress of on-going development and variations in significant industrial processes.

(viii) Verification/Tuning

Verification involves loading the model with the storms recorded during the gauging period and comparing the model's predictions of rate of flow and level with the sewer gauges.

It should be bourne in mind that a verified simulation model which appears to be not very accurate in engineering terms is a considerable step forward from an unverified design model as shown in Fig.2.

(ix) Improvements

Where improvements are shown to be necessary, the verified model may then be used for testing various feasible options to:-

(a) alleviate existing troublespots
(b) accommodate committed growth
(c) accommodate structure plan growth

The model may subsequently be used to assess the hydraulic adequacy of sewer relining schemes and to indicate the hydraulic loading on inlet works, pumping stations and overflows.

It is important to appreciate that the simulation model will not design improvement schemes, but it is perhaps a retrograde step to fall back to traditional design methods. The recommended method is to artificially increase pipe sizes or introduce retention tanks and orifices into the model and to test their adequacy using the built-in design storms. At this time no specific guidelines comparable to those currently used to design new sewerage systems can be given and it may be argued that specific return frequencies for surcharge or flooding are not relevant to the design of improvements. The exercise is rather one of improving the performance at troublespots up to at least the level of the rest of the system and fully utilising reserve capacity.

(x) Final Report

This should be a formal document presenting the results of the analysis and recommendations for improvements or further investigations. It should be addressed principally to the non-specialist who may not wish to comprehend the jargon associated with urban hydrology or computer models. The inclusion of computer print-outs is not recommended and tabulated data should be reduced to a minimum. A section proving the validity of the computer model will be required and some selected hydrographs of measured and simulated flows should be included. A plan showing how closely the model reproduces existing troublespots will be useful in this section.

The results of the various simulations are best presented on catchment plans as surcharged sewers and flooding points at specified return frequencies. Details of flows and surcharge levels are probably unnecessary.

9 Application in Severn-Trent Water Authority

9.1 The inaccuracy of design methods as applied to existing sewerage systems is illustrated in Fig.2. This is a measure of the difficulty in surveying the catchments in instances where records are poor and under these circumstances the use of design methods for any calculations whatsoever cannot be recommended.

9.2 It is considered that in future more emphasis will be placed on finding short term cost effective solutions to local problems based on flooding frequency. This is beyond the capabilities of design models.

9.3 Fig.2 also shows that survey work for WASSP-SIM is not likely to be much more expensive than for design methods and the additional resources required to produce an effective analysis are associated with verification.

9.4 Verification ranges from crude 'fitting' at troublespots through medium cost in-house techniques to more expensive flow surveys and there is sufficient scope to specify the method to suit the work in hand. Consequently Severn-Trent Water Authority is adopting a flexible approach to meet financial restraints and still make good use of the new technology. Initially this is dependent upon training in three techniques which are new to most engineers. These are, use of computers, WASSP-SIM and sewer monitors, and to reduce the costs of training the Authority has set up centres of expertise which can advise, assist or substitute for local engineers as required.

9.5 With this backing all hydraulic calculations on existing sewerage systems can be based on WASSP-SIM and crude verification will become mandatory.

9.6 Formal verification will not be required in every case and flow surveys by contractors will only be specified on a cost benefit basis or to assist with the analysis of large complex systems. However the Authority is establishing stocks of portable gauges to support the in-house technique described earlier or for carrying out spot checks at particular locations, and taking every opportunity to install suitable gauges at sewage treatment works and major pumping stations.

10 Comments

The approach to modelling and monitoring described in this paper has developed over a period of some four to five years. There are still aspects of simulation which require further research and development and at present, considerable skill and some optimism is often necessary for a number of applications. It is usually the case for instance that simulation is applied initially to complex systems where special difficulties falling outside the range of the various hydraulic sub-models arise. This tends to reduce the confidence of engineers despite the fact that often the particular difficulty could not be solved by any common hydraulic laws.

Experienced users however have learned to accommodate such difficulties and, realising the benefits of simulation to check catchment surveys, would hesitate to accept other unverified calculations. It is always the case that, with complex structures in a system, various extremes can be tested quickly on the computer and the engineer can usually be satisfied with the assumptions made, or specify suitable monitoring for the location.

It is expected that simulation will lead the way towards more cost effective solutions to problems but experience in this field has been limited to date. It is certainly true however that with a more accurate analysis provided by simulation, and the ability to test proposals, premature investment can be controlled.

Looking further ahead, the industry has always had to rely on expert interpretation of calculations to predict flooding. The Wallingford Procedure has already provided a means of designing sewers to a given flooding frequency on green field sites but early attemps to design improvements on this basis have not been particularly successful. By far the majority of properties suffer no flooding whatever and a likely design frequency for flooding is in the order of thirty years or more. At this design frequency simulation models overpredict flooding and the results cannot be accepted. It is hoped therefore that further research and development will solve this problem in the near future.

11 References

1 Sewerage - A Strategic Approach, Severn-Trent Water Authority publication, August 1983.

2 Sewerage Rehabilitation Manual. WRC Engineering, Swindon, 1984.

3 Measurement of the Performance of Sewerage Systems and Associated Hydraulic Structures - Recent Developments. A J Baughen, A R Eadon. Institute of Water Pollution Control Journal, 1982.

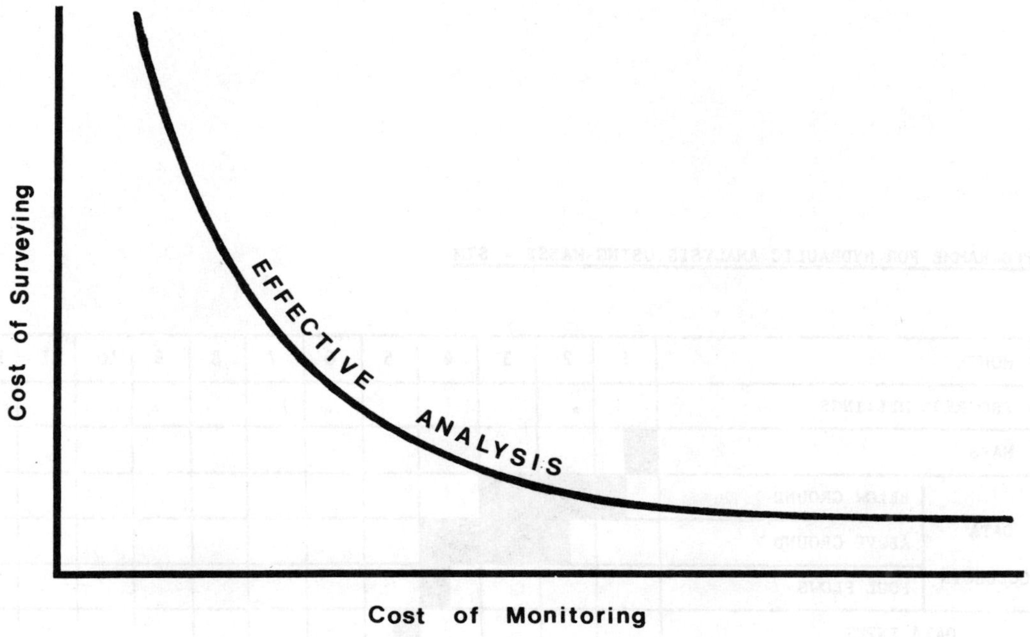
Cost of Surveying
EFFECTIVE ANALYSIS
Cost of Monitoring

Fig. 1

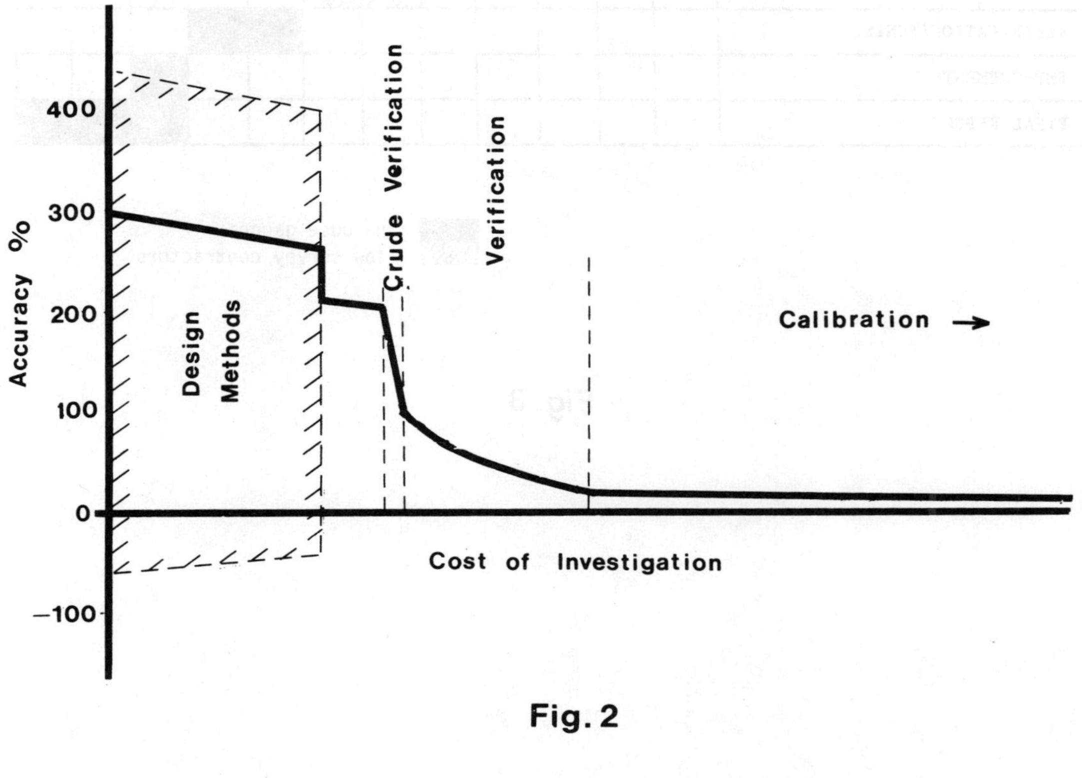
Accuracy %
400
300
200
100
0
–100
Design Methods
Crude Verification
Verification
Calibration →
Cost of Investigation

Fig. 2

PROGRAMME FOR HYDRAULIC ANALYSIS USING WASSP - SIM

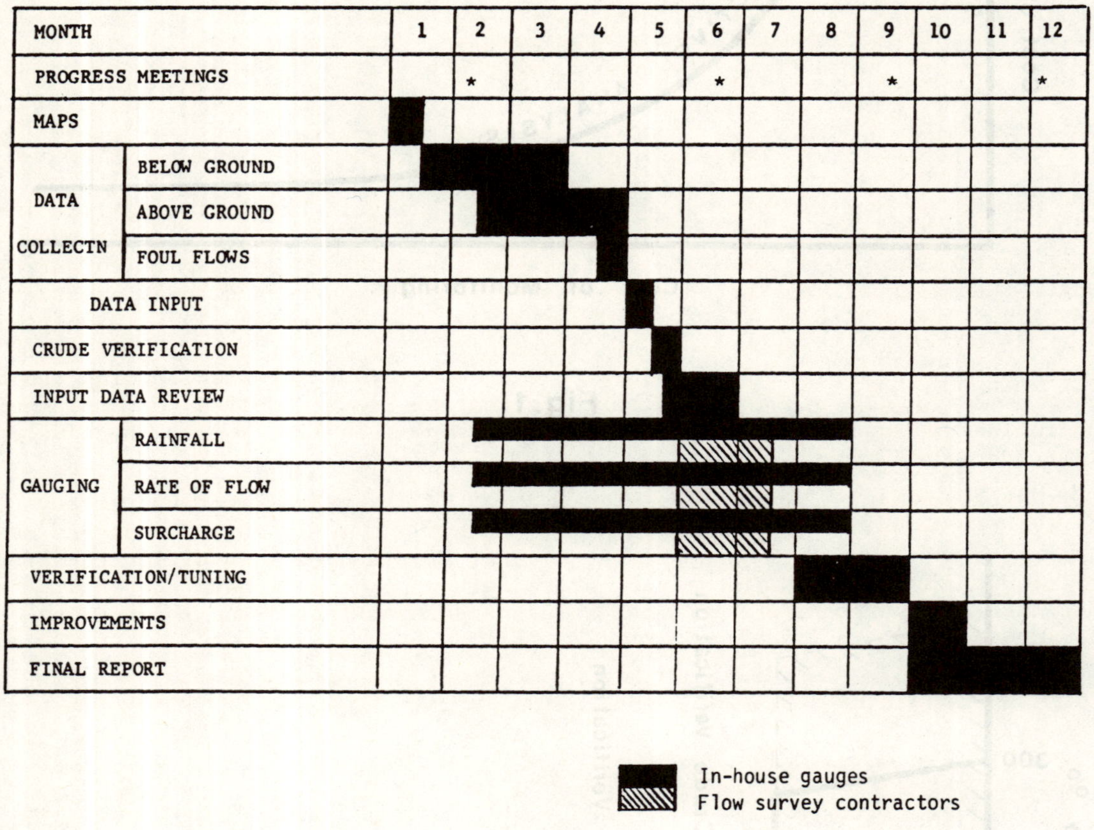

Fig. 3

International Conference on the
PLANNING, CONSTRUCTION, MAINTENANCE & OPERATION OF SEWERAGE SYSTEMS
12-14 September, 1984

PAPER F2

ANALYSIS OF SURCHARGED STORM SEWER SYSTEMS

A.R.D. Thorley* and D.J. Wood†

*Thermo-Fluids Engineering Research Centre, The City University, London, England.
†Department of Civil Engineering, University of Kentucky, Lexington, Kentucky, U.S.A.

Summary

The use of pressurized pipe network hydraulic analysis for analyzing flow in storm sewer systems is investigated. The rationale for this application is that the hydraulic performance of a storm sewer system is satisfactory as long as no surcharging develops. When significant surcharging occurs the pipe network becomes pressurized and the principles governing flow in pressurized piping systems apply. Using this approach analyses can be carried out to determine the degree of surcharging which occurs and, if surface flooding occurs, the degree and duration of surface flooding can be predicted. An advantage of this approach is that it can be applied to complex piping systems including looped systems and systems which contain lift pumps, check valves and other hydraulic devices.

This approach is applied to a representative London storm sewer system to demonstrate the pertinent considerations. Single event analyses and analyses over a period of time are carried out to illustrate this approach. In addition the effect of varying certain critical system parameters such as pipe roughness and loss considerations at manholes and vertical inlets are investigated and shown to be significant.

Held at The University of Reading, England. Organised and sponsored by BHRA, The Fluid Engineering Centre, Cranfield, Bedford in conjunction with the Water Research Centre.

NOMENCLATURE

a	=	exponent for irreversible loss relations
A_s	=	surface area for flooded region
E_p	=	energy input by pump
f	=	number of fixed hydraulic grade nodes in pipe system
j	=	number of junction nodes in pipe system
K	=	constant for irreversible loss relations
ℓ	=	number of closed loops in pipe system
p	=	number of pipes in pipe system
Q	=	flowrate in pipe
Q_e	=	flowrate entering pipe system at junction node
Q_{in}	=	flowrate in pipes flowing toward junction
Q_{out}	=	flowrate in pipes flowing away from junction
Q_s	=	flowrate from pipe system to surface detention area
ΔH	=	difference in H.G.L. between boundary reservoirs
Δt	=	time period for extended period simulation calculations

1. INTRODUCTION

Storm sewer systems are characterized by a series of nearly horizontal pipes which are connected together to form a network of pipes. The pipes are usually 2 to 5 metres below the ground level though sometimes are much deeper, and are connected to the ground by a number of vertical inlet pipes and manholes which provide access to the underground pipes. The purpose of the storm sewer system is to carry off runoff from rain storms. The rainfall runs over the ground through gutters and other channels and enters the storm sewer system through the vertical pipe inlets.

Simply stated the sole purpose of the storm sewer system is to remove the storm runoff and to do so in such a manner that unacceptable surface flooding will not occur. For small inflows into the storm sewer system the flow will occur as free surface flow and under this condition the storm sewer system performs acceptably. However, as the inflow increases portions of the system will be surcharged and the flow will occur as pressurized conduit flow. When this occurs the water rises in the vertical pipe inlets and manholes. Additional increases in inflow may result in a condition where the water in the vertical sections rises to the ground level and surface flooding may occur. In some cases the degree of flooding may be such that disruptions to street traffic may occur and basement or ground level flooding of buildings is possible.

It is necessary to be able to predict the possible occurrence of flooding and the duration and severity of flooding of storm sewer systems due to intense storms. These predictions are necessary to properly design new storm sewer systems or additions to existing storm sewer systems and to properly evaluate the effect of surface developments on existing storm sewer systems. The principal theme of this paper is to demonstrate that an appropriate analytical tool for evaluating surcharged flow in storm sewer systems is a hydraulic analysis model for pressurized pipe network flow. If no surcharging or only limited surcharging occurs, then the storm sewer system is adequate and a pressurized pipe network hydraulic analysis model is appropriate for demonstrating that this situation exists. When significant surcharging occurs with possible flooding, then the presssurized pipe network hydraulic analysis model correctly performs the required hydraulic calculations. In addition, complications such as looped connections, in-line pumping and surface flooding are easily handled by a well conceived hydraulic network model.

Currently there are several models proposed for analyzing flow in storm sewer systems. These include the Storm Water Management Model (SWMM)(1) and the Wallingford Model (2) in addition to models proposed by Song (3) and Yen (4). These models typically account for both free surface and some pressurized flow and include some dynamic effects due to the acceleration of the water in the sewer pipes. Some work has been carried out on the complex phenomenon associated with the transition from free surface to pressurized flow (5). For the most part these models concentrate on simulating the storm sewer flow starting from the initiation of the storm through the free surface phase and into the pressurized phase if this condition is reached. These models are relatively complex and may require certain simplifications and assumptions (such as non looped configurations, no pumping, etc.). Most of the complexities are due to attempts to model the free surface flow and the transition to pressurized flow. Precise modelling of these flow conditions has very little effect on what may be the most significant result of the simulation, the prediction of the degree and duration of any flooding which may occur.

In many instances the only result of substantial interest in a storm sewer hydraulic analysis is the degree of surcharge and, if flooding occurs, the degree and duration of flooding which is calculated for a specified storm. A steady state pressurized pipe network analysis which can account for external storage is proposed as an appropriate procedure for calculating extreme conditions in a storm sewer system. It has been previously demonstrated that a dynamic model of a pressurized storm sewer system yields results which are not significantly different from a steady state model which accommodates external storage (6). This means that even though flow conditions are changing in the storm sewer system, mass-acceleration terms are generally small compared to friction forces and neglecting mass-acceleration terms is acceptable.

ANALYSIS

Figure 1 represents a schematic of a simplified storm sewer system. This system consists of pipe sections which connect junction nodes and boundaries at a fixed hydraulic grade. For all such systems of pipes the following holds

$$p = j + \ell + f - 1 \tag{1}$$

where p is the number of pipe sections, j is the number of junction nodes and f represents the number of nodes at a fixed hydraulic grade. A hydraulic analysis of this system is required to determine the unknown flowrates, Q. The set of equations to determine the unknown flowrates includes the continuity equations for each junction

$$\Sigma Q_{in} - \Sigma Q_{out} = Q_e \text{ (j equations)} \tag{2}$$

where Q_e is the flow entering or leaving the system at that junction. For each loop there is an energy equation which can be expressed as

$$\Sigma_{losses} = \Sigma E_p \text{ (ℓ equations)} \tag{3}$$

where E_p represents the energy put into the system by pumps and losses represents the irreversible losses due to pipe friction and fittings (entrances, exits, junctions, etc.). Between the nodes of constant hydraulic grade one can apply the energy equation in the form

$$\Delta H = \Sigma_{losses} - \Sigma E_p \text{ (f-1 equations)} \tag{4}$$

where ΔH is the difference in the hydraulic grades between the two end nodes. The losses can be expressed as a function of flowrate, i.e.

$$\Sigma_{losses} = KQ^a \tag{5}$$

where the constant K depends on the physical characteristics of the pipe line or fitting where the loss occurs. The pump energy can also be expressed as a function of flowrate

$$E_p = f(Q) \tag{6}$$

This function is determined from a plot of pump head and flowrate (pump characteristic curve).

Equations 2-4 with the relations defined in equations 5 and 6 represent a set of p simultaneous non linear algebraic equations from which the set of p unknown flowrates can be determined. No direct method for solving this set of equations is known. However, a number of algorithms have been proposed for obtaining the solution. The linear method of solution involves applying the Newton-Raphson procedure to remove the non-linearities and solving the resulting set of linear equations. The linearization is repeated using the solution and this procedure is repeated until the correct solution is obtained (7). A recent paper showed this method to be a very reliable procedure for obtaining the solution set while some of the more commonly applied Hardy Cross (8) and nodal methods do not always converge to the correct solution (9). Computer programs utilizing various algorithms and incorporating various features have been developed. A program developed by the authors based on a linearization procedure and incorporating features useful for storm sewer system analysis was utilized for the applications illustrated herein (10).

3. APPLICATION TO STORM SEWERS

Hydraulic network analyses are generally carried out for a specified set of conditions. Flowrates and hydraulic grade lines throughout the system are obtained by solving the basic hydraulic equations. This single event type of analysis is useful for evaluating the storm sewer system's ability to remove specified constant inflows. If a system can remove the various peak inflows expected for a particular storm applied simultaneously without flooding, then the system will be able to handle the actual time variable inflows which occur during a storm.

The significant results for the single event type of analysis are the hydraulic grade lines which occur throughout the storm sewer system. These represent the degree of surcharge which is required to remove the inflows to the pressurized pipe system. Figure 2 illustrates the nature of the results obtained from this type of analysis. The specified inflows are obtained from the storm intensity and storm runoff calculations. If the hydraulic grade lines result in a surcharged condition which exceeds the surface ground level, then surface flooding is indicated. To properly model this condition the hydraulic grade lines in the flooded areas should be maintained near surface ground levels. This can be accomplished by introducing a line connecting a storage area with a water level slightly above ground level with a check valve which only allows flow to go to the surface. This is depicted in Figure 3. This arrangement allows the excessive inflow (over what the system can remove) to go to the surface detention area producing flooding conditions. The magnitude of the flow to the surface may be used to evaluate the potential for flooding. With this arrangement if the inflow does not produce flooding only surcharging occurs and all the specified inflow is removed by the storm sewer system.

If a single event simulation indicates that flooding occurs, an extended period simulation may be carried out to determine the degree and duration of flooding. These time simulations employ single event simulations to determine conditions at a specified time by using storm runoff inflow hydrographs and relations for storage in surface detention areas to project changes in water levels in the detention areas. Figure 4 depicts this simple procedure for handling storage. The change in water level over the time increment, Δt, is approximately given as:

$$H = Q_s \Delta t / A_s \tag{7}$$

where Q_s is the net flow into the flooded area and A_s is the surface area of the flooded region. Equation 7 is used to compute changes in the water surface level of the flooded area and this is used along with inflows determined from storm intensity and surface runoff characteristics. The time increment used for the time simulations must be small enough to reasonably model the inflow hydrographs and also small enough so that the water surface level changes are adequately calculated by Equation 7.

The inflow hydrographs are required input for the hydraulic analysis and these may vary in any prescribed manner for any of the specified storm system inflows. The proposed time simulation does not directly incorporate the underground storage and, as such, is most appropriately applied after the system becomes surcharged. It is possible to introduce additional storage distributed throughout the system to represent underground storage. However, a more precise time simulation (including below ground storage) of conditions prior to surface flooding is unlikely to provide any result more significant than the degree of surcharging throughout the storm sewer system when flooding commences. This result is adequately computed prior to the onset of surface flooding neglecting underground storage. It may be noted that the results for time simulations prior to surface flooding are identical to the result predicted by a single event analysis for the same set of specified inflows. These features are illustrated by the results of example calculations for a typical storm sewer system.

4. AN EXAMPLE SYSTEM

The storm sewer system that has been used in the analyses to be described is a self-contained section of the drainage systems in one of the London Boroughs. It is illustrated schematically in Figure 5, together with a numbering scheme for the 46 pipes and 33 junction nodes. Storage tanks A-H represent detention basins to enable estimates of flooding volumes to be made, if necessary, and 'reservoir' I is the sump at the pumping station to which the storm water is intended to flow.

This system was extensively surveyed in the early 1970s and physical data on pipe lengths and elevations, location of manholes, etc. were available. The system also contained loops, a feature not typical of some sewage schemes.

As the system is some 75-100 years old engineering judgement was required when collating data on sewer diameters, surface roughness and minor loss characteristics. Nominal values were used in the initial studies. Parametric studies followed to assess the importance of the various factors on the surcharge and flooding characteristics of the system.

The inflow to the system was based on rainfall data for the London area for storms having a return period of 1,2 and 3 years. The typical storm lasted 30 minutes and had a rainfall hyetograph as shown in Table I for a storm on a 3-year return cycle. This profile can be normalized and is illustrated graphically in Figure 6. A typical profile for a suburban area is nearly symmetrical (see, for example, Figure 7.3 in ref.2). For the present purposes a symmetrical one has been used. The one- and two-year storm profiles have the same shape and the amplitudes of the rainfall hyetographs are 69.9% and 89.4% respectively.

For the purpose of establishing a baseline for the studies it was assumed that 65% of the rainfall onto paved and roof areas would be input to the system. Surface roughness of the sewers was taken as 1.5 mm and modest values of minor loss coefficients (e.g. 0.3) were used.

5. RESULTS AND DISCUSSION

The analyses were performed on a Sirius 1 16-bit micro-computer. An excerpt of the initial data summary from one of the early runs is shown in Table II. It indicates line sizes, roughness heights, elevation of reservoirs, etc. Following the hydraulic analysis a comprehensive set of results is produced, a small sample being shown in Tables III and IV. These two Tables refer to the flows near the height of the storm having a 2-year return cycle.

The system is surcharged because the hydraulic grades are generally higher than the sewer elevation, giving rise to positive pressures. A limited amount of flooding is indicated, particularly at junctions 16-18 and 24, however it is understood that the manholes will withstand some positive pressure.

As mentioned previously, engineering judgement was required regarding data such as minor loss coefficients,surface roughness, etc. It also applied to some extent to the distribution of the inflow to the system. How important is the accuracy of this data? Is a reasonable approximation adequate or is special care required with some parameters? These factors have been investigated using the storm on a 2-year return cycle as the source of basic inflow data. The effects as felt at selected locations in the system are presented - the 'selected' points being chosen to represent locations both near the extremities and well down the system where the branches converge towards the final leg running into the pumping station sump.

5.1. Frictional Effects

The program used in this study distinguishes between friction losses due to pipeline friction and the minor or local losses associated with junctions, manholes, etc.

Line friction is determined from the Darcy Weisbach equation, for which information on surface roughness is required. An initial assumption was a roughness height of 1.5 mm, with follow-up calculations using 3.5 mm and 5 mm. Figure 7 illustrates the effect on the level of maximum surcharge at junction nodes 4, 12, 18 and 25. It appears that above about 2 mm the effect of surface roughness is small as the graphs flatten out. Similar results occur at other locations.

Minor loss coefficients, however, seem rather more important - see Figure 8. For the initial baseline run a value of 0.3 was used for each line. For later runs some variation was introduced to more closely model junctions and manholes, etc. These individual values were factored up for the test runs and mean values used for the purpose of plotting data out as on Figure 8. That these K-values are more influential should not be unexpected due to the number of inflow points to the system and the number of branches and junctions.

5.2. Partial Blockages

The detailed maps from which the original data were extracted were anotated in several places to the effect that some of the pipelines were partially or totally blocked. The effect of generalised partial blockages was investigated and a set of results for the four junction nodes in the previous discussion is given in Figure 9.

If very significant blockage occured clearly flooding would result in all cases. At more modest degrees of blockage the effects on the system are not uniform. With only small degrees of blockage the surcharge levels increase, but then at higher levels of blockage, e.g. above 20% (i.e. less than 80% of flow c.s.a. is available) the surcharge level drops in some regions - e.g. junction nodes 4 and 12 on Figure 9. The explanation for this is that the storm water is held up in the more remote sections of this particular system, significantly increasing the surcharge and risk of flooding there, whereas lower down (towards the bottom in Figure 5) the system can still cope reasonably satisfactorily.

5.3. Storm Intensity

Very careful judgement has to be exercised when assessing the proportion of rain water which actually enters the sewer system. In the preceding discussion it was assumed to be 65% of the rainfall onto paved areas and roofs. Varying this parameter is analogous to varying the intensity of the storm - the effect of which is illustrated in Figure 10, showing the surcharge levels at junctions 12, 18 and 25 for two mean values of minor loss coefficient.

The data were obtained from a series of steady state analyses at intervals during a storm based on the hyetograph for a 2-year return period. However, this could be extended to cover longer return periods and a subsidiary baseline is incorporated in Figure 10.

The results indicate that at nodes 18 and 20, once it starts, surcharging occurs quite rapidly with storm intensity until ground level is reached, after which it is much slower. The detention basins were represented by tanks having a mean diameter of 10 m which is perhaps on the small size. A larger one would reduce the predicted flood level and then the upper parts of these graphs would be quite flat. At junction 12 only surcharging occurs - no flooding.

The dashed lines represent a parallel study using the higher minor loss coefficients (mean K = 2.6). Surcharging and flooding follow a similar pattern but occur earlier and reach greater elevations.

5.4. Extended Period Simulations

Once surcharging and flooding start to occur features of primary interest are the extent and duration. The mere fact that they occur can be ascertained from steady state analyses but to determine the extent and duration extended period simulations are required. Figure 11 is the first of the specimen results included in the present paper and refers to conditions at junction node 15.

The lower curve shows the extent and duration of surcharging and a short period of flooding for the case when the minor loss coefficients are relatively low. A similar curve is also shown for a similar run with one of the higher values of loss coefficients (average K = 2.6). Earlier surcharging and flooding to a greater depth occurs and they both last for a longer period of time. Also the peak events are delayed.

Superimposed on this graph is the profile of the Storm Intensity Index; the peak of which precedes the flooding peaks.

It should be noted that the depth of flooding predicted is again influenced by the size of the detention basin at H, namely a mean diameter of 10 m.

The extent of the flooding can also be considered in terms of the flooded volume. This is shown on Figure 12 for the same node as that of Figure 11, i.e. junction 15 and detention basin H. The time delay between the height of the storm and peak flooding is again apparent.

6. SUMMARY AND CONCLUSIONS

The primary objective of this paper has been to demonstrate that a pressurized pipe network algorithm can be applied to a storm sewer system. The onset of surcharging can be predicted and the extent and duration of both surcharging and flooding can be investigated.

The accuracy of any computer study is only as good as the model and the basic data with which it has been supplied. This particular model has stood the test of serious investigation (9). When applied to particular systems, as with instrumentation, some degree of calibration is required with any computer study. This arises due to the uncertainties over numerical values for pipe roughnesses, minor loss coefficients, silting up of lines, etc. It has been illustrated that partial blockage and the accuracy of minor loss characteristics are more significant than surface roughness so far as surcharging and flooding predictions are concerned. Storm intensity, and the related parameter of the proportion entering the sewer system, are also of considerable importance. Once surcharging occurs it builds up rapidly, following on the rapid increase in rainfall rate. There is some comfort perhaps in the fact that the flowrates that a surcharged system can handle are greater than one having a free-surface flow.

Ideally, comparisons with field measurements are desirable but these are seldom available in a comprehensive form for storm flows due to the random nature of such events. The next phase of the present project is to effect comparisons with other storm water models in the United Kingdom. The results of this continuation will be reported in due course.

7. ACKNOWLEDGEMENTS

The example application described in this paper would not have been possible without the consent and co-operation of the Thames Water Authority who kindly provided the basic data. This, together with the opportunities to discuss the scope of the parametric studies and their general interest and encouragement is much appreciated by the authors. Particular recognition is due to Mr.W.G. Bedford and Mr.V.Jeyanathan.

8. REFERENCES

1. Roesner, L.A., Shubinski, R.P., and Aldrich, J.A.: "Storm Water Management Model User's Manual Version III, Addendum I EXTRAN," U.S. Environmental Protection Agency, Cincinnati, Ohio, 1981.

2. Dept.of Environment: "Design and Analysis of Urban Storm Drainage - the Wallingford Procedure" Vols.1 and 2. The National Water Council, Queen Anne's Gate, London, 1981.

3. Song, C.C.S.: "Two-Phase Flow Hydraulic Transient Model for Storm Sewer Systems," International Conference on Pressure Surges, Sept.1976, BHRA Fluid Engineering, Cranfield, Bedford, England, 1976.

4. Yen,B.C., and Pansic,N.: "Surcharge of Sewer Systems," Research Report No.149, Water Resources Center, University of Illinois, Urbana, Illinois, March, 1980.

5. Yen, B.C.: "Hydraulic Instabilities of Storm Sewer Flows," Urban Storm Drainage, Proceedings of the First International Conference on Urban Storm Drainage, Ed. by P.R. Helliwell, pp.282-293, Wiley-Interscience, New York, 1978.

6. Heitzman, G.C. and Wood, D.J.: "Dynamic and Steady Flow Modelling of Surcharged Storm Sewer Systems," International Symposium on Urban Hydrology, University of Kentucky, Lexington, KY, 1982.

7. Wood, D.J. and Charles, C.O.: "Hydraulic Network Analysis Using Linear Theory," Journal of the Hydraulics Division, ASCE, Vol.98, No.HY7, July, 1972.

8. Cross, H.: "Analysis of Flow in Networks of Conduits of Conductors," Bulletin No.286, University of Illinois Engr., Exp.Station, Urbana, Illinois, 1936.

9. Wood, D.J. and Rayes, A.: "Reliability of Algorithms for Pipe Network Analysis, Journal of the Hydraulics Division, ASCE, Vol.107, No.HY10, Oct.1981.

10. Thorley, A.R.D. and Wood, D.J.: "BASIC Computer Program for the Analysis of Pressure and Flow in Pipe Distribution Systems Including Extended Period Simulations," Dept.of Mech.Eng., The City University, London, 1983.

TABLE I. Rainfall Hyetograph in mm/hr at 60 second intervals

10.13	10.65	11.18	11.67	12.51	13.53	14.82	16.43	18.69	22.80
29.61	41.50	59.49	81.18	109.00	109.00	81.18	59.49	41.50	29.61
22.80	18.69	16.43	14.82	13.53	12.51	11.67	11.18	10.65	10.13

TABLE II. Sample of the Basic Data for the Sewer System

Pipe Network Analysis - SI Version
Flow Units are Litres/Second

Input Data File Name for this simulation = TW21SSD
Number of Pipes = 46
Number of Junction Nodes = 33

**** Summary of Input Data ****

Pipe No.	Node 1	Node 2	Length (m)	Diam. (mm)	Roughness (mm)	Sum-M Fact.	Pump Type	FGN Grade (m)
1	14	0	174.0	625.0	1.5	0.3	0.0	-1.2
2	1	2	95.0	300.0	1.5	0.3	0.0	
3	2	3	79.0	300.0	1.5	0.3	0.0	
4	3	4	71.0	300.0	1.5	0.3	0.0	
5	4	7	64.5	375.0	1.5	0.3	0.0	
6	7	8	22.4	600.0	1.5	0.3	0.0	
7	8	12	118.4	600.0	1.5	0.3	0.0	
8	12	13	90.8	600.0	1.5	0.3	0.0	
9	13	14	79.0	600.0	1.5	0.3	0.0	
10	1	0	10.0	300.0	1.5	0.3	0.0	6.6
11	11	0	10.0	300.0	1.5	0.3	0.0	5.2
12	11	10	61.8	300.0	1.5	0.3	0.0	
13	10	8	90.8	300.0	1.5	0.3	0.0	
14	8	0	10.0	300.0	1.5	0.3	0.0	5.4
15	10	9	52.6	225.0	1.5	0.3	0.0	
16	9	3	60.5	225.0	1.5	0.3	0.0	
17	3	6	82.0	225.0	1.5	0.3	0.0	
18	6	5	68.5	225.0	1.5	0.3	0.0	
19	5	4	72.0	225.0	1.5	0.3	0.0	
20	7	21	46.0	600.0	1.5	0.3	0.0	
21	21	0	10.0	300.0	1.5	0.3	0.0	5.6
22	30	21	39.5	300.0	1.5	0.3	0.0	
23	31	30	68.4	300.0	1.5	0.3	0.0	
24	32	31	63.0	300.0	1.5	0.3	0.0	
25	33	32	64.5	300.0	1.5	0.3	0.0	
26	33	0	10.0	300.0	1.5	0.3	0.0	6.5
27	22	21	59.2	375.0	1.5	0.3	0.0	
28	23	22	65.0	375.0	1.5	0.3	0.0	
29	24	23	32.0	375.0	1.5	0.3	0.0	
30	25	24	68.5	375.0	1.5	0.3	0.0	
31	26	25	98.7	375.0	1.5	0.3	0.0	
32	27	26	84.2	375.0	1.5	0.3	0.0	
33	28	27	60.5	375.0	1.5	0.3	0.0	
34	29	28	72.4	300.0	1.5	0.3	0.0	
35	29	0	10.0	300.0	1.5	0.3	0.0	8.3
36	15	12	82.9	225.0	1.5	0.3	0.0	
37	16	15	102.0	225.0	1.5	0.3	0.0	
38	17	16	56.6	225.0	1.5	0.3	0.0	
39	18	17	60.5	225.0	1.5	0.3	0.0	
40	18	23	81.6	225.0	1.5	0.3	0.0	
41	19	18	98.7	225.0	1.5	0.3	0.0	
42	19	24	104.0	225.0	1.5	0.3	0.0	
43	20	19	79.0	225.0	1.5	0.3	0.0	
44	20	25	140.8	225.0	1.5	0.3	0.0	
45	15	0	10.0	300.0	1.5	0.3	0.0	4.7
46	25	0	10.0	300.0	1.5	0.3	0.0	6.7

TABLE III. Sample of Storm Water Flows at the Height of a Storm

No.of Trials = 9 - Accuracy = .0008

Pipe No.	Node 1	Node 2	Flow Rate	Head Loss	Minor Loss	Pump Head	Line Velocity
1	14	0	706.18	1.87	0.08	0.00	2.30
2	1	2	48.60	0.24	0.01	0.00	0.69
3	2	3	75.60	0.48	0.02	0.00	1.07
4	3	4	79.48	0.47	0.02	0.00	1.12
5	4	7	138.25	0.40	0.02	0.00	1.25
6	7	8	498.60	0.15	0.05	0.00	1.76
7	8	12	607.88	1.17	0.07	0.00	2.15
8	12	13	706.18	1.21	0.10	0.00	2.50
9	13	14	706.18	1.05	0.10	0.00	2.50
LINE NO.10 IS SHUT OFF							
11	11	0	10.05	0.00	0.00	0.00	0.14
12	11	10	27.75	0.05	0.00	0.00	0.39
13	10	8	84.44	0.68	0.02	0.00	1.19
LINE NO.14 IS SHUT OFF							
15	10	9	-31.85	0.26	0.01	0.00	0.80
16	9	3	-21.05	0.13	0.00	0.00	0.53
17	3	6	2.08	0.00	0.00	0.00	0.05
18	6	5	14.50	0.07	0.00	0.00	0.36
19	5	4	33.94	0.40	0.01	0.00	0.85
20	7	21	-340.91	0.14	0.02	0.00	1.21
LINE NO.21 IS SHUT OFF							
22	30	21	111.24	0.51	0.04	0.00	1.57
23	31	30	82.08	0.48	0.02	0.00	1.16
24	32	31	52.92	0.19	0.01	0.00	0.75
25	33	32	23.76	0.04	0.00	0.00	0.34
LINE NO.26 IS SHUT OFF							
27	22	21	204.83	0.80	0.05	0.00	1.85
28	23	22	167.03	0.58	0.03	0.00	1.51
29	24	23	152.54	0.24	0.03	0.00	1.38
30	25	24	105.47	0.25	0.01	0.00	0.95
31	26	25	150.12	0.72	0.03	0.00	1.36
32	27	26	120.96	0.40	0.02	0.00	1.10
33	28	27	72.36	0.10	0.01	0.00	0.66
34	29	28	48.60	0.18	0.01	0.00	0.69
LINE NO.35 IS SHUT OFF							
36	15	12	60.49	1.46	0.04	0.00	1.52
37	16	15	51.81	1.32	0.03	0.00	1.30
38	17	16	22.65	0.14	0.00	0.00	0.57
39	18	17	22.65	0.15	0.00	0.00	0.57
40	18	23	14.49	0.09	0.00	0.00	0.36
41	19	18	23.11	0.26	0.01	0.00	0.58
42	19	24	12.51	0.08	0.00	0.00	0.31
43	20	19	21.57	0.18	0.00	0.00	0.54
44	20	25	3.27	0.01	0.00	0.00	0.08
45	15	0	20.48	0.00	0.00	0.00	0.29
46	25	0	82.48	0.07	0.02	0.00	1.17

Units: Flow Rate (litres/second)
Head Loss (m)
Line Velocity (m/s)

TABLE IV. Sample of Storm Water Flows and Hydraulic Grade Elevations at the Height of a Storm

Junction No.	Elevation (m)	Demand (ℓ/s)	Pressure (kPa)	Hydraulic Grade (m)	Ground Elevation (m)
1	4.7	-48.6	16.1	6.3	6.6
2	4.2	-27.0	18.8	6.1	-
3	3.5	-27.0	20.0	5.6	-
4	3.1	-24.8	19.3	5.1	-
5	3.7	-19.4	17.7	5.5	-
6	4.1	-12.4	15.0	5.6	-
7	2.4	-19.4	22.8	4.7	5.62
8	2.3	-24.8	21.2	4.5	5.45
9	3.7	-10.8	17.5	5.4	5.44
10	3.2	-24.8	19.5	5.2	5.37
11	3.5	-37.8	17.1	5.2	5.23
12	2.0	-37.8	11.9	3.2	4.82
13	1.8	0.0	1.6	1.9	5.01
14	1.0	0.0	-2.1	0.8	-
15	2.9	-29.2	18.4	4.7	4.73
16	3.2	-29.2	28.6	6.1	4.87
17	3.6	0.0	25.7	6.2	4.90
18	4.1	-14.0	22.8	6.4	6.11
19	4.9	-14.0	17.6	6.7	6.89
20	5.5	-24.8	13.5	6.8	7.45
21	2.8	-24.8	19.6	4.8	5.59
22	2.9	-37.8	27.8	5.7	5.81
23	3.1	0.0	31.0	6.3	6.28
24	3.5	-34.6	30.1	6.6	6.30
25	3.7	-34.6	30.3	6.8	6.74
26	4.1	-29.2	34.4	7.6	7.35
27	4.3	-48.6	35.7	8.0	7.83
28	4.6	-23.8	34.7	8.1	8.06
29	5.5	-48.6	27.2	8.3	8.31
30	2.8	-29.2	24.9	5.4	5.89
31	3.9	-29.2	19.4	5.9	6.41
32	4.3	-29.2	17.6	6.1	6.53
33	4.7	-23.8	14.4	6.1	6.52

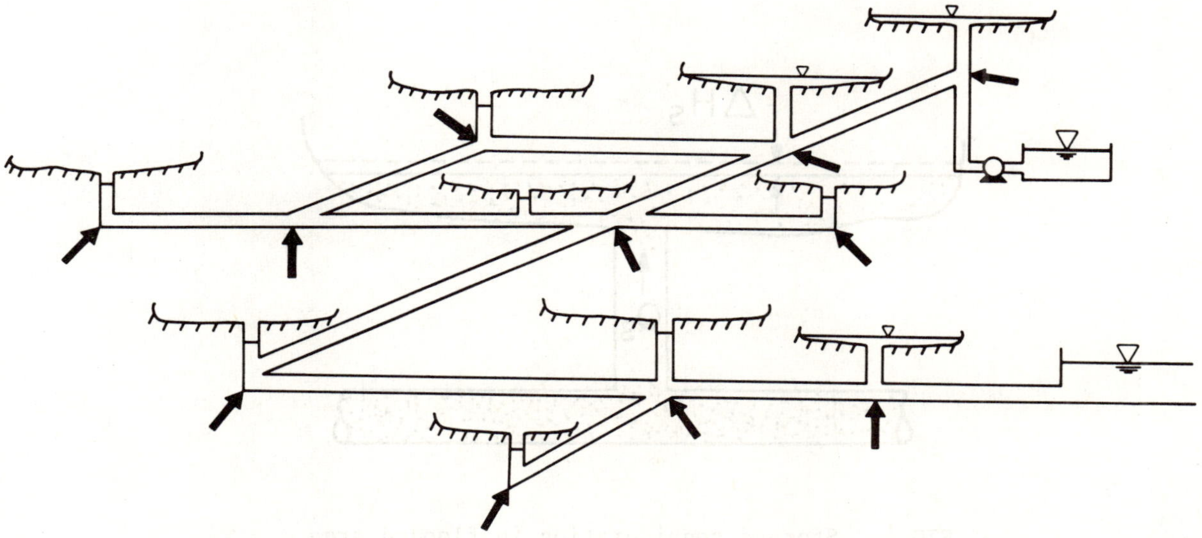

FIG.1. Surcharged storm sewer system

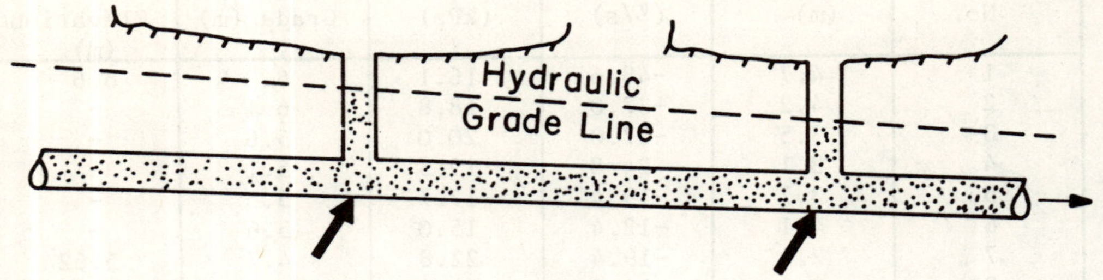

FIG.2. Hydraulic analysis results showing surcharge

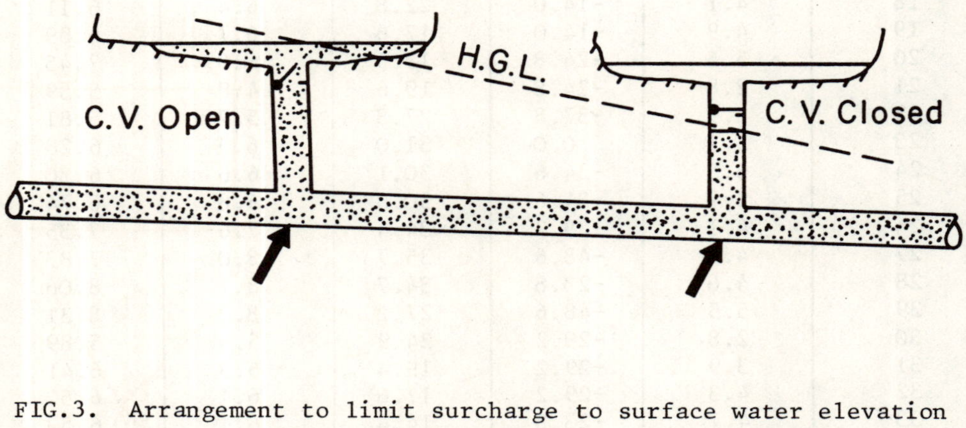

FIG.3. Arrangement to limit surcharge to surface water elevation

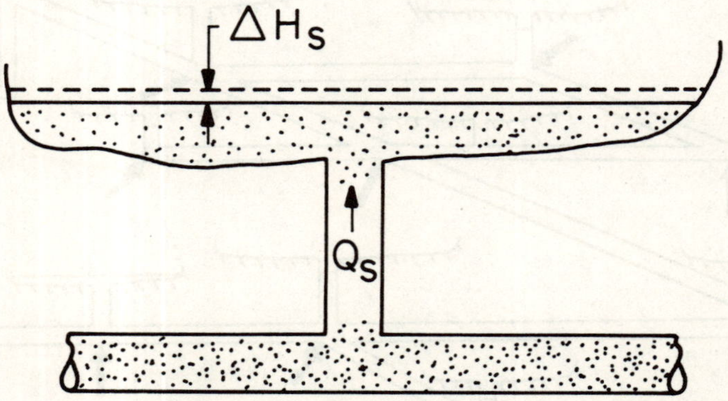

FIG.4. Storage consideration in flooded area

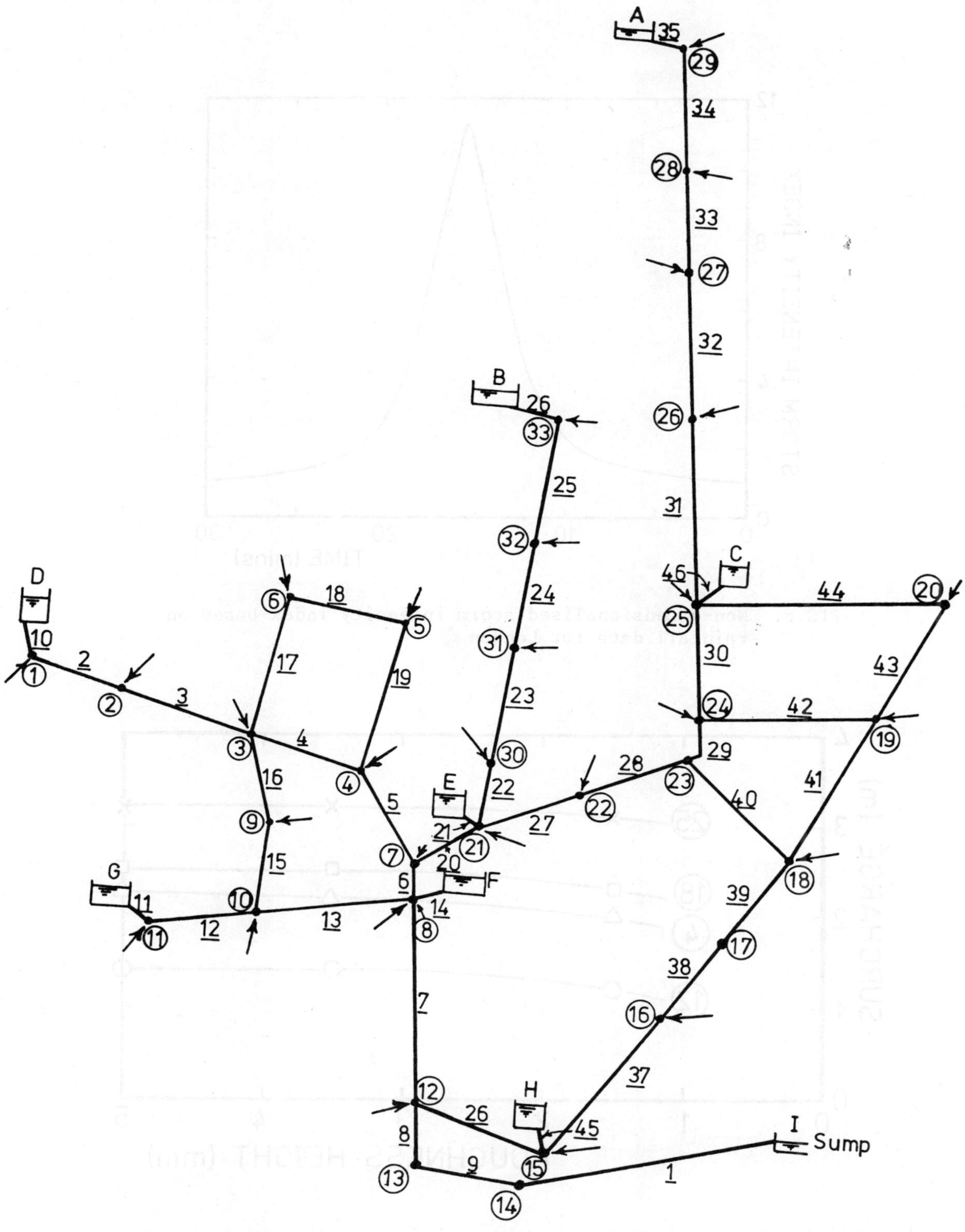

FIG.5. A storm sewer system in London showing the numbering scheme for pipes (underlined) and junction nodes (circled). Storm water inflows are indicated by arrows; letters A-H denote potential detention basins for flooding and I is the sump to which storm water drains.

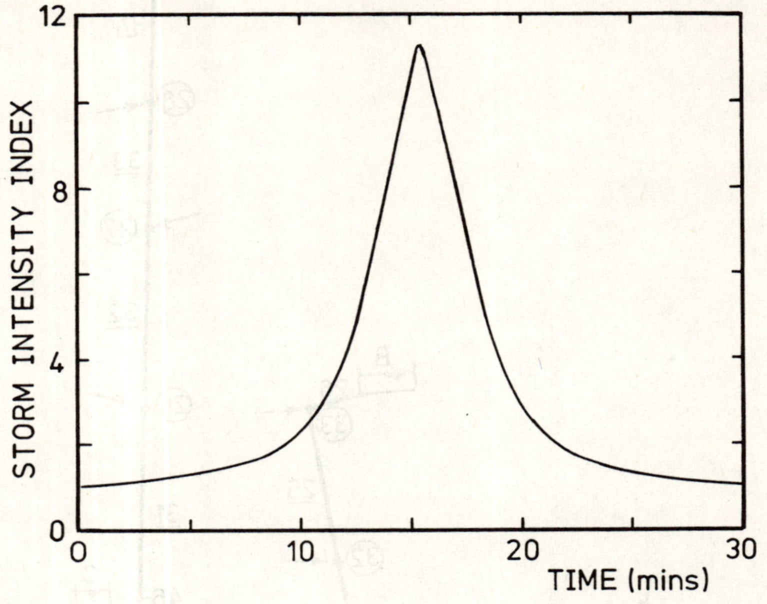

FIG.6. Non-dimensionalised storm intensity index based on rainfall data for London

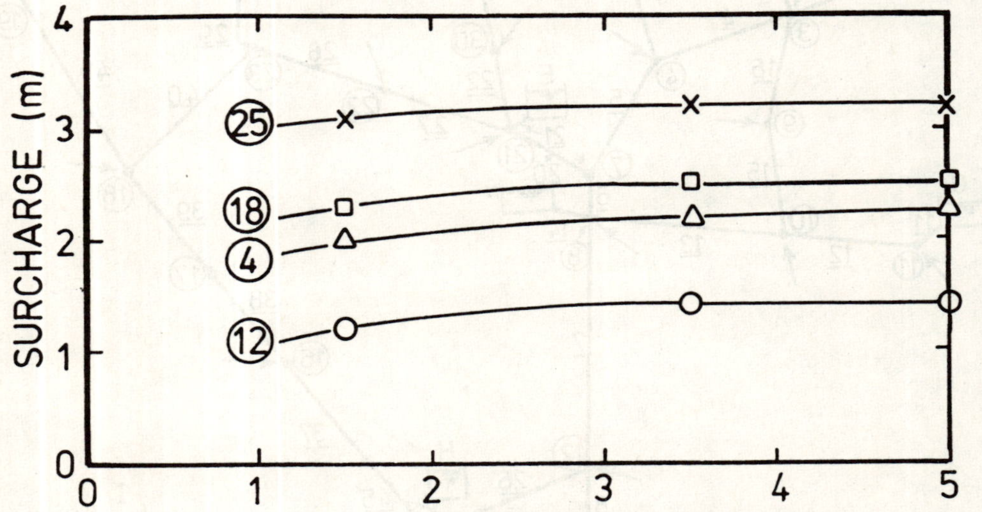

FIG.7. Effect of surface roughness height in the sewers on the maximum surcharge at four representative junction nodes (4,12,18 and 25) due to a storm having a two-year return period.

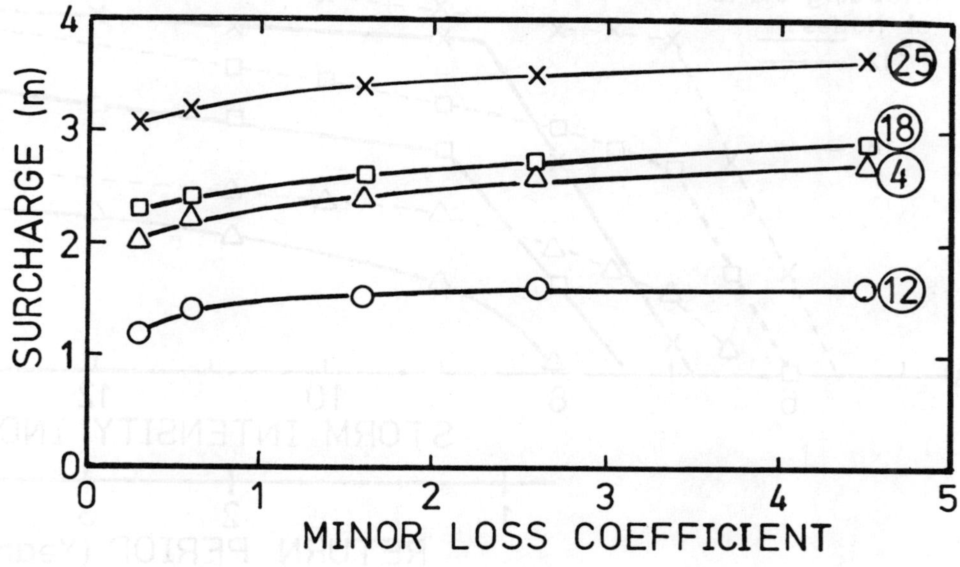

FIG.8. Effect of minor loss coefficient at junctions and manhole connections on the maximum surcharge levels at junction nodes 4,12,18 and 25. The baseline is the average minor loss coefficient in the system.

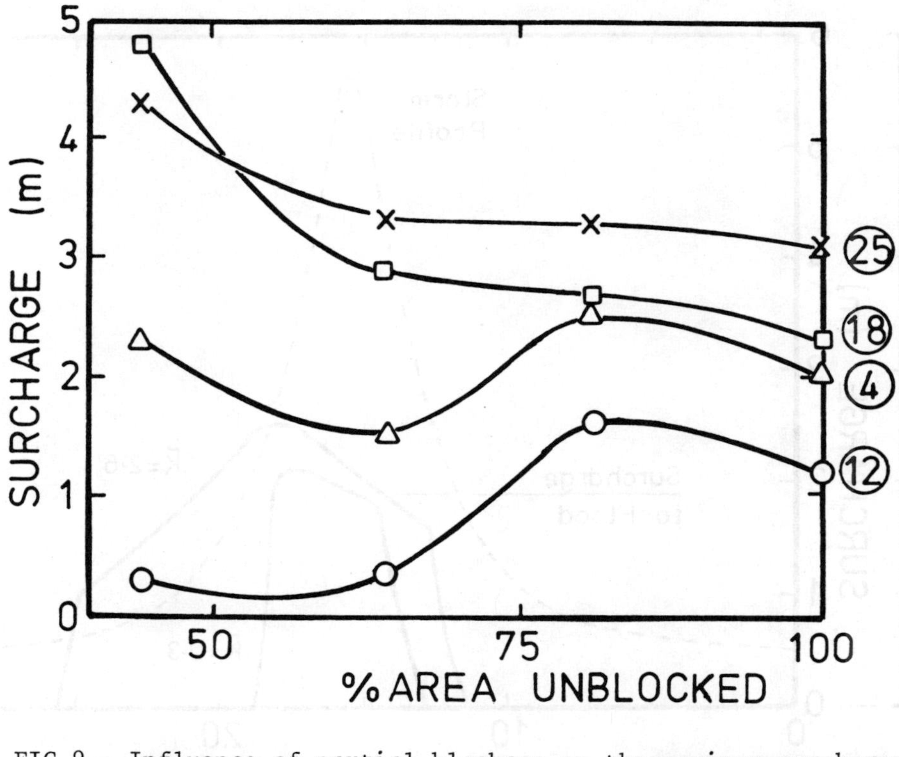

FIG.9. Influence of partial blockage on the maximum surcharge levels at junction nodes 4,12,18 and 25. The baseline represents the average value for the system.

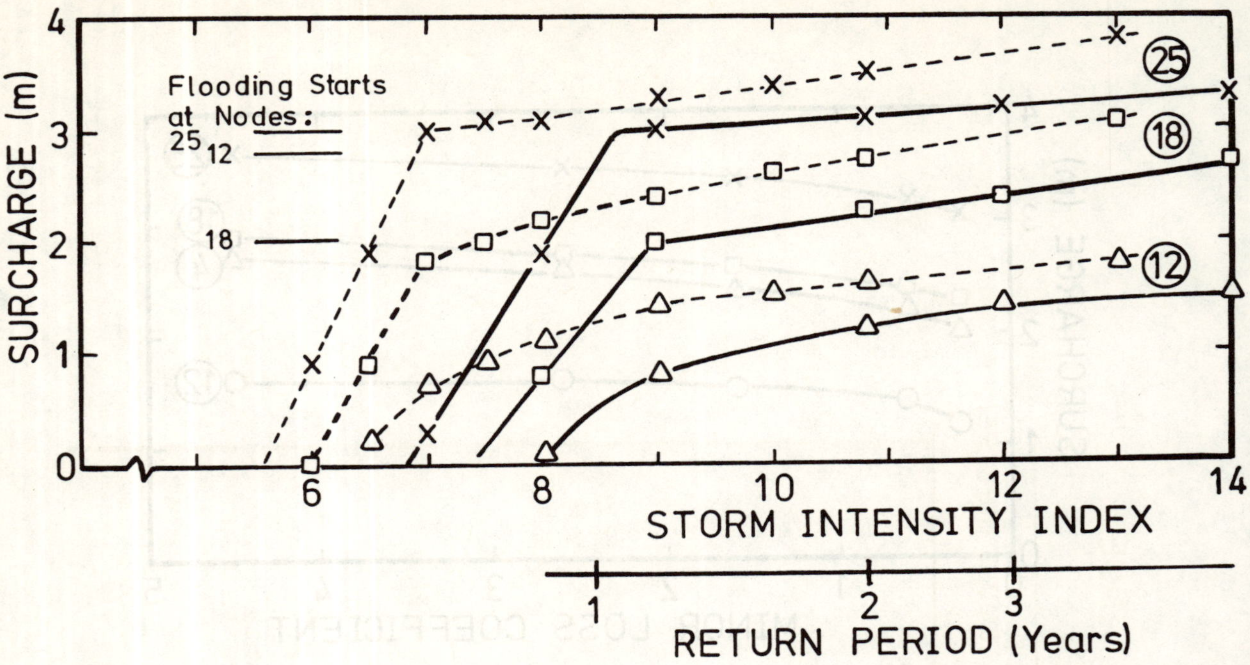

FIG.10. Graph of maximum surcharge versus storm intensity at junction nodes 12,18 and 25. The solid lines are for data including low minor loss coefficients ($\tilde{K}$ = 0.3) and the dashed lines are for high loss coefficients ($\tilde{K}$ = 2.6). Also indicated are the surcharge levels at which surface flooding will occur.

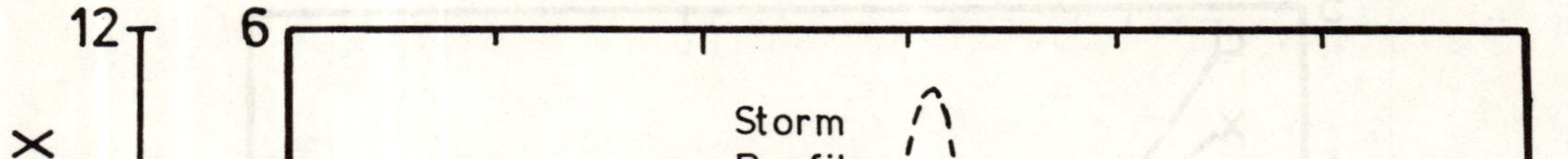

FIG.11. Surcharge levels versus time at junction 15 from an Extended Period Simulation for two levels of average minor loss coefficient. The storm intensity profile is superimposed to indicate time lags.

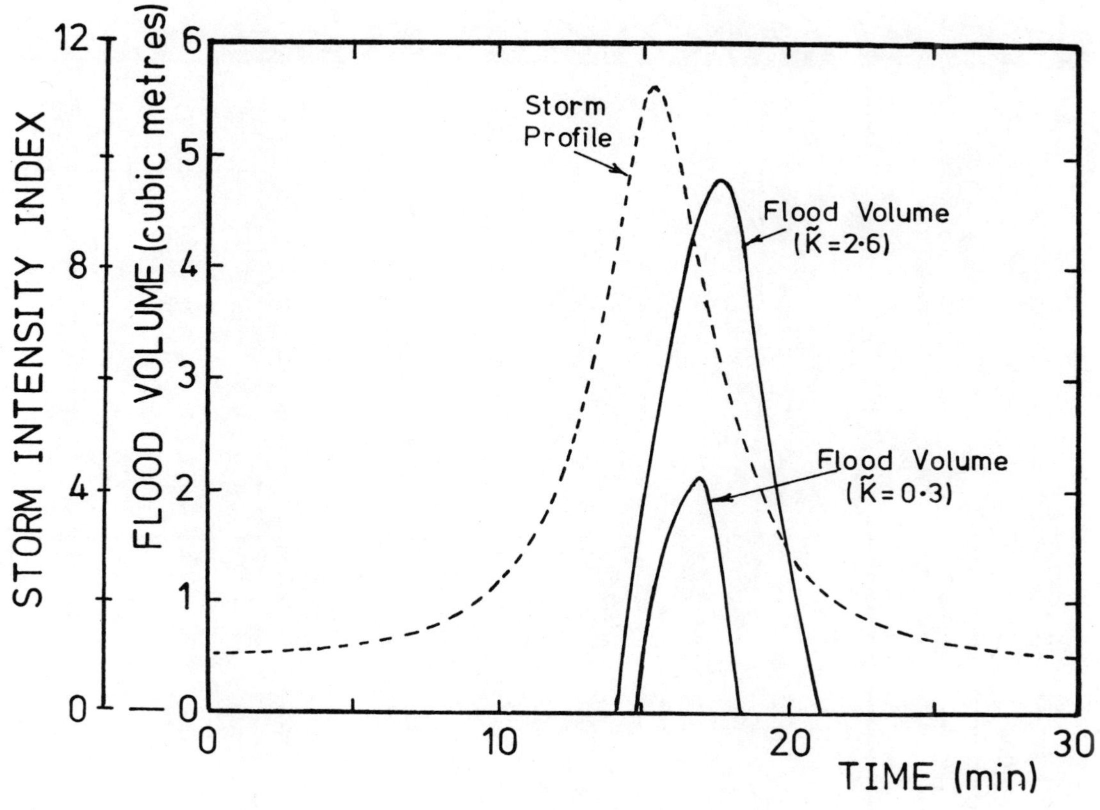

FIG.12. Flooded volume at junction 15, corresponding to Fig.11, from an Extended Period Simulation.

International Conference on the
PLANNING, CONSTRUCTION, MAINTENANCE & OPERATION OF SEWERAGE SYSTEMS
12-14 September, 1984

PAPER F3

APPLICATION OF THE WALLINGFORD PROCEDURE TO THE DESIGN OF URBAN STORMWATER DRAINAGE FOR JEDDAH SAUDI ARABIA

R. Barnsley BSc MICE
Sir Alexander Gibb & Partners
Earley House
Earley, Reading
RG6 1BL
England

Dr. R.K. Price
Hydraulics Research Station
Wallingford
Oxon.
OX10 8BA
England

J.C. Packman
Institute of Hydrology
Wallingford
Oxon.
OX10 8BB
England

Summary

Stormwater drainage facilities in Jeddah are needed less than one day out of every hundred and are utilised even less often for storms which cause extensive flooding. Quantifiable benefits are therefore small compared to the cost of drainage. Various stormwater drainage schemes for South Jeddah have been investigated to a range of criteria. An appropriate scheme was selected and detailed designs completed. The hydrological and hydraulic calculations for the design were carried out using a specially modified version of the Wallingford Procedure for the design and analysis of urban storm drainage. Details of the modifications and the main features of the design are given.

Held at The University of Reading, England. Organised and sponsored by BHRA, The Fluid Engineering Centre, Cranfield, Bedford in conjunction with the Water Research Centre.

NOMENCLATURE

Qp = Peak discharge in litres per second

Cv = Volumetric runoff coefficient

Cr = Routing coefficient

i = Rainfall intensity in mm per hour

A = Catchment area in ha

K = Delay time in minutes

S = Slope %

a = Catchment area in m^2

I = The maximum average rainfall intensity over a 10 minute period in mm per minute.

1. INTRODUCTION

Design philosophies for stormwater drainage works have been developed for areas where rainfall is frequent and urban stormwater drainage is easily justified by the economic and social benefits. In contrast urban stormwater drainage in arid regions is often non-existent and rainfall and runoff data are difficult to obtain both because rainfall is rare and because infrequent flooding is often accepted. Stormwater drainage facilities have been designed for South Jeddah in Saudi Arabia using a specially modified version of the Wallingford Procedure (Ref. 1). This computerised procedure was used to evaluate various schemes and various design criteria to develop an appropriate drainage system. The modifications to the standard Wallingford Procedure and the main features of the design are described in this paper.

2. DESCRIPTION OF THE PROJECT AREA

The project area is located in South Jeddah (Fig. 1) in the coastal strip between the Red Sea and an escarpment. It has an area of 4200 ha and a general gradient of 0.4% towards the coast. The maximum level of the Red Sea is about 0.6m above mean sea level and the normal tidal range is 0.2m. Approximately 20% of the area is undeveloped at present and plans include residential, commercial, industrial and recreational development.

The project area is protected from stormwater outside the area by the Southern Stormwater Channel which is designed to intercept several wadis and divert them to the Red Sea. Within the Project Area a stormwater sewer with length 12 km was constructed in 1979 to drain a narrow strip of land which divides the project area into two approximately equal halves. Its outfall is a 3m wide x 2.5m high culvert which has a longitudinal gradient of 0.04% and a capacity of 10 m^3s^{-1}. Two smaller drainage systems serve 300 ha in the north of the project area and another system in the south of the project area drains about 150 ha into the Southern Stormwater Channel.

3. DESIGN METHODS AND DESIGN CRITERIA

3.1 General Approach.

The design was undertaken in two separate phases; a study phase and a detailed design phase. In the study phase the Wallingford Procedure was first modified to suit the particular conditions in Jeddah and then used for outline design of several alternative drainage schemes. All the schemes were designed to the same storm criteria so they could be compared according to construction cost. The best (least cost) scheme was then redesigned for two other storm criteria and the performance of each version assessed for storms rarer than the design storms. Finally the most appropriate scheme was selected considering construction costs and the level of flood protection provided. The selected scheme was designed in detail for tendering and construction.

3.2 The Wallingford Procedure.

The Wallingford Procedure comprises four main computer programs as listed below, and a program for calculating construction costs:

- Modified Rational Method (analysis and design)
- Hydrograph Method (analysis and design)
- Optimising Method (design only)
- Simulation Method (simulation only).

The original versions of these programs for application in the United Kingdom are described in Ref. 1. Points of particular interest and modifications to the programs for application to Jeddah are described in the following subsections. For this project the main design work was done using the Modified Rational Method and schemes providing different levels of flood protection were compared using the simulation method.

3.3 Rainfall.

Rainfall is very infrequent in Jeddah but intense storms do occur. Over the period 1961 to 1979 the average annual rainfall was 61 mm and the average number of raindays per year was 11 with only 3 having more than 6 mm (Ref. 2). On 16th January, 1979, 80 mm of rain fell in under two hours (Ref. 3). There is little information on short duration rainfall in Jeddah and so data from 15 nearby raingauges in the coastal strip were collected and analysed. Three raingauges, located within 130 km of Jeddah, had autographic equipment and combined to provide a 20 station year record. Annual exceedance series for durations 10, 20, 30, 60, 120, 180 minutes, 1 day and 1 month were obtained from this combined record and intensity-duration-frequency curves compiled graphically. These curves were then transposed to Jeddah (see Fig. 2), making adjustments based on the 1 day and 1 month curves which could be derived for Jeddah itself. Storm profiles for use in the Simulation Method (see Fig.3) were generated from the storm depth-duration-frequency data assuming a symmetrical profile.

3.4 Design Storms.

The 2 year return period intensity-duration curve shown in Fig. 2 was used for the ranking of various schemes. The cheapest of these schemes was then redesigned using the 1 year return period curve shown in Fig. 2 and also the 2 year return period curve truncated at a rainfall intensity of 20 mm per hour, that is any intensities greater than 20 mm per hour were reduced to 20 mm per hour.

The 1 and 2 year return period curves were considered because these frequences are commonly adopted throughout the world for urban stormwater drainage and because longer return periods of 5 or 10 years were thought to be inappropriate for Jeddah where rain days are very infrequent. The truncated curve was used because high intensity rainfall only occurs for very short periods of time (see Fig.3) and so may not cause significant flooding.

3.5 Rainfall Areal Reduction Factor.

An areal reduction factor of 0.9 was adopted in the final design for the main sewers near the outfall. This value was selected, in the absence of any data, with reference to the recommendations in the Wallingford Procedure for the United Kingdom. The reduction was not applied to the truncated section of the intensity-duration curve.

3.6 Volumetric Runoff Coefficient (Cv).

This coefficient relates runoff to rainfall in the Modified Rational Method were:

$$Qp = 2.78 \; Cv \; Cr \; i \; A$$

where Qp is the peak discharge in litres per second
Cv is the volumetric runoff coefficient
Cr is a routing coefficient (taken as 1.3)
i is the rainfall intensity in mm per hour
A is the catchment area in ha

In the original version of the Wallingford Procedure the overall percentage runoff for a catchment is assessed from a multiple regression analysis of data from urban catchments in the United Kingdom, considering the impervious area, the soil type and the wetness of the catchment prior to the storm. Contributions from different parts of the catchment are then determined in accordance with specified rules consistent with the behaviour of the whole catchment. This approach could not be applied to Jeddah because of the lack of data.

For Jeddah volumetric runoff coefficients were ascribed to various types of surface using two sets of indices; one set for fast response areas, that is road lying above the proposed sewers, and the other set for the remainder or slow response areas. Only one fast response index was used corresponding to a coefficient of 0.9.

The volumetric runoff coefficients for slow response areas were estimated by reference to a method devised by the Soil Conservation Service of the United States Department of Agriculture (Ref. 4). This method yields coefficients of 0.25 for permeable areas, assuming the soil to be moderately well drained, and 0.8 for impermeable areas. These coefficients are based on the two year return period storm with duration equal to the time of concentration of the whole catchment i.e. a storm of 30 mm. The volumetric runoff coefficients applicable to various classes of development which were adopted for the design were based on these two extreme cases and are listed below:

- 0.25 for permeable areas
- 0.35 for low density housing (less than 125 persons per ha) and industrial development
- 0.45 for medium density housing (125 to 175 persons per ha)
- 0.55 for high density housing (over 175 persons per ha) and commercial development.

3.7 Time of Entry (for the Modified Rational Method).

The time from the beginning of a storm to the entry of stormwater into the drainage system was assumed to be 20 minutes. This is longer than is often allowed and recognises the generally flat terrain and the large size of individual catchments to each sewer.

3.8 Depression Storage (for the Simulation Method).

The average depth of rain caught on the ground surface in small depressions is subtracted from the rainfall. Three indices are available for different ground slopes but only one was used. This allowed 1 mm and 4 mm depression storage for fast and slow response areas respectively and was used for areas where the gradient was less than 2%.

3.9 Surface Routing (for the Simulation Method).

The Wallingford Procedure for the United Kingdom uses a non linear reservoir to simulate surface runoff from paved and roof areas and any runoff from permeable areas is incorporated implicitly with the runoff from these surfaces. For Jeddah the delay time for the fast response areas was defined by the following equation which is derived from data from catchments in the United Kingdom:

$$K = 0.393 \, S^{-0.264} \, a^{0.333} \, I^{-0.39}$$

where K = Delay time in minutes
S = Slope %
a = Catchment Area in m^2
I = The maximum average rainfall intensity over a 10 minute period in mm per minute.

For the slow response areas a delay time of 20 minutes was adopted.

3.10 Sewer Cost Data.

For pipe sewers the estimated costs of sewers of various diameters at various depths were calculated using budget quotations for pipes and cost data from other construction works in the Middle East. The computer program interpolated for intermediate cases and summed the total. This cost data is shown graphically in Fig. 4.

For culvert sewers cost equations were developed based on the estimated costs of culverts of various heights, widths and depths below ground level. The cost equations are shown graphically in Fig. 5.

Final cost estimates made after completion of the detailed design were prepared from the bills of quantities.

3.11 Data Requirements.

The data requirements for the Modified Rational Method are listed below:

- branch reference number
- reach reference number
- length of reach
- ground level at upstream end of reach
- soffit level of sewer at upstream and downstream ends of reach
- catchment area of reach
- fast response area of reach (%)
- runoff coefficient for fast and slow response areas (2 indices)
- dry weather flow to reach
- sewer shape index (circular or rectangular)
- pipe diameters (allowable sizes)
- culvert vertical dimension
- storm intensity - duration data
- roughness coefficient for sewers
- minimum allowable velocity.

The data requirements for the Simulation Method are similar to those of the Modified Rational Method, except as listed below:

- storm profile data instead of storm intensity - duration data
- surface slope index for depression storage
- number of manholes in reach
- flooded area as % of catchment area of each reach
- head loss coefficient for upstream manhole of reach
- level hydrograph at outfall.

3.12 Output.

The Modified Rational method produced the following information for each reach of the drainage system:

- longitudinal gradient of sewer
- effective catchment area (catchment area x Cv)
- time of concentration
- peak discharge
- pipe diameter or culvert dimensions
- pipe full velocity
- pipe or culvert capacity
- warnings of low velocity or surcharging.

Longitudinal profiles of the sewers were plotted on a drum plotter or a flat bed plotter using a separate drafting program. The specified sewer levels were then adjusted where necessary to avoid interference with other services and the process repeated. This could also be done automatically and construction costs minimised using the Optimising Method.

The Simulation Method produced the discharge in the sewer, the incremental and cumulative discharge out of the sewer onto the ground and flood levels for each time increment. In order to compare schemes designed to different standards the maximum flooded depth was calculated for each reach assuming that flooding occurred with uniform depth along the route of the sewer over a width of 100 m. This is obviously a simplification but it was considered suitable for comparison purposes.

3.13 Computer Costs.

All calculations were done on a PRIME 550-II computer with 2 Megabyte central processing memory.

The Modified Rational method was very quick. Design of a 29 branch sewer system with 130 reaches took only 15 seconds of central processing time. In contrast the simulation method took much longer. To simulate the performance of the same system with a time increment of 1 minute took 10 minutes of central processing time.

4. THE DRAINAGE SCHEMES

4.1 General Features.

All the schemes considered provided an extensive system of large sewers (diameter greater than 500 mm) along selected main roads in the area and included road gullies and gulley connections in those roads. Tertiary sewers were not included in the project except where they were required to reduce the cost of gulley connections for example in wide roads.

The use of open channels and open storage facilities was not generally acceptable within existing development because of lack of space nor within undeveloped areas because of a reluctance to allocate high value land to stormwater facilities which are required very infrequently.

4.2 The Main Schemes.

The topography and existing land use in the Project Area dictated that main sewers to the Red Sea could only be accommodated along two routes: one to the north of Jeddah Oil Refinery and the other along the corridor between Jeddah Oil Refinery and Jeddah Naval Base, alongside an existing stormwater culvert. Only a comparatively small part of the Project Area can be drained into the existing Southern Stormwater Channel because of the relative levels of the Channel and the Project Area.

Two basic schemes developed from these limitations. The first was essentially an enlargement or repetition of an existing stormwater sewer system to serve the whole area with an outfall between Jeddah Oil Refinery and Jeddah Naval Base. The second was similar in the south of the project area but the north of the area was drained along the route of the new elevated Makkah Expressway to a northern outfall. Both schemes were designed using the Modified Rational Method for storms with a return period of 2 years. The first scheme was adopted because it was cheaper by about 10% and only required a single outfall.

4.3 The Level of Flood Protection Provided.

The optimum scheme for the two year return period storms was redesigned for the truncated two year storm curve and the 1 year storm curve for the reasons given in Section 3.4. The Simulation Method was then applied to these three schemes to calculate the flooding which would occur during storms with a return period of 5 years. The results are shown in Fig. 6 together with relative costs of the schemes.

It can be seen from Fig. 6 that during a 5 year storm:-

(a) The scheme designed to the 2 year curve produces only a small amount of flooding.

(b) The scheme designed to the 2 year truncated curve also produces very little flooding. Paradoxically, although designed to a lower standard than the 2 year curve, it produces less severe flooding at its lower extremities. This is explained by the fact that while the sizes of the sewers near the outfall are similar in both schemes, the sewers of the upstream extremities are of lower capacity in the two year truncated scheme. Consequently the rate of entry of stormwater into the system is reduced. This causes a reduction in flooding in the lower extremities and a slight increase in flooding in the upstream extremities.

(c) The scheme designed to the 1 year curve produces considerable flooding to a depth of more than 300 mm.

The scheme designed to the 2 year truncated intensity - duration curve has important advantages over both of the other schemes: it provides better flood protection in the low lying area near Jeddah Oil Refinery than the scheme designed to the 2 year curve and it is some 17% cheaper; it provides much better flood protection than the scheme designed to the 1 year curve. The scheme designed to the 2 year truncated intensity duration curve was selected as being the most appropriate, considering the service provided and the capital costs of the schemes.

4.4 Details of the Final Scheme.

The final scheme, as designed for tendering and construction, is shown in Fig. 7 and physical data are given below:-

Catchment Details (including an existing system).

Catchment area	3051 ha
Effective area (catchment area x Cv)	1280 ha
Mean Volumetric Runoff Coefficient (Cv)	0.42
Time of concentration near outfall	100 minutes
Design discharge at outfall	70 m^3s^{-1}

Construction Details.

Road gullies	2610 No.
Pipe sewer up to 0.4 m diameter (Gulley connections and tertiary sewers alongside main sewers)	51 km
Pipe sewers 0.5 to 2.0 m diameter	30 km
Reinforced concrete culvert	19 km
Open channel	1 km

The majority of the reinforced concrete culvert is single barrel, 2.5m high, and up to 5m wide. Near the outfall a 4 barrel culvert, each barrel 2.5m high and 4.1m wide, is required.

For tendering and construction the works have been shown on plans, scale 1:1000 and longitudinal profiles, horizontal scale 1:1000, vertical scale 1:100.

5. CONCLUSIONS

The Wallingford Procedure has been adapted for application to Jeddah and used for the design of a large drainage scheme. It is now suitable for more general application in other parts of the world providing the necessary rainfall characteristics and runoff factors are available. The Modified Rational Method was particularly easy and quick to use and the results could readily be checked because the method is very simple; it was used for the main design work. The Simulation Method is complicated and is expensive in terms of computing time, but it gives guidance on the level of flood protection being provided, which was very necessary in determining design criteria in this case because of the infrequency of storms and flooding.

The cost program provided immediate cost estimates of the sewers designed and this greatly facilitated the identification of least cost alignments and layouts. Computerised drafting using a separate plotting program allowed small scale preliminary profiles and high quality final drawings to be produced very rapidly.

The volumetric runoff coefficients are clearly most important factors in the design and in the absence of rainfall and runoff data there remains an element of uncertainty about the values adopted. Research is required into this subject and to that end two autographic rain gauges have been installed in Jeddah and pumping records are being kept at a stormwater pumping station in central Jeddah.

6. ACKNOWLEDGEMENTS

The work described in this paper was undertaken for the Municipality of Jeddah by Sir Alexander Gibb & Partners to whom the Authors are grateful for their kind assistance.

7. REFERENCES

1. National Water Council: "Design and Analysis of Urban Storm Drainage The Wallingford Procedure". London, National Water Council, 1981, Vol. 1.

2. El Sayed and Enani: "Some Characteristics of Rainfall in Jeddah". Local Climate Publication No. 1, General Directorate of Meteorology, Jeddah, Saudi Arabia, 1979.

3. Watson Saudi Arabia: "Master Plan and Report on Stormwater Drainage for the City of Jeddah" unpublished report 1974.

4. Soil Conservation Service: "Engineering Manual Part 4". United States Department of Agriculture, 1957.

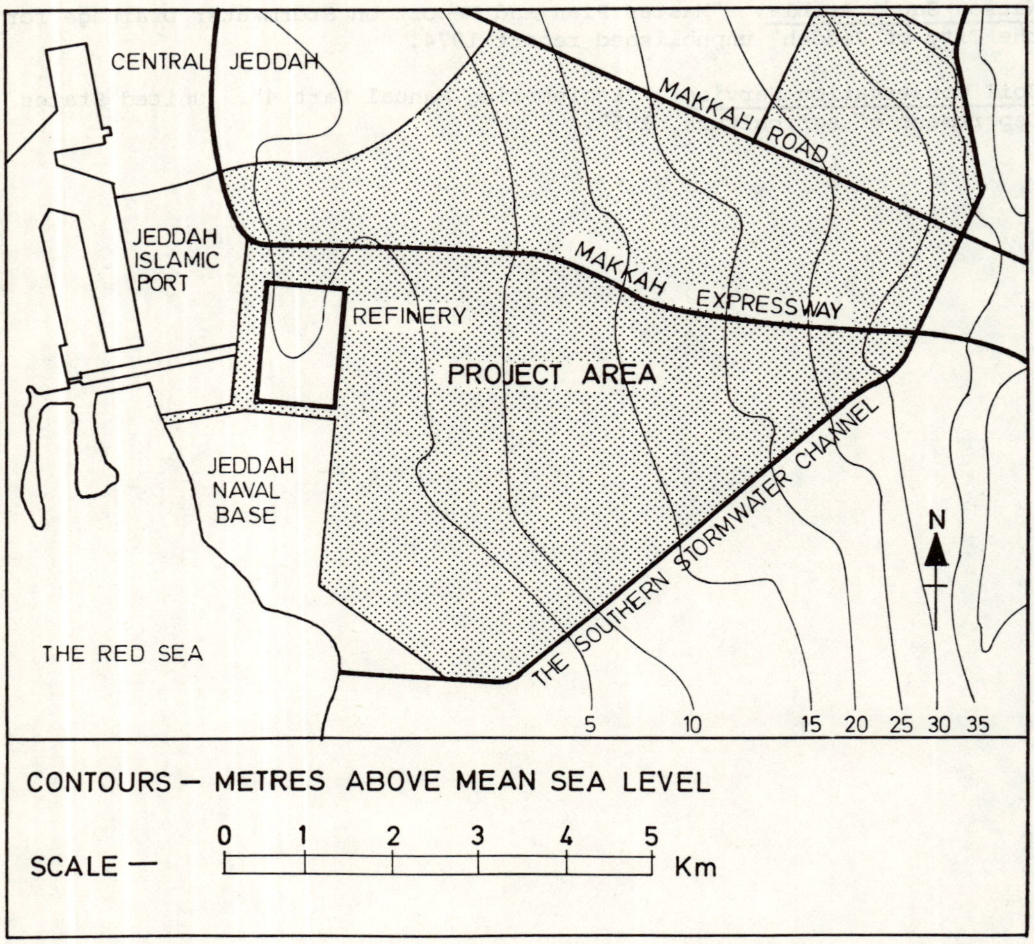

FIG.1 PLAN OF PROJECT AREA

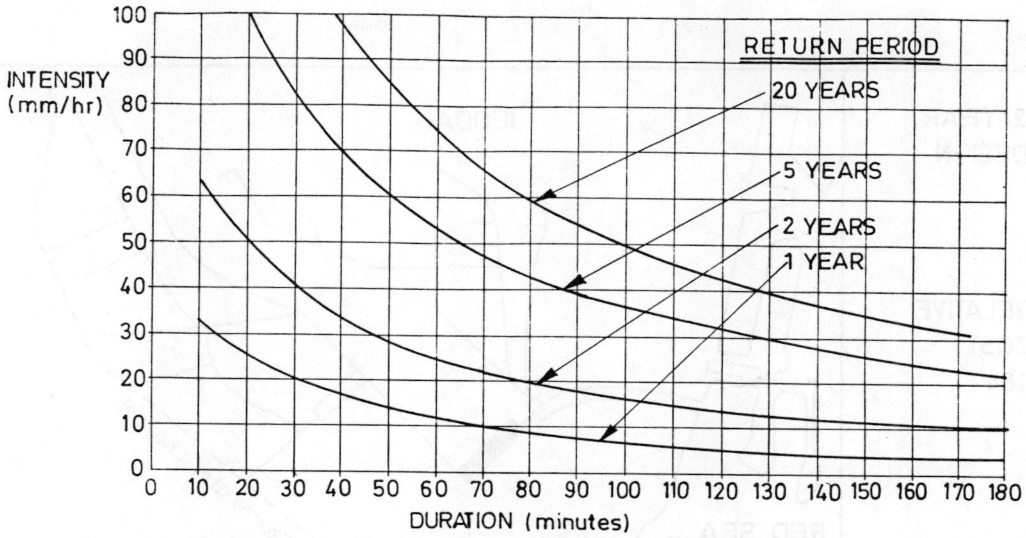

FIG.2 JEDDAH RAINFALL INTENSITY-DURATION-FREQUENCY

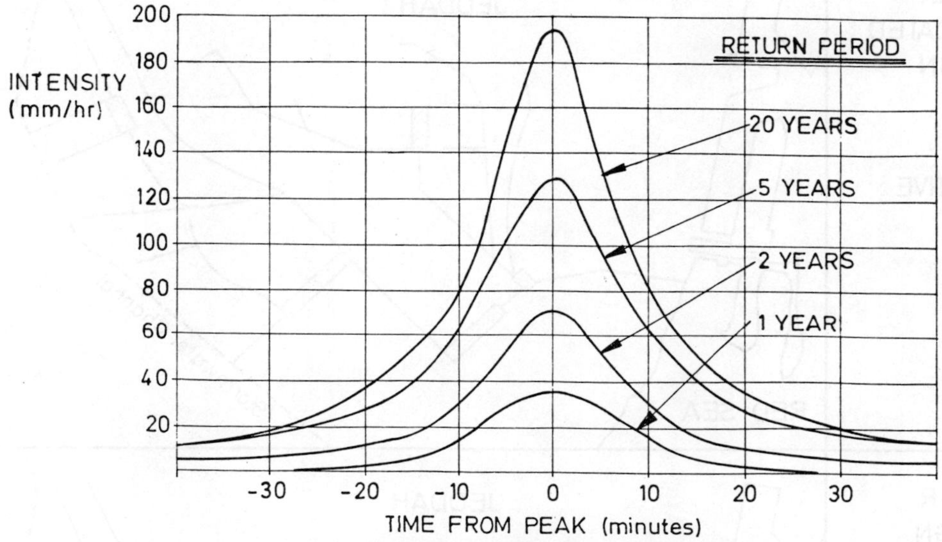

FIG.3 JEDDAH STORM PROFILES

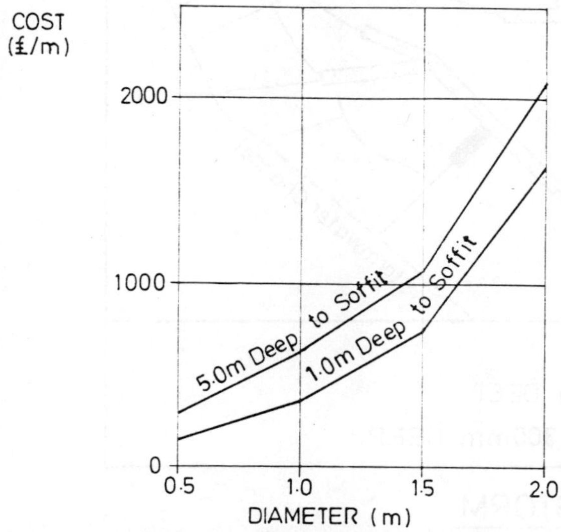

FIG. 4 PIPE SEWER COSTS

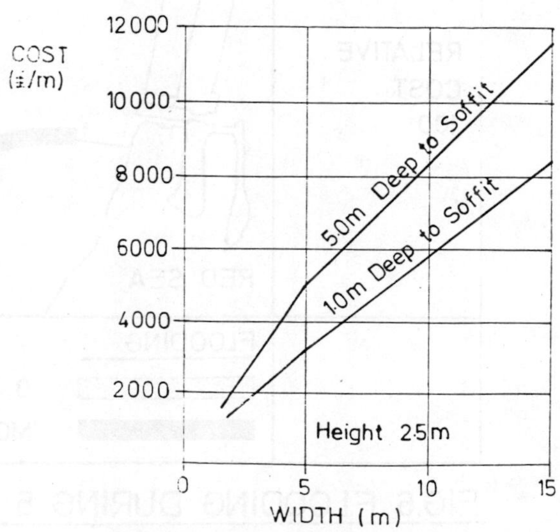

FIG. 5 CULVERT COSTS

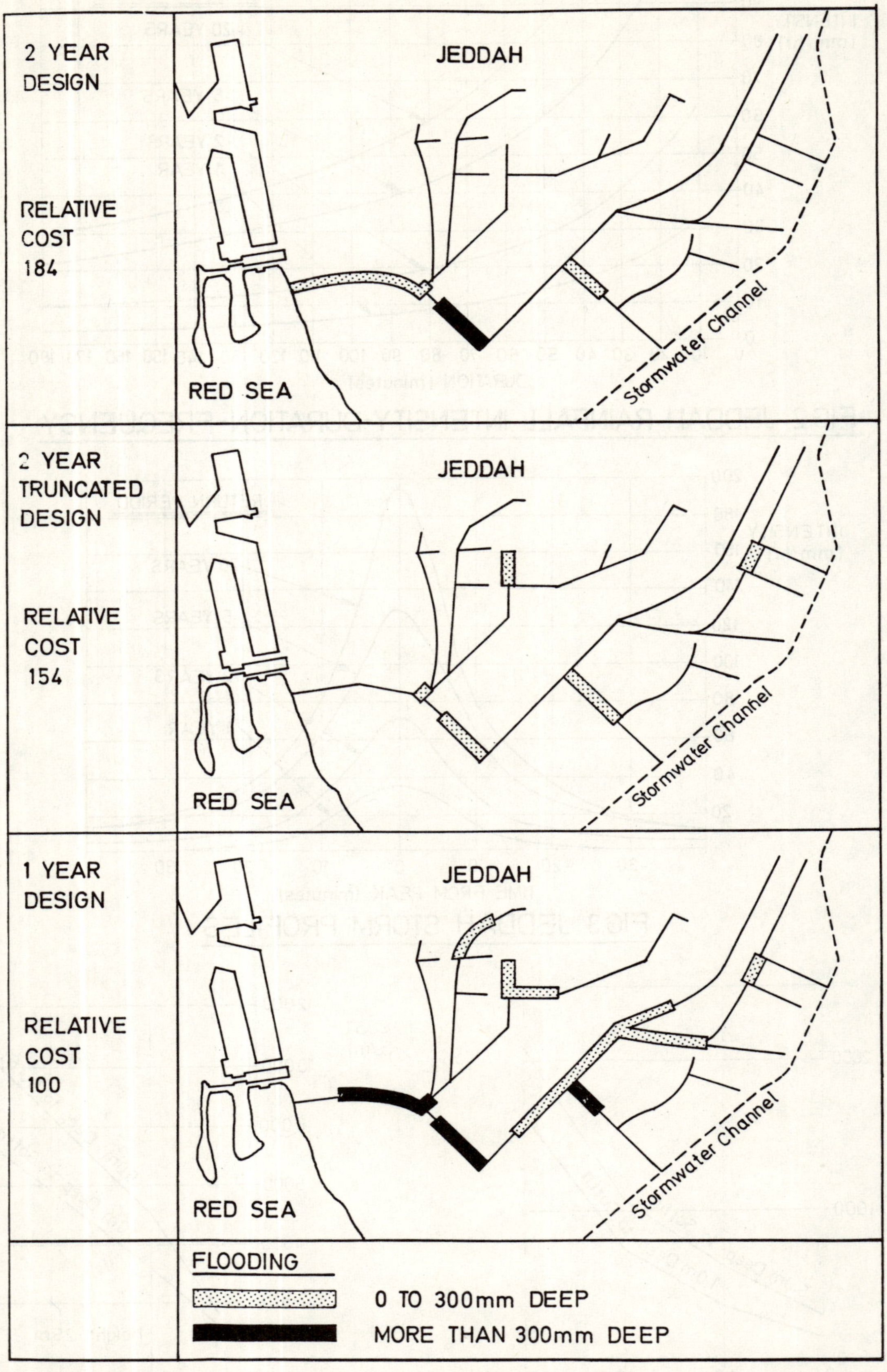

FIG.6 FLOODING DURING 5 YEAR STORM

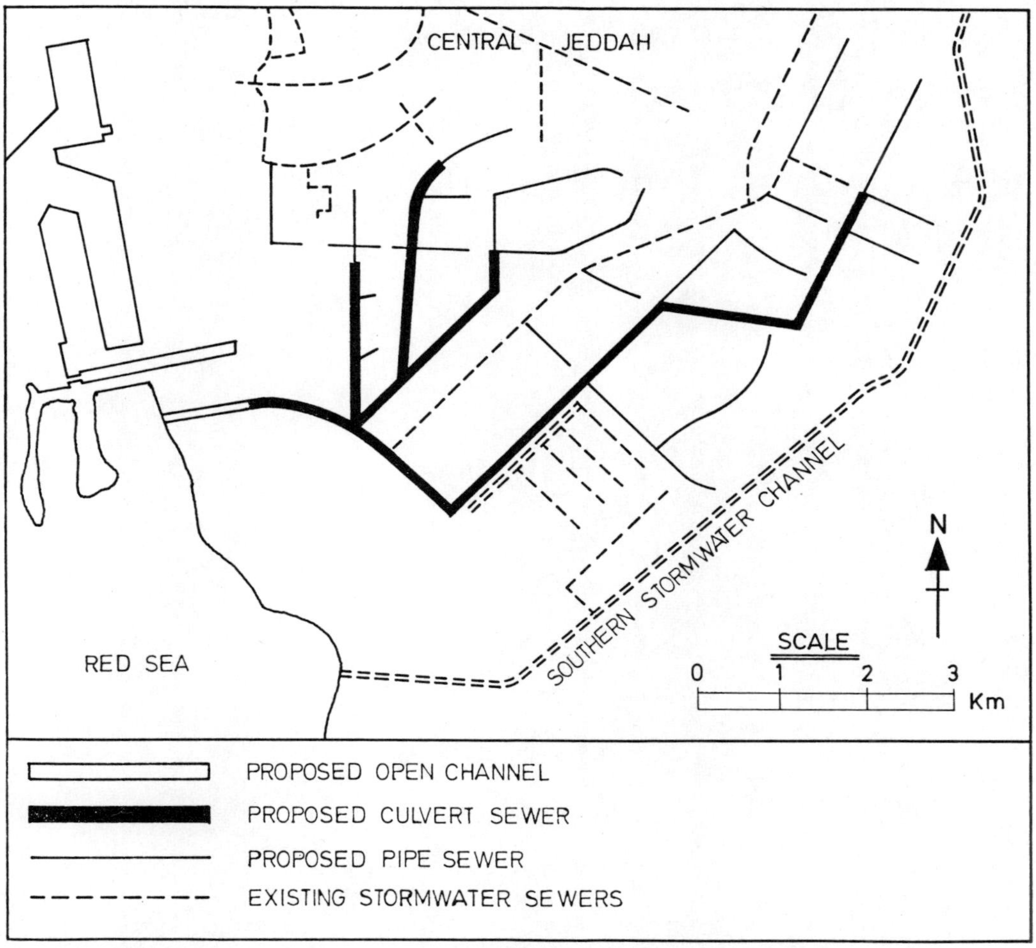

FIG. 7 PROPOSED STORMWATER DRAINAGE SYSTEM

International Conference on the
PLANNING, CONSTRUCTION, MAINTENANCE & OPERATION OF SEWERAGE SYSTEMS
12-14 September, 1984

PAPER G1

STRUCTURAL DESIGN OF BURIED RIGID PIPELINES
A COMPARATIVE STUDY OF INTERNATIONAL PRACTICE

H. A. Allan

Haiste International Ltd.
Consulting Engineers
Leeds, U.K.

SUMMARY

The paper describes the design procedures used in the U.K., Germany, Japan and U.S.A., highlighting the differences which exist, particularly with regard to the degree of support offered by the bedding, imposed loads from traffic, and the influence of lateral soil forces. The effect of design variations on factor of safety and site control is discussed with reference to examples of recently completed installations. The paper concludes by making proposals intended to achieve a more unified approach in the future.

Held at The University of Reading, England. Organised and sponsored by BHRA, The Fluid Engineering Centre, Cranfield, Bedford in conjunction with the Water Research Centre.

NOMENCLATURE

Wc	Load imposed on pipe from trench backfill
We	Total effective external load imposed on pipe
Wt	Maximum crushing test load of pipe
Fm	Bedding factor for pipe in trench
Fs	Factor of Safety

1. INTRODUCTION

Despite the increased use of flexible plastic pipe materials such as uPVC, polyethylene and GRP, in applications where their light-weight, fewer joints and corrosion resistance have proved to be beneficial, the traditional rigid pipe materials of clay and concrete, and to a lesser extent asbestos cement, are still in wide use throughout the world for gravity sewer pipelines. Particularly in the larger size range from, say, 1500 mm dia. and upwards, concrete is by far the most commonly selected pipe material. Where a high degree of corrosion resistance is necessary due to the presence of sulphuric acid arising from sulphide generation, PVC lining may be specified if the calculated sacrificial concrete thickness proves to be uneconomic.

Most design engineers would probably consider their method of structural analysis to be well established and not open to fundamental question, and indeed in a purely national context this would probably be so. However, the export of varying national practices to countries embarking upon major first-time sewerage schemes through international contractors and consulting engineers has served to highlight the wide variation in pipe strength which can result when different design methods are used (ref.1).

The city of Baghdad, Iraq, was first provided with a foul sewerage system in the early 1960's to serve that part of the city lying on the east bank of the River Tigris. Although, following the commissioning of this scheme, there was an on-going programme of new sewer construction, the construction rate failed to keep pace with the population growth (in excess of 3% per annum) and the development of new residential areas. The Iraq government's policy of injecting a substantial proportion of the 1970's oil income into modernising the country's infrastructure resulted in the Baghdad City Authority embarking upon a major sewerage expansion programme in 1979. This was to increase the population served from 650,000 to 2,100,000, all contracts being run concurrently and representing an expenditure of some U.S. $ 2000 million inclusive of pumping stations and treatment plants.

In order to achieve this ambitious objective in the shortest possible time-scale the sewerage authority (Baghdad Sewerage Board) prepared in a matter of months design and contract tender documents for subsidiary sewer networks, trunk sewers, pumping stations, and treatment plants. Flows in the sewerage network were determined by the Sewerage Board as were the drainage areas, routes of trunk sewers and invert levels at key points. The detailed designs and drawings were prepared by the various contractors subsequently appointed by the Board.

Following award of the various contracts in 1980, Haiste was appointed to provide design checking and construction supervision services, and to plan and design the next phase of sewerage construction to extend the sewerage systems and treatment facilities to serve a population of 4,300,000. It was in carrying out the structural design checks on buried concrete sewers that it became evident that there were significant differences in the design approaches adopted by the various international contractors, who in good faith had adopted design practices used in their native countries. For a British consultant it was natural to compare the designs with those which would have resulted from the application of U.K. design procedures and in two cases spot checks revealed that the pipe strength fell considerably short of the U.K. requirement. This necessitated a more rigerous analysis of the various design methods, to determine the reasons for the discrepancies and provide the basis for objective reports to the client.

2. GENERAL COMPARISON OF NATIONAL PRACTICES

2.1 United Kingdom

Concrete pipes manufactured in compliance with B.S. 5911 (ref. 2) are designed to a three edge test load in accordance with the compliance requirements. Manufacturers are free to select the most economic combination of concrete strength, wall thickness and steel reinforcement for any given pipe class and diameter. For pipeline structural design purposes the document in most common use is the Simplified Tables of External Loads on Buried Pipelines (ref. 3), but a more comprehensive design guide has recently been published (ref. 4) which will no doubt gain wide acceptance and supersede the Simplified Tables.

Design charts produced by pipe manufacturers are also used, but should be treated with caution since they are generally applicable to one design case only in respect of trench width, backfill density and factor of safety.

The structural calculation of the pipeline is carried out by the bedding factor method using a pipe manufactured to a known strength in accordance with the British Standard. Having determined the total linear design load applied to the top of the pipe from backfill and traffic loads, the designer selects an appropriate combination of pipe strength and bedding to give the required factor of safety in accordance with the formula:-

$$\frac{Wt\ Fm}{Fs} \geqslant We$$

The bedding factor for pipes in a trench may vary from 1.1 for a pipe laid directly on the trench bottom, to 3.4 for a pipe supported by a reinforced cradle or arch. In all cases where a higher degree of bedding support is assumed in the calculation selected granular material or concrete is specified below the pipe. Whether the designer selects a high strength pipe with a low bedding factor or vice versa is a matter for economic evaluation and the designer's judgement as to the reliability of achieving the necessary pipe support from the bedding under site conditions.

Because a high degree of compaction on the sidefill around the pipe in a trench cannot always be relied upon, it is usual to assume that the whole of the load from the trench backfill, Wc, is carried by the pipe itself, and to neglect the reduction in pipe wall bending moment which would result from the inclusion of lateral forces. However, pipes laid as part of an embankment construction are much more likely to be subjected to lateral loading and an increase in bedding factor may be adopted, up to a maximum of 5.8.

2.2 West Germany

Concrete pipes manufactured to DIN 4035 (ref. 5) are designed in the same way as other buried concrete structures. They are analysed in accordance with DIN 1045 (ref. 6), i.e. bending moment, shear forces etc. are calculated, and concrete and reinforcement provided to suit. Provision is also made in DIN 4035 for strength testing of finished pipes. DIN 4033 (ref. 7) is mainly concerned with construction and is fully comprehensive in its requirements for trench preparation and bedding; general guidance is given in respect of pipeline design and reference is made to other design documents (ref. 8).

Having calculated the total vertical loads on the pipeline, the required pipe strength, i.e. the bending moment resistance of the pipe wall, is determined by a theoretical analysis of the pipe cross-section, taking into account the extent of support reaction provided by the pipe bedding. Under certain circumstances DIN 4033 permits up to 120° of bedding support to be assumed in the calculations when the pipe is placed directly on the prepared trench bottom (with or without pre-shaping), and native soil is compacted in the haunch area.

DIN 4033 emphasises the relationship between the degree of compaction provided to the pipe sidefill and the structural calculation. Lateral earth loads may be included in the calculation, thus reducing the pipe wall bending moment and permitting a lower strength pipe. Proof is required, however, that the design assumption has been realised, by in-situ Proctor density testing.

2.3 Japan

Reinforced concrete sewer pipes are manufactured in accordance with either JIS A 5303 (ref. 9) or JSWAS A-1 (ref. 10) and are classified according to strength under external load conditions. The manufacturer designs the pipe to ensure that the strength requirement is met. The selection of pipe and bedding is determined from a theoretical analysis of the loads on the pipe from trench backfill, live load and bedding support reaction, and the bending moment of resistance of the pipe wall.

Relief of stress in the pipe wall due to lateral soil forces in the trench is not specifically allowed for in the calculations, but recognising that it does exist it appears to be the practice to offset this by neglecting the vertical loads arising from the self weight of the pipe and its contents.

2.4 United States of America

Reinforced concrete sewer pipes in accordance with ASTM C76 (ref. 11) may be manufactured in accordance with the prescribed concrete strengths, wall thicknesses and reinforcement tabulated in the standard, or special designs may be adopted subject to the manufacturer satisfying the purchaser as to the adequacy of his design. External load testing of finished pipe is a requirement of the standard.

The structural calculation of the pipeline is generally carried out by the bedding factor method and is virtually identical to U.K. practice in this respect, the same standard bedding classes being adopted. The A.S.C.E./ W.P.C.F. Manual of Practice (ref. 12) does, however, suggest the use of alternative design methods for larger pipes (above 1200 mm) which would permit lateral soil forces to be taken into account and hence reduce the required pipe strength.

3. VERTICAL LOADS ON THE PIPELINE

3.1 Trench Backfill

Although Marston's equations are considered to give conservative values for vertical soil load in trenches, they seem to remain in general use by most designers. On the one Baghdad contract where a variation on Marston theory was adopted, the load calculated did not differ significantly from the Marston value.

The choice of soil density is likely to have an equal or greater effect than differences in loading theory, since there is a possible range of 1600 to 2200 kg/m^3. In Britain and America minimum design values of 1900 to 2000 kg/m^3 are recommended in the absence of reliable soil information, whereas in Japan lower values in the range 1600 to 1800 kg/m^3 are normally adopted.

3.2 Traffic Loads (Main Highways)

U.K. practice is to allow for 45 units of HB loading in accordance with B.S. 5400 (ref. 13). This comprises a single abnormally heavy vehicle having two pairs of axles, each pair comprising eight 112.5 kN wheel loads inclusive of impact, resulting in a total effective vehicle load of 1800 kN. The loading configuration derives from an experimental system specifically evolved in relation to bridge design, and does not represent an actual vehicle.

The German SLW 60 vehicle has three axles with two 100 kN wheels on each axle. An impact factor of 1.2 is applied to give a total effective vehicle load of 720 kN. The loading does not apply to depths of cover of less than 1.4 m, and in such cases an increased impact factor may be adopted.

The Japanese T-20 vehicle has a total weight of 200 kN of which 80% is exerted through the rear wheels. Only the rear wheels are considered in calculating loads on buried pipelines. With an impact factor of 1.3 applied the total effective load is therefore 208 kN.

The A.S.C.E./W.P.C.F. Manual gives calculation methods for traffic loads, but no recommended vehicle weights are stated. The suggested impact factor of 1.3 applies only to depths of cover less than 3.0 m.

3.3 Temporary Construction Loads

There is general recognition in the various design guides that loads from construction vehicles can be significant, and the designer is required to give due consideration to the possibility of a temporary loading condition being critical. Most designers would, however, be reluctant to increase the project cost to cater for a temporary loading condition, if it can be avoided.

4. EFFECT OF DESIGN VARIATIONS

4.1 Design Factor of Safety

Generalisations are always to be avoided, and one would hesitate to attempt to allocate approximate factors of safety to each national design method since much depends upon the characteristics of the particular installation. Nevertheless, it is evident that in descending order of factor of safety (or ascending order of design economy) the following ranking probably applies in most cases:-

United Kingdom
United States
Japan
Germany

A major factor in achieving the economic German design is in the allowance for lateral soil forces in the calculation, this alone can make a difference of 25% to the required pipe wall strength.

German and Japanese designs also assume a more even distribution of support reaction from the bedding, than is the case with the British and American practice and a reduction in required pipe strength of approximately 20% can be achieved by this means.

If, therefore, the true factor of safety of a pipe installed under controlled site conditions according to German design procedure is, say, 1.25 then the equivalent British design would have a factor of safety of around 2.0, or even greater in shallow burial cases.

4.2 Site Conditions

As previously mentioned DIN 4033 specifies a high quality of site installation in respect of trench width, trench formation, bedding, and side-fill compaction. This is quite consistent with the German design approach which places a high degree of reliance upon the quality of construction. In mobilising lateral soil forces, German design practice for rigid pipes has similarities with flexible pipe design theory in that more of the load is taken by the surrounding soil and less by the pipe itself, if this is not appreciated by the site engineer then the design factor of safety may not be achieved.

Contract drawings and specifications define the maximum trench width which is permissible and this normally corresponds to a "narrow trench" design condition. Whilst every attempt may be made to keep excavation to within the limit, on many sites narrow trench widths are exceeded and low safety factor designs are vulnerable.

Even when granular material is used in controlled conditions, strain measurements made on a flexible pipe installation (ref. 14) would seem to suggest that it is difficult to achieve uniformity of compaction in the haunch area below the pipe. Designs which assume uniform vertical support from the bedding require particular care to be taken during construction, in a part of the trench where working and inspection conditions are least favourable.

4.3 Examples of Completed Installations

Many kilometres of large diameter precast concrete sewer have been laid in Baghdad under the current construction phase. In drawing attention to problem areas for the purposes of this comparison of design practices, it should be appreciated that the number of "failed" pipes represents only a minute fraction of the total.

It is perhaps worth considering initially a contract where an inexperienced contractor encountered considerable difficulty in controlling groundwater in the excavations. As a result the quality of bedding placement was poor and trench width exceeded the design maximum on many occasions. The pipes were manufactured by the contractor according to ASTM C76 and although strength requirements were met, general quality and finish were below average. The design was checked against U.K. practice and was found to comply. Despite the adverse site conditions no significant cracking was found on inspection after completion.

On a second contract, designs based on Japanese practice had been prepared covering pipes in the size range 0.8 to 2.5 m dia. The pipe strengths proposed were significantly less than those calculated on a U.K. basis. The difference was mainly accounted for by lower soil densities and traffic loadings, and the greater degree of support assumed from the bedding. The pipes as manufactured were found to have a higher crushing strength, and this, together with modifications to the bedding detail and increased use of extra strength category pipes in the deeper sections, formed the basis of an agreement with the contractor. On final inspection longitudinal cracking was discovered in some pipelines, the crack widths tended to be greatest at the limiting depth of a given bedding or pipe strength category. Only on eight pipes did the crack width exceed 0.25 mm.

A third example was a contract designed according to German codes, again significantly lower pipe strengths were found when compared with U.K. practice. The difference was attributed to the lateral soil loads and the increased spread of uniform support offered by the bedding in the German design. The 3.0 m dia. pipes as manufactured were of a high quality and average concrete cube strengths were 20% above the design requirement. Some twenty-two pipes in one section had to be removed and replaced however, due to cracking of varying degrees of severity, in the worst case the pipe had visibly deflected and temporary propping was introduced to prevent a total collapse. The design maximum cover for that section was 6.0 m, but only 4.5 m existed. The investigation concluded that (a) the compaction of bedding and sidefill had been inadequate, (b) the specified bedding support arc had not been provided, and (c) a heavy crane operating in the vicinity may have travelled over the top of the pipe when only 0.5 m of cover had been placed.

5. DISCUSSION

The difference in basic design method i.e. bedding factor/crushing strength as against analysis of load distribution/pipe wall bending moment, should not in itself result in significantly different answers to the same problem, since a theoretical relationship exists between the two methods. If it is reasonable to assume that the degree of uniform support from the bedding adopted in German and Japanese designs can in fact be realised under well controlled site conditions then the bedding factors used in Britain and America already include a factor of safety providing the installation is carried out in accordance with the specification. It is of interest to note that whilst the old "Simplified Tables" stated that the recommended factor of safety of 1.25 on ultimate pipe strength was based on first-class workmanship and proper supervision and should be modified (i.e. increased) if considered necessary, the new U.K. design guide takes a less rigid view and recognises that the calculations which lead to the pipe design load tend to err on the safe side.

The strength of the pipeline as a whole is shared between the pipe itself and the bedding/surround. The pipe is a factory made product with good quality control and the risk of falling below the design strength is low, but there will always be less certainty about that element of pipeline strength which is achieved on site. It is apparent that economic designs which have proved to be satisfactory in Europe and Japan are at risk if applied in other countries where there is a less skilled local or imported third world work force, despite supervision by experienced expatriate engineers.

In pipelines laid with shallow cover the use of the British HB traffic loading will result in significantly higher loads than are ever likely to be imposed once the pipeline is complete. HB loading was developed for the purpose of bridge construction and can be regarded as an extreme condition which should be considered where the consequences of a structural member failing would be disastrous. It is arguable, therefore, that for most pipelines under main highways a lower loading such as the German SLW 60 could safely be adopted, bearing in mind the residual strength that remains in the pipeline (ref. 15) beyond the 0.25 mm crack width which is normally taken as the limit in service.

6. CONCLUSIONS

Rigid pipelines designed in accordance with U.K. practice and installed under controlled site conditions are likely to have a true factor of safety on ultimate strength significantly greater than the 1.25 given in the design literature. It would perhaps be preferable if this was recognised by increasing the bedding factors and adjusting the recommended factor of safety.

Economic designs which rely upon a high degree of effective site control should not be adopted without modification in areas where the quality of workmanship required by the design code cannot be reliably maintained.

Designers should be given more guidance on recommended factors of safety. A suggested format is given in the table below, Fs = 1.0 represents the proof load condition, i.e. crack width = 0.25 mm, when the installation is carried out in accordance with the specification and the design assumptions with respect to trench width, bedding support etc. are realised.

		Ground Conditions (as relevant to ease of maintaining design trench width)	
		poor	good*
Site	poor	1.5	1.3
	average	1.2	1.1
Control	good	1.1	1.0

Suggested Values for Factor of Safety, Fs (proof)

*would apply under "wide trench" design conditions

If the lateral soil forces are utilised in the design to reduce pipe wall bending moment, the specifications and drawings should make it quite clear that this condition applies, so that compaction of the sidefill to the specified density is given the requisite degree of attention.

Engineers designing in accordance with U.K. practice should consider whether, in main road loading cases, worthwhile economies might be achieved without sacrificing design life by utilising a more realistic vehicle load.

A more rigorous analysis of the different design methods than is offered by this paper should be undertaken with a view to reconciling the differences and promoting a unified approach to the design factor of safety.

7. ACKNOWLEDGMENTS

The author relied heavily upon the structural engineering knowledge of G. Kronbachs, B.Sc.,M.Sc., M.I.C.E., whose comparative analysis of the various Baghdad designs formed the basis for this paper. Acknowledgements are also made to the various Baghdad contractors whose engineers co-operated with us in a spirit of international friendship, having a shared interest in delving more deeply into the subject.

8. REFERENCES

1. Olliff, J.L. : "Factors of Safety in the Structural Design of large Sewers". Proc. First International Symposium on Urban Drainage Systems, Southampton, U.K., September 1982.

2. B.S. 5911; Part 1 : "Precast Concrete Pipes and Fittings for Drainage and Sewerage. Specification for Concrete Cylindrical Pipes, Bends, Junctions and Manholes, Unreinforced or Reinforced with Steel Cages or Hoops", British Standards Institution, London, 1981.

3. Young, O.C. and Smith J.H., Building Research Station : "Simplified Tables of External Loads on Buried Pipelines". London, H.M.S.O., 1970.

4. Young O.C. and O'Reilly M.P., Transport and Road Research Laboratory : "A Guide to Design Loadings for Buried Rigid Pipes", London, H.M.S.O. 1983.

5. DIN 4035: "Reinforced Concrete Pipes, Reinforced Concrete Pressure Pipes and Suitable Fittings of Reinforced Concrete", (Stahlbetonrohre, Stahlbetondruchrohre und Zugehorige Formstucke aus Stahlbeton), Berlin, Deutsches Institut fur Normung, 1976.

6. DIN 1045 : "Concrete and Reinforced Concrete; Design and Construction".

7. DIN 4033 : "Sewers and Sewer Pipelines", (Entwasserungskanale und Leitungen) Berlin, Deutsches Institut fur Normung, 1979.

8. ATV - Worksheet A127 : "Guidline for the Static Calculation of Sewers and Sewer Pipelines". (Richtlinie fur die Statische Berechnung von Entwasserungskanalen und Leitungen).

9. JIS A 5303 : "Centrifugal Reinforced Concrete Pipes", Japanese Industrial Standard.

10. JSWAS A-1 : "Reinforced Concrete Pipes for Sewerage", Japanese Sewerage Works Association.

11. ASTM C76 : "Reinforced Concrete Culvert, Storm Drain and Sewer Pipe", American Society for Testing Materials, 1979.

12. ASCE/WPCF/Manual of Practice : "Gravity Sanitary Sewer Design and Construction", American Society of Civil Engineers/Water Pollution Control Federation, New York/Washington 1982.

13. BS 5400, Part 2 : "Steel, Concrete and Composite Bridges - Specification for Loads", British Standards Institution, London, 1978.

14. Greenwood M.E., and Vinjirayer P.S. : "Testing of Large Diameter Fibre Glass Reinforced Concrete Pipe", Owens-Corning Fiberglas, October 1973.

15. Trott J.J., Nath P. and O'Reilly M.P. : "Longitudinally Cracked Pipes and their Structural Capacity". In : Proc. Sewerage 81, London 22-24 June 1981, Institution of Civil Engineers, 1981, Paper 22.

International Conference on the
PLANNING, CONSTRUCTION, MAINTENANCE & OPERATION OF SEWERAGE SYSTEMS
12-14 September, 1984

MECHANISED PIPE JACKING AS A METHOD OF SEWER CONSTRUCTION

Jim Washbourne BSc (Eng) Hons MSc CEng MICE

OXFORD POLYTECHNIC
Headington, Oxford

Summary

In the UK manual excavation has been a dominant feature of pipe jacking since the introduction of this tunnelling technique in the 1950's. Hand excavation is both flexible and versatile but it is also astronomically expensive and very slow when compared with machine excavation. Hand excavation effort may cost over £100 per KW hour whilst machine excavation may cost less than £1 per KW hour (excluding the cost of putting the machine in the ground). However, the apparent great cost advantage of mechanisation for pipe jacked tunnel excavation is diminished by the considerable cost of putting a machine into the ground which must include allowances for initial investment, maintenance, handling and fitting. In addition the utilisation of a tunnelling machine is low, usually 5%-50% even with an experienced contractor. In contrast a miner doing manual excavation may be usefully active most of the time and have a very high efficiency rating. Nevertheless, mechanised pipe jacking is proving very competitive because of its capacity to instal lined tunnel at between 4 and 8 times the rate of the corresponding hand excavated version giving the contractor a substantially greater turnover. The cost of the mechanised pipe jack operation is often 1/2 - 2/3 of the cost of the comparable manual excavation operation. The mechanised pipe jack operation costs are usually about 40% of the total cost of the tunnel whilst the pipes themselves represent about 60% of the total cost. To reduce the excavation cost for larger pipe jacking operations and to enable the work to be executed more rapidly mechanisation is becoming an accepted practice. In the last few years several machines have appeared in the UK which will take over from traditional hand excavation mining for many pipe jacking projects. This paper explains the operation of these machines and the attendant pipe jacking developments.

NOMENCLATURE

$\emptyset$ = angle of shearing resistance of soil

c = undrained apparent cohesion of soil

S = crushing strength of rock

Pp = passive thrust for jacking wall

ρ = soil density behing jacking wall

g = acceleration due to gravity

H = height of jacking wall

L = length of jacking wall

M = mass of three dimensional passive wedge behind jacking wall

C = cohesive force developed on three dimensional passive wedge inclined face

R = reaction developed on three dimensional passive wedge inclined face

F = factor of safety applied to jacking wall passive thrust

Fb = net positive buoyancy per metre of tunnel lining

Fd = downward drag force per metre on tunnel lining perimeter

Fp = pressure differential net downward force on tunnel lining perimeter

G = gel strength of lubricating slurry

D = outside diameter of tunnel lining

t = radial overcut for tunnel excavation

Pd = flow induced slurry pressure differential around tunnel lining

1. INTRODUCTION

Since the start of the Industrial Revolution machinery has been developed which has largely emancipated man from the gruelling task of large-scale excavation, transportation and deposition of earth and rock on surface sites. The development of large earthwork machines has become so advanced that nowhere on earth is their use precluded by the availability of cheap human labour. It is startling to compare the efforts of human beings with earthwork machines. A caterpillar D10 bulldozer, for instance, has the same work capacity as 5000 men, and at an operating cost of £100.00 per hour, it means that men would have to work for less than 2p per hour to equal its efficiency. The crude power of machines like the D10 is limited to large scale, general earthmoving and as the size of the earthwork task gets smaller and with increased complexity of the work the yawning gap between the efficiency of machine and man gets less. Nevertheless small scale surface earthwork machines are increasingly being developed as machinery becomes more sophisticated and the cost of human labour increases.

It is reasonable to expect that tunnel excavation should follow the same pattern of development that has occurred in general surface earthwork. Indeed, we are witnessing a steady increase in the proportion of new tunnels which are being mechanically excavated. The general tendency in tunnel excavation has been similar to that of surface earthwork in that the larger jobs were first to be mechanised and the size of job where mechanised excavation can be used is getting smaller. It is probable that for small diameter tunnels (up to 1.5 m OD) and for short lengths (less than 100 m), hand excavation will compete with machine excavation in the UK for some time to come because of the high cost of putting a machine in the ground. Some machines like the Marcon Power Vane, which is pneumatically powered, and the Akkerman, which is hydraulically powered, are light and cheap and may offer effective competition to hand excavation for short tunnels. However, for both of these machines, the powering system precludes their use for long drives because there would be large power losses in the long working fluid delivery pipelines.

2. GENERAL ASPECTS

In the past 4 years in the UK things have changed somewhat for pipe jacked tunnels in the 1.75 m to 2.20 m OD range and several machines have appeared which have made an impact leading one to anticipate further developments. These machines are all electrically powered and can be used on drives in excess of 200 metres. The first machine was the Zoker back-acter machine, first used in August 1981 on the construction of 4600 metres of 2.12 m OD tunnel in London Clay for the West Feltham Surface Water Sewer Contract. The second machine was the Lovat Full-Face Tunnel Boring Machine first used in November 1981 to install 2100 metres of 2.18 m OD tunnel in sands and gravels above the water table for the Ipswich South East Area Drainage Scheme. The first rock-boring machine to be introduced was the Dosco 180P boom cutter machine first used in September 1983 for the construction of 2800 metres of 1.80 m OD tunnel in hard marls and mudstone on the River Bollin Surface Water Outfall (Manchester) Contract. The first bentonite machine to be used in the UK for a pipe jack excavation was the Iseki Poly Tech Crunching Mole machine. This was employed to drive 593 m of 1.78 m OD tunnel through sand and gravel 14 m below the water table where there were boulders in the face on the Eastern Interceptor - (Newcastle-upon-Tyne) Contract. The second rock boring machine to be introduced was the Markam disc cutter machine which was used first in February 1984 in hard sandstone on the Halifax Main Sewerage Reconstruction Scheme Contract.

A brief description of each of these machines, their applications and limitations (authors opinion) and the UK work with which they have been involved are given in Appendices 1-5 respectively. Figure 1 indicates the range of strength of excavation material for which the machines can be usefully employed and Figures 2 and 3 are approximate charts for relating excavation energy output.

Tunnel excavation when carried out with a tunnelling machine is very rapid compared with hand excavation. If a tunnelling machine is used the rate of progress of tunnelling is mainly governed by the speed at which spoil can be removed from the workings and the rate at which linings can be installed. For a small diameter tunnel, lined with bolted segments, progress of 1/3 to 2/3 of a metre per hour would be reasonable because of the time taken to fit and bolt together the segments. If a wedge block lining were to be used in such a tunnel a rate of progress of 2 to 3 metres per hour is feasible. If, however, a jacked lining were to be used for a similar sized tunnel then a rate of 3 to 6 metres per hour might be achieved. Speed of installation is the main advantage of jacked linings for machine bored tunnels. The other big advantage is that the end product is a less jointed and more robust structure. It is only because pipe jacking has been associated with very slow (hand) excavation that it has failed so far to develop into the universal form of tunnelling limited only by the maximum pipe diameter that can be feasibly handled.

When the change is made from short, small hand-excavated pipe jack tunnels to larger diameter longer machine excavated ones, certain things have to be changed. More substantial jacking walls are required to give the large jacking reactions required to push the linings into place. The use of slurry is usually necessary to provide lubrication and maintain clearance between the bore and the outside surface of the pipes (Ref 2). Belt loaders are necessary (for other than bentonite machines) to enable large quantities of excavated material to be loaded quickly. The spoil removal skips have to be of appreciable capacity and sometimes, it is necessary to employ multi skip muck trains on the longer runs to increase the rate at which spoil can be transferred to the jacking pit for removal.

3. JACKING WALLS

Jacking walls are usually required to produce between 500 and 3000 tonnef of reaction depending upon the size of the tunnel installation and are usually formed from heavy section steel piles with a substantial vertical reinforced concrete load distribution slab bedded against them. The efficient design of this item is of considerable importance and two dimensional earth pressure theory has proved to be unsuitable for this purpose. A three dimensional wedge passive pressure theory was developed by the author for short deep walls (Ref 3) and is illustrated in Figures 4, 5, 6 and 7. Using this theory, a jacking wall was analysed for a Lovat Machine drive at Poole in Dorset as follows:

$$\frac{Pp}{F} = \frac{1}{F} \times \rho g H^2 \frac{(L+H)}{2\ \ 3} \tan^2 (45^o + \emptyset/2)$$

$$+ 2cH\ (L+H) \tan (45^o + \emptyset/2)$$

For the installation

Passive thrust is	Pp
Factor of safety	$F = 1{,}5$
Soil density	$\rho = 1890\ kg/m^3$
Wall height	$H = 8.0\ m$
Wall length	$L = 4.0\ m$
For sand	$\emptyset = 35^o$
	$c = 0$

$$\frac{Pp}{F} = \frac{1}{1.5} \times \frac{1890}{1000} \times 9.81 \times 8^2 \ \frac{(4+8)}{2\ \ 3} \times 1.921^2 + 0$$

$= 13622$ kN

$= 1{,}389$ tonnef

The corresponding two dimensional analysis gives a working load of only 595 tonnef. It is worth noting that the wall was loaded to 1500 tonnef when the tail-can of the tunnelling machine jammed and no perceptible distress was observed in the wall. The sheet pile manufacturer and a consultant engineer, using two dimensional earth pressure theory, both separately rated the ultimate passive resistance of the wall at about 1000 tonnef. Two dimensional passive analysis can be regarded as being conservative in pipe jacking wall design and wasteful of resources.

4. SLURRY LUBRICATION

The maximum end thrust which can be used to slide a reinforced concrete pipe run into the ground is usually that which generates 15 N/mm^2 of compressive stress in the lining's cross-sectional area, which have jack loads evenly distributed by a heavy steel thrust ring. To maximise the distance that can be jacked with a given maximum end-thrust, it is standard practice to overcut by about 20 mm the bore (on the diameter) into which the pipes are slid, and to introduce a lubricating slurry under pressure into the overcut space (Ref 4). Using this technique it has been shown that it is possible to jack pipes over a kilometre distance in any type of soil. The author believes that lubrication techniques used in pipe jacking are as yet underdeveloped and in the next decade jacking distances of several kilometres could become possible without the need for intermediate jacking.

Lubricating slurry is pumped under pressure around the outside of the jacking linings just behind the tunnelling machine and, to compensate for leakage and consolidation of the slurry, it is supplemented by injections of slurry through the top of the lining at intervals all along the run throughout the tunnelling process (Ref 5). As the lining is normally lighter than the weight of the volume of slurry which it displaces, injection of slurry through the top of the lining counteracts the tendency of the lining to be displaced upwards due to its net positive buoyancy.

The net positive buoyancy Fb per metre giving a tendency for floatation of a tunnel lining in a slurry is given as follows:

$$Fb = \text{Wt/m of displaced slurry} - \text{Wt/m of lining} \qquad (1)$$

The tendency for a relative downward flow of slurry in the space around the tunnel lining, when the tunnel lining tends to float, or when injection is carried out close to the top of the lining, gives downward forces which counteract the net positive buoyancy. If G is the gel strength of the slurry and D is the outside diameter of the tunnel lining, the downward drag force Fd per metre on the lining perimeter will be

$$Fd = 2\ G\ D \qquad (2)$$

In addition there is a net downward force Fp per metre available because of the pressure differential in the slurry generated by its tendency to flow around the lining where

$$Fp = \frac{2\ G\ D^2}{t} \qquad (3)$$

Where t is the radial clearance between the outside of the tunnel lining and the surrounding ground.

Formula (3) for Fp is obtained by integrating the vertical components of load due to the pressure differential in the fluid Pd caused by the relative downward flow of slurry around a tunnel lining. Pd increases uniformly from 0 at the bottom of the lining to a maximum value of $\frac{G\ D\ \pi}{t}$ at the top.

Figure 8 indicates the relative downward flow of slurry around a tunnel lining and Figure 9 indicates the maximum pressure and drag effects generated by that relative movement. Figure 10 shows the summation of drag and pressure effects for Fd and Fp. It follows that if

$$Fd + Fp > Fb \qquad (4)$$

then there will be no appreciable uplift of a jacked tunnel lining surrounded by a slurry. This is ensured by a reasonably small slurry thickness even for the largest jacked lining.

For example a 2.03m external diameter "Spunthrust" pipe weighs 2.510 tonnef/m and it displaces 3.327 tonnef/m of 1.026 tonnef/m slurry giving the right hand side of equation (4) as 0.817 tonnef/m net upward buoyancy force. If the gel strength of the slurry is 1Q N/m and the thickness t of the slurry layer is 10 mm then the left hand side of equation (4) becomes,

$$2 \times 10 \times 2.2 + \frac{2 \times 10 \times 2.2^2}{0.010} = 9724 \text{ N/m}$$

$$= 0.991 \text{ tonnef/m}$$

Thus the lining will be maintained in position by the gel strength of the slurry in spite of the net upward buoyancy force. The stabilising influence of the gelled slurry will, of course, increase with time as it consolidates and the gel strength consequently increases.

5. CONCLUSION

Mechanised pipe jacking is undoubtedly the tunnelling system of the future. The speed at which tunnelling can take place using this technique and the distances that can be jacked would have been unbelievable when the system was first introduced. What must be remembered is that the system is still being developed and much is yet to be done before its full potential is achieved. For use in the UK this must be a comforting thought considering the extent of the work that needs to be done on our sewerage network infrastructure. This paper refers to the first machines of their type to be introduced into UK Pipe Jacking. They will be the first of many.

6. ACKNOWLEDGEMENTS

The author wishes to thank the following for the assistance rendered in preparing this paper.

Zokor International
Fairclough Civil Engineering
Lovat Tunnelling Equipment
A Streeter & Co
Hawker Sidderley Dosco
Iseki-Poly-Tech
Lilley Group
Tunnels & Tunnelling Journal
Markham & Co

7. REFERENCES

Ref 1 "A Strata Strength Index for Boom Type Roadheaders". V P Aleman. Tunnels & Tunnelling, March '81, pp 53-55.

Ref 2 "Thixotropic Slurries for Civil Engineering". J Washbourne. Civil Engineering Technical. April '83, pp 9-13.

Ref 3 "Three-dimensional Passive Analysis for Jacking Walls". J Washbourne. Tunnels & Tunnelling, March '81, pp 13-17.

Ref 4 "Sliding Resistance of Jacked Tunnel Linings". J Washbourne. Tunnels & Tunnelling, November '82, pp 17-18.

Ref 5 "Slurry Lubrication for Jacked Tunnel Linings". J Washbourne. Civil Engineering Technology (Cover Story), February '84, pp 7-9.

APPENDIX 1 - Zokor Back Acter Machine (2.12 m O D)

Manufacturer; Zokor International - Aurora, Illinois, USA
First Operator; Fairclough Civil Engineering - Stone, UK
Length of UK Tunnel installed as of 1.1.84 - 4600 m
Maximum UK drive distance - 460 m
Maximum UK production per shift - 2 2 m (approx)
Power supply to face; 440 V 3 phase
Maximum power of driving head - 20 KW
Speed of excavation unit; 6 cpm for a 0.12 m heaped capacity bucket
Spoil removal; duplicate single 1.5 m^3 heaped capacity skip belt loaded
Maximum pit jacking force; 1200 tonnef (machine is not self thrusting).

Advantages

1. Can be used in harder material than can the soft earth full face machine because it can exert greater breakout forces. It can remove large rocks and access to the face is easy.

2. Can excavate ahead of the lining without the need for a telescope joint in the tail can.

3. The machine can supply high concentration of excavation effort selectively at the hard patches of an uneven face (which may be supplemented with pneumatic chisel work). This should enable the machine to be kept on line in variable ground.

Disadvantages

1. Cannot be used with a head of water at the face unless compressed air is used.

2. The excavation is interjected with the withdrawal of material back from the face onto the conveyor so that the full power of the machine cannot be continuously applied to excavation.

3. The bore cannot be regarded as neat dig because of the very nature of bucket excavation.

APPENDIX 2 - Lovat Full Face Tunnel Boring Machine (2.18 m OD)

Manufacturer - Lovat Tunnelling Equipment; Rexdale Canada
First UK Operator - A Streeter - Godalming, UK
Length of UK Tunnel installed as oo 1.1.84 - 3260 m
Maximum UK drive distance; 520 m
Maximum UK production per shift 42 m
Power supply to face - 440 V 3 phase
Maximum power of boring head - 110 KW
Speed of boring head - 6 revs per minute
Spoil removal - duplicate 4 no 1.5 m^3 heaped capacity skip muck train belt loaded
Maximum jacking force - 1500 tonnef (machine is self thrusting with 400 tonnef jacks)

Advantages

1. With the unimpeded uninterrupted excavation process large quantities of material can be dug out very quickly if the face conditions are right.

2. The adjustable face doors of the machine allows soft and waterlogged ground to be fully supported during and after excavation.

3. The machine can excavate ahead of the lining and can thereby apply a large thrust to the boring head.

Disadvantages

1. Use is severely handicapped by the presence of large stones on the face as access to the face is limited.

2. The telescopic joint in the tail of the machine can give serious problems in collapsing ground.

3. With uneven conditions at the face, ie large hard and soft areas the machine may be deflected off line.

APPENDIX 3 - Dosco 180p Boom Cutter; Tunnelling Machine (1.80 MOD)

Manufacturer - Hawker Siddeley Dosco - Newark UK
First UK Operator - Fairclough Civil Engineering - Stone UK
Length of UK Tunnel installed as of 1.1.84 - 500 m approx
Maximum UK drive jacking distance - 450 m
Maximum production per shift - 10 m approx
Power supply to face - 440V 3 phase
Maximum power of cutter - 67 KW
Speed of cutter - 61 revs per minute
Spoil removal - single 1 m^3 heaped capacity skip belt loaded
Maximum pit jacking force - 1200 tonnef (machine is self thrusting with 200 tonnef jacks)

Advantages

1. Can be used on a hard and variable face and harder patches can have high concentrations of excavation effort selectively applied to them.

2. The machine can excavate ahead of the lining and the open front of the machine allows the bore to be inspected and picked clean of loosened material.

3. The cutter guide enables a neat excavation to be performed.

Disadvantages

1. The cutting head can be easily damaged by boisterous use.

2. The pace of excavation in good uniform conditions is slower than with the disc cutter machine.

3. The open face may necessitate the use of compressed air where water is a problem.

APPENDIX 4 - Iseki Crunching Mole (1.78 m OD)

Manufacturer - Iseki Poly - Tech - Tokyo Japan
First UK Operator - Lilley Group - Nuneaton UK
Length of UK Tunnel installed as of 1.1.84 - 593 m
Maximum UK drive distance - 307 m
Maximum UK production per shift - 7 m approx
Power supply to face - 440V 3 phase
Maximum power of boring head - 30 KW (includes crusher)
Speed of boring head - 2 1/2 revs per minute
Spoil removal - 75% spoil suspended in slurry through 100 mm pipes
Maximum pit jacking force - 1200 tonnef (machine is not self thrusting)

Advantages

1. Can excavate a face where there is a large head of water without the need for compressed air.

2. Can deal with large stones in the face which are passed through a crusher built into the machine.

3. The spoil removal process is continuous along with the excavation process.

Disadvantages

1. The face is inaccessible and obstructions are difficult to deal with.

2. The machine cannot excavate ahead of the lining this means that thrust to the boring head has to be supplied together with overcoming sliding resistance.

3. Bentonite transport pipes as well as electricity supply have to be broken and reformed at each pipe addition.

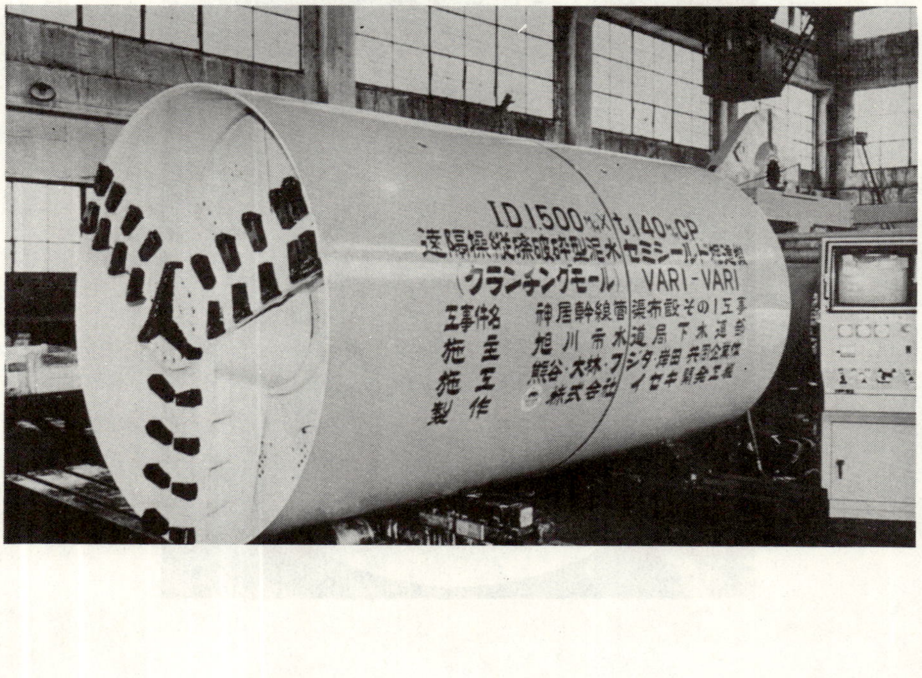

APPENDIX 5 - Markham Disc Cutter Machine (2.12 m OD)

Manufacturer - Markham and Co, Chesterfield, UK
First UK operator - Fairclough Civil Engineering - Stone, UK
Length of UK Tunnel installed as of 1.1.84 OM
Maximum UK drive distance - 800 m to be attempted in 1984
Maximum UK production per shift
Power supply to face - 1100V 3 phase
Maximum power of boring head - 250 KW
Speed of boring head - 5 revs per minute
Spoil removal - single 1.5 m heaped capacity skip belt loaded
Maximum pit jacking force - 1200 tonnef (machine is self thrusting with 500 tonnef jacks)

Advantages

1. Can excavate hard materials which would otherwise demand blasting to be carried out.

2. Full face disc cutting is the most rapid machine excavation in hard material.

3. The full face machine imports support to the face and to the bore.

Disadvantages

1. The machine does not excavate ahead of the lining and therefore the bore cannot be picked clear of loose material. This means that the machine and/or the pipes can jam in fissile material.

2. With uneven conditions at the face, ie large hard and soft areas the machine may be deflected off line.

3. Face and cutting head is less accessible than those for boom cutter machines.

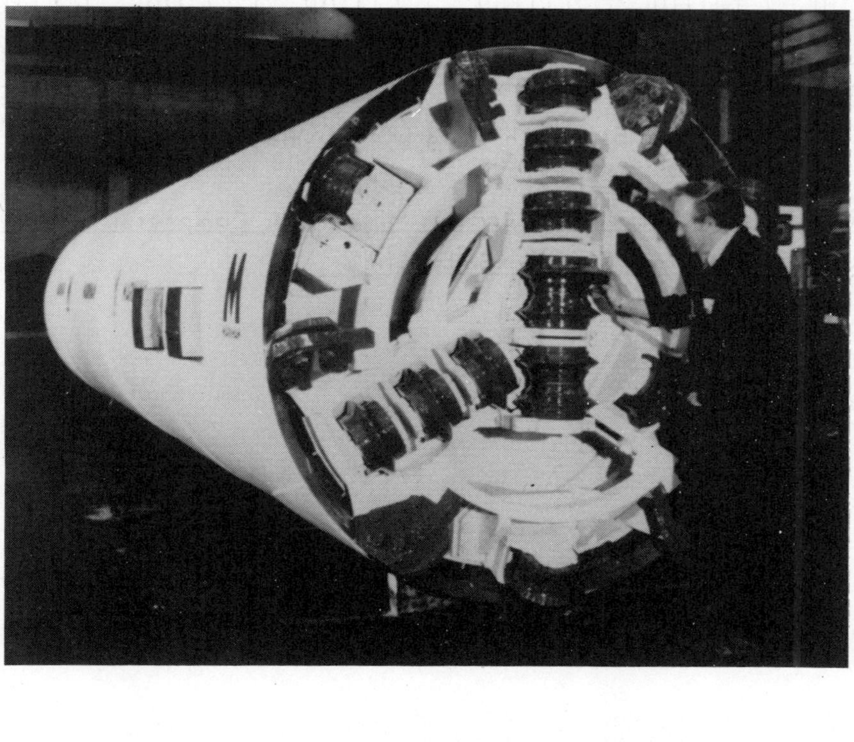

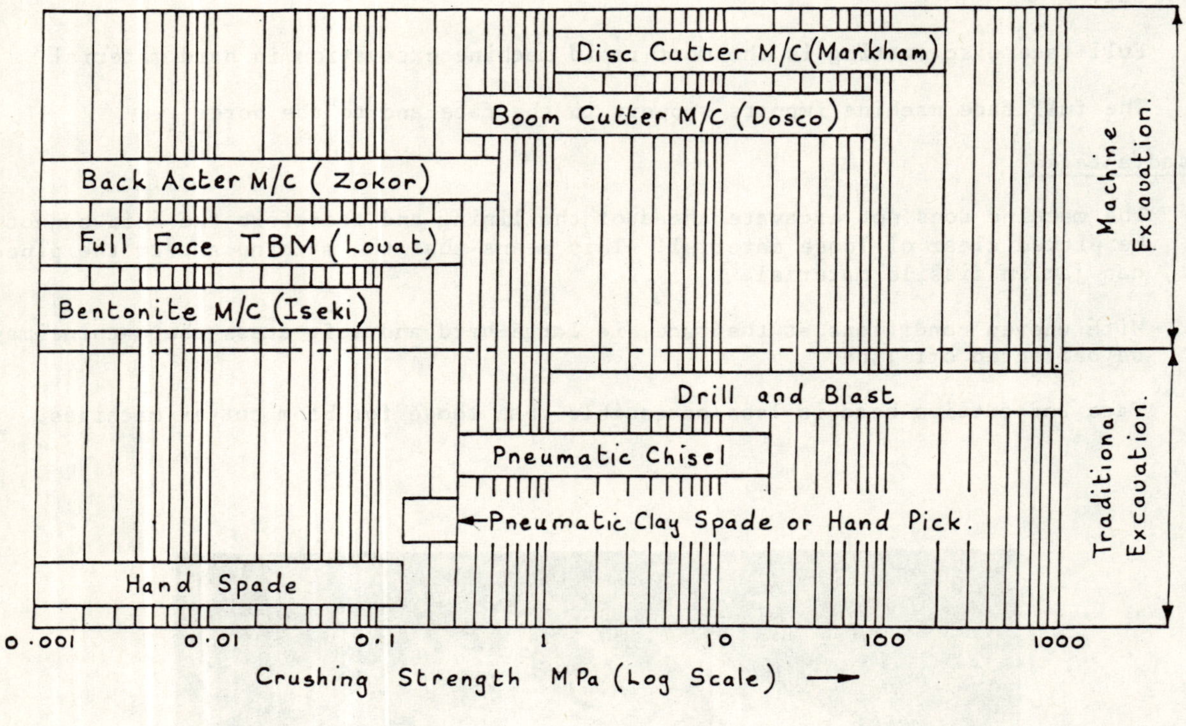

Figure 1. Relationship Between Material Unconfined Compressive Strength and Excavation Process Suitability.

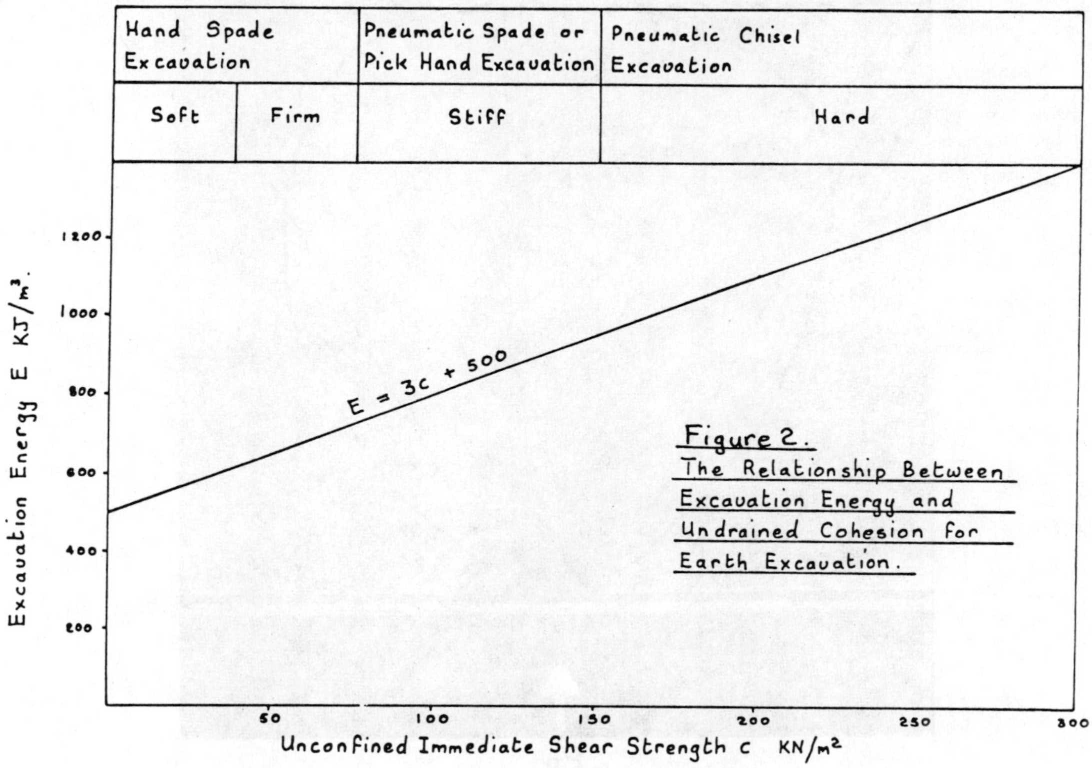

Figure 2. The Relationship Between Excavation Energy and Undrained Cohesion for Earth Excavation.

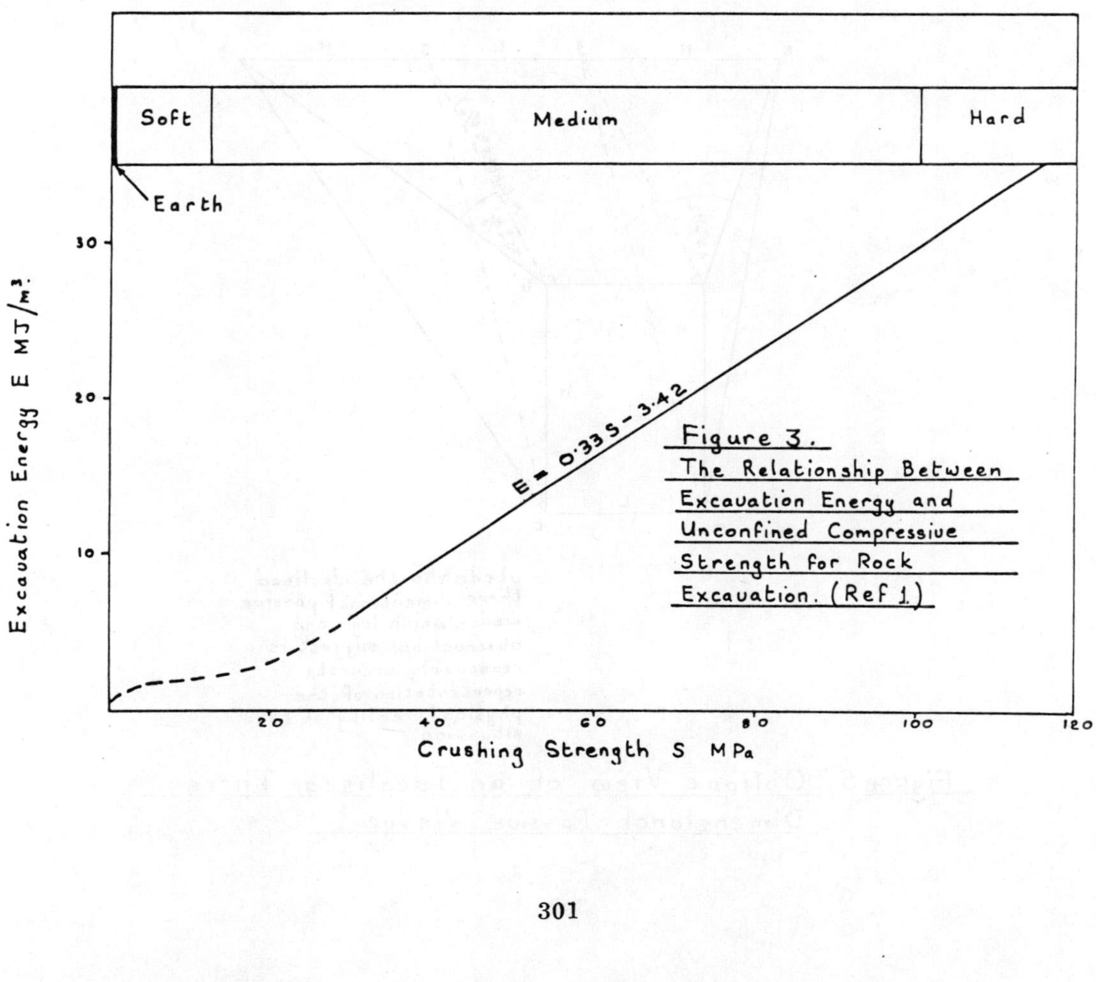

Figure 3. The Relationship Between Excavation Energy and Unconfined Compressive Strength for Rock Excavation. (Ref 1)

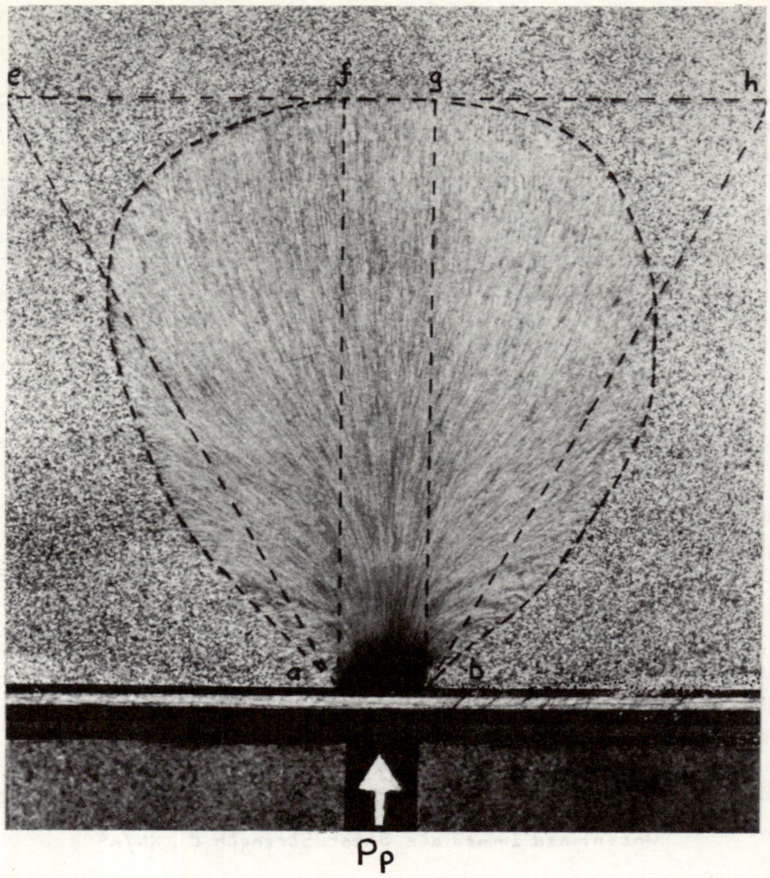

Figure 4. Plan View Showing the Area of Disturbance for a Passive Wedge in a Model Test.

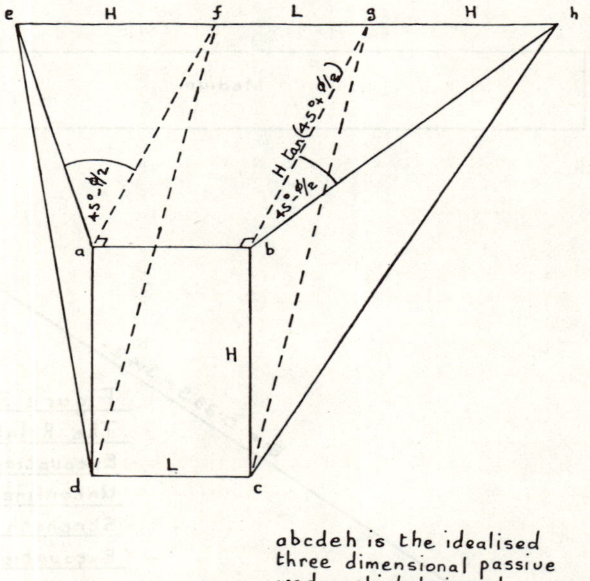

Figure 5. Oblique View of an Idealised Three Dimensional Passive Wedge.

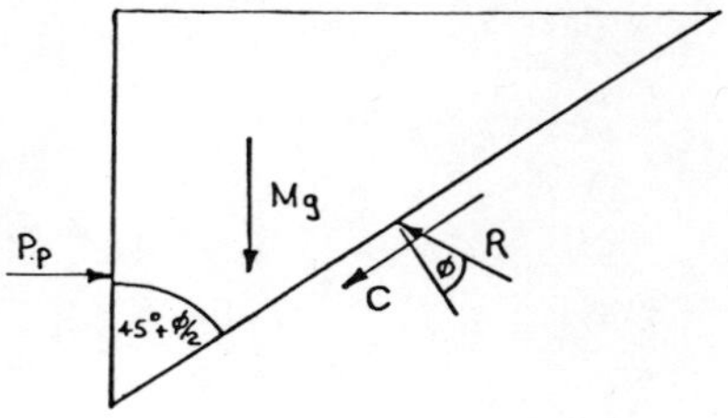

M_g = Wedge Volume x ρg

C = c x Inclined Shear Area cdeh

The forces on the splayed faces ade and bch of the three dimensional wedge are ignored in the analysis.

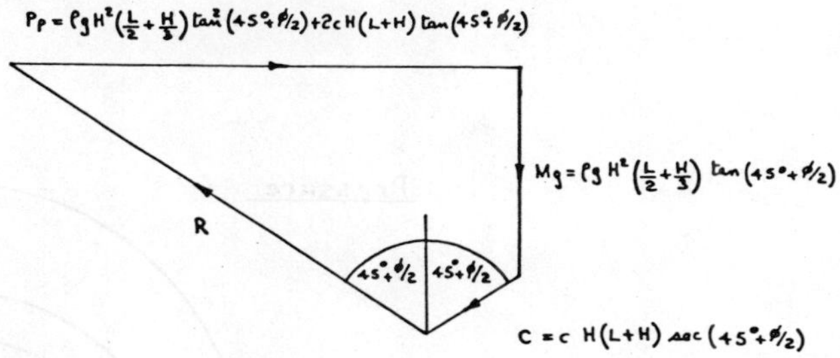

Figure 6.
Cross Section Perpendicular to Jacking Wall for the Three Dimensional Passive Wedge.

Figure 7. Force Diagram for the Three Dimensional Passive Wedge.

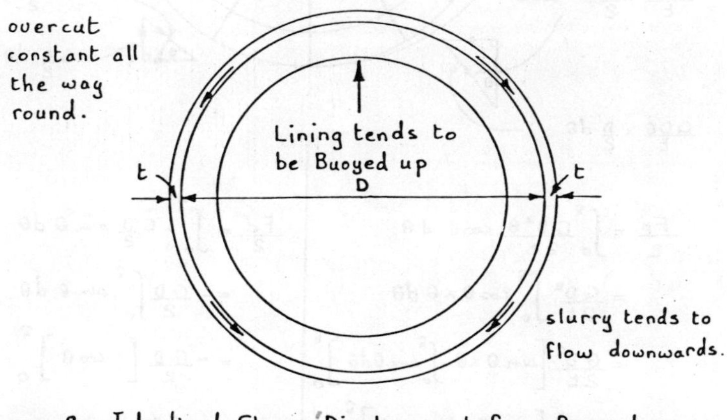

Figure 8. Idealised Slurry Displacement for a Buoyed up Tunnel Lining.

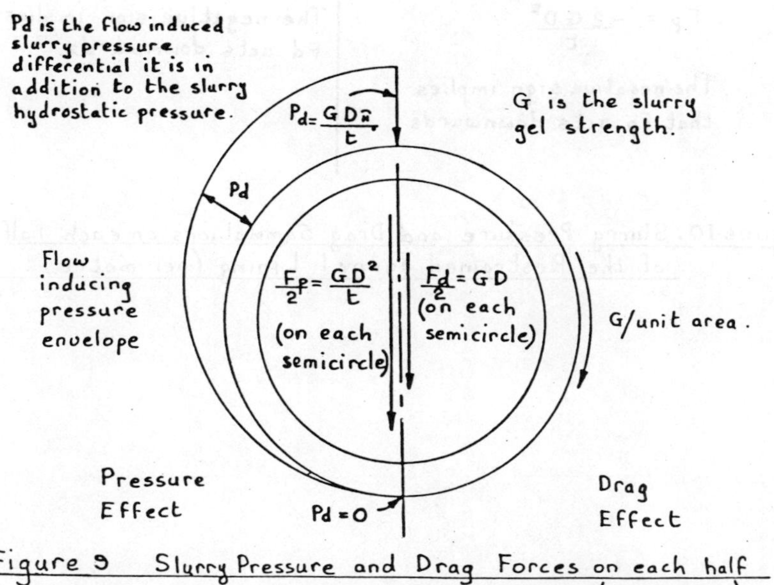

Figure 9 Slurry Pressure and Drag Forces on each half of the Restrained Tunnel Lining (per metre)

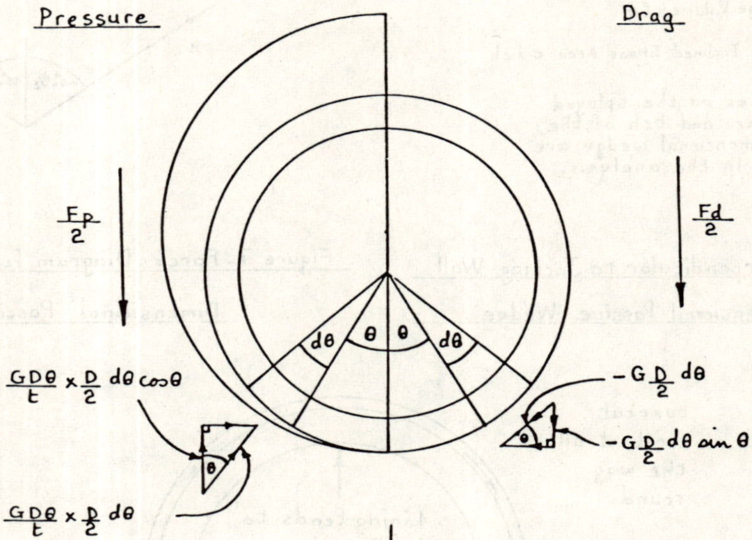

$$\frac{F_p}{2} = \int_0^{\pi} \frac{GD^2\theta}{2t} \cos\theta \, d\theta$$

$$= \frac{GD^2}{2t} \int_0^{\pi} \cos\theta \times \theta \, d\theta$$

$$= \frac{GD^2}{2t} \left[\sin\theta \times \theta - \int_0^{\pi} \sin\theta \, d\theta \right]_0^{\pi}$$

$$= \frac{GD^2}{2t} \left[\sin\theta \times \theta + \cos\theta \right]_0^{\pi}$$

$$= -\frac{GD^2}{t}$$

$$F_p = -\frac{2GD^2}{t}$$

The negative sign implies that Fp acts downwards

$$\frac{F_d}{2} = \int_0^{\pi} -G\frac{D}{2} \sin\theta \, d\theta$$

$$= -\frac{GD}{2} \int_0^{\pi} \sin\theta \, d\theta$$

$$= -\frac{GD}{2} \left[-\cos\theta \right]_0^{\pi}$$

$$= -GD$$

$$F_d = -2GD$$

The negative sign implies Fd acts dowmwards.

Figure 10. Slurry Pressure and Drag Summations on each half of the Restrained Tunnel Lining (per metre).

International Conference on the
PLANNING, CONSTRUCTION, MAINTENANCE & OPERATION OF SEWERAGE SYSTEMS
12-14 September, 1984

PAPER G3

PIPE JACKING IN JAPAN

D. B. Downey

ARC Slimline Limited
The Ridge, Chipping Sodbury,
Bristol BS17 6AY
England

Summary

Pipe jacking in Japan is a well-established technique for sewer construction and is significantly advanced in comparison to the practice of the art elsewhere. Methods and practices are described from the author's discussions with manufacturers, contractors and users over the past three years and their development and relevance are considered with a view to suitability in the U.K. scene. There is a clear trend of increasing mechanisation and sophistication at the root of the evident popularity of this method of construction and those methods developed to deal with the problems of pipelaying in congested urban locations may have some interesting implications for urban renewal here.

Held at The University of Reading, England. Organised and sponsored by BHRA, The Fluid Engineering Centre, Cranfield, Bedford in conjunction with the Water Research Centre.

1. INTRODUCTION

It is often considered that Japan lags behind the West in the provision of sewerage and drainage. The statistics tend to support this view for where the U.K. can boast 97% of population served by a public sewer, Japan can only claim to have so far made provision for the needs of 32% of its staggering population of 117.06 million people. This situation is remarkable considering Japan's history in respect of public health; in the Edo period when London and Paris reeled under repeated epidemic diseases the citizens of Japan's cities were sustained by an efficient night soil disposal system and surprising in view of Japan's exceptional growth in the post war period to its current economic status with the world's second largest G.N.P.

However as many modern western innovators have been made painfully aware the Japanese are not slow learners; they installed an activated sludge treatment plant in 1913 only 13 years after its British invention at a time when the practicalities of technology transfer must have been far more formidable than at present. It is simply that protected by their own highly efficient and conservationist system the Japanese did not suffer the same desperate urgency to involve themselves in a 19th century flurry of sewer building and when later the breakdown of this system occurred through the introduction of modern fertilizers and increased urbanisation, the capital expenditure required was committed to other priorities.

The first sewer construction reported in Japan was at Kanda, one of the oldest districts of Tokyo, other municipalities like Kyoto, Yokohama, Nagoya and Osaka quickly followed with similar pilot schemes. In 1900 the first sewerage law was enacted but progress was very slow, public health was a relatively low priority and the major upheavals of successive international conflict influenced to the extent that by 1958 only 6% of the population enjoyed the sophistication of a public sewerage service.

Since that time a major programme of public expenditure has been underway to provide sewerage for the bulk of the population. Annual public expenditure on sewerage currently exceeds £4,500 million and data provided by the Japan Sewage Works Agency (Ref. 1) shows that most urban connurbations with population exceeding 300,000 already have sewer networks and have further infrastructure under construction. Although nationwide a very substantial volume of construction has yet to be tackled the infrastructure in the main centres is nearing completion and this is a clear demonstration that the Japanese have developed considerable expertise in urban construction techniques and therefore have a great deal to offer to Western countries who face massive expenditure in urban renewal in the years to come.

The Japan Sewage Works Association figures for 1981 (Fig. 1) indicate that half the sewer pipelines were constructed from concrete. Obviously in the very small diameters clay and PVC pipe make up a substantial proportion of the length installed, however above 300 mm centrifugally spun concrete accounts for 74% of the total length and concrete segments add a further 11% to make concrete the major sewer material. Some weighting of the distribution to take into account diameter and associated cost implications clearly demonstrates the importance attributed to concrete pipe in Japanese sewer construction. It is worth noting that the Japanese have improved the original Hume style of centrifugal spinning to such an extent that high quality pipe can be manufactured at production rates which would compete with many of the inferior vibratory or mechanical methods. In the congestion common in Japan's urban connurbations the benefits of pipe jacking offering minimal surface disturbance are of particular benefit and the techniques have made great progress since their first use at Amagaseki City in 1948. In 1981 jacking techniques were used in 30% of all concrete pipeline construction and 1982 figures show a 25% growth in the popularity of these methods with a total length of 502 kms of pipe jacked. (Ref.2). It is evident that the extent of technical development that can be sustained by a sewer building programme on this scale must outstrip the best efforts of our indigenous industry and a review of methods and products and trends can at least provide some stimulus as we enter a period of reconstruction of our much envied asset.

2. METHODS

2.1 General

Updating the published analysis of pipe jacking for the years 1977-80 with additional data (Ref 3)it is possible to indicate two important and fairly clear trends (Fig. 2). There is a substantial growth in pipe jacking at the expense of traditional open cut and segment tunnel methods and there is a steady trend towards increasingly mechanised techniques. There is steady growth in the use of full face shield machines, in small diameter jacking which is exclusively automated and in the use of sophisticated systems of automated interjacks.

2.2 Hand Excavation

A substantial volume of pipe jack is executed using hand excavation. In 1980 there were 229 kms installed by this method, 5210 thrusts of an average 45 ms. Methods used are not dissimilar to those employed in the U.K. A variety of shields are available ranging from simple cutting edges extended or enclosed to provide an appropriate degree of protection for the miners to more complicated double chamber shields for use with compressed air. Spoil removal is accomplished by wheeled muck skip or conveyor belt where circumstances merit the investment.

There are a large number of small pipejacking contractors active in Japan and their long term involvement is to some extent secured by policy in allocating work to ensure continued employment and keen competition. Many of these smaller contractors specialise in hand excavation techniques because they are limited in their capacity for capital exposure. However there are schemes offered by machinery manufacturers for leasing of equipment and it may be reasonably expected that the steady decline in simpler techniques will continue.

Much of the construction work has been undertaken in the Kanto and Kansai regions where soils are frequently unstable soft clays and silts with a higher water table adding complications. There are many specialist companies to provide dewatering and chemical stabilisation though as an alternative the use of compressed air methods is widespread. Compressed air methods though often essential in difficult ground produce specific problems of noise and miner safety and these drawbacks have greatly assisted the popularity of closed shields. Use of compressed air systems is time consuming,expensive and hazardous.

2.3 Mechanical Excavation

In many circumstances conditions are unsuitable for miners to work directly at the face particularly where there is abundant ground water and risk of instability. There is increasing use of mechanical methods for excavating usually manoeuvrable screw conveyors or backhoe equipment which can be controlled remotely from a secondary chamber thus avoiding some of the dangers and improving markedly the rate of spoil removal. A sealed conveyor is usually employed to tranfer spoil from the pressurised primary chamber to a skip or conveyor system. These more elaborate shields often employ steering jacks to control the direction of the cutting edge.

2.4 Closed Shields

A further development designed to take the worker further from the face involves a fully closed shield. These machines, like Kidoh Construction's Helical Blind Shield employ an adjustable rotary full face cutter within the shield which transfers spoil through doors in the cutter face to a sealed screw conveyor through an air lock and onto a conveyor belt for transport to the thrust pit. The method is particularly suited to clays, silts, sand and aluvial gravel. It provides full protection for the work force in difficult conditions and is particularly suited to the soft wet ground found in river beds. It has been used for pipes upwards from DN 900.

2.5 Slurry Shields

Another alternative to compressed air systems is the Bentonite shield machine, believed to be an originally British development, now taken on by the Japanese and substantially improved. It was first used in Japan in July 1967.(Ref. 4) Its use was reported by Tohyama and Ishibashi crossing the Arakawa River with a 2.6 metre diameter concrete pipe. The 1.6 km contract was completed in 16 drives using a shield remarkably like those popular today. The basic system involves a mud water slurry pumped to the face at a pressure just above ground water pressure to sustain the face ahead of the cutter and hold back ground water. Spoil is admitted to a chamber behind the cutter through adjustable vents in the cutter face and is pumped as a slurry SG 1.1-1.2 to the thrust pit to a settlement system. Slurry is also injected through holes fabricated in the pipes to provide lubrication. This was shown at Arakawa to substantially reduce the frictional forces.

Bentonite shield systems require a substantial investment in separation plant and are complex in respect to maintaining flow, SG and pressure. Errors in measurement can lead to face collapse and surface heave. Notwithstanding the problems the practical benefits of the slurry shield method have led to considerable popularity and its use has grown from 1.4% of pipe jacks in 1977 to a current penetration of more than 20%. Much of this growth has been due to the efforts of machinery manufacturers Okamura, Kidoh and particularly Iseki Kaihatsu who have extended the basic technology signficantly and whose equipment was used on more than 50% of all shield method pipe jacks last year.

Iseki have developed a range of new shield machines stemming from their own invention the Mechanical Earth Pressure Counterbalanced Bentonite Shield (Fig. 3). This equipment uses the cutter face to balance the earth pressure thus separating the functions of the Bentonite in the basic slurry shield. This simplifies the monitoring and control with continuous contact from the adjustable cutter face to hold up the ground and Bentonite solely to counteract water pressure and transport the spoil to the surface. The machine is particularly effective in water saturated clay and silt. The machine is articulated for steering through powerful jacks and its progress is monitored by laser to give an accuracy as good as ± 10 mm on line and level. The use of shield machines has been extended to mixed soils by the introduction of a crushing capability in another Iseki machine. The Crunching Mole contains a rock crusher which can reduce boulders as large as 500 mm to pea gravel for transport to the surface in the slurry.

2.6 Small Diameter Jacking

The jacking of small diameters is an important component in the Japanese scene. It currently makes up just over 10% of the nation's pipe jacking. The first small bore pipe jack machine was the Komatsu Iron Mole developed in 1973. The system involves an auger which is fixed within a pilot pipe from thrust pit to arrival pit prior to the rotating excavation head which completes the preparation of the bore to the appropriate diameter sometimes using additional water to soften the ground to an appropriate consistency for transfer by auger to the waiting skip in the arrival pit. The systems accuracy ± 50 mm on line and ± 20 mm on level is monitored using a theodolite aimed at two targets located in the cutter head. Steering is achieved by use of a level jack which adjusts the position of the excavation head to the pilot line.

A compatible system is Sanwa Horizonga devised in 1975, this machine excavates with a rotary cutting head and transfers spoil along an auger running back through the pipeline to the thrust pit. Like the Iron Mole the machine itself not only excavates but also applies the thrust through a long stroke jack. The cutter head is equipped with steering jacks and controlled by a laser theodolite. A similar machine more recently developed is the Kagatagumi sand worm auger.

Bentonite shield machines have also been developed for small diameters. The Iseki family of machines includes the Telemole for pipes DN 600 upwards and the Telemouse for pipes DN 350 upwards. An even smaller machine is under development. These machines involve close circuit television transmission of gauge readings at the cutter head and control from the thrust pit via an operator monitoring the TV screen in complete security. Small diameter jacking tends to be restricted to relatively short drives. There is generally insufficient sectional area to sustain the thrusts required for drives longer than 50 metres. The Crunching Mole concept has been extended down to DN 450.

2.7 Interjacks

Published figures show a marked increase in the use of interjacks, the benefits are very clear, in 1980 63.1 kms were driven in 720 thrusts, an average jacked length of 94 metres. A fairly typical arrangement is shown in Fig. 4 with an appropriate number of jacks regularly spaced around the circumference of an intermediate chamber formed from a bolted segment housing located within an external steel sleeve. The jacks are actuated as pressures applied by the main rams approach a critical maximum and thus extend very significantly the length of pipe jack. Lengths of 2-500 m are common using sophisticated linked interjacks which can be operated automatically and in sequence. The use of these systems is particularly beneficial in conjunction with closed steerable shields suitable for curved jacking.

2.8 Curved Jacking (Ref. 5)

Curved jacking was undertaken in Minami Urawa by Kidoh Construction and reported in detail in the Monthly Sewage Journal by Ishibashi,Yabe and Nishimura in 1981. The contract involved two curved drives of radius 500 m and 80 m and was accomplished using a Bentonite shield machine. Curved jacking requires careful control of thrust forces and overcuts and the published theoretical analysis indicates that for curves of radius 50 - 200 m jacking pressures must be increased by a factor of 1.07 - 1.33 and excavation increased by 2 - 15 mm to generate the curve action. Soil stabilisation is required to provide sufficient thrust resistance for the pipeline to follow the curve. The opening of joints is unavoidable in curved jacking and it is usual to reseal the pipeline with plastic cement or fibre reinforced mortar. The use of automated interjacks and curve spacers is particularly helpful to reduce concentrated loads on the pipe.

2.9 Jacks and Thrust Rings (Fig. 5)

The range of jacks available is extremely wide reflecting the plethora of contractors at differing levels of capitalisation. Jacks from 50 - 200 tonnes are commonly used and increasingly sophisticated long stroke (up to 3 m) thrusting machines like the Mole Meister are in use. Carefully designed thrust rings are frequently lined to equalise thrust pressures around the periphery of the pipe. Test work undertaken by the Japan Hume Pipe Association has shown that the thrust capacity of pipes can be improved by as much as 66% by use of appropriately designed thrust rings. (Ref. 6).

3. PIPES

3.1 Historical Development

Early jacking pipes were specified in Tokyo Municipal specifications covering the diameter range 600 - 2400 with diameter to wall thickness ratio ranging from 7.7 - 9.4 : 1. The pipes were centrifugally spun concrete tubes with a 6 or 8 mm rebate to form a simple double spigot. The pipes were butted together with a 300 mm long 'T' shaped thin steel band loosely located on the spigot with no rubber gasket. The root of the T shape ring was about 20 mm less than wall thickness and the cavity was grouted after laying. (Ref.7).

It is not clear whether this design preceded or was concurrent with a thinner pipe (D/T = 12) described in the first national specification JSWAS A-1 published in 1974 (Ref 8)This specification for pipes DN600 - 3000 designated strengths as prescribed in JIS A5303 Class 2. the thrust capacity of these pipes was indicated as 1300T/sq m of pipe wall. The joint was even simpler, a thin external steel sleeve with a central T.

3.2 Current Pipe Design

The Japan Sewerage Works Agency revised its specification again in 1975 (Ref 9)with some fairly radical changes to the pipe design. Wall thicknesses were increased; D/Tratios range from 9.2 - 12. Provision has been made for holes for Bentonite injection and bolts for fixing pipes together during jacking to avoid twisting when used with powered closed face cutters and to resist deviation in line and level. Double rebated spigots with a groove for rubber seal location are provided and various finned gaskets for use as sliding seals are specified. These finned gaskets are particularly suited to resist infiltration. The specification also includes details of interjacking sleeves approximately 1.2 metres in length. In addition to specifying tolerances, strengths and an external hydrostatic test on the joint to demonstrate pressure tightness (1 kg/cm² for three minutes) the specification details jacking methods, pipeline design and practical and laboratory tests on the basic design of pipe to resist jacking forces. The specification was under revision again in 1983.

3.3 New Products

A new style of jacking pipe was introduced by a major consortium of pipe makers including Nippon Hume, Daido Concrete and Teikoku Hume who joined together with the larger contractors Maeda, Kidoh and Nanno, calling themselves the Semi Shield Technical Association. In the larger diameters severe competition (see Fig. 2) has been experienced from shielded segment construction. For this reason this Association was formed to develop an improved alternative in the diameter range 1200 - 3000 offering additional flexural and axial strengths for use in larger thrusts and low pressure applications. The 3S type semi shield pipe includes a fixed steel socket with very small clearance between spigot and socket providing tight compression of seals and a very square joint when constructed.

Another variant on traditional steel reinforced jacking pipe has been introduced by Fujimura Hume. It involves a traditional style steel ring with double rubber seals on each spigot. A hole formed in the wall between the rubber seals provides the means to pump grout into the joint chamber after construction to provide additional resistance to leakage and corrosion. The joint formed is particularly flexible and is suited to curved jacking. The pipe is provided in three strength classes and there is a choice of two concrete grades to provide additional thrust capacity.

The most recent innovation has been launched by the Japan Superline Pipe Association, a group of eight major companies, including the six largest concrete pipe producers, who exercise rights granted by licence to manufacture the glass reinforced concrete pipes developed by ARC in St. Ives, Cambridgeshire. This group, in addition to manufacturing Superline pipe for open trench construction, have introduced their Long Jacking Pipes (LJP). This pipe, a thick-walled glass reinforced spun concrete pipe was first explored by the Kurimoto Hume Pipe Company (Fig. 8) on behalf of the Association and confirmed the findings of earlier laboratory tests conducted by ARC Slimline (Ref10)The presence of glass fibre in the pipe provides the required flexural strength and augments the compressive strength of good quality spun concrete to a remarkable degree. Tests carried out at Nihon University in September 1982 showed that the LJP offered extra compressive strength, as much as 50% higher than equivalent tests on conventional jacking pipe. The pipe is offered with several alternative joint forms. The JSLPA specification 2 covering jacking pipe indicates the standard loose collar style in two forms with or without Bentonite injection holes. Kurimoto Hume and others also offer special fixed joints and pressure joints with a capacity up to 10 atmospheres. The pipe had been used in over ten cities in its first three months of manufacture and has proved very successful. A notable contract recently completed is the crossing under the River Daido, a 197 metre curved thrust of a special 4K pressure pipe for irrigation purposes.

4. DISCUSSION

It is not the intention of this paper to form conclusions; the foregoing paragraphs represent the impressions of the author from a lengthy dialogue over the past three years with manufacturers, contractors, local government and Ministry officials. It is however sensible to summarise the main observations and to question their relevance to sewerage and drainage construction in the U.K.

It is very evident that the abundant construction work in Japan encourages innovation at a rate unparallelled by other nations and the Japanese have a reputation for exploring new technology, modifying and in many circumstances improving it to meet their specific needs. There is substantial growth in sewer construction despite an international downturn in this sector of public expenditure and pipe jacking can be seen as a major factor in increasingly mechanised and efficient construction. The increasingly sophisticated techniques and products developed in recent years are probably more advanced in Japan than anywhere else in the world and these techniques are beginning to make headway in South East Asia and have just been introduced in Europe.

In the U.K. we are privileged to enjoy a sewerage and drainage system unequalled elsewhere. Our system developed over the past 150 years catalogues the history of technology and know-how in this field. However its longevity is its weakness, the faults in our system whether reflecting limited state of the art knowledge or simply old age are becoming increasingly manifest. They are one of the few areas of our infrastructure showing real growth. NWC estimate that current repair costs run out in excess of £100 million annually and that replacement of our systems which must be regarded as inevitable at some time in the future will cost £31 billion. The majority of this cost, some 80%, is attributed to excavation. The Lords Select Committee (Ref. 11) reporting on the Water Industry surmised that this sum may be reduced by new techniques and noted that the requisite impetus for the necessary research and development may lie with legislation disallowing excavation of city streets. This is just the kind of stimulus that has been applied in Japan when vast urban connurbations on a scale that dwarf our cities were built with no sewer network. It may not be necessary for us to repeat their learning curve if we take note of the established techniques and innovative products. It is evident that pipe jacking offers an economic and attractive method for urban renewal particularly in poor ground conditions and in installation of small diameter pipe.

There are obvious differences to be considered, it would be wholly unreasonable to assume that jacking techniques developed in the soft loam of the Kanto plain are instantly appropriate to all our requirements. However great strides have been made in mixed soil conditions and many of our older cities have developed from historic sites in river valleys and our situation is not so different from that prevailing in the urban areas of Japan currently being explored by pipe jackers. We must also consider practices selectively, the use of steel joints requires detailed consideration of corrosion aspects. Steel joints have been used for more than 30 years in Japan and no significant adverse reactions have been detected, it is however an area of opportunity to identify an alternative that would be universally acceptable and take the pipe jacking option one stage further.

5. REFERENCES

1. Tohyama, S. "The Current Situation and Problem of Japan's Sewage Works" MIPRO lecture Dec 1983.

2. Journal of Sewerage, Vol. 6 No. 3, March 1983, p4 (in Japanese)

3. Kurosi, S. Private communication

4. Ishibashi, N. and Tohyama, S. "Arakawa River Scheme". Japan Journal of Sewage 1968 (in Japanese)

5. Ishibashi, N., Yabe, T. and Nishimura, S. "Curve Jacking Method of Mud Water Type for Construction of Sewer Pipelines in Urawa City". Dobku Seko Vol. 22 No. 8 1981 (in Japanese).

6. Japan Hume Pipe Association "Report on Tests of Jacking Pipe". Dec 1982 (in Japanese).

7. Iseki Poly-Tech Inc. "Standard Pipe Jacking Construction System"

8. JSWAS A-1-1974 (in Japanese)

9. JSWAS A-2-1975 (in Japanese)

10. Japan Superline Pipe Association "Outline report on performance tests on the glass fibre reinforced pipe". June 1983. (in Japanese)

11. House of Lords Select Committee on Science and Technology The Water Industry Vol. 1 report. HMSO London, Dec 1982.

Fig. 1 Sewer Length by Material and Size Constructed in Japan in 1981

Size (mm)	Length by Materials (km)						
	Vitrified clay pipe	PVC pipe	Reinforced concrete pipe	Special reinforced concrete pipe for jacking method	Segment	Others	Total
Less than 200	657.4	1,084.4	191.3	0.6	–	46.8	1,980.5
250 - 300	460.8	963.2	2,077.6	24.1	–	156.5	3,682.2
350 - 450	7.0	49.6	378.6	26.2	–	61.8	523.2
500 - 900	–	22.0	370.7	186.3	1.2	41.8	622.0
1,000-1,850	–	3.2	151.5	147.6	86.6	63.7	452.6
2,000-2,800	–	1.5	22.6	23.3	63.5	29.7	140.6
3,000-4,800	–	–	0.5	0.9	24.1	6.0	31.5
More than 5,000	–	–	–	–	8.6	1.1	9.7
TOTAL	1,125.2	2,123.9	3,192.8	409.0	184.0	407.4	7,442.3

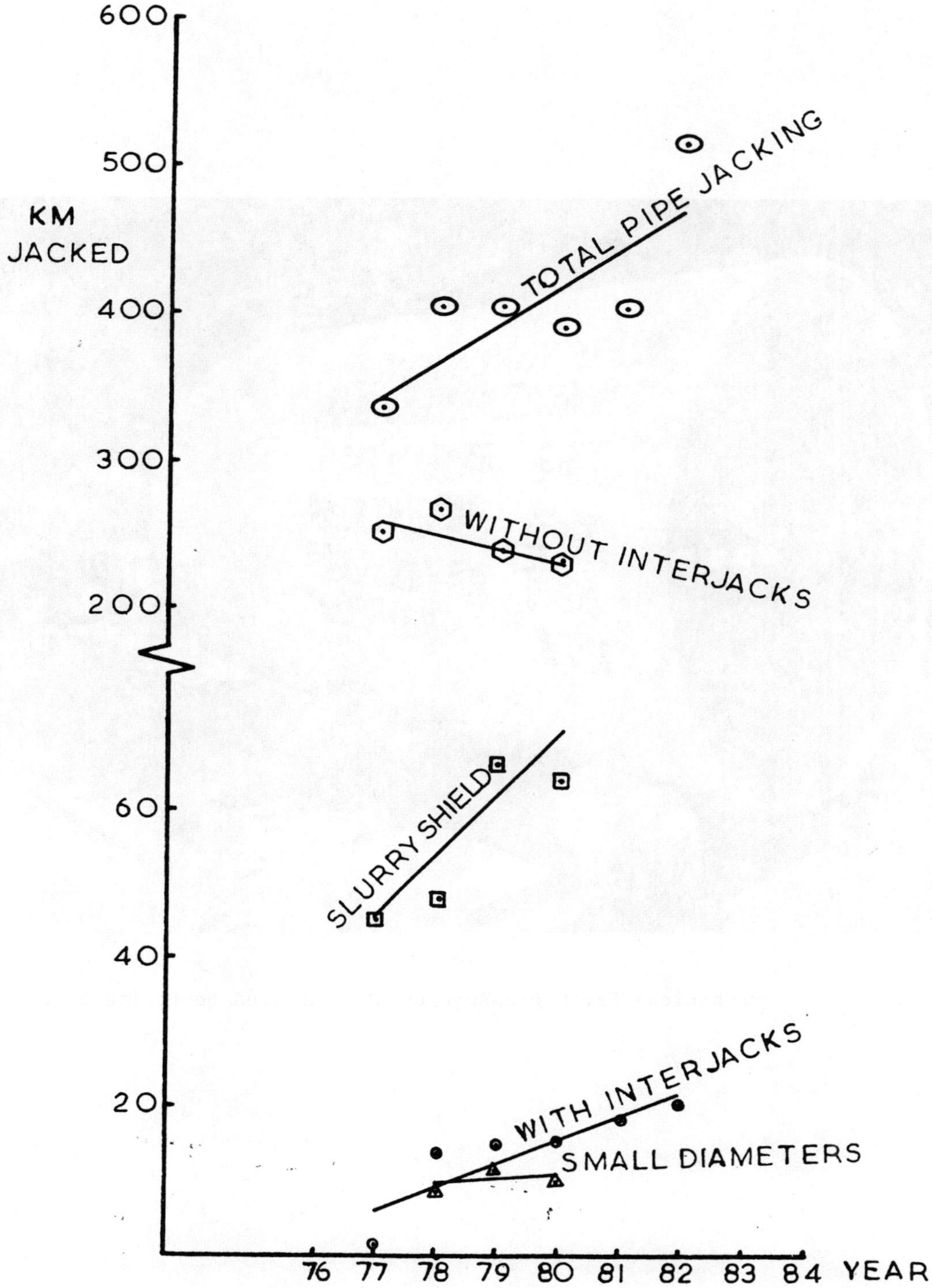

Fig. 2 Trends in Pipe Jacking in Japan (Ref. 2 & 3)

Fig. 3 Mechanical Earth Pressure Counterbalanced Bentonite Shield

Fig. 4 Typical interjack arrangement

Fig. 5 Thrust ring and long stroke jacks in use at Urawa City

Fig. 6 Glass reinforced Long Jacking Pipe under Test

International Conference on the
PLANNING, CONSTRUCTION, MAINTENANCE & OPERATION OF SEWERAGE SYSTEMS
12-14 September, 1984

PAPER G4

VACUUM SEWERAGE

G.J. Stanley C. Eng. MI Mun E. FIPHE. M Inst. WPC.

D. Mills BSc. ACGI. MICE. FIPHE.

Anglian Water
Ambury Road, Huntingdon, PE18 6NZ
England

Summary

The provision of conventional gravity sewers in the flat fenland areas of East Anglia is difficult and costly. Poor trenching conditions, high water levels, deep excavations to achieve gradients and continuous pumping deters the provision of public sewerage facilities. Many properties are served by cesspools or septic tanks which fail to operate because of water levels and structural deterioration in the shifting ground. Infiltration is high and sewers are prone to silting, blockage and structural failure. The advent of vacuum sewerage in the United Kingdom provides a viable alternative sewerage system which overcomes many of these obstacles.

Held at The University of Reading, England. Organised and sponsored by BHRA, The Fluid Engineering Centre, Cranfield, Bedford in conjunction with the Water Research Centre.

1. BACKGROUND

The origins of vacuum sewerage can be traced to Amsterdam in 1868. Modern two pipe systems have been installed in a number of countries but special vacuum traps were required on all connections and appliances.

The development, in 1970, of a 75mm. vacuum interface valve heralded the one pipe system evacuating all foul waste from strategically sited collection sumps to which conventional gravity drains and sanitary fittings could be connected.

Since then the one pipe system has been used in the USA to serve industry, small townships, leisure homes and yacht marinas.

In 1980 it was introduced into the UK and used to resewer St. John's Hospital, Chelmsford for the Essex Area Health Authority. A more detailed history of vacuum sewerage and the particular problems associated with the hospital installation are described by Cressey (Ref. 1)

2. THE ONE PIPE SYSTEM

The collection system comprises one or more pipelines, which are maintained under vacuum, radiating from a central collection station and separated from the gravity drains by pneumatically operated interface valves situated at each collection sump.

Sewage, from individual properties, is conveyed through gravity drains and sewers to the collection sumps. At predetermined sewage levels the interface valve opens for a preset time and the sewage, mixed with air, is drawn into the vacuum main after which the valve closes.

Transportation takes place at a velocity of 4.5 to 5.5 m/sec into the central vacuum collecting tank from where the sewage can be pumped by conventional means into existing sewers or to a treatment works.

Plastic pipes are normally used and range in size from 75mm. to 200mm. diameter although larger diameters are now being contemplated. The applied vacuum is between 400mm. and 500mm. of mercury with a minimum of 175mm. and the maximum lift for each main is about 7.5m.

3. CHARACTERISTICS

Vacuum mains are relatively independant of gradient such that depth of cover can be reduced to that required for protection, i.e. 1 to 1.5m.

Shallow depth, totally sealed flexible pipes and small diameters make vacuum sewerage an attractive alternative to gravity sewers in the following conditions.

1. Flat terrain
2. High ground water levels
3. Unstable soils
4. Rock
5. Narrow trench requirements
6. Underground obstructions.

However, there is a constant energy requirement.

4. FULNEY ESTATE, SPALDING, LINCOLNSHIRE

The estate comprises 61 horticultural holdings scattered over 60 ha of flat fenland on saturated silt about 3m. above sea level with a ground water level less than 1m. below the surface. Cesspools and septic tanks were provided some 45 years ago and these had deteriorated with the result that ground water entered the tanks which had to be constantly emptied at considerable cost. It was this that prompted the owners to requisition public sewers from Anglian Water through their agent South Holland District Council.

Several conventional sewer options were considered and the most cost effective required the construction of three small below ground pumping stations fed by deep gravity sewers and pumping sewage to a fourth collection station before pumping into the existing sewerage system at nearby Spalding.

In comparison the vacuum option required three shallow vacuum mains radiating out into the estate from one above ground collection/pumping station; 22 collection sumps were required to serve the area, see Fig. 1. The estimated costs of the two alternatives are shown in Table 1.

5. HIGH STREET, SPALDING

In contrast, the High Street area of Spalding was served by public sewers which, because of slack gradients and deteriorating pipes, frequently blocked and caused sewage flooding inside private properties. Mechanised maintenance was impossible because the sewers were inaccessible.

The provision of new sewers laid to steeper gradients had to take account of several factors.

New sewers would be laid in running silt within a narrow, heavily trafficked street fronted on one side by 3 and 4 storey listed buildings and on the other side by the River Welland. In places the sewers would be 4m. deep to pick up basement connections and the river water level rises to 2.5m. below roadway level. A brittle cast iron gas main would be endangered and there were underground chambers and buried tie rods securing the river bank piling to be avoided.

The characteristics of shallow depth, narrow trench and flexibility in routing made vacuum sewerage an attractive alternative. The final vacuum main layout is shown in Fig. 2 and the estimated cost comparisons are given in Table 2.

6. DESIGN AND CONSTRUCTION

6.1 Design

Major design parameters were as follows:

Fulney Estate

Domestic population	214
Estimated dry weather flow (daily)	48038 l.

High Street Spalding

Domestic population	61
Commercial population	71
Estimated dry weather flow (daily)	18340 l.

Appendix 1 shows the method used to size the principal components in the Fulney Estate Scheme.

The design parameters for dry weather flow and maximum flow are in widespread use for the design of gravity sewers with inbuilt allowances for infiltration and peak daily rates of flow.

In the past, those who have argued for reduced parameters for foul sewer design, see Stanley (Ref. 2), have been unable to overcome the fact that small decreases in pipe diameters in gravity sewers do not materially reduce overall costs. Such small cost savings can rarely be justified having regard to the 80 year sewer asset life, future infiltration increases, unforeseen future long term development and illegal connections.

However, the opportunity for infiltration into the vacuum system is much reduced; the collection sumps attenuate peak rates of flow; illegal connections are more quickly and easily detected and much of the cost is associated with 20 year asset life equipment. Over design has a greater impact on the costs of mechanical and electrical plant and running costs.

It is significant that detention periods in the vacuum system at Fulney are sufficient to produce septic sewage requiring treatment before discharge. Undoubtedly the gravity/pumping option would have produced longer detention periods and more acute problems but there is scope for producing more suitable design parameters for future vacuum systems.

For example it might be more sensible to measure the water inflow to the catchment, derive the average daily per capita allowance and allow for a peak flow of no more than 2 to 3 times the average. More flow data would be required to provide design parameters for general usage.

6.2 Construction

No problems were encountered during construction. Both schemes were completed well within the six month contract periods. The 2400m. of vacuum main at Fulney was laid in 7 weeks. High Street Spalding was closed to traffic for 4 weeks during which time the vacuum main was completed and the street reopened. Shallow trenches and small above ground structures ensured swift and trouble-free construction. The simple, compact layout of the collection/pumping station at Fulney is shown in Fig. 3.

7. COSTS

Actual capital costs and running costs for 1982/83 for both schemes are shown in Table 3. Detailed breakdowns are given by Stanley and Mills. (Ref. 3).

True operating costs at Fulney are obscured by a number of extraneous factors.

The actual maintenance crew wages were higher than expected at £409 compared with the estimate of £260.

However these costs included familiarisation courses and a higher than usual visiting frequency to check on correct operation. In addition wages were increased by regular sampling for septicity testing and the overheads on wages increased proportionately.

As yet, the operating situation is too unsettled to derive true operating costs but first year costs are close to the original estimate. It is likely that operating costs for both systems will reduce in subsequent years.

8. OPERATION

Both systems have operated reliably.

The sensing open and close device on one of the 22 interface valves at Fulney failed and was replaced.

Scale has built up on the vacuum pumping impellers due to the hardness of the local water supply but softening of the seal water presents no difficulty.

In August 1983 a large volume of super heated steam was accidentally discharged to one of the collection sumps at High Street. The short length of UPVC pipe joining the collection sump to the interface valve chamber softened and deformed. It has since been replaced with ABS pipe which has a higher softening temperature. High velocity and separated flow conditions in the vacuum mains prevented further damage.

No leaks or blockages have occurred in the vacuum mains and simulated incidents have been produced to train personnel in the fault tracing procedures.

9. CONCLUSIONS

1. Both installations proved cost effective for the two situations described.
2. Initial doubts about the reliability of the interface valve and the sensor appear groundless.
3. Operating personnel have confidence and are enthusiastic in the maintenance of the systems.
4. Operation is less onerous, dirty and hazardous than that associated with many gravity systems in flat fenland areas where deep manholes, septic sewage, gases, odours and frequent blockages occur.
5. Maintenance costs of stations are no more than those associated with traditional sewage pumping stations.

6. With the absence of manholes and with mains laid to curves it is very important to maintain accurate locational records for swift location of division valves should leaks occur.

7. Operating instructions and fault finding procedures should be clearly displayed and operatives require training in both aspects. In view of the infrequent occurrence of faults refresher courses with simulated breakdowns may be required.

8. Further economies in capital and operating costs may be achieved by developing design parameters specifically related to the characteristics of vacuum sewerage.

10. ACKNOWLEDGEMENTS

The opinions expressed are those of the authors who wish to thank their colleagues in Anglian Water and South Holland District Council and also Messrs Cowells for assistance in providing data and illustrations.

11. REFERENCES

1. Cressey, E.G.: "New Vacuum Drainage at St. John's Hospital Chelmsford". Institution of Public Health Engineers, Eastern Area Nov. 1980.

2. Stanley, G.J.: "Design Flows in Foul Sewerage Systems". Department of the Environment, Directorate General Water Engineering, Research and Development Division 1974.

3. Stanley, G.J. and Mills, D.: "Vacuum Sewerage". Institute of Water Pollution Control, Scottish Branch Symposium, March 1984.

APPENDIX 1 DESIGN PARAMETERS

Population	214
Per capita sewage production (daily)	225 litres
Operating limits of vacuum	400-600mm. mercury
Number of interface valves	22
Vacuum consumed at each interface valve	$0.225m^3$ per hour
Minimum retention time in collection tank at DWF	0.5 hour
Length of Mains: 100mm. diameter	195 metres
75mm. diameter	2495 metres
Sewage pumping main	1690 metres

Nomenclature

DWF	Estimated daily dry weather flow (24 hours)
Q_{max}	Estimated peak rate of flow
F	Vacuum pump sizing factor dependant on network length 0 - 900m F = 5 901 - 2135m F = 6 2136 - 3050 F = 7 over 3051 F = 8
Q_{VP}	Capacity of vacuum pump
Q_{VV}	Total vacuum consumed by interface valves
Q_{SP}	Sewage pump delivery rate
V_O	Operating volume of sewage collection tank
V_T	Total volume of sewage collection tank
V_R	Total volume of vacuum reservoir
R_T	Sewage retention time in collection tank at DWF
V_M	Total volume of vacuum mains

Calculations

Assume sewage discharged over 16 hours

$DWF = 0.84\ ls^{-1} =$

$Q_{max} = 4 \times DWF = 3.36\ ls^{-1}$ ($12.10m^3$ per hour)

$Q_{VV} = 22 \times 0.255 = 5.61m^3$ per hour

$Q_{VP} = (F \times Q_{max}) + Q_{VV}$

$= (7 \times 12.10) + 5.61$

$= 90.3m^3$ per hour

Nearest standard vacuum pump size = $127.5m^3$ per hour
Therefore provide 2 No. (1 duty, 1 standby)

Assume a time (t) of 90 seconds to raise vacuum in system from 400-600mm of mercury ($53\text{-}800\ kNm^{2}$)

$V_M = 11960\ l$

$V_R = (3Q_{VP} \times t) - (\frac{V_M}{3})$

$= \frac{(3 \times 127.5 \times 1000 \times 90)}{3600} - (\frac{11960}{3})$

$= 5576.0\ l$

Assume R_T = 1800 seconds at DWF

Then $V_O = 1800 \times 0.84$
$= 1512\ l$

$$V_T = V_R + V_O$$
$$= 5576.0 + 1512$$
$$= 7088 \text{ l}$$

Nearest collection tank size = 6820 litres

$$Q_{SP} = 6 \times DWF$$
$$= 6 \times 0.84$$
$$= 5 \times ls^{-1}$$

Nearest standard pump size 5.5 ls^{-1} at 22m total lift

Assume 100mm diameter UPVC main with solvent joints

k_s	=	0.030mm
Velocity	=	$0.7ms^{-1}$
Static Lift	=	3.66m
Friction	=	10.00m
Negative	=	7.62m
		21.28m

TABLES

Net Present Value (NPV) has been calculated for a 40 year period using a 5% discount rate assuming 2.5% inflation on energy costs and takes account of residual values at year 40. The following asset lives have been assumed.

Gravity Sewers	80 years
Pressure & Vacuum Mains	40 years
Mechanical & Electrical Plant	20 years

TABLE 1. FULNEY ESTATE : ESTIMATED COST COMPARISON

Description	Gravity System £	Vacuum System £
Capital	280800	205700
Energy (pa)	331	438
Maintenance (pa)	1300	475
NPV	315980	248430

TABLE 2. HIGH STREET : ESTIMATED COST COMPARISON

Description	Gravity System £	Vacuum System £
Capital	187400	118400
Energy (pa)	33	153
Maintenance (pa)	520	360
NPV	189890	137150

TABLE 3. ACTUAL CAPITAL AND OPERATING COSTS

Scheme	Capital Cost Estimate in (£)	Operating Cost 82/83 Estimate in (£)
Fulney Estate	207000 (205700)	927 (913)
High Street	112000 (118400)	651 (513)

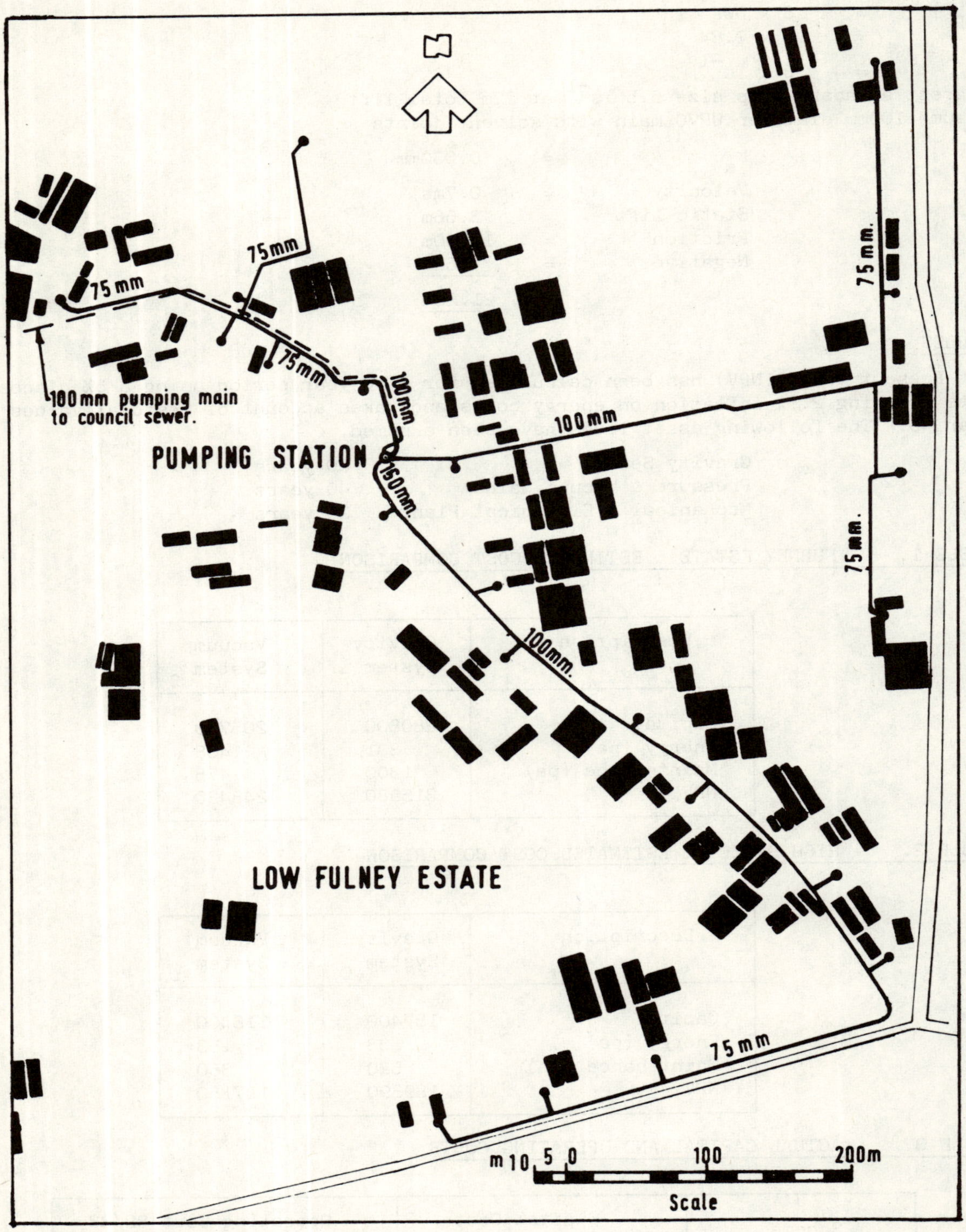

FIG. 1. FULNEY ESTATE VACUUM MAIN LAYOUT

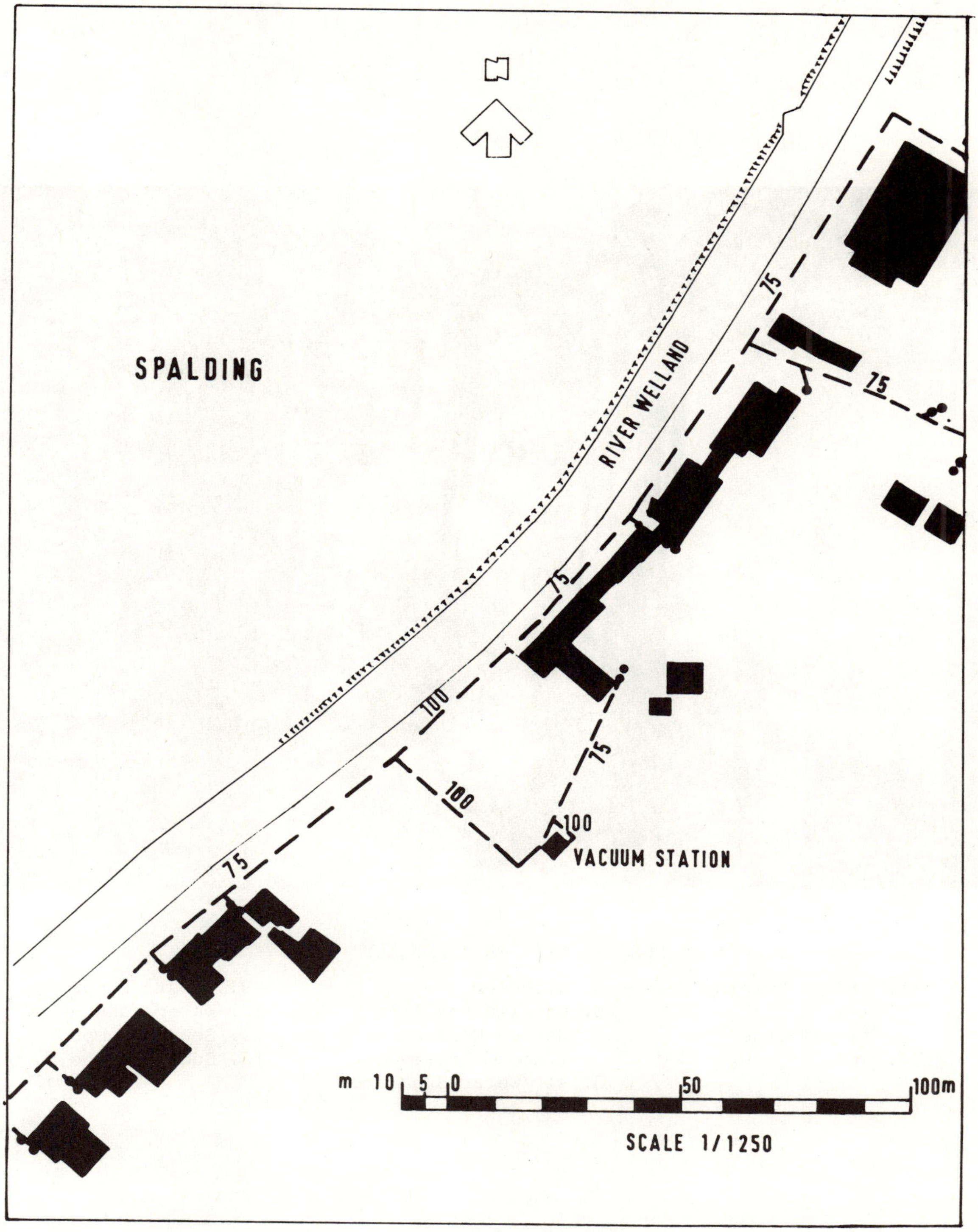

FIG. 2. HIGH STREET SPALDING VACUUM MAIN LAYOUT

FIG. 3. COLLECTION/PUMPING STATION AT FULNEY

1. Sewage Pumping Main
2. Vent
3. Vacuum Line
4. On-line gauges
5. Silencer
6. Equalising Line
7. Sewage Collection Tank (Vacuum Reservoir)
8. Air/Water SEparator
9. Vacuum Pumps
10. Water Tank
11. Sewage Pumps
12. Suction Pipe

International Conference on the
PLANNING, CONSTRUCTION, MAINTENANCE & OPERATION OF SEWERAGE SYSTEMS
12-14 September, 1984

PAPER H1

SEGMENT LINED TUNNELS
FIELD INVESTIGATION OF ROUGHNESS

P. Ackers
and
J. D. Pitt

Binnie & Partners
Artillery House, Artillery Row,
London, SW1P 1RX, England

Summary

The hydraulic performance of tunnels lined with precast concrete segments was investigated as part of the research programme of the UK Construction Industry Research and Information Association (CIRIA).

Fifteen tunnels were surveyed to provided data on geometry covering the four main types of lining currently in use, with special attention to details such as the joints between segments. Hydraulic tests were also carried out on five of the tunnels.

A good correlation was found between the measurements of joint displacement and the hydraulic roughness factor of the Colebrook-White equation. It is thus possible to recommend k_s values for design, taking account of segment type, tunnelling method and ground conditions.

Held at The University of Reading, England. Organised and sponsored by BHRA, The Fluid Engineering Centre, Cranfield, Bedford in conjunction with the Water Research Centre.

NOMENCLATURE

A	=	cross-sectional area of tunnel
D	=	diameter of tunnel
g	=	gravitational acceleration
i	=	hydraulic gradient (head loss per unit length)
k_s	=	equivalent roughness coefficient
l	=	segment length, i.e. distance between principal joints in direction of flow
m	=	hydraulic mean depth (A/P = D/4 for circular tunnels flowing full)
N	=	number of segments per ring
p	=	wetted perimeter
Q	=	discharge
s	=	step height
V	=	mean velocity of flow (Q/A)
Re	=	Reynolds number
ν	=	kinematic viscosity of fluid

1. INTRODUCTION

Segmental linings of precast concrete are being used increasingly for construction of tunnels to carry surface water and sewage, yet there was no definite information about the most important measure of hydraulic performance - hydraulic roughness. As a result, some authorities have imposed apparently arbitrary penalties on the use of such linings. Because of this lack of adequate data for hydraulic design, the Construction Industry Research and Information Association (CIRIA) commissioned an investigation by Binnie & Partners, London. It was anticipated that the hydraulic roughness would depend on the degree of lipping between segments, and hence the accuracy of construction. In turn, this might depend on segment type and tunnelling conditions. Four main types of segmental lining are currently in use in the U.K, details of which are given in Table 1.

The research had two main components: surveys of 15 tunnels to establish standards of construction, alignment and lipping at the joints; and hydraulic tests on 5 tunnels (2 flowing full-bore as water supply conduits; 3 flow part-full as drainage conduits)

Factors affecting the choice of tunnels for survey and testing included:

- the need to make preliminary inspections
- availability, ease of access
- ease of making surveys on newly completed tunnels before commissioning
- the need for hydraulic tests to be carried out in tunnels in service

2. TUNNEL SURVEYS

2.1 Method

General information was provided by the authorities responsible for the tunnels, and information on segment dimensions was obtained from the lining manufacturer. Most of the survey effort was therefore directed to obtaining data on the geometry of the tunnel, and particularly on joints and surface condition of the lining.

A purpose-made extending gauge, accurate to 1 mm, was used to measure diameter at the mid-point of segment length. Measurements were taken for two orthogonal directions, usually horizontal and vertical, but 10 to 15° off vertical for Tattersall wedge block linings, and 45° off vertical in the large-diameter tunnels.

Tunnel alignment was measured by taking the offset between tunnel centre and a reference line provided by a laser beam. A target with graduated scale was set up at each measurement location, and the deviation both horizontally and vertically from the zero setting was recorded. Precision was in the range ± 1 mm close to the laser to ± 10 mm at a greater distance, but values were generally well within ± 5% of maximum deviation from line. When laser measurements could not be obtained, data on invert levels were taken from records of surveys during construction.

A purpose-made gauge was used to measure the steps, s, between adjacent segments. Four equidistant points around the circumference were scraped clean for this purpose. The were usually defined by diametrical lines at 45° to the vertical, but additional measurements at points defined by horizontal and vertical diameters were sometimes taken for comparison. In all measurements, the distinction was made between 'positive' steps, where the downstream segment protruded into the flow, and 'negative' steps, where the downstream segment was recessed relative to the upstream segment.

Information on joints, including type of filler and whether or not this protruded into the tunnel, was also recorded. From 20 to 50 rings were surveyed in each case.

2.2 Survey results

Full details are given in the Project Record (ref. 1). Table 2 lists the tunnels surveyed, and indicates the segment type and other relevant details. Table 3 summarises the joint measurement, in the form of the step height, averaged disregarding sign.

Data for Type A lining (Wedge block) include one anomalous set, where a geological fault caused major construction problems. The more usual low value of step reflects the high standard of erection possible in good ground for this lining type, when expanded against an accurate cut. Segment of Type D (reinforced, bolted) are often difficult to place, and there is evidence to suggest that step values between the invert segments can be greater than elsewhere.

The results show that steps at joints are influenced less by tunnel size than by lining type and construction method. In turn, choice of lining type and accuracy in erection are largely dependent on ground conditions.

Measurements of tunnel diameter indicated that a higher than nominal diameter in a given direction (usually not more than 2%) tended to be balanced by a correspondingly lower value in the direction at right angles. On average, differences in area of cross section did not exceed 2%, and they were unlikely to have influenced hydraulic roughness. Values of standard deviation for level and, where available, line (measured about a best-fit line) also turn out to be small compared with tunnel diameter, so their direct effect on hydraulic roughness is unlikely to be significant.

Visual inspection of surface finish of concrete segments revealed occasional minor flaws, including small cracks, leaks, grout holes or pitting of the surfaces. However, most panels were in excellent condition and had a high quality finish. Only one of the tunnels surveyed was over 10 years old, so there was limited information on the effects of long-term use, but there was no significant abrasion or chemical attack.

There was a deposit of slime or silt in several tunnels, though these rarely exceeded 1 mm, and the slime was limited to part of the circumference. Deposits were thickest in the newly constructed Littlebrook tunnels (through chalk), where calcite deposits were up to 10 mm thick. However, this material was soft, and would have been expected to scour rapidly after the tunnels were commissioned.

Joints, too, were mostly of good quality, with the surface of the joint-sealing compound, in general, flat. Where bulging or hollowing was evident, it seldom exceeded 5 mm.

3. HYDRAULIC TESTS

3.1 Methods of test

Details of the hydraulic tests will be found in refs 1,2 and 3. The hydraulic roughness is a function of the head loss (hydraulic gradient), flow velocity, cross-section of flow (hydraulic mean depth) and, to a relatively insensitive degree, the viscosity of the fluid being conveyed. For tunnels flowing full, velocity is readily calculated from accurate measurements of discharge and diameter: for tunnels flowing part-full, it is preferable to measure velocity directly by assessing the time of travel of a tracer over a known distance. In fact several methods of discharge and velocity measurement were used, including a differential-pressure pipe meter, volumetric assessment (using a reservoir), radio-active tracer, chemical tracer and also a gauging weir. The choice of method depended on whether a permanent installation was available, or temporary facilities had to be provided, as well as on flow conditions.

It was feasible to carry out hydraulic tests on five sites only (indicated in table 2), within the financial and time constraints for the project.

3.2 Analysis and results

The Colebrook-White equation was used as the basis of analysis, as this is a widely applicable function covering smooth, transitional and rough-turbulent conditions.

Expressed in terms of the engineering variables, m, i and V, the equation can be written:

$$V = \sqrt{32gmi}\,\log\left\{\frac{k_s}{14.8m} + \frac{1.255\nu}{m\sqrt{32gmi}}\right\} \quad (1)$$

This equation can be solved directly for k_s given the other variables. Results are shown in Fig. 1, in the form of friction factor, λ, plotted against the Reynolds number of the test, Re,

where

$$\lambda = 8gmi/V^2 \quad (2)$$

and

$$Re = VD/\nu = 4Vm/\nu \quad (3)$$

Fig 2 shows k_s plotted against $\overline{|s|}^2/l$, where $\overline{|s|}$ is the mean step height irrespective of sign. A satisfactory correlation emerged, confirming that the main source of hydraulic resistance in segment lined tunnels comes from the joints between segments.

4. CONCLUSIONS AND RECOMMENDATIONS

The physical condition of 15 segmentally lined tunnels was surveyed and hydraulic tests were conducted in five of these. In interpreting the results of the work, it should be noted that joint spacing was 0.6 m (the segment length) in almost all tunnels examined. The main findings were:

1. The average values of hydraulic roughness, k_s, in the Colebrook-White equation for the five tunnels tested lie in the narrow range 0.5 to 2.1 mm.

2. A correlation exists between hydraulic roughness and mean absolute joint step height and spacing.

3. Mean absolute values of step in the surveyed tunnel were between 1.0 and 5.3 mm, and depended more on lining type and construction method than tunnel size (The higher values were usually recorded in shield-driven tunnels; most steps at joints were less than 10 mm).

4. Using the correlation between step height and roughness, values of k_s for the 15 tunnels surveyed lie in the range 0.4 to 3.1 mm. These values are only marginally higher than those commonly accepted for in situ concrete lined tunnels, i.e. 0.3 to 1.5 mm.

5. Standard deviations of departures in level and line (relative to the best-fit line) were between 9 and 61 mm for level, and between 16 and 244 mm for line (six tunnels only).

6. Mean diameters were generally within 1% of nominal diameter.

7. The surface finish of precast concrete segments and the condition of the jointing compound were generally good and, therefore, unlikely to contribute significantly to hydraulic roughness.

8. neither ageing nor deposition of thin (<2 mm) layers of slime or sediment was found to affect hydraulic roughness. However, thicker deposits might reduce the difference in hydraulic performance between segmental and in situ linings.

The prospective use of a segment line tunnel frequently involves comparison with in situ lining. The comparison of hydraulic performance may point to the need for different diameters in the two cases, although this will depend on assessment of appropriate roughness values. In good ground conditions, segment lining may achieve as high a capacity as an in situ lining unless it is particularly smooth and remains free of slime. Table 4 indicates the sensitivity of tunnel diameter to the roughness, k_s.

The following recommendations apply in this context:

1. Recommended design values of k_s (mm), based on tests carried out as part of this project, are as follows:

Ground condition	Hydraulic quality		
	Good	Normal	Poor
Ground permitting the use of expanded wedge block linings	0.5	1.0	1.5
Other conditions	1.0	2.0	3.0

2. Provided there is not extensive sliming or sediment, the value of k_s is primarily dependent on the steps at the joints, and k_s can be estimated if measurements are made after construction, from:

$$k_s \simeq 0.3 + 60\ \overline{|s|}^{\,2}/l \ \text{(mm)}$$

3. The above expression should be used with caution, as it is based on limited data and only for l = 600 mm. However, it can form the basis for specifying or checking the construction tolerance required to meet design criteria.

4. Control of tolerances in diameter and line and level does not appear to be as important as control of the quality of erection exemplified by the steps at the joints, but large deviations of level create sediment traps, and therefore should be avoided (as for in situ linings).

5. If extensive sliming (greater than 2 mm thick, say) is expected, hydraulic design should take into account the results of work in which k_s is shown to be related to operating velocity e.g research by Hydraulics Research, Wallingford, results of which are incorporated in ref. 4. Under these conditions, the difference in k_s between segmental and in situ linings is unlikely to be significant.

5. ACKNOWLEDGEMENTS

The work was carried out as part of a project under contract to CIRIA. In addition, the Authors wish to express thanks to the many authorities who supported the research. The paper is presented with the permission of the Director of the Construction Industry Research and Information Association, London, and the Authors are grateful to the Association and to members of the Project Steering Committee for their support and guidance.

6. REFERENCES

1. Pitt J.D. and Ackers P, Hydraulic roughness of segmentally lined tunnels, project record, CIRIA, 1982, RP277.

2. CIRIA, Hydraulic roughness of segmentally lined tunnels, CIRIA Report 96, 1982.

3. Ackers P and Pitt J.D., Segment lined tunnels: field scale test to determine hydraulic roughness, 3rd Int. Conf. on Urban Storm Drainage, Goteborg, Sweden, June 1984.

4. Hydraulics Research, Tables for the hydraulic design of pipes, fourth edition, 1983.

TABLE 1 - Types of segmental lining currently in use in UK (after CIRIA, ref 2)

Type	*Description of lining*	*Examples*
A. Expanded wedge block (see Figure 1)	Unreinforced concrete segments held together by passive earth pressures mobilised by expanding the lining circumferentially against the smooth circular profile of excavation. Constructed using hand shield or soft rock tunnelling machine, usually in clay	Tattersall wedge block, Donseg
B Smoothbore knuckled (see Figure 2)	Unreinforced segments knuckled along longitudinal joints to provide articulation during loading. Circumferential joints are located by interlocking tongues and grooves. Voids between lining and excavation filled by grout. Constructed with or without a shield, generally in good cohesive soils	Charcon Universal, Charcon Rapid, Spun Concrete Flexilok, Mini
C Smoothbore bolted (see Figure 3)	Reinforced segments connected longitudinally and circumferentially by a hidden connecting system. Voids between lining and excavation filled by grout. Constructed in all ground conditions	Charcon One Pass
D. Reinforced bolted with infill panels (see Figure 4)	Flanged reinforced concrete segments bolted together. Voids between lining and excavation filled by grout. Constructed in all types of ground. Segment recesses later filled by precast panels shaped to tunnel radius	Buchan bolted with Bucfil pellets, Charcon bolted

TABLE 2 Summary of tunnel surveys

Reference location	Diameter (mm)	Use (see notes)	Year Built	Ground condition, method, direction, tested.	Joint dimension and step N	l (mm)	s (mm)	s (mm)
A.WEDGE BLOCK								
Ely Ouse-Essex	2540	RWS/F	1971	Gault clay, Sh, WF, HT	10	600	0.4	1.4
Ely Ouse-Essex (a)	2464	RWS/F	1971	Gault clay, Sh, WF, HT	10	600	0.0	1.8
Ely Ouse-Essex (b)	2464	RWS/F	1971	Gault clay, Sh, AF, HT	10	600	0.3	2.3
Ely Ouse-Essex (c)	2464	RWS/F	1971	Chalk gault, Sh, AF, HT	10	600	-0.4	4.6
Three Valleys	2540	RWS/F	1972	London clay, Sh, AF	10	600	-0.1	1.0
Datchet (North)	2540	RWS/F	1970	London clay, Sh, AF, HT	10	600	-0.6	2.7
Datchet (South)	2540	RSW/F	1970	London clay, Sh, AF, HT	10	600	-0.0	2.2
B. SMOOTHBONE - KNUCKLED								
Roding Valley	1500	FS/PF	1974	Clay, Sh, AF, HT		600	0.0	4.3
Margate	1372	CS/PF	1976	Chalk, H, AF		600	0.1	1.5
Margate	2134	CS/PF	1977	Chalk, Sh, WF		600	2.1	2.7
Margate	2057	CS/PF	1961	Chalk, H, WF		600	-0.2	2.3
Enfield (a)	1000	SS/PF	1977)	Water bearing	3	600	-2.2	4.0
			)	sand and gravel				
Enfield (b)	1000	SS/PF	1977)	Sh, AF, HT	3	600	-1.9	4.1
C. SMOOTHBORE-BOLTED								
West Ham	1800	FS/PF	1979	Soft clay, peat, Sh, AF	5	600	-0.5	5.3
D. REINFORCED BOLTED with infill panels								
Stoke-on-Trent	2590	CS/PF	1977	Mudstone, H, AF		600	-0.5	4.1
Littlebrook Intake	4270	CW/F	1979	Chalk, Sh, AF		600	-1.9	3.0
Littlebrook Outfall	4270	CW/F	1979	Chalk, Sh, WF		600	2.5	4.1
Walsall (B-C)	3660	RD/PF	1980	Siltstone, shale, H, AF, HT	7	600	-0.2	3.0
Walsall (E-C)	3660	RD/PF	1980	Sand, gravel, Sh, WF, HT	7	600	-0.9	4.1

Notes:

CS = Combined sewer
CW = Cooling water
F = Full bore
FS = Foul sewer
PF = Part full
RD = River diversion
RWS = Raw water supply
SS = Storm water sewer

Sh = Shield driven
H = Hand driven
AF = Driven against flow direction
WF = Driven with flow direction
HT = Hydraulic test carried out

See ref 2 for information on diameter measurements and alignment.

Table 3 Step values for the main types of segmental lining (after CIRIA, ref 2)

Lining type	Mean step, regardless of sign, s	No. of data sets
A. Expanded wedge block	1.8 ± 0.9 mm	6
B. Smoothbore knuckled	2.4 ± 1.3 mm	6
C. Smoothbore	5.3 mm	1
D. Reinforced bolted	3.5 ± 0.6 mm	5

Table 4. Required tunnel diameter as a function of k_s for various flows and gradients (after CIRIA, ref 2)

	Diameter (m) for:			
Flow rate (m^3/s)	$k_s = 0.3$ mm	$k_s = 0.6$ mm	$k_s = 1.5$ mm	$k_s = 3.0$ mm
$i = 10^{-4}$				
3.0	2.51	2.56	2.65	2.73
10.0	3.98	4.07	4.20	4.32
30.0	6.07	6.20	6.40	6.57
100.0	9.65	9.84	10.15	10.43
$i = 10^{-3}$				
3.0	1.60	1.64	1.71	1.76
10.0	2.54	2.60	2.70	2.78
30.0	3.87	3.97	4.11	4.23
100.0	6.16	6.30	6.51	6.70

Note: Values are for tunnels flowing full.

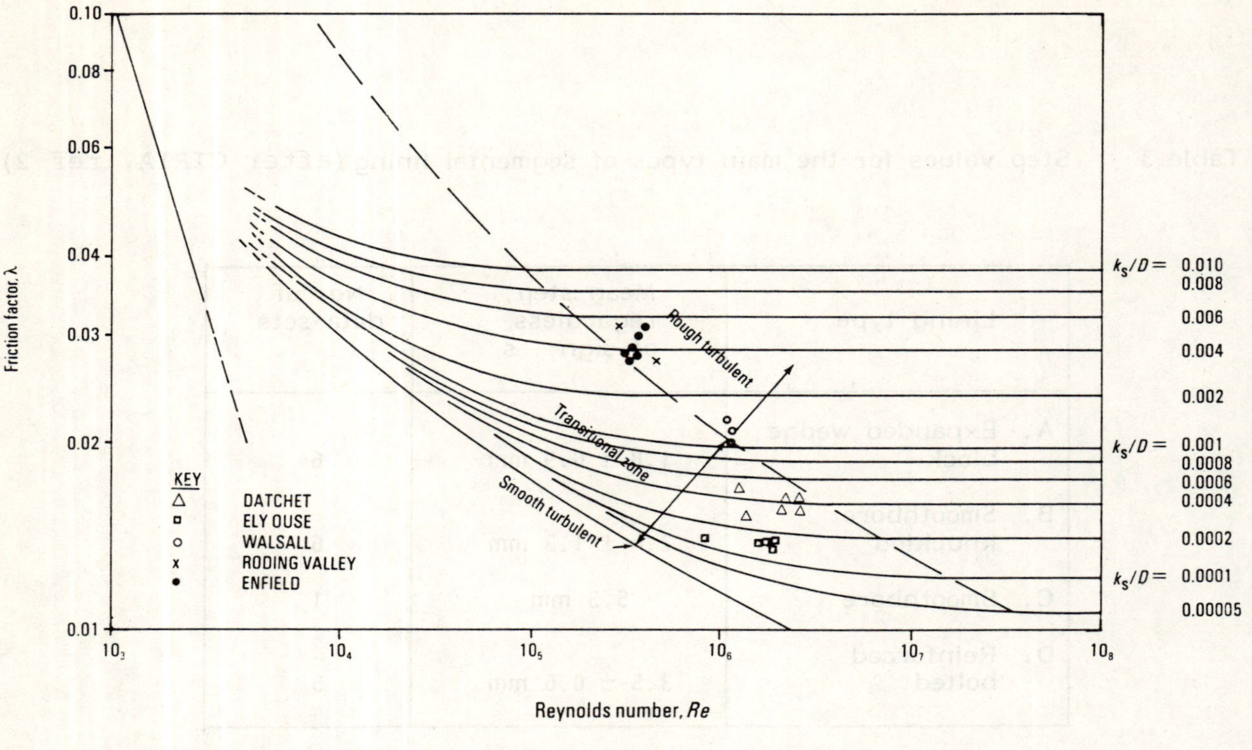

Figure 1. Hydraulic test data (after CIRIA, ref 2)

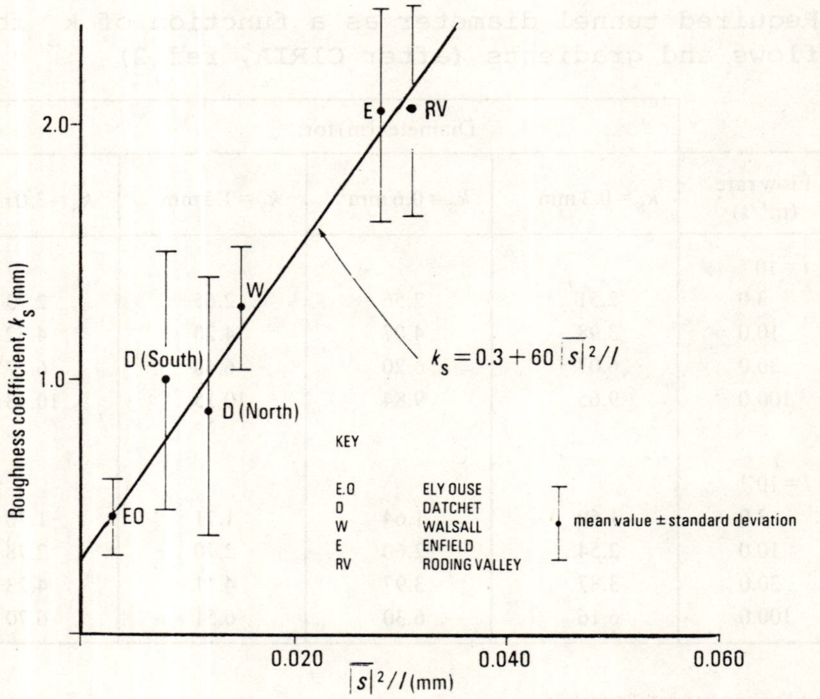

Figure 2. Roughness coefficient, k_s, as function of steps between segments (after CIRIA, ref 2)

International Conference on the
PLANNING, CONSTRUCTION, MAINTENANCE & OPERATION OF SEWERAGE SYSTEMS
12-14 September, 1984

THE HYDRAULIC ROUGHNESS OF USED SEWERS

R J HENDERSON BSc AMIWES

WRC ENGINEERING

PO BOX 85

BLAGROVE SWINDON SN5 8YR
ENGLAND

SUMMARY

The Colebrook-White equation is commonly used in calculations of pipeline capacity and is embodied in hydraulic models used for design and analysis of sewer systems in the UK. Alternative flow formulae are considered and the resistance term in these is related to the hydraulic roughness (k_s).

The Water Research Centre recognised the need to provide guidance in assessing roughness values for combined and storm sewers in service, and flow survey results have been used to calculate k_s values for 30 brickwork and pipe sewers covering a variety of age and condition in 8 UK towns.

In many sewers the roughness was observed to decrease significantly with increasing depth of flow; causes and implications of this effect are considered. Pipefull k_s values of between 0.2mm and 36mm have been observed and an attempt is made to relate these to sewer condition where possible. The results are in broad agreement with those derived from earlier research.

The factors contributing to roughness are considered and methods suggested for estimating k_s values arising from each source. It appears that, in pipe sewers, the material of construction exerts far less influence over k_s than does the condition of the pipe structure and jointing or the presence of slime and silt deposits. Presence of deposits and deterioration of the fabric are the main sources of roughness in brickwork sewers.

An example of the improved modelling accuracy achieved by using pipe-specific roughness values is described.

Held at The University of Reading, England. Organised and sponsored by BHRA, The Fluid Engineering Centre, Cranfield, Bedford in conjunction with the Water Research Centre.

NOMENCLATURE

A = cross-sectional area of flow (m^2)

D = pipe diameter (m)

F_r = Froude Number

P = Wetted perimeter (m)

R = hydraulic radius (m) = A/P

R_e = Reynolds Number

V = flow velocity (ms^{-1})

g = acceleration due to gravity ($m^2 s^{-1}$)

i = hydraulic gradient

k_s = hydraulic roughness (m unless specified in mm)

n = Manning friction coefficient

ν = kinematic viscosity ($m^2 s^{-1}$)

1. INTRODUCTION

When undertaking the hydraulic analysis of an existing sewer system the engineer is required to estimate the resistance imposed by the component pipes. Likewise, when designing new systems or upgrading measures he must anticipate the capacity and therefore maximum roughness the sewer is likely to attain during its useful life.

The dearth of information on the roughness of used sewers provided stimulus for a series of field tests at some 20 sites by the Hydraulics Research Station (HRS) in the early 1960's(1). Since this time however, little new information has been derived from in-sewer tests and there remain many conditions where confident selection of appropriate roughness values is impossible. The Water Research Centre (WRC) recognised the need to provide guidance and initiated research to analyse both published information and newly available data derived from its programme of sewer flow surveys. This paper presents these data and draws general conclusions regarding the roughness values most appropriate to a range of sewer conditions.

2. HYDRAULIC RESISTANCE EQUATIONS

The subject of hydraulic resistance in pipelines has been reviewed elsewhere(2) and it is concluded that the **Colebrook-White equation** is the most appropriate for sewerage design since it has a sound theoretical basis and applies over a wide range of flow and roughness conditions. In engineering terms the equation is most usually written:

$$V = -\sqrt{32\ gRi}\ \log_{10}\left[\frac{k_s}{14.8R} + \frac{1.255\ \nu}{R\sqrt{32\ gRi}}\right]\ ms^{-1} \ \ldots.(1)$$

In sewers flowing full the Reynolds Number ($R_e = 4VR/\nu$) will invariably exceed 10^4 hence the flow will be in the fully turbulent rough zone. The second term in the bracket of equation (1) is therefore negligible when compared with the first for commercially available sewerage pipes and this allows solution for the hydraulic roughness k_s when the other variables are known. Thus

$$k_s \simeq 14.8\ R \times 10^{-A} \times 1000\ mm \ \ldots..(2)$$

where $A = \dfrac{V}{\sqrt{32\ gRi}}$

Charts and tables(3) have been published to simplify use of the equation for pipefull and free-surface flow in conduits having k_s values specified in increments between 0.006mm and 15mm. The Colebrook-White equation is also embodied in hydraulic models currently used in the UK for design and analysis of sewer systems, for example the Wallingford Procedure(4) and SPARROW(5).

The **Manning equation**, frequently used for open channels may also be applied in the rough conditions encountered in sewers.

For part-full flow the equation is

$$V = \frac{R^{2/3}\ i^{1/2}}{n}\ ms^{-1} \ \ldots\ldots...(3)$$

and for a circular pipe flowing full

$$V = \frac{0.397\ D^{2/3}\ i^{1/2}}{n}\ ms^{-1} \ \ldots\ldots..(4)$$

The relationship between Manning's 'n' and k_s is:

$n \simeq 0.012\ (k_s)^{1/6}$ ....(5) where k_s is in mm

The Manning equation gives a reasonable approximation to the Colebrook-White equation when the relative roughness (k_s/R) is between 0.008 and 0.14 and the flow is fully turbulent. A number of hydraulic models developed in North America and Europe, for example SWMM[6] and HVM[7], employ the Manning equation for pipe routing and require specification of ´n´ values.

The **Crimp and Bruges** equation corresponds to the Manning equation with a fixed value of n = 0.012, a somewhat conservative value for used sewers.

3. REVIEW OF EXISTING INFORMATION

A literature search revealed several references detailing hydraulic tests conducted in sewers during the last 80 years in the UK and overseas. Wherever possible the data from these have been evaluated to express roughness as a Colebrook-White k_s value in mm. Full details of the construction, condition, age and hydraulic parameters of the tested sewers, together with calculated k_s values are to be found in a WRC report[8].

The majority of existing data relate to sewers in fair condition at the time of examination. The field tests employed a variety of flow velocity measurement techniques which may account in part for the wide range of roughness observed. Furthermore almost all of the tests were made in pipes flowing part-full, a limitation imposed by the normal flow regime, available instrumentation and difficult operating conditions. It has been observed that hydraulic roughness varies with depth of flow due to the non-uniform distribution of sewage slimes and sediment around the pipe perimeter. Measurements of roughness in sewers flowing at different depths may not therefore be comparable and those taken at low flow would not be representative since the k_s value at greater flows (eg pipefull discharge) may be appreciably lower.

4. DETERMINATIONS OF SEWER ROUGHNESS BY WRC

4.1 Sewer flow surveys

Sewer flow surveys are recommended as a valuable aid to the effective hydraulic analysis of sewer systems, particularly those having any unusual features.[9, 10] A flow survey has 2 principal roles:

i) to verify any assumptions made in constructing a hydraulic model and to identify unknown features such as overflows and throttles,

ii) to observe the performance of the sewer and its ancillary structures under observed rainfall events.

Flow surveys are not a new concept but, until recently, the instrumentation necessary for reliable operation has been unavailable. The WRC has mounted over 40 flow surveys using modern instrumentation in the UK during the last 2 years and it is from some of these that the roughness data described below have been obtained.

Broad objectives when planning positioning of in-sewer instrumentation (flow velocity and depth monitors) are to deal with trouble-free areas with a minimum of equipment leaving the bulk of the capability for problem areas. Thus monitors will be located on the principal sewer runs to assess inflow from large component areas and also at junctions, bifurcations, overflows or other specific points. The sites for which flow data are available were thus selected on grounds other than pipe roughness; the information is however, a valuable ´by product´ of hydraulic investigations in a range of combined and surface water sewers.

4.2 Flow survey methodology

A full account of the methods and instrumentation used in WRC flow surveys is described elsewhere.[11] Flow depth is measured using a pressure transducer and flow velocity by means of a Doppler shift system. Both sensors are mounted in a small wedge-shaped housing installed on the sewer invert, generally well upstream of the access manhole. Sites are selected to avoid clogging or unstable hydraulic conditions and where silt is present on the invert the housing is mounted above the silt level to prevent it being covered.

The sensors are connected to a solid state unit which logs readings at 2 minute intervals for up to 2 weeks. The measured Doppler velocity is related to average flow velocity by means of a laboratory flume calibration and weekly in-situ calibration using a portable velocity meter.

Survey duration is normally planned to be 5 weeks but this may be terminated or extended as rainfall conditions dictate. Data are transferred to a mini computer for quality control checks during the survey with final processing being undertaken on the mainframe computer at WRC.

4.3 **Determination of k_s values**

Any sites displaying unusual hydraulic characteristics were eliminated from the data base for pipe roughness analysis, thus sites immediately upstream of overflows, at changes of gradient or direction and those suffering serious backwater effects were not used. Pipe roughness was evaluated using a computer program which solves the Colebrook-White equation for k_s, given depth and velocity data, using equation (2) above. The program can accommodate circular and egg shaped pipes only, analysis of odd-shaped pipes required modification of inbuilt depth-hydraulic radius-area functions. Output consists of depth-velocity data, proportional depth (flow depth/vertical pipe diameter) and hydraulic roughness. At each site k_s was determined over a range of flows, sometimes including surcharge, depending on storm severity during the survey period. Figure 1 shows calculated roughness values plotted against the proportional depth observed during a single storm for one site, a brickwork trunk sewer in York.

It is recognised that a drawback of using velocity-depth data derived from a single monitor is that the hydraulic gradient will not necessarily equal the pipe slope, as taken from sewer records. Ideally, monitors would be placed at both ends of a sewer length allowing assessment of hydraulic head difference, however such an arrangement would conflict with flow survey objectives and add to the cost. It is reasonable to assume that hydraulic gradient varies relative to pipe slope during the rising and falling limbs of the hydrograph and therefore, by plotting calculated k_s against proportional depth for a full range of flow any systematic variation in the calculated roughness caused by this effect would be revealed. It was found that in almost all cases there was little difference in k_s values on the rising and falling hydrograph for a specified depth and that the degree of scatter does not warrant use of two monitors to define hydraulic gradient.

4.4 **Site Surveys**

During site visits the size and internal condition of the sewer was noted and measurements made of the depth of any deposits present. Colour photographs were also taken of the sewer reach upstream of the monitor at most of the sites and these form the basis of a visual guide to sewer roughness, published as Appendix J of the WRC/WAA Sewer Rehabilitation Manual[9]. The photographs allow estimation of joint displacement between adjacent pipe segments and assessment of mortar loss in brickwork sewers. An indication of the extent of sliming and encrustation may also be gained by this means in non-accessible sewers.

5. DISCUSSION OF RESULTS

Calculated roughness values and details of the 30 sewers examined are presented in Tables 1 and 2. The k_s values quoted are the observed or estimated pipefull, or, in cases where proportional discharge was less than 0.7 they correspond to the k_s value at peak flows.

Of particular note is the considerable range in k_s values for a given sewer material; these findings support Ackers' suggestion that, for used sanitary sewers the differences in roughness between materials of construction is less than commonly accepted[1]. Table 3 summarises available data for sewers of brick, clayware and concrete construction and shows that the recent results are broadly consistent with those of earlier research, although the range in k_s is narrower. Table 3 (iii) suggests that used clayware and concrete pipes would typically occupy the same roughness increment (3.0mm). However, it is clear from the range in k_s for pipe sewers of 0.2 to 12.0 (WRC data) or 0.1 to 120.0mm (all UK data) that selection of a roughness value on grounds of material alone and without knowledge of the sewer condition could give a very misleading indication of capacity.

It is appropriate here to briefly consider some of the factors affecting sewer roughness and relate these to the sites examined.

5.2 Age of sewer

Observations suggest a general trend to high roughness at greater ages although this is often due to factors other than wear of the internal surfaces of pipes. The age of a sewer gives no direct indication as to likely roughness. Analysis of the data in Tables 1 and 2 relating to brickwork sewers, supplemented by information from 15 brick sewers in London (for which no condition information is available) shows there to be no correlation between age and roughness[8]. This is because deterioration rates vary from site to site and re-pointing may reduce the roughness of an ageing sewer to that of a new one.

Several researchers have, however reported a progressive roughening of concrete and cement-lined pipes, believed due to chemical interaction between pipe surface and effluent. This phenomenon has been observed in both foul and storm sewers; the rapidity and eventual degree of roughening being dependent upon pipe composition, temperature and aggressiveness of the effluent. Vant[12] suggests the following relationship.

$$e \simeq 0.56 \log_{10} \text{Age} + 0.02 \text{ mm} \quad .....(6)$$

where e is a depreciation coefficient (mm) to be added to the k_s value of the pipe when new and the age is measured in years. Any increase in roughness due to concrete deterioration will normally be small when compared with that arising from other sources.

5.3 Sliming

In systems conveying foul sewage the rapid development of slime over the perimeter wetted by the daily cycle of flow will significantly increase the roughness over that of a clean pipe. Horton[13] gives examples of two brickwork sewers in which k_s was observed to increase from 0.3 to 12.0mm and from 1.2 to 3.7mm during the first 5 years of use primarily due to sliming. Clay[14] quotes tests on a concrete sewer in which k_s when new was 0.6mm increasing to 1.8mm after 9 months usage. Tests by the HRS[15] also confirm the rapid establishment of slime accompanied by an increase in roughness. In these tests k_s increased from 'as new' values of 0.07 and 0.09mm to fully slimed roughness of 2.5 and 4.6mm for clayware and pre-cast concrete pipes respectively within 4 months. The slimed roughness is governed partly by pipe material, with the smoother surfaces tending to shed slime more readily, and the flow velocity, which acts as a control over the depths of slime able to form under the shearing action of the flow. This explains the tendency towards thicker slime deposits in the flatter clayware sewers of York and Swanage (Nos. 6, 7, 16, 17, 18 and 20 in Table 1) whilst in comparable steeply sloping sewers of Accrington the sliming was minimal.

Figure 2 shows a suggested relationship for pipes slimed to half depth, between the pipefull roughness and normal flow velocity for 5 sewer materials. This relationship whilst derived from experiments is generally supported by field observations in those sewers where roughness is primarily due to sliming. For example daily flows in the concrete sewer in Torquay (No 23) generally exceed 2.5 ms^{-1} and cover only 20% of the wetted perimeter. Thus sliming is limited in both depth and extent, resulting in a k_s of 0.2mm.

5.4 **Pipe jointing**

Displacement at pipe joints, a common problem in ageing sewers, introduces energy loss additional to that due to the texture of the pipe surface. Clearly the extent of displacement affects the k_s value over a length of sewer and, based upon the results reported here and earlier work on drainage pipes[2], the following table is suggested as a guide for use in pipes of 225 to 525mm diameter.

Estimated mean joint step (mm)	**k_s due to joint eccentricity (mm)**
<10	0.15
10 - 13	0.30
14 - 18	0.60
19 - 27	1.50
28 - 40	3.0
>40	6.0

These values should be added to the roughness value estimated for a well-aligned pipe, making due allowance for sliming or deposits.

5.5 **Sediment**

The presence of sediment is acknowledged to seriously impair hydraulic performance although its effects upon flow resistance are difficult to predict. In extreme cases k_s may exceed 150mm due to apparent development of bedforms under certain flow conditions, or alternatively sediment may have little effect on flow capacity other than to reduce flow cross-section.

Where the sediment forms a reasonably plane bed an appropriate k_s value for the deposit is given by

$$k_s \simeq 1.5 \ D_{65} \quad ...... (7)$$

Where D_{65} is the diameter of the sediment for which 65% of the sample is finer. In storm sewers D_{65} will typically be in the range 0.5 to 5mm. The lateral extent of the deposits should be measured and a composite k_s value for the whole pipe calculated.

Available evidence from sewers suggests that bedforms and consequently the highest roughness values are associated with part full flow when the Froude No is between 0.25 and 0.5. 'Washing out' of bedforms occurs when F_r is between 0.55 and 0.9. This effect was observed in a silted sewer at Littlehampton (No 13); roughness declined to 12mm as proportional depth increased but then began to rise again, peaking at k_s = 72mm at Froude Nos around 0.3 and a proportional depth of 0.5 (actual depth $\simeq$ 500mm). As F_r increased beyond 0.4 the roughness declined, believed due to washing out of bedforms, resulting in a minimum k_s of 16mm, extrapolated to 9mm pipefull. The subject of roughness due to sediments and cohesive silts is beyond the scope of this paper but is considered in greater depth elsewhere[8, 16].

5.6 **Mortar loss in brickwork sewers**

The highest k_s values obtained in the latest research are associated with brickwork sewers in which mortar is missing to some extent and where repointing is required. Such sewers would fall into roughness increments of 15 and 30mm. As a very approximate guide, for eroded brickwork of typical construction the roughness value is likely to be between half and two thirds the estimated depth of mortar loss measured in mm. Where sections of brickwork protrude or are missing the roughness may be considerably higher.

5.7 **Variation of roughness with depth of flow**

Previous work(1) has indicated that k_s is not fixed but varies with hydraulic conditions, particularly depth of flow. This finding is reinforced by the latest research. Figure 1 is typical of depth-roughness plots where deposits of cohesive silt were present; at lower flows the relative roughness (k_s/R) is high and the sediment occupies a greater proportion of the wetted perimeter. At pipefull flows some degree of bed-smoothing may occur but the principal reason for the reduction in k_s is that sediment affects a smaller section of the flow. A similar effect was observed in sewers suffering joint eccentricity.

In several clean and well aligned sewers the maximum roughness was observed at proportional depths of 0.2 to 0.4, corresponding to the maximum daily flow level at which slime growths are observed to be thickest. Whilst pipefull roughness is generally required for design purposes, an assessment of part-full k_s values may be necessary, as when estimating peak daily flow velocities for self-cleansing, and should take effects such as sliming of the lower part of the pipe into account. Figure 3 shows the rating curve for a slimed sewer with deposits in the invert and illustrates the variation in k_s with depth. Variable roughness may well result in pipefull velocities close to the maximum, usually observed at around 80% full for a pipe of uniform roughness.

6. **APPLICATIONS IN HYDRAULIC MODELLING**

As an illustration of the effects of pipe roughness upon the predictions by hydraulic models of real sewer systems two examples are presented here. Both were produced using the Wallingford Procedure Simulation Package (WASSP-SIM)(4), one for a design event and one during a model verification exercise.

Figure 4 shows depth and level hydrographs resulting from a 1 year 30 minute design storm for 3 roughness values applied globally to the sewers of a 70 ha catchment in York. Overestimate of roughness values in a model will attenuate simulated flows and suggest greater surcharge depth and duration than would be observed in the field. Figure 4 reinforces the need to select appropriate k_s values to produce valid hydraulic models.

The second example, shown in Figure 5, indicates how model verification results may be improved by careful assessment of individual sewers. In the absence of specific information regarding pipe material and condition the Oswaldtwistle sewerage system was initially modelled using a global k_s value of 1.5mm. The response of a subcatchment within this model is shown in Figure 5(a), the modelled peak discharge is early and is overestimated by about 18%. Subsequent analysis of flow survey data allowed estimation of roughness values at 3 points within the catchment and, by inspection of in-sewer photographs and use of the visual guide(8), specific roughness values were assigned to each pipe. Some of the brickwork sewers were seen to be in poor condition (k_s = 15mm or 30mm) and many of the clayware pipes were misaligned (k_s = 3mm or 6mm). The results of a second simulation using revised roughness values are shown in Figure 5(b) and illustrate the improved fit.

7. CONCLUSIONS

The results reported here are broadly consistent with those of earlier research and confirm that roughness is generally governed more by the condition and flow regime of a sewer than by its′ age and pipe material. Degree of joint displacement contributes significantly to observed roughness in pipe sewers and may result in k_s values exceeding 6mm. The results for brickwork cover a wide range and may be related to the extent of mortar loss. Badly worn brickwork attains k_s values of 30mm; continuing the use of 15mm for worn brickwork and 3 or 6mm for well pointed brickwork seems justified.

An accurate assessment of roughness is important for realistic appraisal of existing sewer capacity and should be incorporated in investigations prior to hydraulic model development. Flow survey techniques may be used for local determinations of roughness at all stages of flow; alternatively a visual guide to assist engineers in estimating pipefull k_s values is available[(9)].

The subject of roughness in used sewers is not yet fully researched and there remains a particular need for data from sewers in very poor condition. To this end it is intended to continue analysis of existing and future WRC flow survey results.

8. REFERENCES

1) Ackers P, Crickmore M J and Holmes D W. "Effects of use on the hydraulic resistance of drainage conduits". Proc. Inst. Civ. Engineers **28** 1964 pp 339 - 360. Discussion: **34.**

2) Ackers P. "Hydraulic resistance of drainage conduits" Proc. Inst. Civ. Engineers **19** July 1961 pp 307 - 336.

3) Hydraulics Research Station. "Tables for the hydraulic design of pipes". Third Edition 1977 HMSO.

4) "Design and analysis of urban storm drainage - the Wallingford Procedure". NWC/DoE Standing Technical Committee Report No 28, NWC September 1981.

5) Thompson J L and Lupton A R R. "A method of assessment of piped drainage systems taking account of surcharge and overground flooding". Urban Storm Drainage, Proceeding of the Internationl Conference, Southampton April 1978. Ed Helliwell P R. Pentech Press.

6) Huber W C **et al** "Storm Water Management Model User´s Manual Version II. USA EPA Report EPA-670/2-75-017, March 1975.

7) Klym H **et al** "Urban hydrological processes computer simulation". Dorsch Consult, Munich. ETH Seminar, Zurich 1972.

8) Henderson R J. "A guide to hydraulic roughness in sewers" WRC External Report No 131 E. March 1984.

9) "WRC/WAA Sewerage Rehabilitation Manual". WRC Engineering 1984.

10) Williams D W. "Hydraulic analyses - the role of sewer flow surveys". WRC External Report - to be published. WRC Engineering 1984.

11) Drinkwater A R. "Sewer flow surveys; a manual of guidance". WRC Report - to be published. WRC Engineering 1984.

12) Vant R J. "Depreciation coefficients in pipelines". Civil Eng. and Public Works Review (Australia) **58** 1963.

13) Horton T. "Flow in the sewers of the North Metropolitan System of Massachusetts". Trans. Am. Soc. Civ. Engs. **56** 1901.

14) Clay W V in Discussion of reference(1). Proc. Inst. Civ. Engineers 1964.

15) Perkins J A and Gardiner I M. "Measurements of the hydraulic roughness of slimed sewer pipes". Report No IT 237. Hydraulics Research Station. July 1982.

16) May R W P. "Sediment transport in sewers". Report No IT 222. Hydraulics Research Station. February 1982.

17) White K et al. "Absolute flow measurements to obtain depth against discharge for sewers". Conference on Fluid Flow Measurement. National Engineering Laboratory, East Kilbridge, Glasgow. April 1975.

TABLE 1 SUMMARY OF ROUGHNESS DATA OBTAINED BY WRC

N°	Location	Sewer material	Diameter (mm)	Slope	Proportional depth (max)	k_s (mm)
1	York	Brickwork	915 x 610	.0013	0.96	6.0
2	York	Brickwork	840 x 560	.0015	0.92	10.0
3	York	Brickwork	940 x 760	.0021	1.00	12.0
4	York	Brickwork	1675 x 1120	.0007	0.92	13.0
5	York	Brickwork	760	.0040	1.00	18.0
6	York	Clayware	380	.0033	1.00	2.3
7	York	Clayware	525	.0014	1.00	1.5
8	Accrington	Clayware	380	.0526	1.00	0.7
9	Accrington	Clayware	380	.0833	0.90	2.0
10	Accrington	Clayware	380	.0588	1.00	8.2
11	Littlehampton	Brickwork	1140 x 850	.0008	0.80	3.3
12	Littlehampton	Brickwork	1100 x 830	.0099	0.70	5.0
13	Littlehampton	Brickwork	940 x 620	.0009	0.75	9.0
14	Littlehampton	Brickwork	1140 x 850	.0012	0.74	35.0
15	Swanage	Clayware	380	.0525	0.29	3.3
16	Swanage	Clayware	475	.0032	1.00	3.8
17	Swanage	Clayware	700	.0012	1.00	3.9
18	Swanage	Clayware	305	.0161	1.00	4.5
19	Swanage	Concrete	610	.0044	1.00	11.0
20	Swanage	Clayware	380	.0032	1.00	4.0
21	Torquay	Brickwork	1240 x 850	.0058	0.48	3.0
22	Torquay	Brickwork	525	.0069	0.69	26.0
23	Torquay	Concrete	1140	.0069	0.36	0.2
24	Torquay	Clayware	525	.0408	0.68	12.0
25	Enfield	Brickwork	2010 x 1830	.0300	0.59	35.0
26	Enfield	Concrete	1250	.0960	0.12	5.5
27	Oswaldtwistle	Brickwork	610	.0641	0.61	36.0
28	Oswaldtwistle	Clayware	450	.0257	0.64	7.5
29	Oswaldtwistle	Clayware	450	.0550	0.37	3.3
30	Stevenage*	Concrete	990	.0031	1.00	0.9

* Earlier WRC test (17).

TABLE 2 INTERNAL CONDITION OF SEWERS EXAMINED BY WRC

No	Approx age (years)	Description	Depth of silt (mm)
1	90	Trunk sewer. Structurally sound, well pointed. Slimed to 3mm. Fine silt on invert.	170
2	90	Structurally sound, encrustation on walls. Silted invert.	70
3	90	Structurally sound, well pointed. Slimed to 3mm. Silt deposits form a rough bed.	150
4	90	Trunk sewer, good condition. Slimed to 4mm. Incrustation and debris on wall and soffit	100
5	90	Structurally sound but pointing showing signs of deterioration.	60
6	90	Good condition. Slimed to half depth. Minimal joint misalignment.	-
7	90	Good condition. Slimed to half depth. Debris on walls and soffit.	-
8	90	Good condition, minimal sliming, no deposits (fast flowing). Joint eccentricity typically 8mm.	-
9	90	As (8) above but with more marked joint displacement, typically 15mm.	-
10	90	Extensive joint displacement (20 to 25mm) causing standing waves under part full flow.	-
11	80	Good condition. Fine sand and silt on invert. Slimed.	140
12	80	Brickwork in good condition. Fine sand and silt on invert. Slimed.	160
13	80	Brickwork in fair condition. Fine sand on invert. Slimed.	150
14	80	Structurally sound but brickwork may need re-pointing. Silted and slimed.	50
15	80	Clayware. Thin film of slime to $^1/_3$ depth. Slight displacement of up to 10mm at pipe joints.	-
16	80	Clayware, normally slimed to $^1/_3$ depth. Displacement at pipe joints of 15 - 20mm.	
17	80	Clayware. Well aligned and structurally sound. Thick coating of slime throughout.	60
18	80	Clayware. Slight displacement at joints, typically 10mm. Thick coating of slime (5mm) on invert.	-
19	80	Concrete. Rough internal finish, no displacement at joints. Thickly slimed and with debris on soffit.	-

Nº	Approx Age (years)	Description	Depth of silt (mm)
20	80	Clayware. Good condition, slight displacement at joints, typically 10mm. Slimed.	-
21	100	Trunk sewer in good condition. Glazed brickwork, all mortar intact. Minimal sliming.	-
22	100	Brickwork worn and irregular. Repointing needed. Clean and with minimal sliming.	-
23	50	Well aligned concrete pipe with smooth finish. Thin film of slime on invert.	-
24	50	Pipe free of slime but with rough cement deposits on invert. Joint eccentricity of 15mm.	-
25	80	Brickwork, circular arch. Worn but in reasonable condition. Culverted watercourse containing rubble and coarse gravel.	-
26	10	Pre-cast concrete box culvert. Watercourse, free of sediment. Joint steps typically 20mm, some incrustation.	-
27	100	Brickwork in poor condition. Mortar worn or totally lost.	-
28	100	Clayware. Displaced pipes - estimated joint eccentricity of 30 - 35mm.	-
29	100	Clayware. Displaced pipes - joint eccentricity of 20 - 25mm.	-
30	19	Well aligned concrete pipe, close joints. Smooth wall but slimed at water level.	-

TABLE 3. SUMMARY OF ROUGHNESS VALUES CATEGORISED BY SEWER MATERIAL.

Material	Range in k_s (mm)	Mean	Std. Dev.
(i) WRC Results			
Brickwork	3.0 -- 36.0	16.2	12.5
Clayware	0.7 -- 12.0	4.4	3.1
Concrete pipe	0.2 -- 11.0	4.4	5.0
(ii) All UK data			
Brickwork	0.3 -- 61.0	15.7	15.9
Clayware	0.3 -- 18.0	5.0	4.7
Concrete pipe	0.1 -- 120.0	10.5	30.4
(iii) All UK data excluding sites with significant sediment deposits			
Brickwork	2.2 -- 36.0	15.2	12.4
Clayware	0.3 -- 18.0	3.5	2.3
Concrete pipe	0.1 -- 11.0	2.7	3.2

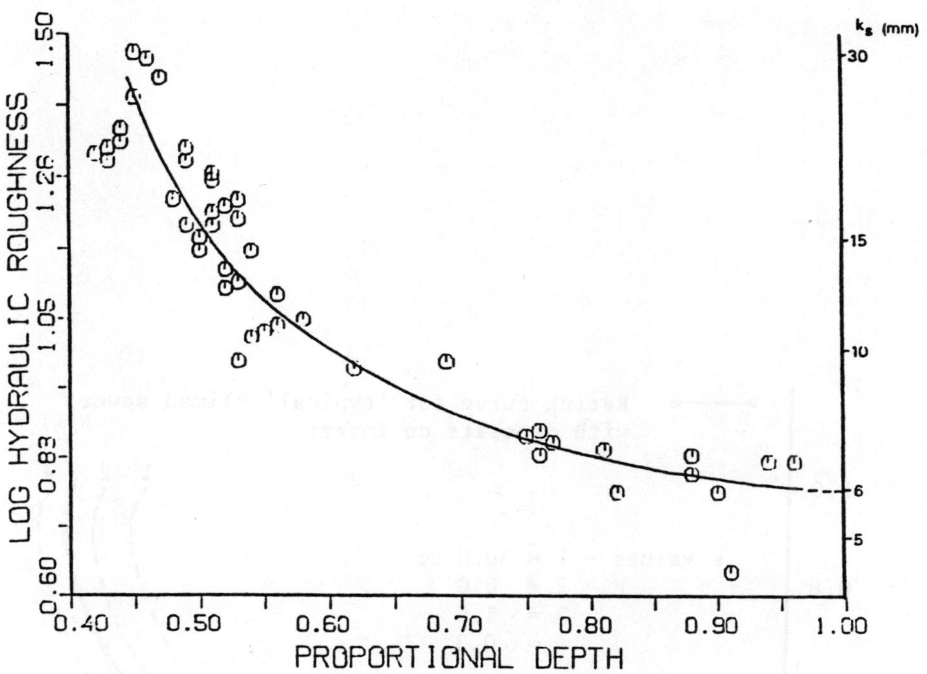

Figure 1 Calculated hydraulic roughness as a function of proportional depth. Results for site 1 during one storm.

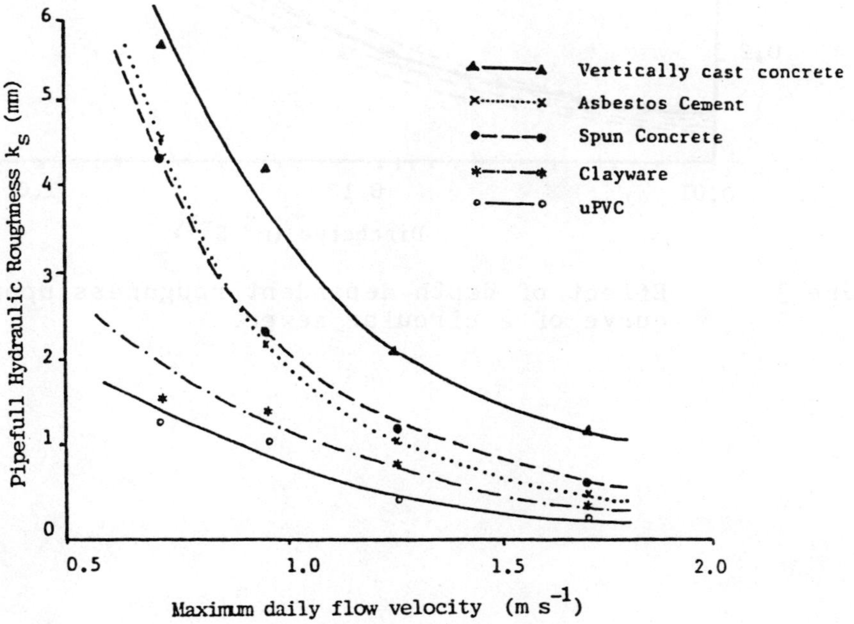

Figure 2 Roughness due to sliming as a function of flow velocity for 5 pipe materials. (Data from HRS tests).

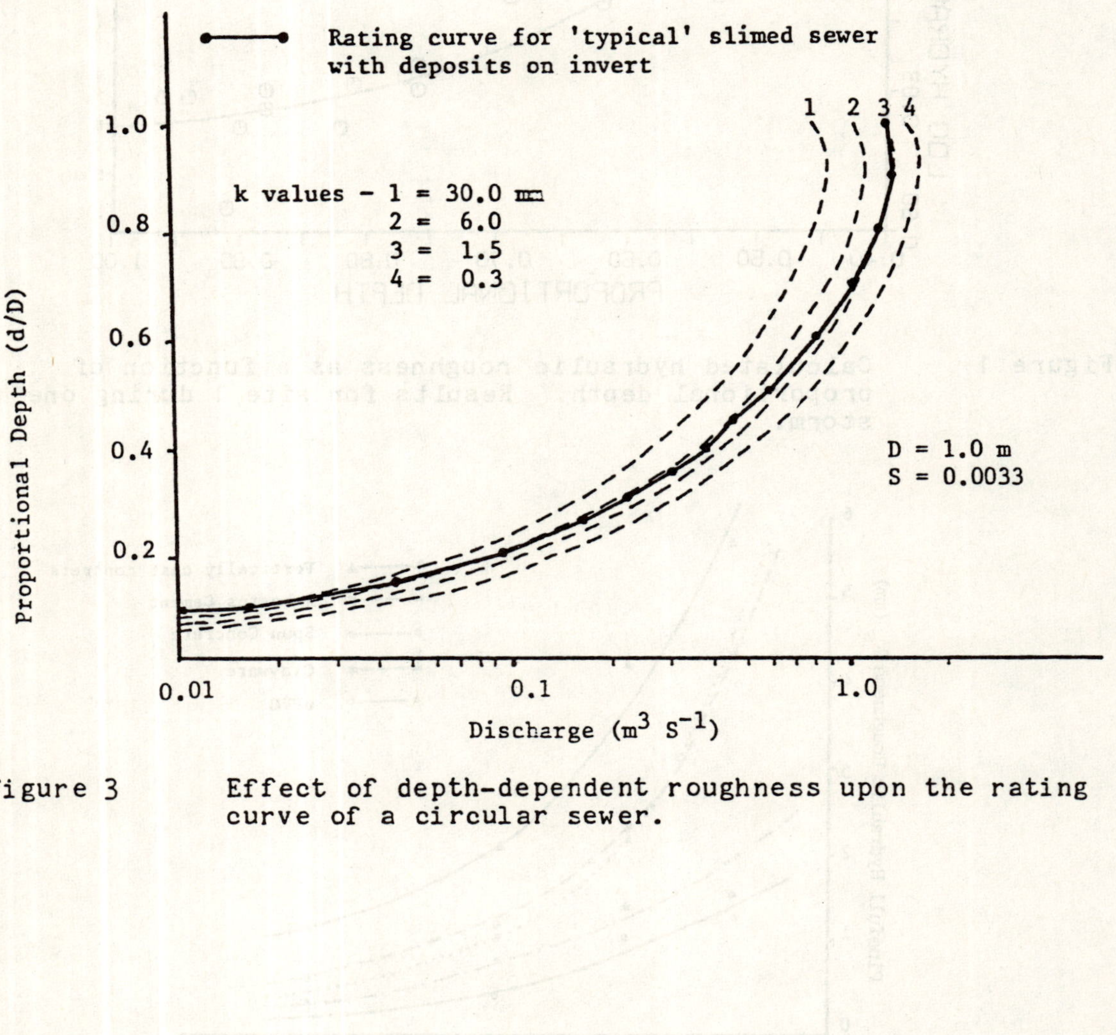

Figure 3 Effect of depth-dependent roughness upon the rating curve of a circular sewer.

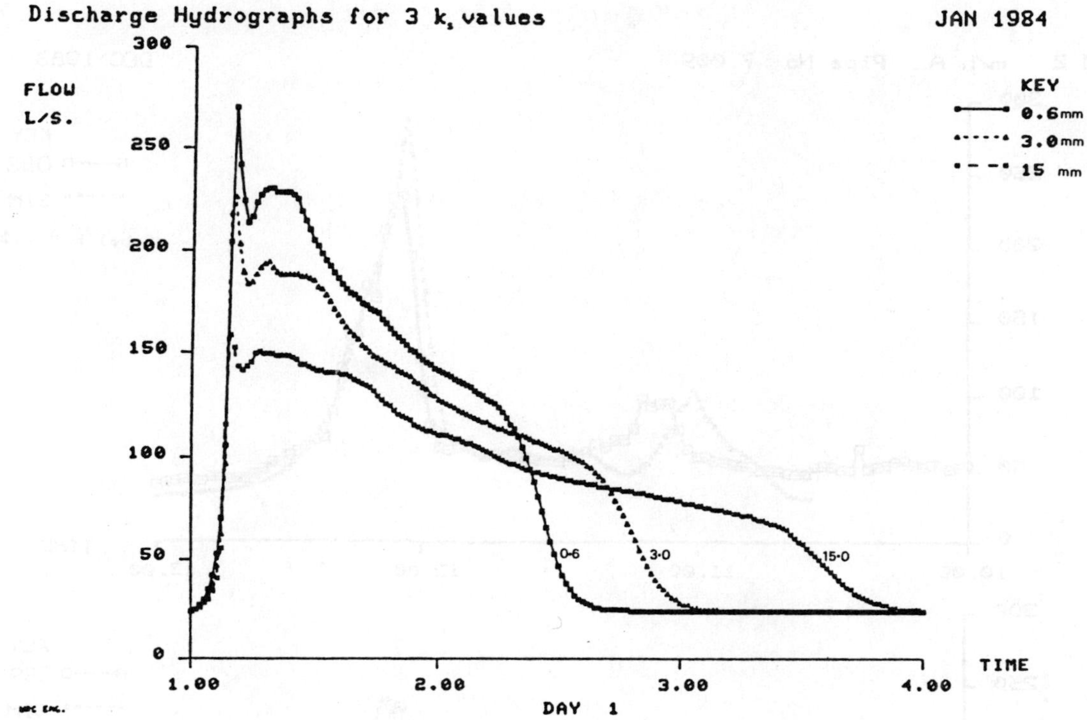

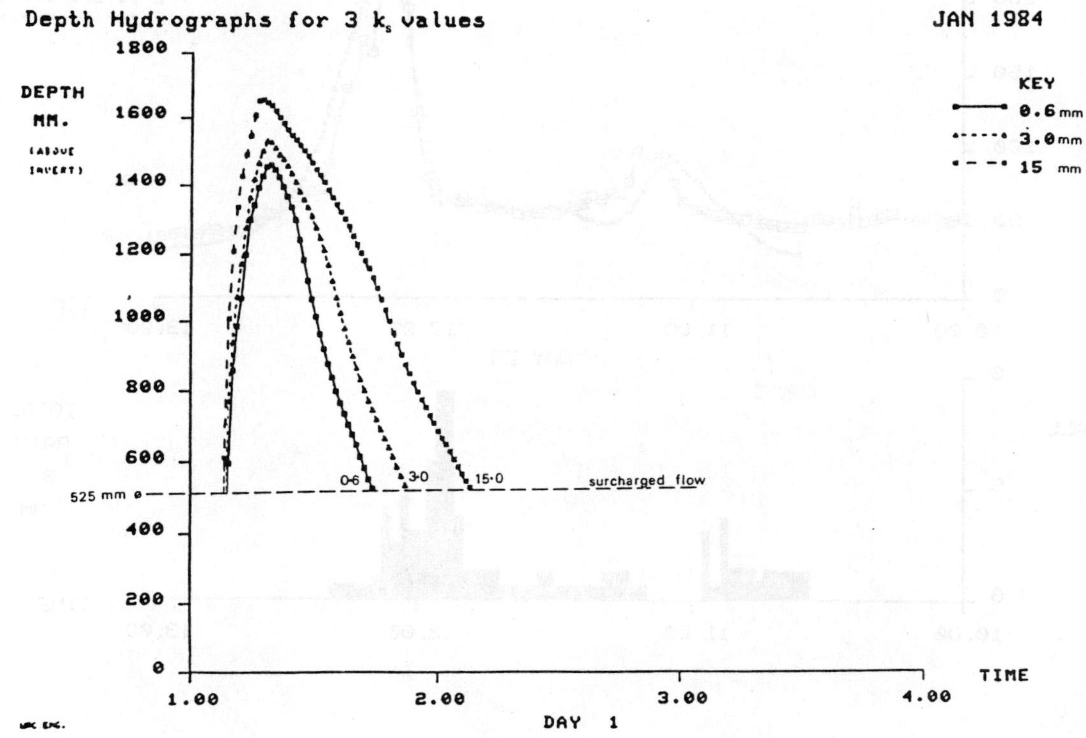

Figure 4 Discharge and level hydrographs for a sewered subcatchment showing the effects of applying 3 global roughness values.

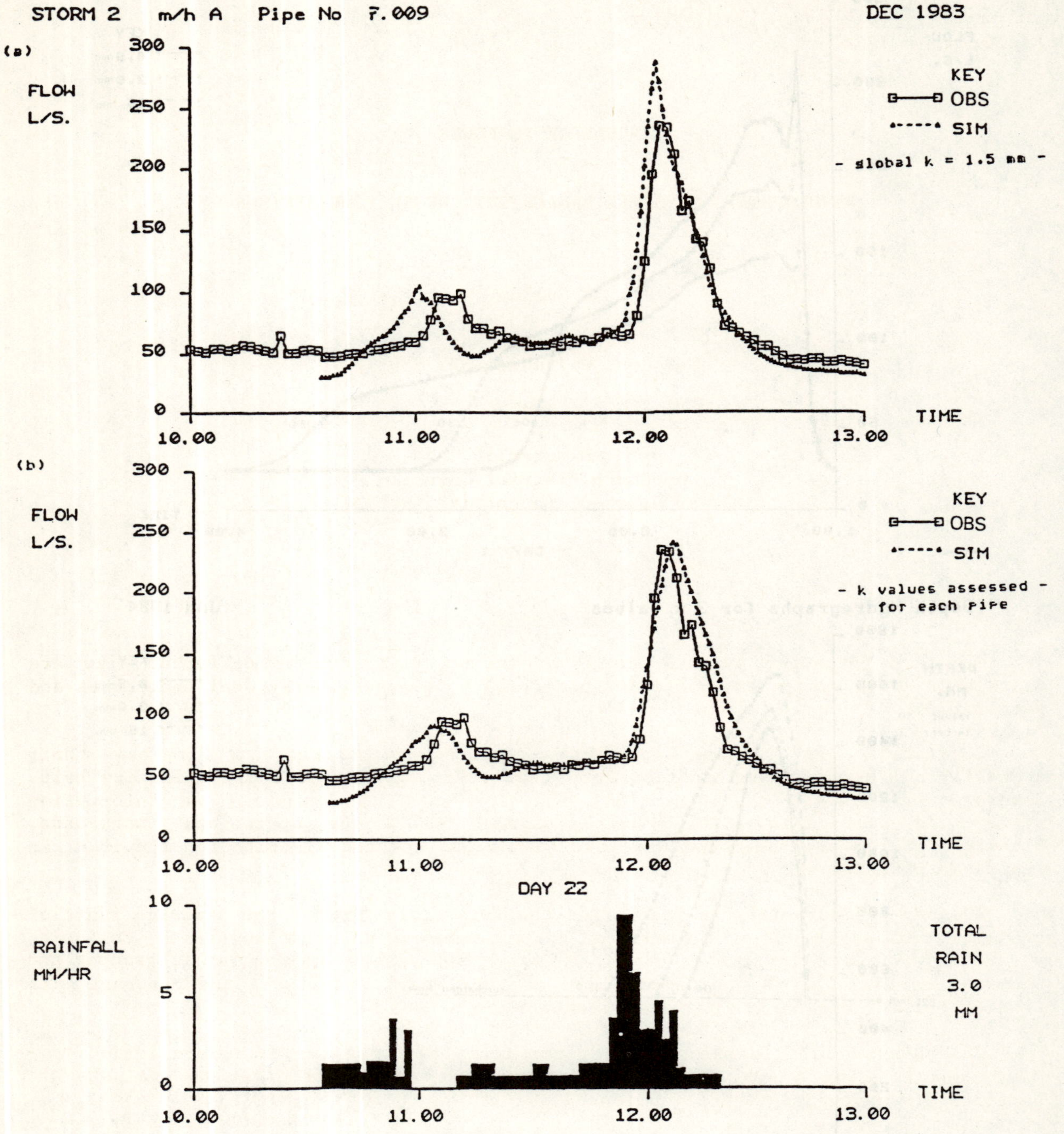

Figure 5 Discharge hydrographs for subcatchment in Oswaldtwistle showing improved fit resulting from pipe-specific assessment of roughness.

International Conference on the
PLANNING, CONSTRUCTION, MAINTENANCE & OPERATION OF SEWERAGE SYSTEMS
12-14 September, 1984

PAPER J1

KNOWLEDGE IS POWER

DEVELOPMENTS IN SEWER MANAGEMENT INFORMATION SYSTEMS

N Cullen

WRC Engineering
Swindon, Wiltshire SN5 8YR

and

M Murrell
Hyndburn District Council
Great Harwood, Lancashire BB6 7DA

SUMMARY

Simple information systems installed in two districts of different size are described. Each system was part of a programme to improve operational procedures and performance.

Government and consumer pressure requires authorities to improve their acquisition and use of information, and new technology is available which may help. Sewerage has special characteristics which affect the design of an information system, in particular a high crisis/planned ratio for operational workloads. Blockages and other random failures may follow the Poisson distribution and this can assist in planning the crisis response.

A study in Yorkshire has identified the main activities of the sewerage function and has itemised the major elements of a comprehensive information system, following an analysis of activity. Area Files, in which all system information is grouped and stored geographically by catchment, are a novel and key feature, and contrast with traditional methods of filing by subject.

Following a full locational sewer survey in Hyndburn the data were loaded on a microcomputer, for a small addition to the survey cost, in order to evaluate the benefits of improved data utilisation.

Trial implementation of computerised digital mapping and data base methods is proceeding but requires and includes a thorough analysis of data requirements and relationships. Studies are needed, to demonstrate the opportunities and benefits arising from increased information availability.

Held at The University of Reading, England. Organised and sponsored by BHRA, The Fluid Engineering Centre, Cranfield, Bedford in conjunction with the Water Research Centre.

1. INTRODUCTION

There is a growing interest in the subject of information systems for sewerage management. There are three main reasons for this:-

(i) the steady accumulation, particularly over the past 5 years, of system and condition data,

(ii) pressure from Government and consumers on Water Authorities and their agents to demonstrate and improve efficiency, and

(iii) the availability of new technology, particularly microcomputers and digital mapping.

WRC Engineering has maintained a watching brief in this area since its foundation in 1980. Much of the early work was concentrated on overcoming the "severe lack of information relating to the lengths, diameters, age, material and condition of both sewers and water mains" documented in the "National Assessment" (Ref. 1). Large scale and sample surveys (Refs. 2, 3) were used in the development of a Sewerage Strategy (Ref. 4). This strategy states that survey and rehabilitation effort should be concentrated in the most important 20% of sewers and the rest left under a reactive policy. Both halves of this approach require good long term information in order to be efficient and effective. This reason, coupled to parallel needs in the many other functional areas of a Regional Water Authority (RWA), has helped to initiate WRC activity on Management Information Studies.

Whilst not ignoring the potential of the more glamorous products of Information Technology now on the horizon, this paper describes the basic work undertaken by two sewerage departments in setting up simple management information systems. Although further advances are possible, what has been done to date may be adequate for many organisations. The steps taken are an essential foundation for any subsequent enhancements.

The paper outlines the background to the problem, describing the need, the development philosophy and the nature of sewerage operations. It then picks out key elements of the field studies at Kirklees by Yorkshire Water Authority (YWA), WRC and the Local Government Operational Research Unit (LGORU), highlights the need to improve operational performance and install an efficient manual information system, outlines the generalised information system proposed by LGORU and comments on implementation of a simplified system by Hyndburn DC, including locational sewer surveys and the use of a microcomputer. Finally, current developments at WRC in digital mapping and database system development are noted.

2. BACKGROUND

2.1 The Need.

Significantly, the National Assessment did not say "there is a serious problem with the condition of our assets" - it said "there is a serious lack of information ..." which prevented formulation of realistic policies and estimates. The Severn Trent Water Authority (STWA) Director of Finance states "the more we have got to know about the state of our assets, the more we have reduced the perception of what is needed in renewal terms" (Ref. 5).

STWA has been encouraged (Ref. 6) to "press ahead with its engineering survey to establish the physical state of its underground assets .... to provide a firmer base for the calculation of financial target returns ..." and depreciation charges. The data handling and analysis implications of such studies are large.

Similarly, it has been suggested that a major cause of uncertainty in North West Water Authority (NWWA) about whether the correct priorities are being set is "ignorance of the extent, capacity and condition of the sewerage system". The RWA must also "be able to require and get prompt and accurate information on complaints, flooding and other aspects of service" (Ref. 7).

The problem of getting the right information to the right people at the right time has been recognised. STWA has been advised (Ref. 6) to "set up a project team to review in a systematic manner the need for management information at all levels and to provide a strategy for providing divisional and regional managers with management systems which are appropriate to their responsibilities and duties".

It is still remarkably difficult to collate regional figures on a consistent basis into national estimates of how money is spent. In 1981/82, RWA sewerage revenue expenditure was £176m and capital £192m (Ref. 8). Taking the NWWA figures (Ref. 7) as a guide, only about 30% of revenue is spent on "repair and maintenance", and pumping requires possibly another 30%. The technical and administrative effort involved in controlling and directing the flow of sewerage information and activity requires 20% of revenue.

2.2 Development Philosophy.

Information services need to be aimed at efficient use of capital, of operations revenue, and of the time and abilities of the managers, engineers and administrators themselves. Information is used to control and direct resources; in other words knowledge **is** power.

Information has been called "the life-blood of an organisation"; increasingly it is viewed as a major asset or resource to be managed and channelled as rigorously as the water which the Industry collects, treats and distributes. Mere collection is insufficient; it must also be directed to the point of need.

In setting up systems to achieve this, it is unwise simply to "automate today's bad habits" (Ref. 9). Shortcomings in operational procedures need to be remedied prior to installing an information system.

In Kirklees and in Hyndburn, following a study of existing procedures, simple paper-based management information systems were installed concurrently with organisational changes needed to streamline operations.

The next stage is to use a computer for the retrieval of some information needed for day-to-day operations and also to archive, retrieve, analyse and review records of activity. The design and implementation of such systems relies heavily on already having a well organised department.

2.3 Nature of the Operation.

The structure of an information system will mirror the needs of those it serves.

The "capital" part of expenditure is the subject addressed by the WRC Sewerage Strategy. Most of the "operational" part of revenue is devoted to reactive work. While it is a "fire brigade" operation, there is no cost effective alternative.

An acceptable reaction time to problems must be sustained, and a system instituted for identifying recurrent failures so that the cause can be identified. Some tasks need to be carried out routinely, but without good long term data it is difficult to establish appropriate frequencies. Pumping is necessary in some areas, and suitable maintenance and pump scheduling policies must be devised and monitored.

The sewerage manager is responsible for both direct labour and capital budgets. Tight cash limits with penalties for over or under spending in any year place a premium on timely and accurate information for monitoring and forecasting. Financial services are frequently provided by another department and beyond the manager's direct control.

In addition to new works design and minor works, a sewerage department has to deal with planning applications, building over and trade effluent agreements, and similar matters. Much of this work is closely interlinked with Local Authority responsibility and most of it deals extensively in maps and plans.

3. FIELD STUDIES

3.1 Kirklees.

On 1 August 1977, nearly three and a half years after Local Government Reorganisation and formation of the RWAs, Kirklees Metropolitan District Council exercised its right to relinquish the sewerage "agency". Yorkshire Water Authority assumed direct control of an area previously administered by eleven District Councils and serving 375,000 people. This is still the largest operation of its kind to be administered directly by a Regional Water Authority.

3.1.1 Internal O & M Study.

The information system emerged as one outcome of an internal YWA Organisation and Methods study, which had a brief to prepare proposals for a gradual change from a crisis system of operation and maintenance to a more planned system, and also to investigate staffing levels, existing methods of working and safety requirements, possible introduction of further mechanical aids, operational frequencies, depot locations and facilities.

During the study, senior managers, sewerage technical staff, productivity services staff and two external research agencies worked together, aiming to develop new operational policies on a sound technical foundation.

On takeover, there were two separate bases 10 km apart, one in Huddersfield and the other in Dewsbury. "Crisis" work accounted for 85% of workload at one centre and 72% at the other. Available information was analysed and short term studies of work patterns undertaken.

A number of key issues were discussed. These and similar questions are common, and in the design of any information system one should consider whether and how it should be structured to make data about these topics readily available for periodic policy reviews:

1. Differing operational practice in the two areas.
2. Differing standards of routine maintenance within the two areas.
3. Low performance of the operating gangs.
4. Overtime levels.
5. Need for a safety code of practice.
6. Need for a technically based policy on operational frequency.
7. Better maintenance arrangements for the jetting machines.
8. Pumping stations being over-maintained.
9. Better purchasing arrangements for materials.
10. Radio communications.

Operational decisions were made, to: (i) work from one base rather than two, (ii) segregate "crisis" and "routine" work and implement a job scheduling policy, (iii) restructure the organisation, one section handling day-to-day operations and another identifying and reviewing problems requiring more extensive work and (iv) use heavy goods vehicle jetting machines (Ref 10) as the first line of response to blockages.

The paperwork system was then designed around these proposals, to facilitate (a) efficient "crisis" response, (b) identification of recurrent problems, and (c) servicing the "planning" function.

3.1.2 **Planning the Response to Complaints.**

It is sometimes suggested that because so much sewerage work is "crisis" based, i.e. a response to random events, work planning is impossible and should not be attempted. Blockage studies have shown that these events follow simple statistical "laws" which can be used in planning staff levels and work patterns.

A comprehensive analysis of about 15,000 blockage reports from many districts, collected during 1979 and 1980, has shown a wide range of blockage rates for different districts (Table 1). The frequency of blockages per day follows closely the Poisson distribution (Table 2) although local reporting procedures can create anomalies. The larger the area served, the more predictable the daily, monthly or yearly workload. In Kirklees, there were 7 "incidents" - mainly minor blockages - per 1000 population per annum. Fortunately, blockages are much cheaper to deal with than collapses (Ref 11).

Two defensive measures can be taken. These are (i) "Queue" the work, so that all except the most urgent problems are tackled the day after they are reported; this permits one to (ii) Plan the route taken by the maintenance vehicle. A simple desk study shows savings of 30% to 70% on journey length (and hence time) over the unplanned case with between 5 and 10 points of call per day. Since travelling can occupy over half the working day, this is important.

The short term procedures and paperwork associated with such measures are easy to devise and operate (e.g. Ref. 12 and Fig 2) and will not be discussed further.

3.1.3 **Area Files.**

Many problems on a sewerage system are intermittent, with a return frequency of 6 months or longer. An information system to monitor system performance is needed which can be easily kept up to date and be scanned to indicate problems which need attention, periodically and also in response to complaints, enquiries or planning applications. If information is required about a specific problem the most usual form of initiating a search is from location. The system must provide summaries, cross-keyed with correspondence files and with sewer maps.

To meet these needs the Area File system was devised. The network was divided into catchments, sub catchments and sub-sub catchments. A catchment was defined as "an area of land which drains to a single point in the sewer system". A hierarchical numbering system was devised so that e.g. 11.06.27 means "Sub-sub catchment No. 27 of sub catchment No. 06 of catchment No. 11". Catchment boundaries were drawn on the maps and all correspondence, summaries, etc. numbered and filed according to the location to which they referred.

This treatment varies from the more traditional approach of filing information in terms of subject matter. It offers the advantage that all items relating to any problem are contained on a single file.

3.1.4 **Local Government Operational Research Unit.**

Simultaneously with the YWA O & M study, the Water Research Centre commissioned LGORU to undertake a wider study of the information needs of a sewerage department. Although it would be largely based on the Kirklees system, it was to be sufficiently general to be of use to the whole Industry.

A comprehensive report (Ref. 13) resulted, based on detailed analysis of the objectives and activity of a sewerage agency. The report was not widely publicised at the time partly because of the more pressing need for development of the Sewerage Strategy, partly because it was ahead of its time, being couched in terms more appropriate to computerised system design, and partly because basic system information was not available at that stage.

The report provides a checklist of activities and flow charts which provide a useful reference point for anyone concerned with design of sewerage management information systems. It identifies five main groups of activity:

Group	Activities
System Planning	Long term development policy (system capacity and community needs).
Planning Services	Main interface between the developer and the sewerage system. Searches, planning consent, adoptions, connections and PUSWA.
System Maintenance	Determine system condition. Review system performance. Design minor remedial works. Plan new operational cleaning/maintenance frequencies.
Maintenance Engineering	Day-to-day operations. Execution of minor works.
New Works	Formulate strategic plans for network. Design and supervise construction of new sewers/rehabilitation.

Operational Control is exercised by a line manager over System Maintenance and Maintenance Engineering, to coordinate the operation and maintenance functions of the organisation.

The proposed information system has three major types of element:

(a) Reference Files. These describe the sewer system, its condition, the external loads, and planning constraints. They will change relatively infrequently and are commonly in the form of maps, with record cards or bound files giving more detailed information.

(b) Work Directories. These are lists which chart the progress of each task through the administrative system. There is one row per task and one or more columns for each stage in the task and for cross-referencing. There are Work Directories for planning applications, adoptions, requisitions, maintenance tasks, investigations and capital works schemes (large and small).

(c) Work Record Files. These contain detailed historical data, used either where a reservoir of information has to be accumulated over a period pending decision, action, or management analysis of activity and performance. They are: Previous Planning Applications, Blockages Summary, Systems [Investigation] Reports, Work Preparation Pool and [As Constructed] New Works [Records]. (Words in square brackets added for clarity).

The report proposed that the whole information system should be geographically indexed using a map coloured up with appropriate symbols, appropriately cross referenced to the files and directories, to give a current at-a-glance view of work in progress.

One of the many diagrams is given in Fig. 1, showing how the flow of activity might interrelate with the various files and procedures. Eventual computerisation of key aspects of the manual system was envisaged. A simple computerised complaints system is currently under joint YWA/WRC development.

3.2 Hyndburn.

Hyndburn District Council was formed in 1974 from the amalgamation of the former Borough of Accrington, and the Urban Districts of Altham, Great Harwood, Church, Oswaldtwistle and Rishton. It covers an area of 7300ha with a population of 80,000.

Prior to the commencement of the study in 1980 the sewerage system was operated on a day to day crisis basis. There was no adequate system of recording complaints, messages were passed to the Sewer Inspector through various channels and he would mobilise the two sewer gangs accordingly. A report sheet was filled out by the Sewer Inspector giving brief details of jobs undertaken in the previous week.

3.2.1 Locational Surveys

After a detailed appraisal of the existing records it was decided to undertake a full resurvey. At that time the sewerage system was thought to contain 4,500 manholes and approximately 200km of public sewers. Based on these figures and utilising a 3 man team of in-house staff it was estimated that the survey would take approximately 3 years to complete.

By the end of the survey period in April 1984 approximately 6,300 manholes have been surveyed out of the current estimated total of 7,000. There are thus 56% more manholes than originally estimated. The estimated length has also been increased to 240km but the final total may be considerably higher.

Survey work was undertaken in three separate phases. Initially all the manholes on a single 1/1250 ordnance survey sheet would be plotted and flow routes verified. The individual manholes would then be surveyed and the information logged on site copies of the standard manhole cards which would be later transferred in the office onto the final manhole cards. Finally the manhole covers were levelled. Reference would be made to all existing information but this was not relied upon to identify all existing public sewers.

New sewerage record sheets were produced as overlays in accordance with Ref. 15, utilising standard Al negatives. The amount of information that is required to be printed adjacent to each sewer was found to be excessive in most urban areas and could make the drawing almost unreadable. Any information omitted from the overlays can be tabulated adjacent to the map. Minor refinements to the manhole cards were suggested.

3.2.2 Information System.

Once the sewer survey was under way, attention was directed to possible improvements in the management of the sewer maintenance function. A simplified version of the information system proposed by the "LGORU" study and implemented in Kirklees was developed. There were three principal elements: the area file system, the complaint cards and the job cards.

Each area file contains three standard forms:-

(a) Sewer Maintenance Record. This contains details of all complaint cards and job cards relating to each sub-area together with an analysis.

(b) Planning Control. This contains details of all applications submitted for planning and building regulations, together with details of all connections made to the public sewerage system and information regarding the adoption of sewers.

(c) New Works and Trade Effluents. This combines details of all new works constructed in each sub-area together with information relating to trade effluent agreements. Special regard is given to details of any toxic wastes and their discharge location.

Copies of the complaint/enquiry form and job card appear in Figure 2.

It was necessary to ensure that all reportable incidents were incorporated into the system. For this purpose a central contact person was established in the Revenue Section. This post was graded as a clerk/tracer and the duties included dealing with complaints, maintaining the Area Files, and general tracing with special reference to upgrading sewer records.

To assist in the planning and programming of the work a simple visual display board was devised for work in progress and complaints received.

A two-way process of upgrading of information is essential. Any information gained from the operations side, buried manholes uncovered, alterations to manholes, etc. must be passed to the Sewer Records Section and all new information on the sewerage system must be passed to the Revenue Section. In certain instances other sections or departments are involved, e.g. Highways or Environmental Health.

3.2.3 Computerisation.

The high cost of survey (typically £10 - £15 per manhole) makes it important to store the resulting information in a way which affords maximum usefulness. Computerisation can now be accomplished for a small fraction of the cost of the site work and its feasibility was therefore investigated.

A pragmatic decision was taken, based on the circumstances at the time, to use a microcomputer. Care was taken to protect the "upgrade path" to a more advanced system later, by seeking ease of bulk data loading and extraction. It was decided that WRC should write the software from scratch (in BASIC) and support it throughout the study, rather than seek and adapt an existing management information "package". Compatability with WRC´s own computer allowed more rapid development of software and bulk transfer of data.

The cost of key punching each manhole record by an external bureau was £0.22 and the first 3,000 were completed inside four weeks. The data were transferred to WRC´s minicomputer on magnetic tape and validated by WRC before transfer to the micro.

At present the system will: (i) retrieve, display and print manhole records, (ii) allow records to be modified, (iii) provide backup records of all additions and alterations to the data file, (iv) search for manholes by single or multiple field selections, (v) sort manholes and give listings (either displayed or hard copy printout) by either single or multiple fields, for simple routine work scheduling.

It is stressed that the present system is being used for evaluation of a particular application only. The machine is proving to be a valuable management aid but problems in capacity, versatility and support are already evident and should be taken seriously by others contemplating a similar route. Despite these limitations, the micro option may have a useful place particularly in smaller drainage offices.

Trials by other users e.g. Thames Water Authority and at Hounslow, Bristol and Kennet are mainframe-based and this appears more appropriate if major extensions to other sewerage applications are envisaged.

4. CURRENT DEVELOPMENTS

4.1 A Glance at the Horizon.

New technology currently offers some exciting possibilities for the storage, analysis, retrieval and presentation of sewerage information.

At one end of the spectrum is the widespread availability of the low-cost microcomputer. At the other end the combined database/digital mapping capability of systems like "Intergraph" or "ViewMap" (Ref. 14) provides the equivalent of the LGORU work index map linked to all the supporting data at any desired scale with printed output and multi-coloured true-scale drawings. Some systems offer colour TV. Trial evaluation of these systems and their implications is being promoted by WRC.

What can be shown to date for these systems is feasibility, not desirability. Benefits can be listed but not quantified. Costs may be high, or hidden; e.g. the cost of software, or of simply obtaining and entering/updating data can be extremely large. Further programming may be necessary to tailor a system for particular applications. Despite this, there is a groundswell of optimism, borne out by experience elsewhere, that once systems are fully operational, many new and beneficial uses will be conceived. The Industry needs to make an enlightened but cautious approach.

The ultimate possibility now being mooted, appropriately in 1984, is the fully computerised database integrated across all the functional areas of a Local Authority or Water Authority. It need scarcely be stated that full scale trials would be a large and daunting task.

Nevertheless it should be possible to implement efficient information systems in stages, progressing the work in blocks and stopping at appropriate intermediate points with a workable arrangement. A logical and systematic approach is needed. The technique of Logical Data Analysis, currently being used in a collaborative project by Wessex Water Authority and WRC, is part of such an approach. Because of the inherent similarity of the physical tasks involved, wherever in the world they are performed, the information model resulting from this analysis should theoretically be independent of both organisation and computer system, and hence only one model should need to be prepared for each RWA function. One early outcome of this work will be Data Dictionaries for Water Distribution and for Sewerage.

4.2 **Logical Data Analysis.**

Logical Data Analysis has developed as a tool for the efficient design of any large data handling procedure, (manual or computerised). It is a necessary design stage for any database application. The idea of a database is that you have a large reservoir into which many people pour information and from which others suck what they need. Under a database management system (DBMS): (a) duplication (and hence multiple updating) of information is controlled, (b) a flexible conversational language is provided for the manager to ask for what he wants, when he realises he needs it. In order to achieve these features, together with acceptably small data storage requirements and adequately fast response times for multiple users, one needs a careful analysis of the data relationships - i.e. which data are commonly entered or used together.

The analyst works from a comprehensive set of the formatted paperwork used in a current system (see also Ref. 15). He then rechecks with users to ensure that he has accurately represented the true situation. (This is not always what the manager thought in the beginning!) The main user requirements for retrieval are then agreed and documented.

Part of the Logical Data Model for the sewerage function is illustrated in Fig. 3. Once the data analysis is complete, it is open to the designer to decide how each information need can most appropriately be met. If computerisation is to be staged, the analysis makes it easier to draw the boundaries and assess implications.

5. **CONCLUSIONS**

There is a good case for improving the information systems used in sewerage management. Understanding of procedures to cope with the "crisis" nature of much of the function has improved as a result of blockage studies. Successful implementation of improved operational procedures with streamlined paper-based information systems has been accomplished in two districts, one large and one small, and the key elements listed. In one case a complete manhole survey was undertaken and the data loaded on a microcomputer to improve its usefulness. A theoretical analysis of the information needs of a sewerage agency has been undertaken and the results are available as a starting point for the design of real systems.

Current developments include a thorough analysis of data requirements and relationships as an essential precursor to possible implementation of computerised data base/digital mapping information management for which the hardware is now on the market. The way is now open for studies to seek and demonstrate opportunities and tangible benefits arising from the increased availability of and ability to manipulate sewer system data.

6. ACKNOWLEDGEMENTS

The authors gratefully acknowledge the advice and assistance of colleagues at WRC and in the other organisations mentioned in the text. In particular the work at Kirklees was largely undertaken by Mr D A L Clark of YWA Productivity Services, Mr K Robinson, (Principal Engineer) and his successor, Mr B Ellis.

The permission of Mr W Jollans, (Director of Operations, YWA), Mr M J Rouse, (WRC Director of Engineering) and Mr J Hayden, (Director of Engineering Services and Transportation, Hyndburn D C), to publish this paper is gratefully acknowledged. The opinions expressed are those of the authors and not necessarily those of their employers or of the Water Authorities involved.

7. REFERENCES

1. Standing Technical Committee on Sewers and Water Mains: "Sewers and Water Mains - A National Assessment." NWC/DoE Standing Technical Committee Report No. 1. National Water Council, 1977.

2. Cullen, N.: "The Sewer Dereliction Problem - Evidence from Collapse Studies." In: Restoration of Sewerage Systems, International Conference organised by the Institution of Civil Engineers (London, June 1981), Thomas Telford Ltd 1982, Paper 2, 8pp.

3. Balmer, R. and Meers, K G.: "Surveying Severn-Trent's Sewers". As Ref 2, Paper 14, 7pp.

4. Water Research Centre. "Sewerage Rehabilitation Manual" WRC Swindon 1983.

5. Severn-Trent Water Authority. "Sewerage - A Strategic Approach". STWA August 1983, p4.

6. Monopolies and Mergers Commission.: "Severn Trent Water Authority etc." HMSO, HC 339, 1981. p 297 recommendation No. 1 and p 306 recommendation No. 28.

7. Monopolies and Mergers Commission.: "Anglian Water Authority and North West Water Authority". HMSO, Cmnd 8726, 1982. p 133, Clause 14.22, Clause 14.24 and pp 190 - 192.

8. National Water Council.: "Annual Report and Accounts 1981/82". NWC 1982. p75.

9. Institution of Municipal Engineers.: "Information Technology - Can we afford not to have it?" Editorial Comment, Municipal Engineering, 110 Jan 1983, p4.

10. Cullen, N.: "Performance and Utilisation of Sewer Jetting Plant". The Public Health Engineer, 9, 1, Jan 1981, pp 18 - 23.

11. Cullen, N. and Whittaker, J.C.: "Sewer Blockages in South West Glasgow". WRC External Report No. 80E, WRC September 1982, p 16 and 36.

12. Priel, V.Z.: "Systematic Maintenance Organisation". London, MacDonald and Evans Ltd, 1974, pp 50 - 88.

13. Houghton, A.G. and Jackson, D.R.: "Sewer Records - A Description of an Information System for a Sewerage Agency". Local Government Operational Research Unit, Royal Institute of Public Administration. LGORU Report No. C283. Private report to WRC, May 1979.

14. Middleboe, S.: "Offshore Firm Maps Services Future." New Civil Engineer, 25 Aug 1983, pp 22 - 23.

15. Standing Technical Committee on Sewers and Water Mains.: "Sewer and Water Mains Records". NWC/DoE Standing Technical Committee Report No. 25, NWC 1980, pp 38 - 64.

Table 1 - Sewer Blockage Rates in North West England

Rate	0 - 1	1.1 - 2	2.1 - 3	3.1 - 4	4.1 - 5	TOTAL
Population	2.49	3.05	0.89	0.62	0.05	7.1

NOTES:
1. Rate is in incidents per 1,000 population per year.
2. Population is total population (millions) living in all districts with rate in range given.
3. Average rate for study = 1.4 incidents/1,000 population/year.

Table 2 - Typical Daily Variations in Workload

(a) Glasgow Study A = 2.71

	n = 0	1	2	3	4	5	6	7	8	8+
Theory	67	180	244	221	150	81	37	14	5	1
Actual	140	170	200	180	170	60	40	20	18	2

Number of days out of 1000 on which exactly n incidents occur.

(b) Theoretical values for various annual numbers N

N	n = 0	1	2	3	4	5	6	7	8	8+
50	819	164	16	1						
100	670	268	54	7	1					
200	449	360	144	38	8	1				
400	202	323	258	138	55	18	5	1		
800	41	130	209	223	178	114	61	28	11	5
1600	2	11	34	73	116	149	159	145	116	195

NOTES:
1. Number of days with exactly n incidents is taken from Poisson distribution $P(n = r) = (e^{-A} A^r)/r!$
2. 1000 day sample assumed for ease of calculation.
3. A is approximate average daily rate taken as N/250 (5 day week assumed).

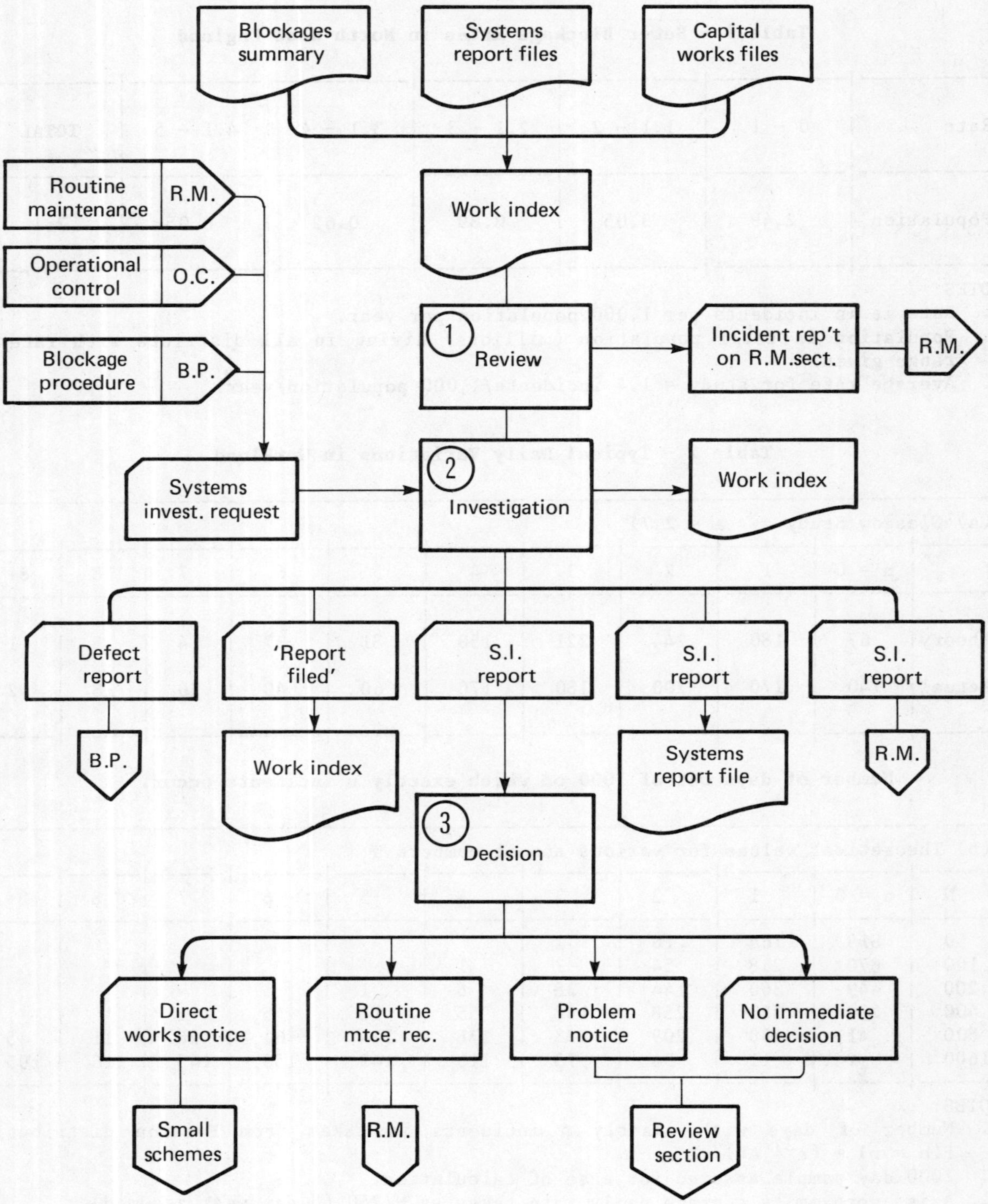

Figure 1. Review/investigation/decision procedures

HYNDBURN BOROUGH COUNCIL

SEWERAGE COMPLAINT AND ENQUIRY RECORD

DC Nº 679
DATE. / /
TIME.
TAKEN BY

NAME. Mr. / Mrs. / Miss.

ADDRESS.

Telephone / Letter / Personal call / Other (Specify)

DETAILS.

ACTION.	COMMENTS.	Analysis of Complaint.
Investigation. ☐		1 Blockage ☐ 2 Collapse ☐ 3 Sewer surcharge ☐ 4 Flooding ☐ 5 Reinstatement ☐ 6 M H Cover ☐ 7 Watercourse ☐ 8 Pollution ☐ 9 Enquiry ☐ 10 Other ☐ Remarks:–
Job Card Number ☐		
Observe. ☐		
Other. ☐		PRIORITY. Immediate. ☐ Next day. ☐ A.S.A.P. ☐

COMPLAINT CLEARED.
Signed
Date / /

HYNDBURN BOROUGH COUNCIL

SEWERAGE JOB CARD

DJ Nº 2166
DATE
ORIGIN

LOCATION

JOB DETAILS

CHARGE

COST CODE

PRIORITY
1 2 3

Received	Started	Completed	Operative

REMARKS

ACTION TAKEN
1 INVESTIGATION ☐
2 JETTED ☐
3 RODDED ☐
4 REPLACE M.H C&F ☐
5 BREAKDOWN ☐
6 REMOVE DEBRIS ☐
7 REPAIR ☐
8 DEFECT CLEARED ☐
9 DEFECT NOT CLEARED ☐
10 FURTHER ACTION REQUIRED ☐
11 OTHER (SPECIFY) ☐

Materials	Quantity	Plant	Time

Signed.....................

Figure 2. Complaint and Job Cards (Reproduced at reduced scale)

Data items from a Standard CCTV Inspection Report

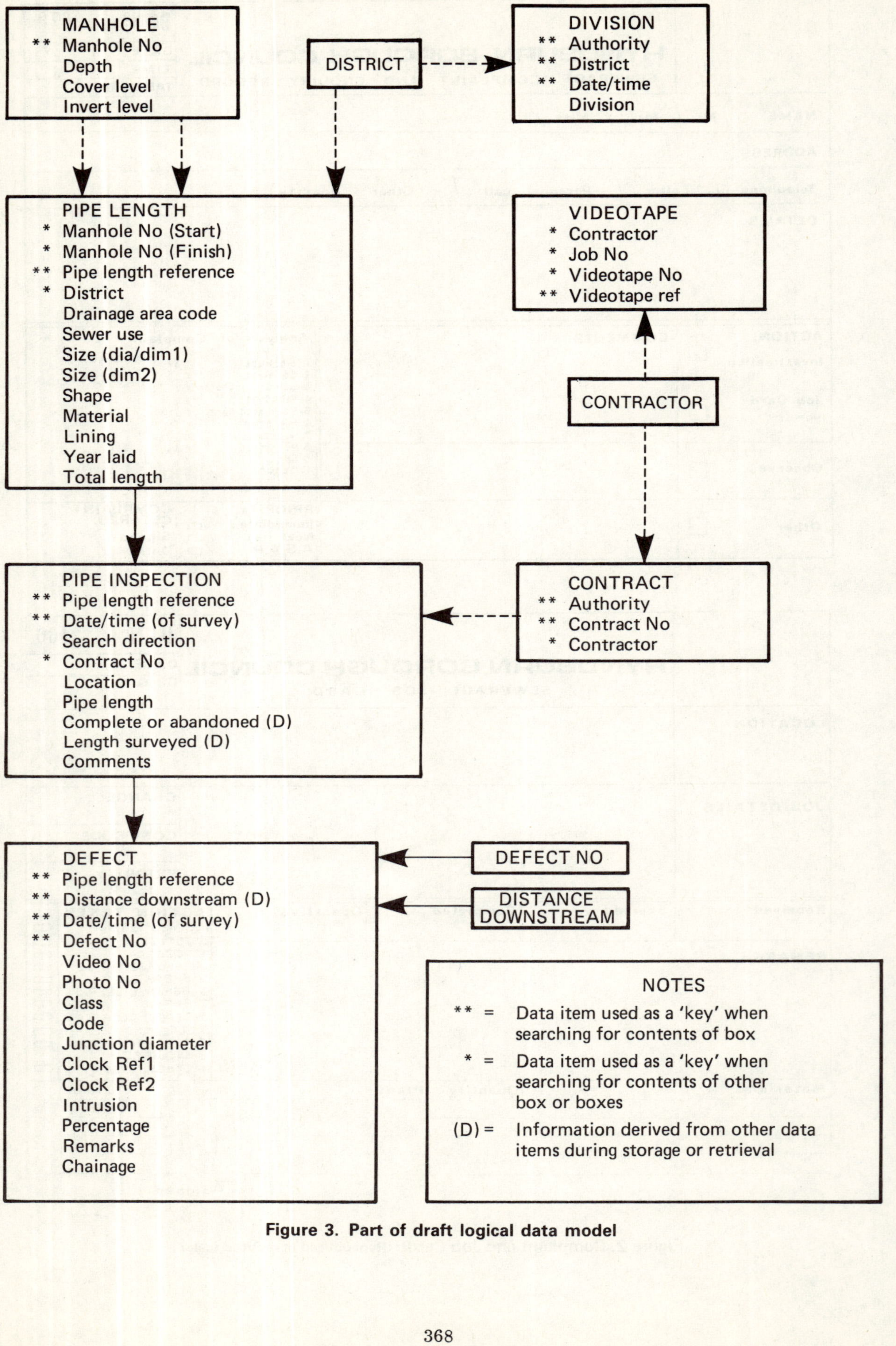

Figure 3. Part of draft logical data model

International Conference on the
PLANNING, CONSTRUCTION, MAINTENANCE & OPERATION OF SEWERAGE SYSTEMS
12-14 September, 1984

PAPER J2

WASTEWATER COLLECTION MAINTENANCE MANAGEMENT SYSTEM

Robert J. Hansen

R. J. Hansen Associates, Inc.
1745 Markston Road
Sacramento, CA 95825
U.S.A.

Summary

This paper describes the WASTEWATER COLLECTION MAINTENANCE MANAGEMENT SYSTEM which is a comprehensive microcomputer/data base information systems package for scheduling and tracking both preventive and regular sewer maintenance work.

The Wastewater Collection Maintenance Management System is one member of a family of management information systems designed and developed to provide support to the wastewater management environment. A sampling of other decision support systems include: treatment plant and pump station scheduling and maintenance; capital improvement resource management; and modeling.

The data base used in this system is an extended network approach to data structuring and includes manhole to manhole sections, street address, and parcels. The paper describes how the system provides access to underground scheduled maintenance, special line history, condition index, TV history, main line history, and parcel/street cross reference.

The paper will cover the strategies used to increase preventive maintenance while decreasing emergency repair work. The process was designed as a decision support system to generate work orders, answer "what if" questions and, above all, to help manage the wastewater collection network. Also described is the report generator for on-line requests that gives users control for initiating their reports and work orders.

Held at The University of Reading, England. Organised and sponsored by BHRA, The Fluid Engineering Centre, Cranfield, Bedford in conjunction with the Water Research Centre.

1. INTRODUCTION

This management and planning tool for Wastewater Collection Systems is designed to enhance the scheduling and tracking of both preventive and regular maintenance work. The system is designed to handle both small and large organizations, and is stand-alone microcomputer-oriented software. It is scoped to provide real-time informational visibiity for all significant underground activities.

The system provides the automatic generation of work orders, a tracking system of completed and non-completed work, assistance in strategic planning for maintenance versus rehabilitation, records manpower resources expended, instantaneous inquiry and update capabilities, and an on-request report generator.

The system was first developed for the Sacramento County Public Works Department as a joint effort between R. J. Hansen Associates, Inc. and the Sacramento County Systems Development and Programming Department in 1979 and 1980, where it is used to control and track the maintenance of 4,000 kilometers of sewage lines.

The system provides management with a control and work planning tool which greatly enhances the daily operation of the Department and is used primarily as a decision support system.

The primary goal of the software is to provide all levels of management a tool for monitoring the condition of Wastewater Collection Systems, formulating a planning process for preventive maintenance policy, and resolving priority conflicts for maintenance and rehabilitation.

The achievement of this tool requires that not only must resources be scheduled and monitored, but improvement strategies must be forecast and simulated to insure that the desired results can be achieved.

The need for quick and frequent visibility of work to be done, system condition, and needs mandate that this approach to managing be automated and operate in an interactive real-time environment. The need to schedule and maximize preventive maintenance while minimizing non-scheduled work can be accomplished if fingertip control of information is provided.

2. HOW THE SYSTEM WORKS

The primary design criterion for the Wastewater Collection System software throughout its development was that the user totally control his information resource, and that the operation of the system to provide the resource be simple. Therefore, the system user's interaction with data processing has virtually been eliminated and replaced with the user's interaction with information he requires to do his job. The maintenance planner/scheduler is not concerned with how the computer works; rather, he is vitally interested in getting relevant and up-to-date information quickly, simply, and in the format he requires. Further, the manager/user can preset exception situations that if and when they occur, the software will automatically notify him.

The system works for the end-users via three information components: the Data Base; the Inquiry and Update resource; and the Hard Copy Exception Reporting and Report Generator. The system is visually depicted in Figure 1. Figure 2 also depicts how the system must react to changing conditions.

The phrase "data base system" is often used in a very loose sense to refer to any approach (computerized or manual) for managing data. Since this loose usage of the phrase covers a multitude of techniques without distinguishing among them, it is not particularly meaningful. In its strict sense, and in the context of this paper, "data base system" refers to a computerized system for managing data in which records of many different types can be integrated into a single organized structure. This single organized structure is called the Wastewater Collection Maintenance Management System data base.

This approach provides the users of the system a much more flexible approach to managing the information and, consequently, easier and more complete ad hoc inquiry capability.

The Data Base is a central repository for all information available from the system. It is subdivided into function subdivisions such as the Main Line Inventory, Main Line History, TV History, Service Line History, and Street Address/Parcel/Manhole to Manhole Cross Reference. These files may be considered as a single, dynamically-interactive entity, with the information being passed and accessed automatically between them to respond to information demands. Figure 3 is shown as the user might visualize a data base but internally is structured somewhat differently.

The data is stored on hard disk files which are directly accessed by the computer in either a sequential or random approach. The on-line disk files can be copied to magnetic tape files to provide backup and security and also the archival of past history files. The disk space required to support the system will vary depending on the number of main lines (manholes) and service lines that are being maintained by the department. Most cities or sewer districts will fall into a range of twenty to eighty megabytes. A megabyte is a million characters of data.

The Inquiry and Update resource is an interactive display process and is the main mechanism to drive the system. When the Main Line Inventory and History and TV History data is placed into the system, a dialog can begin with the planner/scheduler. A maintenance interval plan for preventive maintenance can be inputted for each manhole to manhole section and for area-wide work activities. During this dialog, the system can begin providing work load plans for future months. This can be viewed as an initial start-up simulation mode to see where conflicts may arise. At this point, overall management intervention and/or some policy changes may take place to resolve contention for a resource, or to start getting the P.M. practice to operate in a real-world mode.

At any rate, by evaluating the information available through the inquiry capability, a conflict can be isolated and new direction can be given.

As the system grows with more recent updated activities and information overall, sewer line status can be viewed to observe where each event stands. If the status is satisfactory, no action is taken; however, if some events are occurring more often than planned, or not at all, then changes should be made. These changes are made on an iterative basis to see if the proper mix of needs and resources can be resolved into an acceptable solution.

The Hard Copy Exception Reporting and Report Generator - in general, the hard copy reporting capability of this software falls into two types: query-generated exceptions and standard report requests.

These reports, due to the large amount of data that must be processed to develop them, are not available on the video display but are printed on a hard copy computer printer.

Exception query reports are used to "pop up" red flags to indicate trouble or at least a situation that bears investigation without slowing down the on-going productive efforts. These reports generally help answer three management questions, i.e., what is likely to go wrong? how and when will you know? and what will you do to correct it?

When an unplanned problem has occurred, it can be reflected in red flag reports showing excess or insufficient manpower, targets not achieved, or some work stopped or accelerated.

The question of "what will you do? is answered outside of the system by the planner/scheduler and managers, and the results of management's response entered into the system. This could take the form of shifting preventive maintenance intervals, reducing or adding staff, and/or adjusting the overall planned program.

Most of the exception/query reports will be generated by defining problem condition sets within the system. As these condition sets are monitored on a periodic basis so that when the problem occurs or an unusual pattern is found, the report can be generated and addressed by the appropriate management level.

The reporting and inquiry capability, coupled with an on-line data base, provide the planner/scheduler and managers with a powerful feedback resource to produce operational visibility and control to attain a true management by objectives environment.

3. SYSTEM FUNCTIONS

The Main Line Scheduled Maintenance function provides the user the ability to specify which sewer lines are scheduled for maintenance and when the maintenance is scheduled.

The following information can be entered:

+ Manhole numbers: these manhole numbers define the location of the line

+ Date and interval of maintenance for each type of activity

+ Line description information such as service address, depth, line size, line location, etc.

+ Special maintenance instructions: any special information needed by maintenance personnel can be recorded here

+ Last TV date: the last time the line was TV'd.

The TV History function gives the user the ability to record the information obtained from a TV report.

The information is recorded for a pair of manholes and can have variable footage readings. The system allows the recording of water levels, root rating, grease presence, and a line grade. The user may also indicate if a work order should be produced because of line conditions determined from this report.

The Main Line History function gives the user the capability to track what work was performed on a main line and when it was done.

Such information as was it rodded?, stoppage or other problem information, and footages can be recorded. This information can be queried on-line to see a history of the maintenance performed on a line. The system automatically produces main line history records as scheduled work orders are generated. The user may also enter his/her own records for unscheduled activities.

The Parcel/Street Address/Manhole to Manhole Cross Reference is used. Since the service line information is kept by parcel number, a cross reference of addresses and parcel numbers must be maintained. This function allows the user to acquire a parcel number for a particular address and vice versa. Additionally, information is available to provide reference between parcel numbers and/or street address and the related main line sections.

4. PREVENTIVE MAINTENANCE SCHEDULING STRATEGIES

The Wastewater Collection Maintenance Management System capabilities are designed to respond to managers in a planner/scheduler environment. The logic behind this approach is that scheduling single activities and balancing manpower in a work program is important, but effective scheduling of the entire work program based upon needs and condition ratings is most important. In other words, the emphasis is on maintaining fixed assets versus tracking individual manhours in an accounting system. That means we are looking at the business as a capital-intensive business versus a labor-intensive business.

The system features assist management in developing the scheduling strategies for preventive maintenance, unscheduled maintenance, repair work, and rehabilitiation. It also monitors the progress of these activities to provide management with control and performance information to react to "activities out of control" or to "unrealistic plans".

The scheduling concepts rely on needs, conditions, and intervals, and since the demand for work activities can vary, the concepts had to be simple, easy to use, and without some of the so-called automatic computer complexities.

Management sets the overall scheduling objectives, but again, the concepts are keyed to the several levels of management. In the Sacramento County system, preventive maintenance is scheduled for the next six years. This is reviewed and approved by management but is set up in such a way that the planned work programs can be modified and fine-tuned. Then as problems occur, or do not occur, the strategy is modified either by increasing or decreasing the time intervals set to automaticaly trigger work orders; or activities can be added or completely dropped. It was found that some P.M. activities were overdone, i.e., repeatedly cleaning lines that did not require cleaning.

Just as setting objectives is a management responsibility, so is monitoring progress with respect to those objectives. Through the inquiry, data base, exception reporting and on-request reporting, the system provides the complete information tool for evaluating status and progress.

Again, it should be stated that the scheduling concepts employed in this system place primary importance on responsiveness to preventive maintenance work programs - management's main objective.

The main advantages then to these system concepts are as follows:

+ Management sets the objectives within the framework of the total program.

+ Maintenance scheduling can be correlated to provide a visibility to the total program.

+ Schedules are detailed only to a level consistent with needs - not overly complex and burdensome in data required.

+ Update and feedback requirements are minimized.

5. COMPUTER PROGRAM SUMMARY

To illustrate how the system is structured, the following is a listing of each of the screen programs. The main functions of these computer programs would be to: (1) inquire; (2) update; (3) create; and (4) delete. The screens are also shown in Figures 4. thru 9.

SCREEN PROGRAM NAME

Main Menu
Main Line History
TV History
Service Line History
Main Line Scheduled Maintenance
Report Request Menu
Main Line Scheduled Maintenance Report Selection
Service Line Report Selection
Area Report Selection
Table Maintenance Menu
Table Maintenance
Table Report Selection
Archival Procedure
Service Line Cross Reference
Main Line Stoppages
Prepared Reports on File
Main Line Inventory
Area Scheduled Maintenance
Miscellaneous Report Selection
Main Line Unscheduled Maintenance Report Selection

As previously mentioned in system functions, there were query report capabilities as well as pre-defined reports processed upon request by providing parameters. Each of the following listed reports are parameter-driven to minimize the amount of paper one can get out of a computer. This method also follows the philosophy of working with the computer in a dialog mode.

REPORT PROGRAM NAME

Generate Main Line Scheduled Work Orders
Main Line Incomplete Scheduled Work Orders
Main Line Pending Unscheduled Work Orders in Priority-Date Sequence
Service Line Pending Unscheduled Work Orders
Main Line Completed Scheduled Work Orders
Main Line Completed Unscheduled Work Orders
Service Line Completed Work Orders
Area Completed Work Orders

Selected Scheduled Main Line Work Orders
Selected Scheduled Area Work Orders
Main Line Stoppages Summary
Main Line Stoppages
Area Incomplete Scheduled Work Orders
TV History Hardcopy
Main Line Pending Unscheduled Work Orders in
Priority-Line Sequence
Generate Area Work Orders
Generate Main Line Unscheduled Work orders
Generate Service Line Work Orders
Selected Main Line Stoppages
Table Report
Archive Retrieval Report

And finally, there should be some mention of the specific features of the system to emphasize the dialog concept of talking with your data. These features are:

+ Help screens for immediate on-line assistance.

+ Menu-driven and transaction-driven. Users may select from a menu, but may also use transaction IDs to switch from screen to screen without returning to the menu.

+ Data Base Query Language provides capability of producing ad hoc reports for management or operations personnel.

+ Security provided for any data or set of data in the system.

+ Tables allow the users to tailor codes and descriptions to their own needs.

+ Stand-alone Archival Program allows the user to selectively archive data to another medium, and thus free up space on the primary device.

6. CONCLUSION

Although the following was not originally contemplated, administrative management has found the system extremely useful for answering questions posed by the public, managers, and other government officials. It is reported that this function alone has saved countless hours of professional time formerly spent researching these questions.

The initial implementations of the system have been judged to be successful by the only peer group that really counts - the planner/schedulers and managers (users). In the opinion of those managers interviewed, the most significant aspects of the system are the direct man/machine interface between them and the complete information file through the video screens and the simplicity of use. This initial start into decision support systems will no doubt be expanded upon in the future.

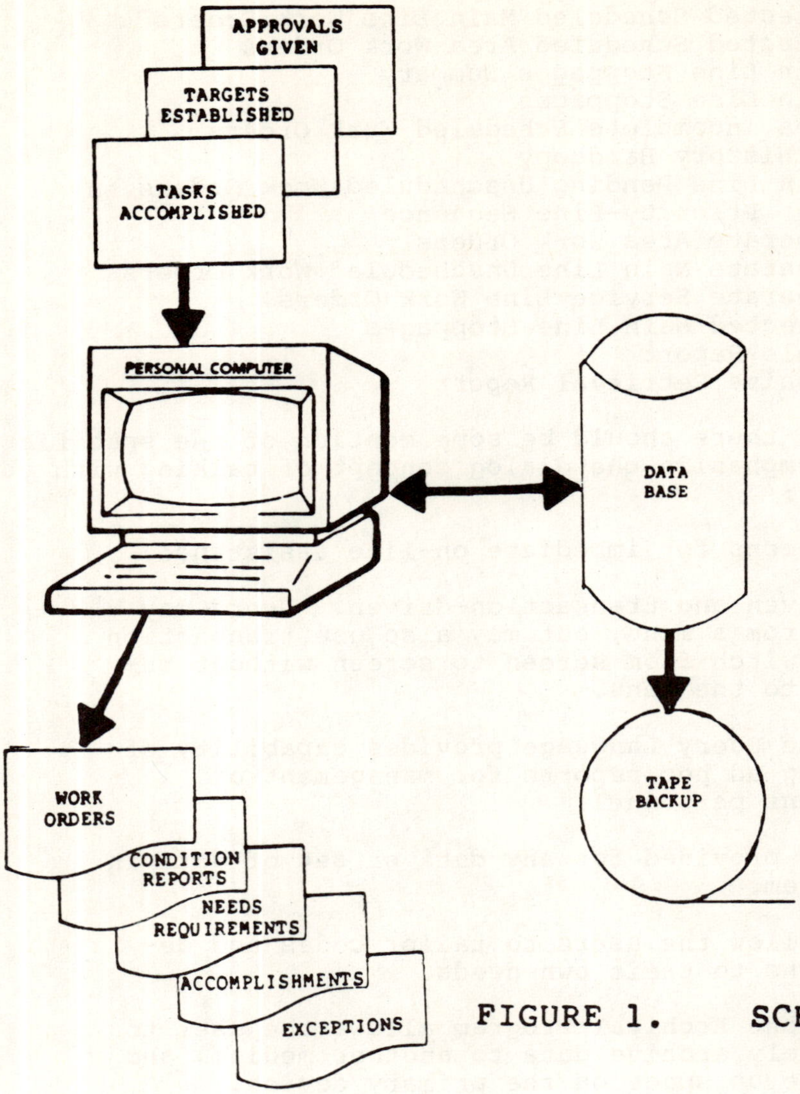

FIGURE 1. SCHEMATIC OF THE SYSTEM

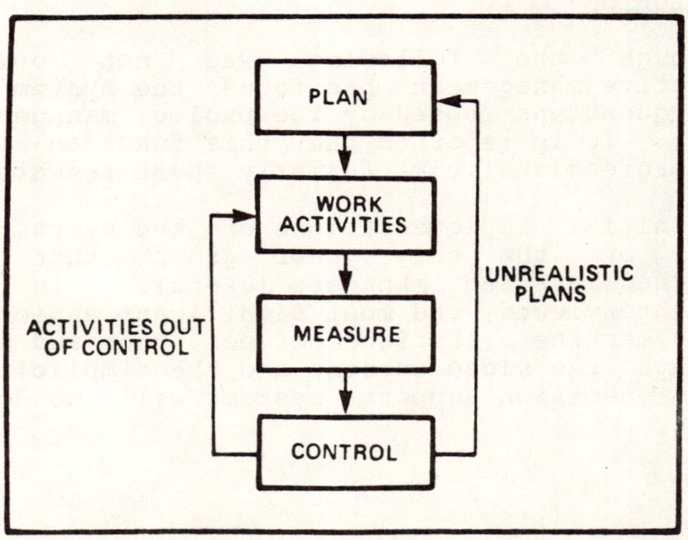

FIGURE 2. INQUIRY AND UPDATE PLANNING AND CONTROL PROCESS

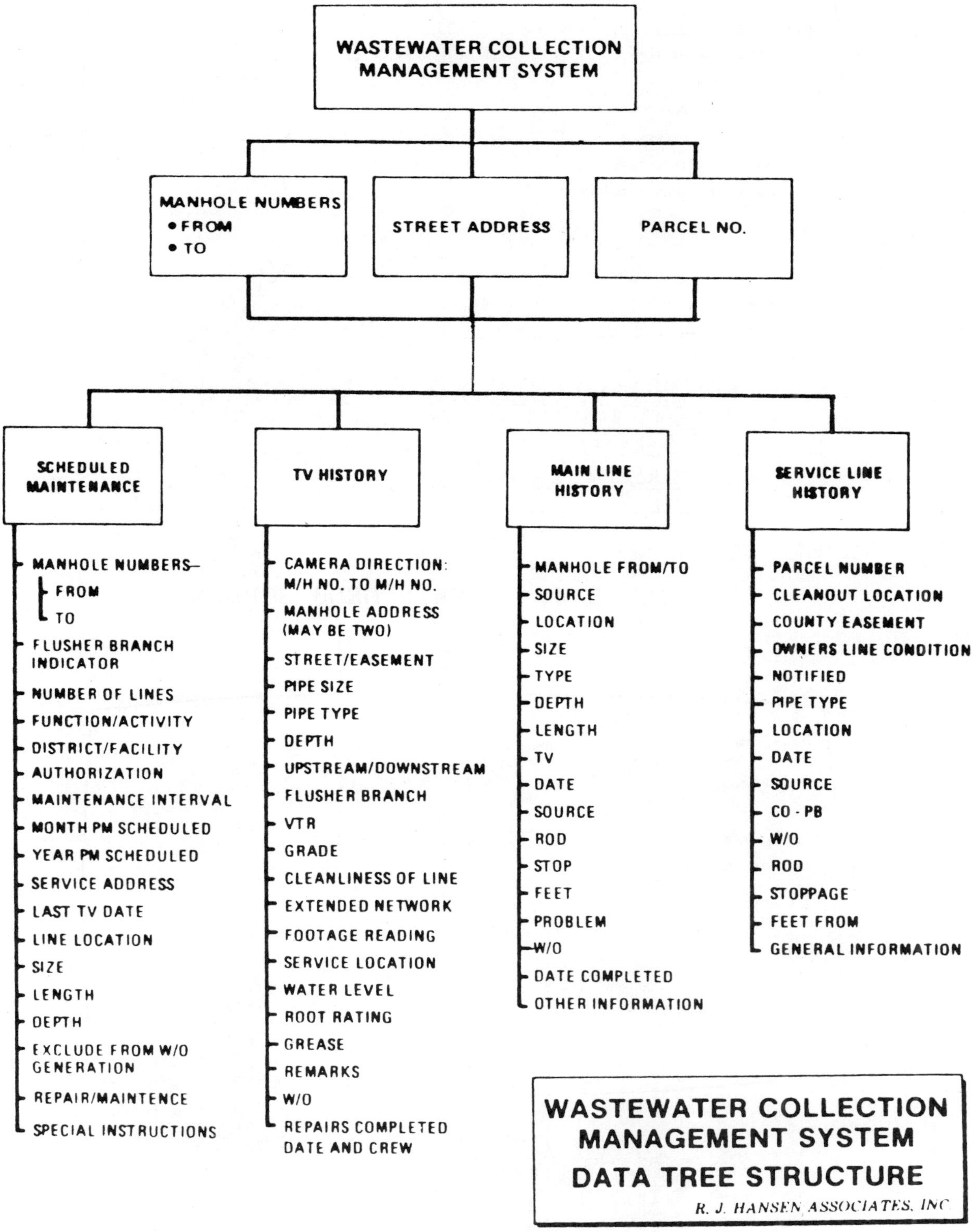

FIGURE 3. DATA TREE STRUCTURE

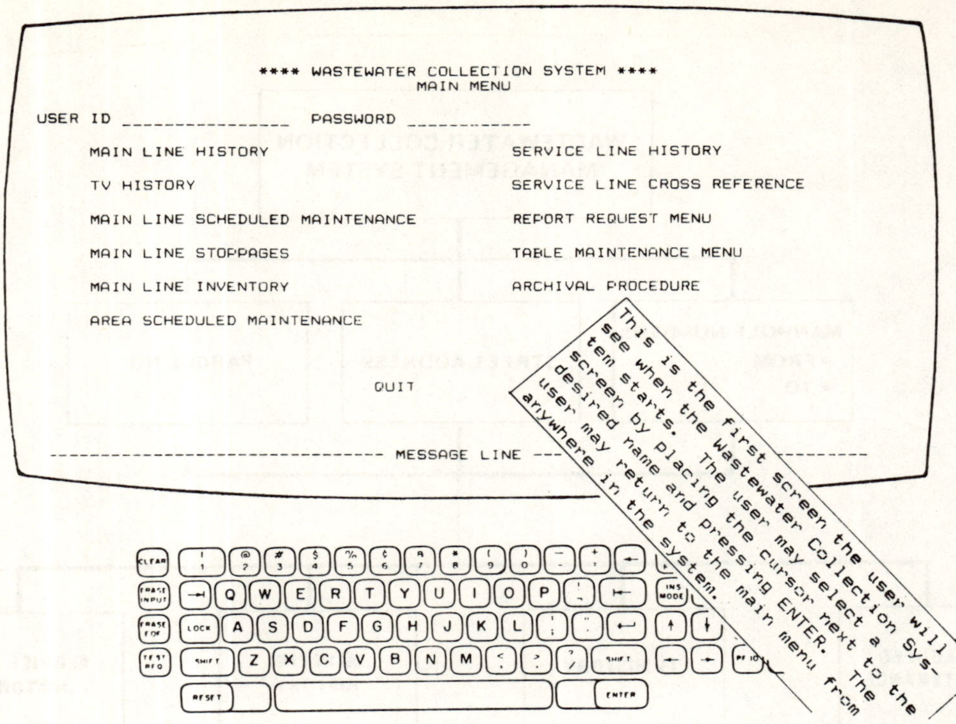

FIGURE 4. MAIN MENU SCREEN

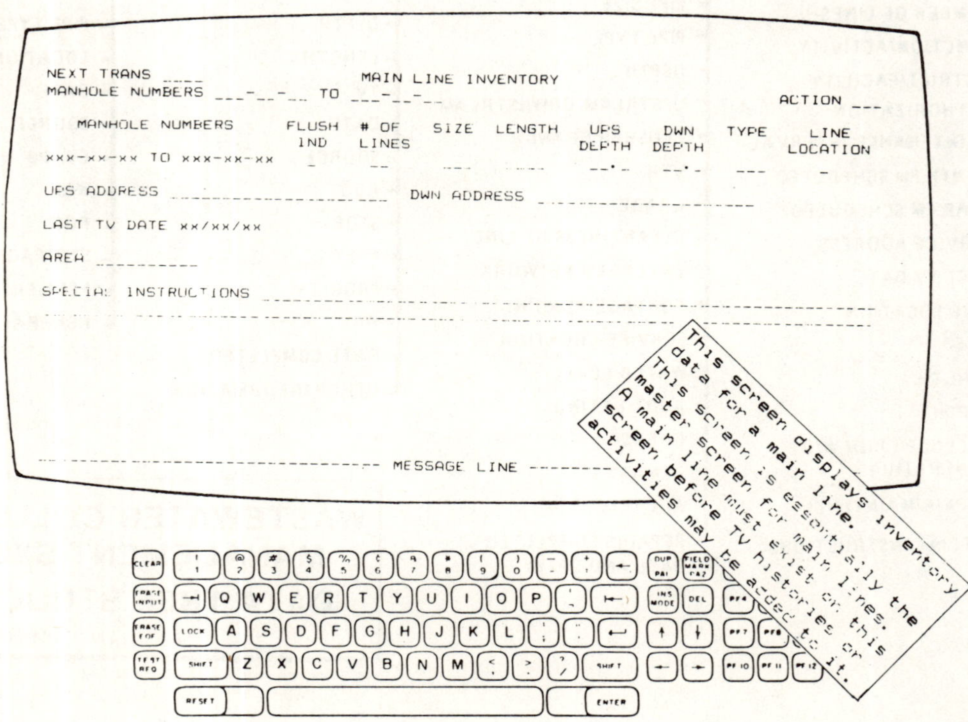

FIGURE 5. MAIN LINE INVENTORY SCREEN

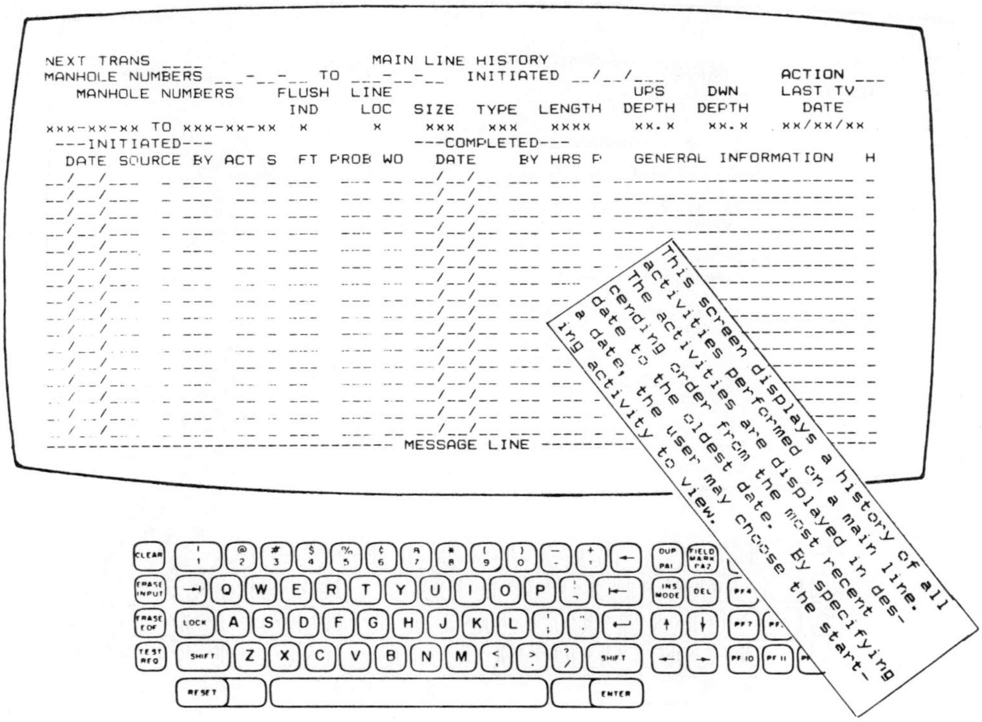

FIGURE 6. MAIN LINE HISTORY SCREEN

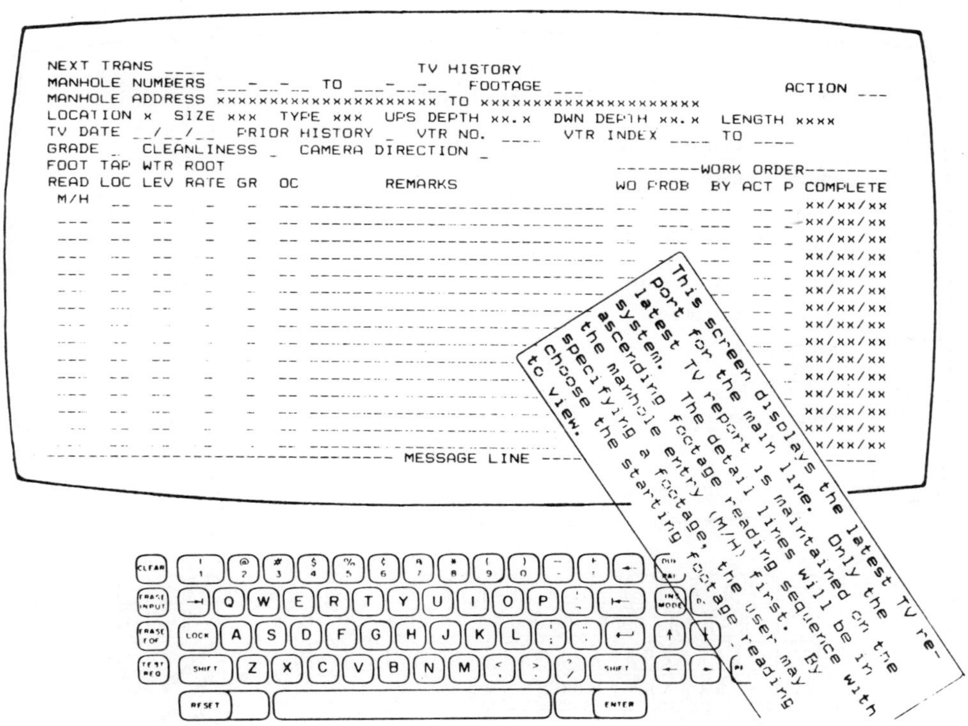

FIGURE 7. TV HISTORY SCREEN

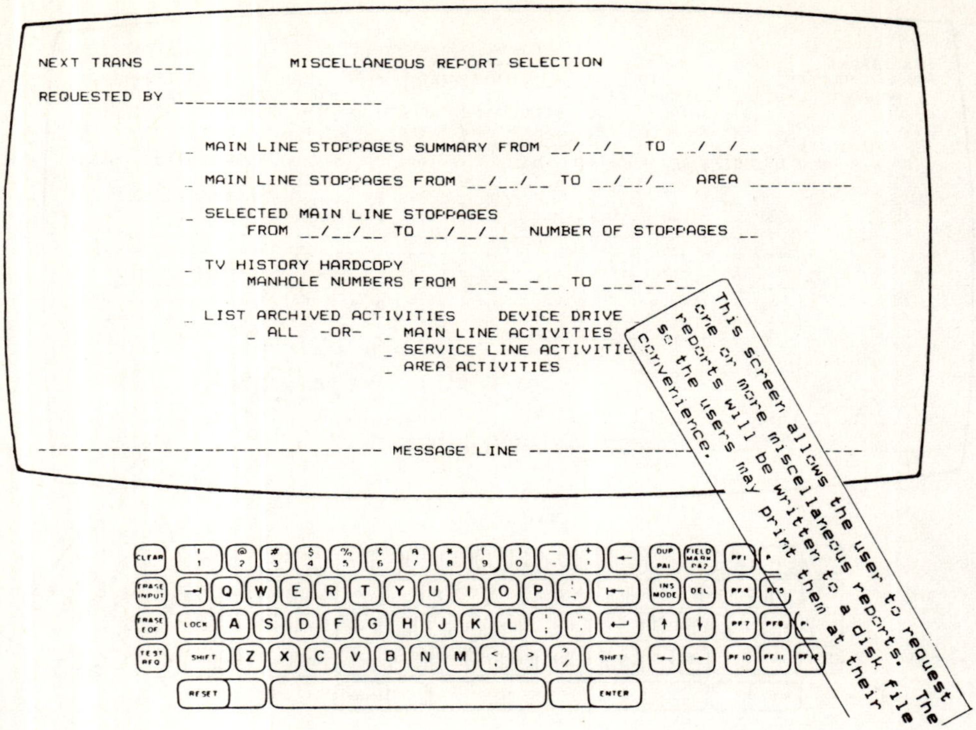

FIGURE 8. MISCELLANEOUS REPORT SELECTION SCREEN

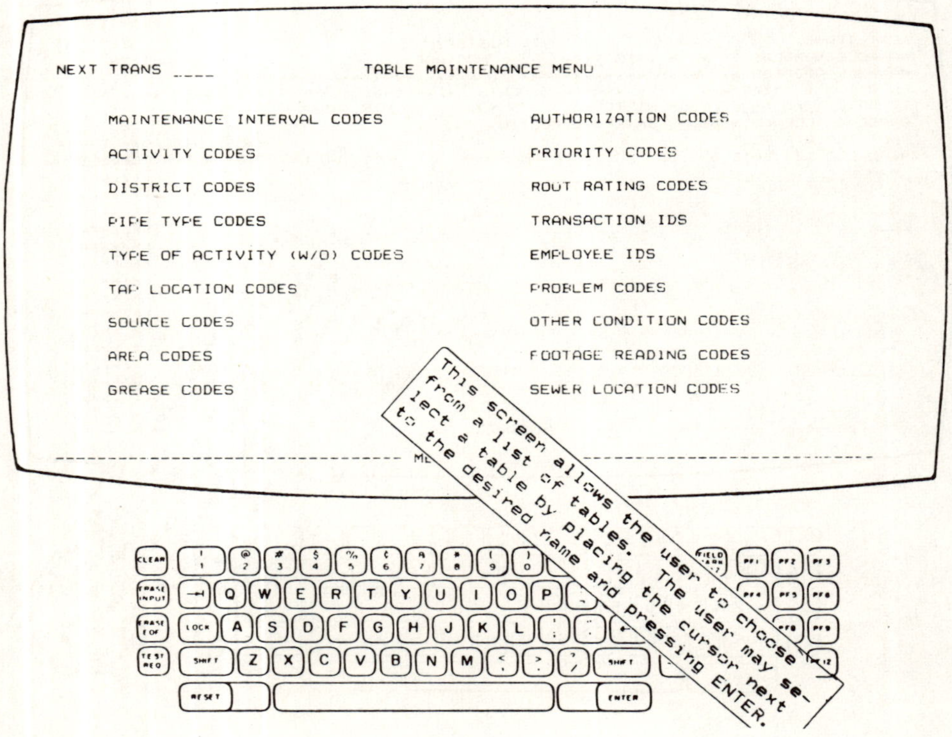

FIGURE 9. TABLE MAINTENANCE MENU SCREEN

International Conference on the
PLANNING, CONSTRUCTION, MAINTENANCE & OPERATION OF SEWERAGE SYSTEMS
12-14 September, 1984

SEWAGE PUMPING OPTIMISATION IN WESSEX WATER AUTHORITY

R Tyler, BSc, CEng, MICE, MIMunE, MIWES
Divisional Engineer (Design and Contracts)
Avon and Dorset Division

A D Plews, BSc, CEng, MIEE, MIElecEE
Assistant Divisional Engineer (Mechanical and Electrical)
Bristol Avon Division

Summary

The paper describes how the use of industrial computer based control systems will optimise the pumping operations of two pumping stations in the Wessex Water Authority.

The first system at a major sewage pumping station for Weymouth, Dorset, depends on knowledge of rainfall information and the characteristics of the sewerage system in order that time-area diagrams and storage in the sewerage system can be calculated to predict rates of flow to the pumping station and hence optimise pump operations to minimise electricity costs.

The second system, installed at an existing pumping station in the Bristol main drainage area, assesses the variable inlet flow conditions by time-level measurements and output control commands to pumping plant to match the variable flow conditions.

In addition to optimising the pumping operations, the control systems are used to control other equipment in the pumping stations, and provide the basis for developing and refining further facilities in these and other pumping stations.

Held at The University of Reading, England. Organised and sponsored by BHRA, The Fluid Engineering Centre, Cranfield, Bedford in conjunction with the Water Research Centre.

1. INTRODUCTION

This paper compares the use of microprocessors for optimising pump control sequences at two different sewage pumping stations in Wessex Water Authority.

The first is a new sewage pumping station, called the Radipole Lake Pumping Station, constructed at Weymouth, Dorset, in the Authority's Avon and Dorset Division, as part of the Weymouth and Portland Main Drainage and Long Sea Outfall Scheme. It has been constructed to redirect sewage from the major catchment area of the town to the headworks of the recently completed long sea outfall, and the opportunity has been taken to incorporate the latest technology available to provide as far as practically possible, a fully automatic failsafe operational station with minimum running and energy costs.

The second is a sewage pumping station called Blackrocks Pumping Station in the Authority's Bristol Avon Division, which was constructed before the Authority's formation in 1974. The station is required to lift sewage flows from the Northern Stormwater Interceptor to the main trunk sewer to the Avonmouth treatment works. Existing pump control equipment was very limited, which resulted in high operational and energy costs. The basic approach here has been to install a simple programmable controller system which will considerably reduce the daily maintenance costs of the station and the annual energy consumption for the minimum outlay.

2. THE WEYMOUTH AND PORTLAND MAIN DRAINAGE AND LONG SEA OUTFALL SCHEME

The recently completed long sea outfall provides a modern marine sewage treatment system for the Borough of Weymouth and Portland. From a headworks at Wyke Regis, pre-treated sewage is discharged by gravity into West Bay through a 2700 m tunnelled pipeline that has been designed to provide for expected population growth into the twenty-first century.

The Radipole Lake Pumping Station delivers sewage from most of Weymouth, and has been operational since January 1984. A new sewerage scheme for the Wyke Regis area will connect this catchment to the headworks in 1985 and the Portland connection will be complete by 1986. Further extensions to the Portland trunk sewerage system will complete the entire project by 1990. Fig. 1 shows the outfall and trunk sewerage system, and Fig. 2 shows the configuration of the Radipole Lake Pumping Station.

Wessex Water Authority employed Consultants, John Taylor & Sons, for the design and construction of the Headworks and Outfall. The Main Drainage works have been carried out under the direction of the Borough Engineer, and L G Mouchel & Partners, Consulting Engineers, have been responsible for main drainage feasibility studies and detailed design and construction of the Radipole Lake Pumping Station, (including the computer control system), as well as the rising mains. Expenditure to date is £24.8 million and when complete the scheme is expected to have cost £34.7 million.

3. RADIPOLE LAKE PUMPING STATION

Sewage arrives at the station through two trunk sewers, each 1.50 m diameter, and sewage is delivered to the headworks through parallel 0.80 m and 1.10 m diameter ductile iron pumping mains 1700 m long. From this summit, sewage gravitates to the headworks through a 1.90 m diameter and 1.40 m diameter outfall partly in tunnel.

Two centrifugal sewage pumps, each of 0.375 m^3s^{-1} capacity, pump either alone or in combination through the 0.80 m diameter main, which is sufficient for dry weather conditions and small rainfall events. Two centrifugal pumps, each rated at 0.725 m^3s^{-1}, pump through the 1.10 m main in a similar manner for storm events, giving a total delivery capability to the headworks of 2.20 m^3s^{-1}. A third pump of 0.725 m^3s^{-1} capacity will eventually be provided as a standby.

The output of 2.20 m^3s^{-1} is the estimated catchment Formula A flow for the year 2010. Flows in excess of 2.20 m^3s^{-1} arriving at the station are discharged to the Backwater in Weymouth by two axial flow pumps each of 1.60 m^3s^{-1} capacity through a 1.50 m diameter low pressure concrete pumping main.

Sewage arriving at the station passes through three bar screens, and screenings are macerated and returned to flow. Before entering the pumping wet well, the coarse grit fraction is removed from the sewage by means of aerated spiral flow grit channels. The screens and grit channels are constructed underground.

Because the 1.10 m diameter main will only be operated infrequently, it is necessary to drain it down after each storm event to prevent septicity in the main. Due to the topography, a considerable proportion of the sewage in the main will not drain back to the Radipole Lake Pumping Station by gravity and a subsidiary pumping station has been provided at the low point, sewage from the 1.10 m main being pumped into the 0.80 m main, which will normally remain full. This subsidiary pumping station is known as the Knightsdale Road Pumping Station.

3.1 Requirements of the Control System

As for any pumping station, the prime objective is to start and stop pumps to ensure that sewage in the wet well is kept within safe minimum and maximum levels, which has traditionally entailed approximately matching pumping output to sewage inflow. Conventional pump control sequences bring in more or larger capacity pumps on rising levels in the wet well, and it is well known that frequently large pumps are operated for relatively short periods to deal with the peak run off rate in storm conditions. The Radipole Lake Pumping Station system attempts to predict likely flows and flow rates into the station and maintains the minimum pumping capability necessary, thus saving energy by reducing the friction head to be overcome and reducing maximum KVA demand charges.

In addition to controlling the 7 pumps in the station, the control system monitors the operation of the Knightsdale Road Pumping Station, and controls the operation of the three bar screens, two aerated grit channels and associated equipment in the station.

3.2 Background to Pump Control Optimisation

Between 1974 and 1976 a comprehensive sewer survey was carried out by L G Mouchel & Partners to produce up to date records, and hence determine by analysis deficiencies in the system. The consultants used the Transport and Road Research Laboratory (TRRL) hydrograph method, which generated hydrographs of rate of flow against time for each of the sub-catchments and hence enabled a total hydrograph to be developed for the rate of flow into the pumping station.

If rainfall intensity is measured in representative sub-catchments, it is possible to calculate the likely inflow from each drainage area up to 2 hours in advance, and hence with knowledge of the storage in the system predict likely pumping requirements.

3.3 Pump Control Optimisation System

Three automatic rain gauges, (Rimco tipping bucket type 0.2 mm capacity), have been installed in suitable locations in the three drainage areas which comprise the Radipole Lake Pumping Station catchment. Rainfall information is transferred in digital form into the microprocessor, which calculates the likely run off to the station. The system allows reasonably accurate predictions to be made for flows up to 15 minutes in advance, and makes increasingly less reliable predictions for up to 2 hours in advance by comparing rainfall behaviour with historical data. Fig. 3 shows predicted hydrographs for 1 year and 5 year storms with estimated error factors.

Knowledge of the sewerage system enables the available storage within the station and sewers to be determined for a given depth measurement. This relationship is shown in Fig. 4.

By deducting the maximum storage in the system from the inflow curve, the minimum outflow curve is derived, and the limits of the two curves define the pumping requirements to avoid flooding of the station and sewers.

Optimum pump sizes to be installed can thus be determined, and installed pumps can be operated to give minimum running costs. The principle is demonstrated in Fig. 5.

The above principles were used in 1976 to make preliminary selections of dry weather pumping rates and to determine likely benefits from the best rainfall information then available, from Hurn, (Bournemouth). During 1978 rainfall information for Weymouth itself was obtained and subsequently further simulations carried out. Fig. 6 shows a comparison of pumping operations on level control and prediction control for rainfall recorded on 2nd August 1978. Fig. 7 shows a 1978 data cumulative volume profile with pump sequencing under optimum and level controls.

3.4 Mathematical Modelling and Updating

The run off to the pumping station is derived from mathematical models developed by L G Mouchel and Partners based on a TRRL output hydrograph from 3 sub-catchments, which were selected based on topography and local knowledge. Although the calculations may not be completely valid due to inaccurate assumptions, (e.g. non-uniform rainfall intensity over each sub-catchment, surcharging in sewers, varying impermeability factors, etc.), they will produce a time-area diagram and hence a cumulative inflow profile for a given rainfall event. The processor will store the necessary data for extraction and analysis by main frame computer at periodic intervals, which will enable revised time-area diagrams to be inserted into the model when the existing predictions are considered too inaccurate. Thus the effects of development in the catchment will be taken into account, as well as basic inaccuracies in the modelling.

The prediction model for optimum pump starting has been developed by Dr G Evans of the Water Research Centre, Medmenham. By reference to Figs. 5 and 7, it can be seen as using fixed slopes (pump outputs) to find the optimum path through a tunnel, the shape of which is known with reasonable certainty for some time ahead, although the accuracy will deteriorate with time.

The installed system represents a fully instrumented catchment, with possibilities of further research into the prediction of storm flows from sewerage systems, as well as facilities for updating the mathematical modelling used for pump optimisation.

3.5 Additional Facilities

Having decided to incorporate a computer control system, it becomes economical to use it for all the control functions in the station. Thus the computer will control:-

(i) three inlet screens and inlet penstocks. Either one, two or all three screens will operate according to flows, and the rotation of screen duties will be varied several times per day.

(ii) two spiral flow grit removal channels and inlet penstocks. Either one or both channels will be in operation according to flow and duty will be rotated. The associated air compressors, air lifts, grit pumps and screw lift classifier will also be controlled by the computer.

Outfall Flushing Sequence. Although all sewage passing to the long sea outfall will have been de-gritted, fine particles will be present, and a flushing sequence is necessary to remove the risk of solids compacting in the outfall invert. This consists of the maximum 2.20 m^3s^{-1} station output delivered to the headworks for about 9 minutes each week. To achieve this, the maximum volume of sewage must be accumulated in the sump and sewers before commencing the sequence. The 1.10 m diameter main is first filled and then both dry weather and both storm pumps are run to deliver the flush. Clearly this would negate the objective of the pump optimisation routine if powered by mains electricity, and hence this will be carefully sequenced to coincide with the weekly test runs of the standby generators to ensure that they are fully operational on load.

Draining Sequence. The 1.10 m diameter main must be drained down after each use. If the computer predicts that it will not be required again, a washout valve at the station allows part of the main to drain back into the station sump. The remainder is drained by opening a valve at the low point enabling the main to drain down into Knightsdale Road Pumping Station, from where it is pumped into the 0.80 m diameter main. These sequences are controlled by the computer, which also makes the necessary adjustments to the measurements of flow in each pumping main.

Telemetry and Fault Monitoring. Comprehensive status reports of key items of plant are available at the station terminal. Similar information about the headworks equipment is also available, a direct land line having been laid with the pumping mains. This information is available to Weymouth and Portland Borough Council staff either at the Radipole Lake Pumping Station or elsewhere, a portable terminal having been provided which enables the computer to be accessed by public telephone line. An interface with Wessex Water Authority's main telemetry system allows for constant status monitoring, although responsibility for maintenance of both station and headworks is being undertaken by the Council. An automatic call out unit has been provided to give the Council immediate notice of key faults developing.

3.6 Possible Future Uses

The use of the standby generators to avoid incurring maximum demand tariffs has already been reported in respect of the flushing sequence. It is expected that this operation will take up the full monthly time allowance for test running of the generators permitted by Southern Electricity Board. If present S.E.B. restrictions are removed in the future, it will be possible to use the standby generators for peak lopping, the generators being started under computer control.

The basic computational facilities exist to monitor possible siltation in the long sea outfall. The water level in the outfall shaft at the headworks at zero flow will equate approximately to sea level, and the change in head needed to discharge a given flow rate will be a measure of the pipe friction. After an initial period to establish stable slime conditions on the tunnel walls, changes in friction can be an indication of siltation.

3.7 Cost Benefit Analysis

When first proposed, the likely savings in maximum demand charges alone, based on the 1976 Bournemouth rainfall data, were estimated at £9,500 per annum for the optimised control scheme compared to conventional level control, (£18,500 per annum against £28,000 per annum). Savings in energy costs would increase thus saving still further, and the development of the optimised control system was approved based on an estimated system cost of the order of £40,000 to £50,000, including software development.

Tenders were received in September 1979 and the lowest tender of Computronic Services accepted in the sum of £57,154. A revised cost benefit analysis basis on the 1978 Weymouth rainfall data indicated reduced savings of £4,542 per annum (£20,211 against £24,753). These savings would be achieved by a fully optimised pump control system at an additional cost of £11,620 over an alternative tender of £45,534 which would have provided for a fully automatic station with the same telemetry facilities, but excluding the prediction routines.

Although the estimated savings were lower, the additional cost of the prediction routines was calculated to have a pay-back period of 2.5 years, and the proposals were accepted.

There have been delays for various reasons in completing the scheme, and inevitably costs have risen as a result. Based on the same 1978 rainfall data and the latest electricity board tariffs, the savings are now expected to be of the order of £5,800 per annum (see Fig. 8), against a revised total cost of the computer control system of about £90,000. Some of these cost increases are attributable to the basic scheme as well as inflation, and therefore the updated additional cost of the optimisation system is £16,100. The pay back period is now estimated to be 2.8 years.

It is interesting to note that the cost savings of the reduced number of kilowatt hour units represent sightly less than 6% of the total savings. Thus the system will only be viable where the tariffs for electricity consumption are similar to those currently in use in the United Kingdom, which impose cost penalties for incurring high kilovolt-ampere maximum demands, especially in the winter months of highest electricity consumption.

4. BLACKROCKS PUMPING STATION

Blackrocks Pumping Station was constructed as part of the Bristol Main Drainage Scheme some twenty years ago, the works being undertaken in two stages. The first stage consisted of installing four single entry mixed flow pumps in a circular plan chamber which lifted the foul water received from the Northern Stormwater Interceptor to the main trunk sewer.

Subsequently the foul water from the Northern Stormwater culvert was diverted through an oval plan chamber fitted with a channel containing a coarse screen and a vertical raked fine screen, discharging into the pump sump. Motorised penstocks are fitted to the inlet channel. Foul water in the main trunk sewer gravitates to Avonmouth Sewage Treatment Works for subsequent treatment and disposal. The pumping station is constructed almost completely underground approximately 1 km downstream of the Clifton Suspension Bridge. Fig. 9 shows a simplified view of the station.

The site was visited by shift personnel from Avonmouth Sewage Treatment Works once every 8 hours. On every visit, the flow conditions were manually assessed on the basis of whether the pump selection made at a previous visit was able to meet current demands. By experience and intuitive reasoning connected with prevailing meteorological conditions, a pump duty selection was made, such that at rising sump levels detected by electrodes, pumps would start and run in a desired sequence. It was found that flow conditions rapidly varied on numerous occasions following the routine visit, resulting in the pumps making frequent starts, especially under low flow conditions where once per minute has been recorded. This was obviously not an efficient way of running the plant.

Immediate solutions to the problem could have been effected by increasing the level of surveillance either by more routine visits or via telemetry. However, the Authority has been implementing an optimisation and automation programme both at Avonmouth Sewage Treatment Works and the Bristol Main Drainage area which will eventually result in the system being supervised during normal working hours only, and manual intervention either in the control room or on site would still have been required.

The pumping station was constructed such that structural modifications for flow measuring devices, for example a flume, could not easily be realised, and therefore a pumping regime whereby the pump sequence and operation could be regulated by a direct flow proportional control system was discounted. Further, space was extremely limited in the electrical plant room so that the use of variable frequency inverter drives to continuously modulate pumping rates and thus minimise the number of starts and stops of pumps was judged impractical. The problem resolved to the question; "by what means can the pumping plant and ancillary equipment be automatically operated to compensate for variable inlet flow conditions and to minimise mechanical and electrical stress on the equipment". The solution was to be at minimum cost with little or no structural modification.

4.1 Requirements of the Control System

The definition of the word "Automation" suggests a possible approach:-

> "Those processes in which mechanisms are used to perform tasks that previously required the attention and control of human beings".

The shift personnel visiting the station observed the variation of level in the sump and formed some judgement as to what the likely conditions would be for the next 8 hours and selected pump duties accordingly.

If the level in the pump sump could be continuously measured, and an algorithm to select the most suitable pump(s) set up, then this could perform the task which previously required the attention and control of the shift personnel. It is necessary for the algorithm to be repeated at relatively high frequency to respond to the sometimes sudden fluctuations in inlet flow conditions which are a feature in this system.

4.2 Level Measurement

For a number of years Wessex Water Authority has been successfully applying ultrasonic level measurement techniques for measuring foul water and sewage levels and this technique was selected for this application, the transducer output being a 4 - 20 mA analogue of the 0 - 6.1 m range of level in the wet well. The earlier probe system also suffered from the disadvantage that attendance of the confined spaces team was essential whenever probe cleaning was required, thus use of the ultrasonic level measurement equipment allowed more reasonable access for maintenance, etc.

The analogue signal was looped through two dual channel trip amplifiers which provided discrete adjustable level trip signals, the adjustable trip settings being an important facility for tuning the final automatic control system arrangements.

The volt free contacts driven from the trip amplifiers were set up initially to change state at discrete levels above the sump invert. For the purposes of this paper, an example might be as follows:-

Level (m)		Analogue Signal Level (mA)
h_1:-	1.27	7.3
h_2:-	1.5	7.93
h_3:-	2.0	9.25
h_4:-	3.12	12.2

Table 1

The output contacts are connected to an industrial programmable controller. The analogue signal is looped through a chart recorder and the telemetry analogue/digital converter for transmission to Avonmouth Sewage Treatment Works.

An ultrasonic level transducer was provided upstream of the screen and both pump sump level and screen upstream levels were compared to provide a high differential level signal.

The pump sump level equipment provides an additional low level pump protection signal. The circuitry for this system is segregated from and independent of the normal operation control system. Fig. 10 shows a block diagram of the level measurement system.

4.3 The Pump Control Programme

The main criterion adopted in the control algorithm is to use the pumping capacity of various combinations of pumps to maintain the level in the pump sump between the example limits of 1.5 m and 2.0 m metres above sump invert, for inlet flows between almost zero flow and 2.55 m^3s^{-1}.

The pump combinations indicated below give nominal delivery flow as shown:-

Pump Combination + indicates OR . indicates AND	Nominal Delivery Flow m^3s^{-1}
1+2	0.425
(1.2)* + 3 + 4	0.85
(1+2) . (3 + 4)	1.275
(1.2.3)* + (1.2.4)* + (3.4)	1.7
(1+2) . 3 . 4	2.125
1.2.3.4	2.55

Table 2

In all, fourteen combinations are possible to give 7 delivery flow states from 0 to 2.55 m^3s^{-1} although those combinations marked with an asterisk are undesirable for continuous operation as they are potentially less energy efficient.

The volume of the pump sump and adjacent culvert was taken off drawings and is shown in fig. 11. The volume contained between example levels h_2 and h_3 is 16 m^3, and therefore the time taken to fill and empty this volume can be calculated, and plotted in Fig. 11. Thus by timing increasing or decreasing level between h_2 and h_3, the rate at which flow into the sump is increasing or decreasing relative to pumping rate can be deduced.

By comparing this time taken for the level to rise and fall between two levels, with time taken by the various combinations of pumps to deliver the same volume of liquid, it is possible to switch pumps on and off to counteract the tendency to rise and fall. Above the 2.1 m level, the effective volume increases considerably. This is the level of the invert of the Northern Stormwater Interceptor and it is the intention to raise level h_3 following trials to provide necessary damping action on the control system produced by the volume of liquid in this culvert. Fig. 13 shows the flow diagram of the pump control algorithm.

To illustrate the sequence of operation, with an inlet flow condition of 1.0 m^3s^{-1} present and one of the larger pair of pumps running together with one of the smaller pair of pumps, the nominal pumping rate is 1.275 m^3s^{-1}, so that the level in the sump will fall between h_3 and h_2 in approximately 58 seconds. On reaching level h_2, the time taken to fall between the two timing levels will be input to the algorithm, and in this example, the smaller pump will stop. The pumping rate will then be 0.85 m^3s^{-1} and the sump level will rise at a slow rate. The time taken to rise between levels h_2 and h_3 will be approximately 107 seconds. This will exceed the time setting TA to restart a small pump, and the pump will restart and the sequence will be repeated.

Variations in inlet flow will cause increasing or decreasing rates of change of levels which will be timed between h_2 and h_3 as in the above example, and pumps started and stopped as appropriate. The pump control logic ensures that operation of undesirable pump combinations detailed in Table 2 are minimised.

4.4 Station Shut Down Control Sequences

4.4.1 Station Shut Down on high screen differential level

The level measurement system described in section 4.2 provides a high screen differential level facility by means of a trip amplifier, connected such that the pump sump level signal is input as reference level, and the level upstream of the screen signal is input as variable. When the difference between the two signals is greater than 0.5 m, the trip amplifier provides a contact closure which is input to the programmable controller, and also initiates an alarm to Avonmouth Sewage Treatment Works.

The programmable controller then outputs commands to fully close both inlet penstocks. Those pumps which are running continue to do so until the low level h_1 is reached when all pumps are required to stop. The rake motor then is stopped by command from the programmable controller. With the pump sump at low level, manual clearance of the screen blockage is facilitated.

4.4.2 Station Control on high level Northern Stormwater Inceptor

In the event of a high level in the Northern Stormwater Interceptor being detected, the two inlet penstocks are required to close to 50% of their normally open position to reduce the flow into the pumping station. Should a further increase in level be detected, then the inlet penstocks will close fully, the station proceeding to shut down as above. The station is reinstated automatically in reverse sequence when the level in the Northern Stormwater Interceptor reduces to below the very high level and high level settings respectively.

4.4.3 Station Shut Down in the event of high level in the Main Trunk Sewer

In these conditions, the main trunk sewer will overflow into the Northern Stormwater Culvert, which may itself be at high level. The station in these circumstances serves no useful purpose and both inlet penstocks will fully close, the station pumping down to level h_1 as before. Flaps over the pump discharge branches prevent backfilling of the pump sump from the trunk sewer. The pumping station will reinstate when the main trunk sewer level has fallen sufficiently to allow normal pumping to be resumed.

4.5 Override Control Facilities via the Telemetry System

Two override facilities are provided, controlled from Avonmouth Sewage Treatment Works via the telemetry system.

(i) Individual pumps may be excluded from the automatic control sequence.

(ii) The position of the two inlet penstocks may be adjusted by means of a desired percentage open (0-100%) signal transmitted via the telemetry from Avonmouth Sewage Treatment Works. To save the costs of replacing the penstock actuators, potentiometers were fitted to monitor penstock positions and to provide feedback to the penstock position control system. The control system was designed in the conventional manner to move the actuator motor in the correct sense to give zero offset between desired position and measured position.

By remote positioning of the penstock with reference to data received at Avonmouth from the telemetry system, some adjustment of inlet flow is possible to allow pumping to continue when the Northern Stormwater Interceptor approaches or is above the very high level shut down point, providing that the main trunk sewer has sufficient spare volume.

4.6 System Design and Software

The detailed design of the system was undertaken by staff of the Mechanical and Electrical Engineering Section in Bristol Avon Division, including detailed panel layout and cabling diagrams for the station modifications.

The programmable controller software was developed in conjunction with the hardware by the same engineers, and loaded and tested off line in an identical unit to that installed at the station. Provision was made in the program to damp inputs likely to suffer from transient effects, for example ripple in liquid surfaces.

4.7 Cost Benefit Analysis

The capital cost of the hardware, including installation and commissioning for the system described at Blackrock Pumping Station were approximately £12,000. The associated design and software costs were £3,000.

The annual savings expected are almost entirely due to elimination of shift visits, the station now being monitored by the telemetry system and visited routinely once per week. Savings in electricity charges are anticipated but are difficult to quantify by prediction.

Savings in shift visits alone are estimated at £15,000 per annum, and even ignoring electricity savings, the payback period is approximately ten months.

5. CONCLUSIONS

Comparison of the two stations shows that modern technology can substantially reduce the operational involvement at sewage pumping stations with a corresponding increase in overall management efficiency.

The Weymouth station allowed for a fully automated system to be developed including prediction routines to minimise energy consumption. To maximise the benefits of this approach it is necessary to have a comprehensive understanding of the station catchment and to provide full instrumentation coupled with efficient data retrieval. The processor required for this application will therefore be several orders larger than that required for the simple event driven system installed at Blackrocks Pumping Station.

The Blackrocks Pumping Station scheme shows that a considerable improvement over traditional pump control systems can be achieved at moderate cost by the utilisation of a comparatively small capacity processor. This system, however, does not fully realise the potential savings in electricity charges arising from the use of predictive techniques coupled with maximum usage of storage available in the sewerage system.

6. ACKNOWLEDGEMENTS

We are grateful to the Directors of the Wessex Water Authority for their permission to publish this paper. Particular acnowledgement must be made to Colin Martin of L G Mouchel and Partners who has progressed the development of the Radipole Lake Pumping Station Computer Control System from inception to completion, and to all fellow engineers in the Authority, the Weymouth and Portland Borough Council and L G Mouchel and Partners associated with the Weymouth project.

7. REFERENCES

(1) Martin C and King D "Control of a Pumping Station by Mini Computer"

"Pumps 79" Sixth Technical Conference of the British Pump Manufacturers Association Canterbury 28-30 March 1979

(2) Martin C and King D "Control of a Pumping Station by Mini Computer - A Progress Report"

"Pumps - The Developing Needs" Seventh Technical Conference of the British Pump Manufacturers Association York 31 March - 2 April 1981

(3) Evans G P "Optimisation of Sewage Pumping TR170" WRC Environmental

(4) Pittman J F H, Martin C and King M W "Weymouth and Portland Marine Treatment Scheme - Inland Sewerage Works" Proceedings Institution of Civil Engineers Part 1 Design and Construction February 1984 Volume 76 page 95

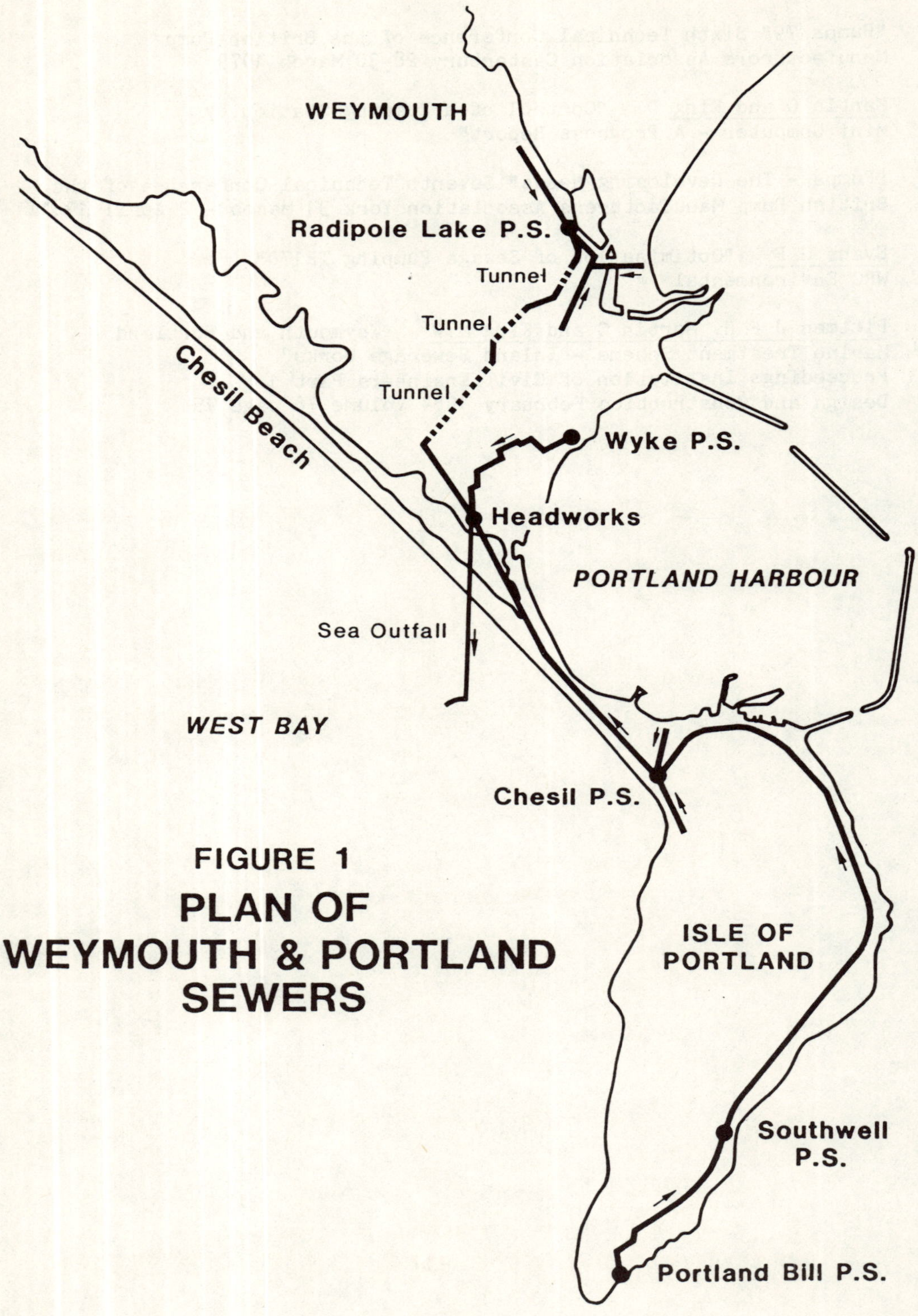

FIGURE 1

PLAN OF WEYMOUTH & PORTLAND SEWERS

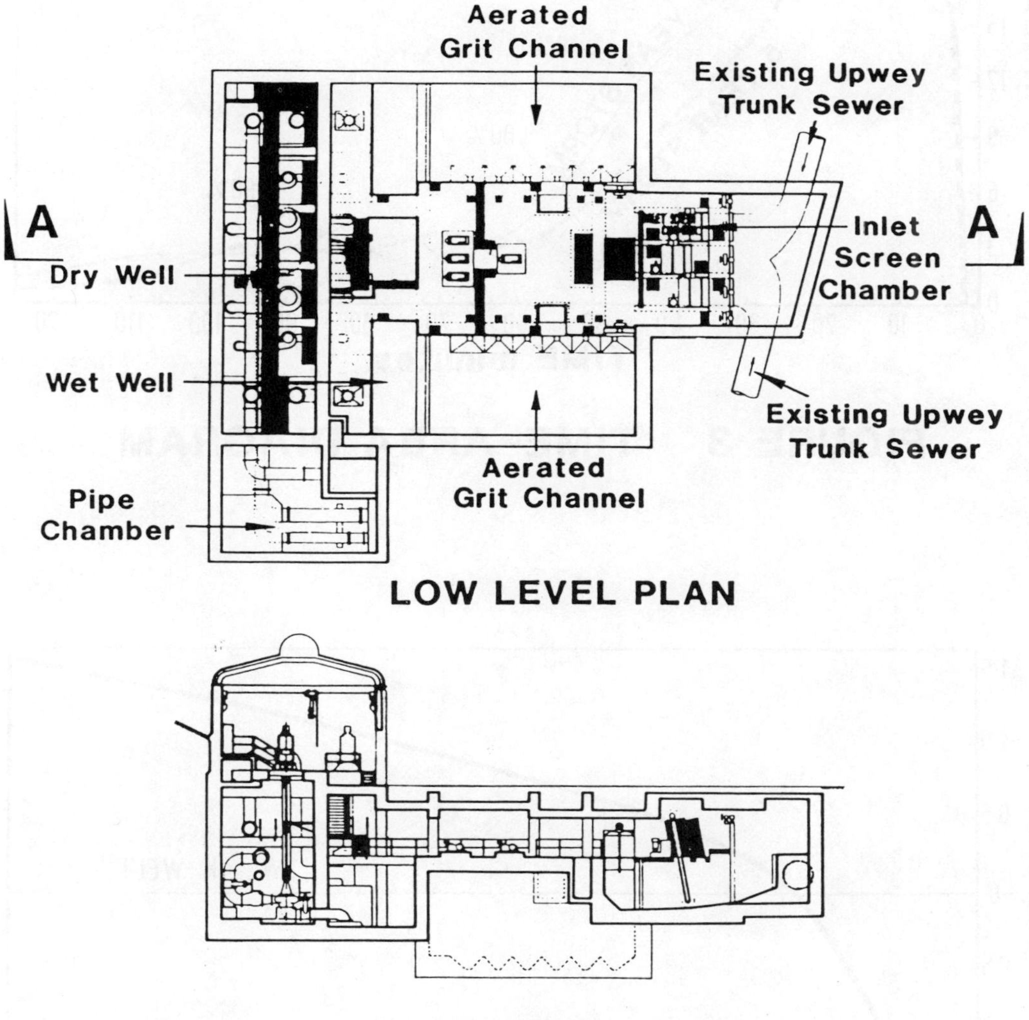

FIGURE 2
RADIPOLE LAKE PUMPING STATION

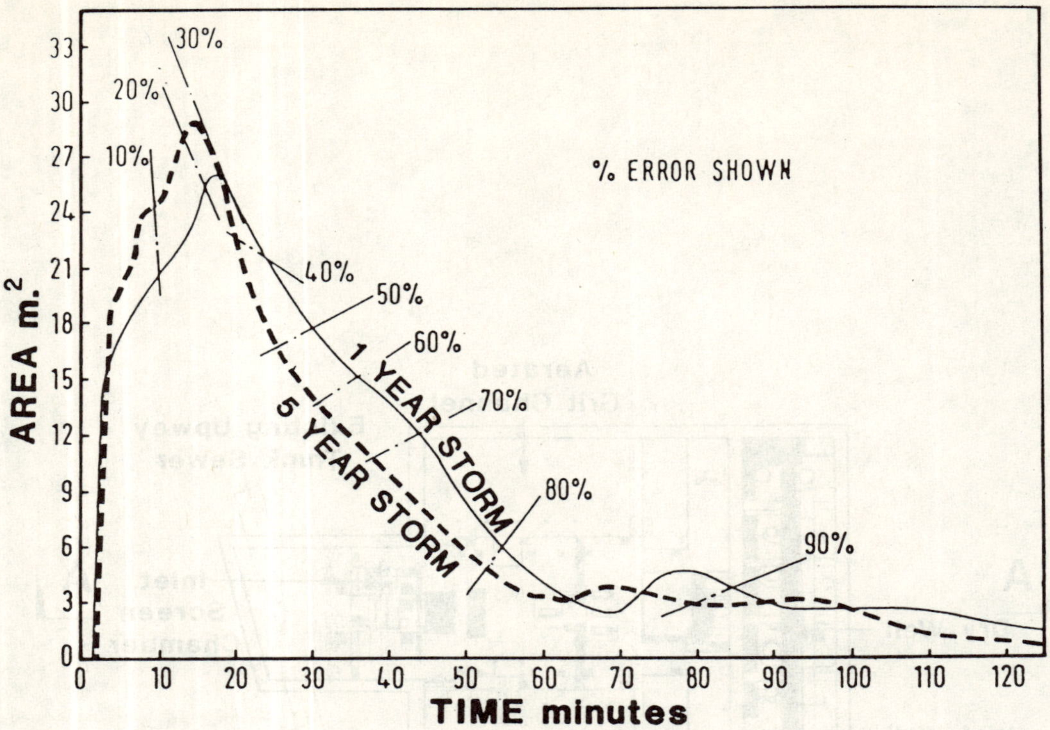

FIGURE 3 TIME-AREA DIAGRAM

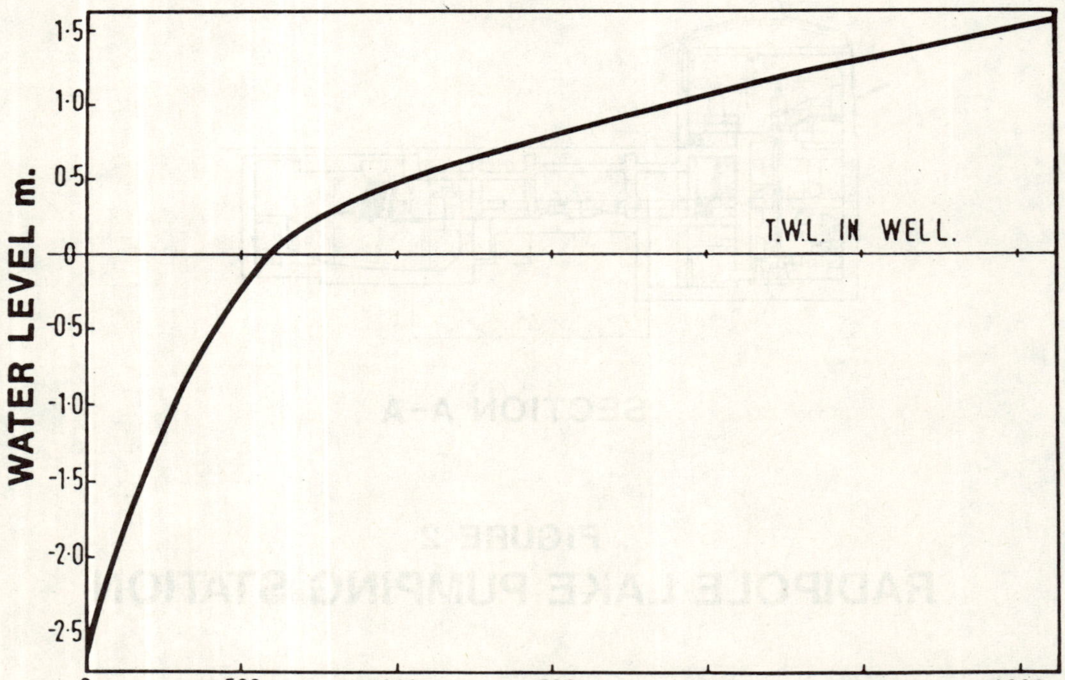

FIGURE 4 DEPTH-STORAGE CURVE

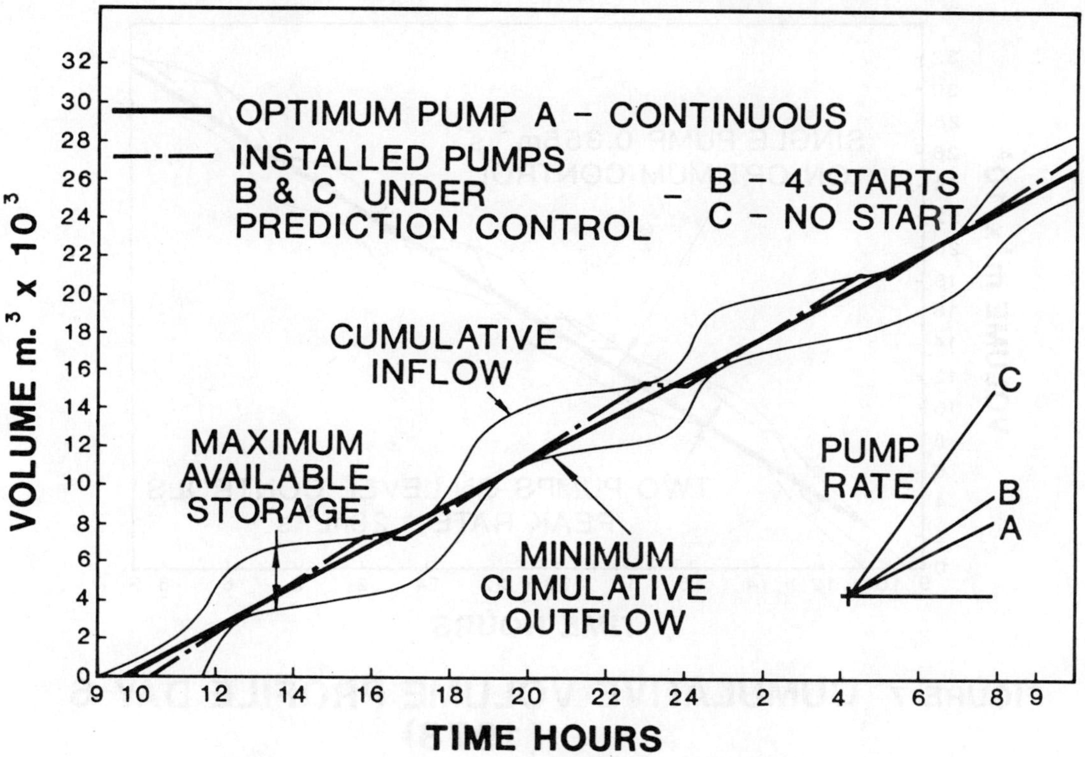

FIGURE 5 **INFLOW STORAGE OUTPUT CURVE**

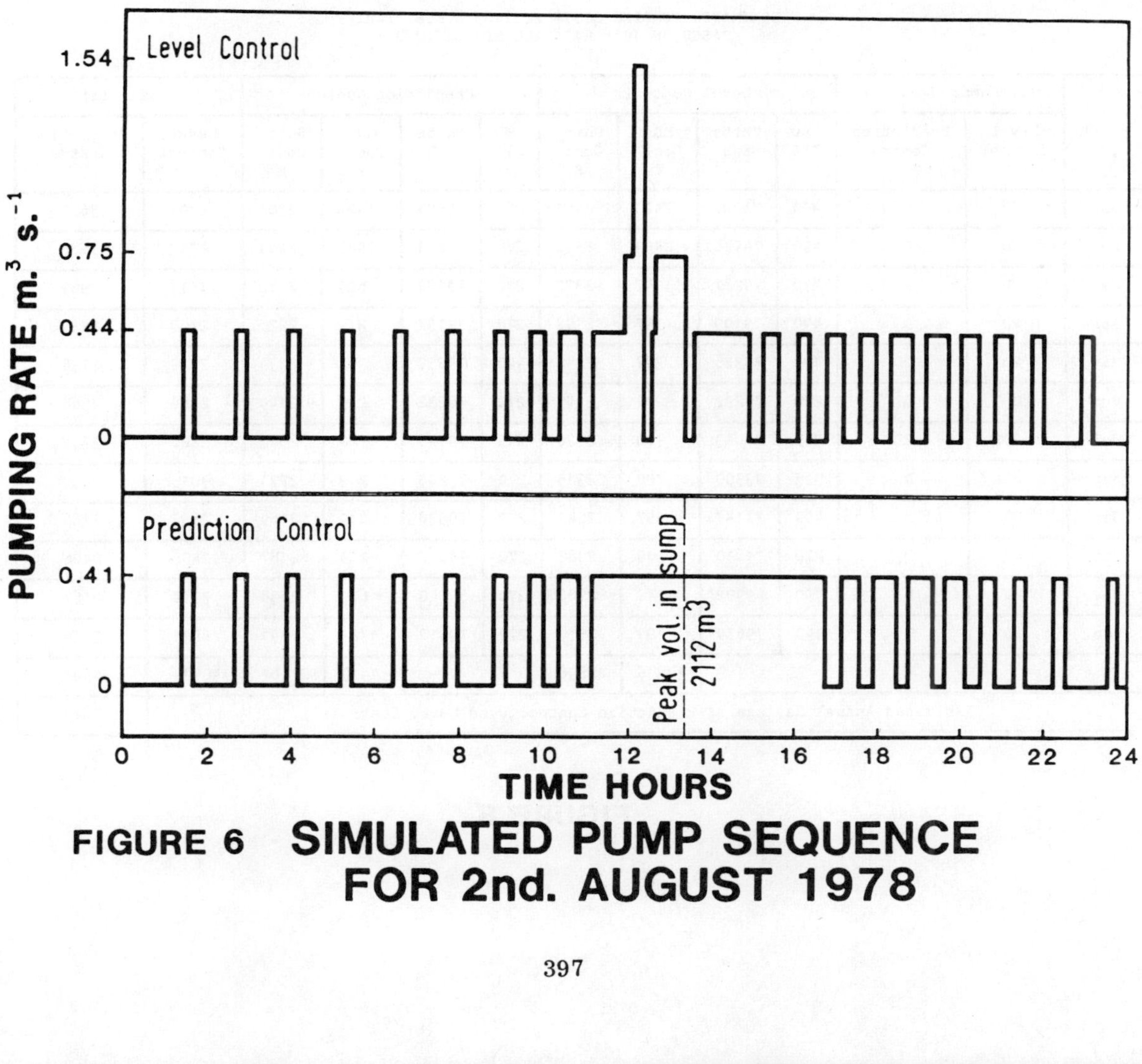

FIGURE 6 **SIMULATED PUMP SEQUENCE FOR 2nd. AUGUST 1978**

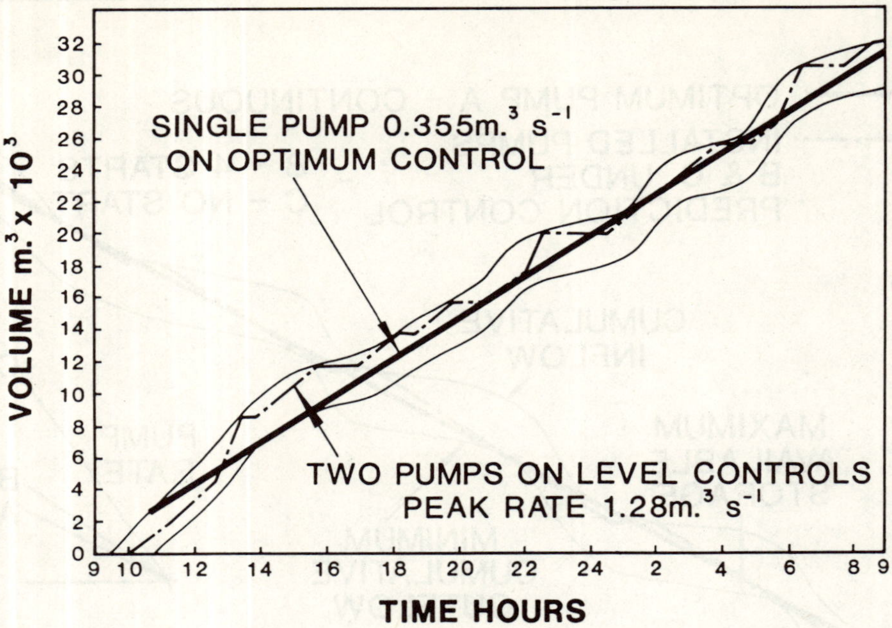

FIGURE 7 CUMULATIVE VOLUME PROFILE DAY 6 (1978)

ESTIMATED PUMPING COSTS AT 1983 S.E.B TARIFF
BASED ON 1978 RAINFALL SIMULATIONS

Month	Pumps Used		Level Control				Prediction Control				Total Cost	
	Level Control	Prediction Control	MD KVA	Units KWH	MD Cost £	Unit Cost £	MD KVA	Units KWH	MD Cost £	Unit Cost £	Level Control	Prediction Control
Jan	2D	D	490	76228	2401	2401	270	75904	1323	2306	4802	3629
Feb	2D	D	490	74786	2401	2356	270	74241	1323	2260	4757	3583
Mar	2D	D	490	75287	1142	2372	270	74971	629	2280	3514	2909
Apr	2D	D	490	73109	397	2303	270	72797	219	2220	2700	2439
May	2D	2D	490	74326	397	2342	490	74259	397	2339	2739	2736
Jun	D	D	270	70232	219	2149	270	70232	219	2149	2368	2368
Jul	2DS	2DS	915	80013	700	2520	915	79746	700	2512	3220	3212
Aug	2DS	D	915	73500	700	2315	270	73206	219	2231	3015	2450
Sep	2D	D	490	71147	397	2241	270	70938	219	2169	2638	2388
Oct	D	D	270	71610	219	2187	270	71610	219	2187	2406	2406
Nov	2D	D	490	69902	1142	2202	270	69840	629	2138	3344	2767
Dec	2D	D	490	79679	2401	2510	270	79010	1323	2391	4911	3714
					12516	27898			7419	27182	40414	34601
Estimated Annual Savings of Prediction Control over Level Controls												5813

FIGURE 8

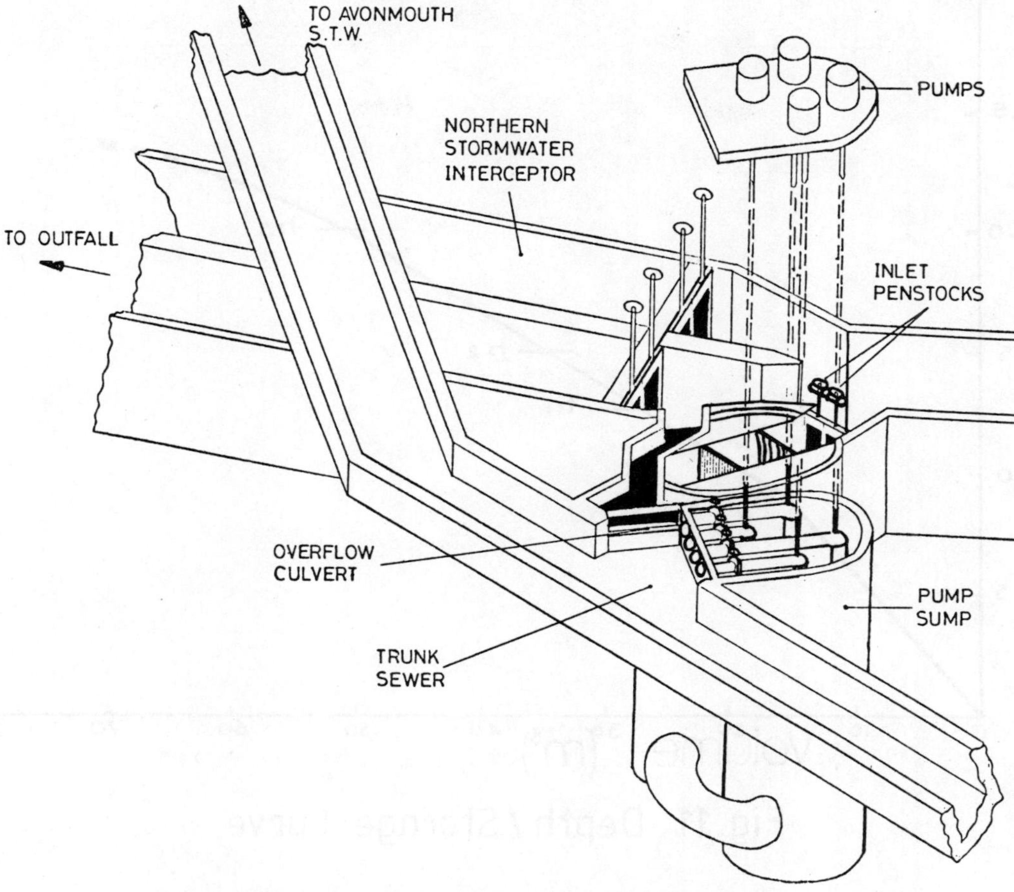

Fig. 9 SIMPLIFIED VIEW - BLACKROCKS P. STN.

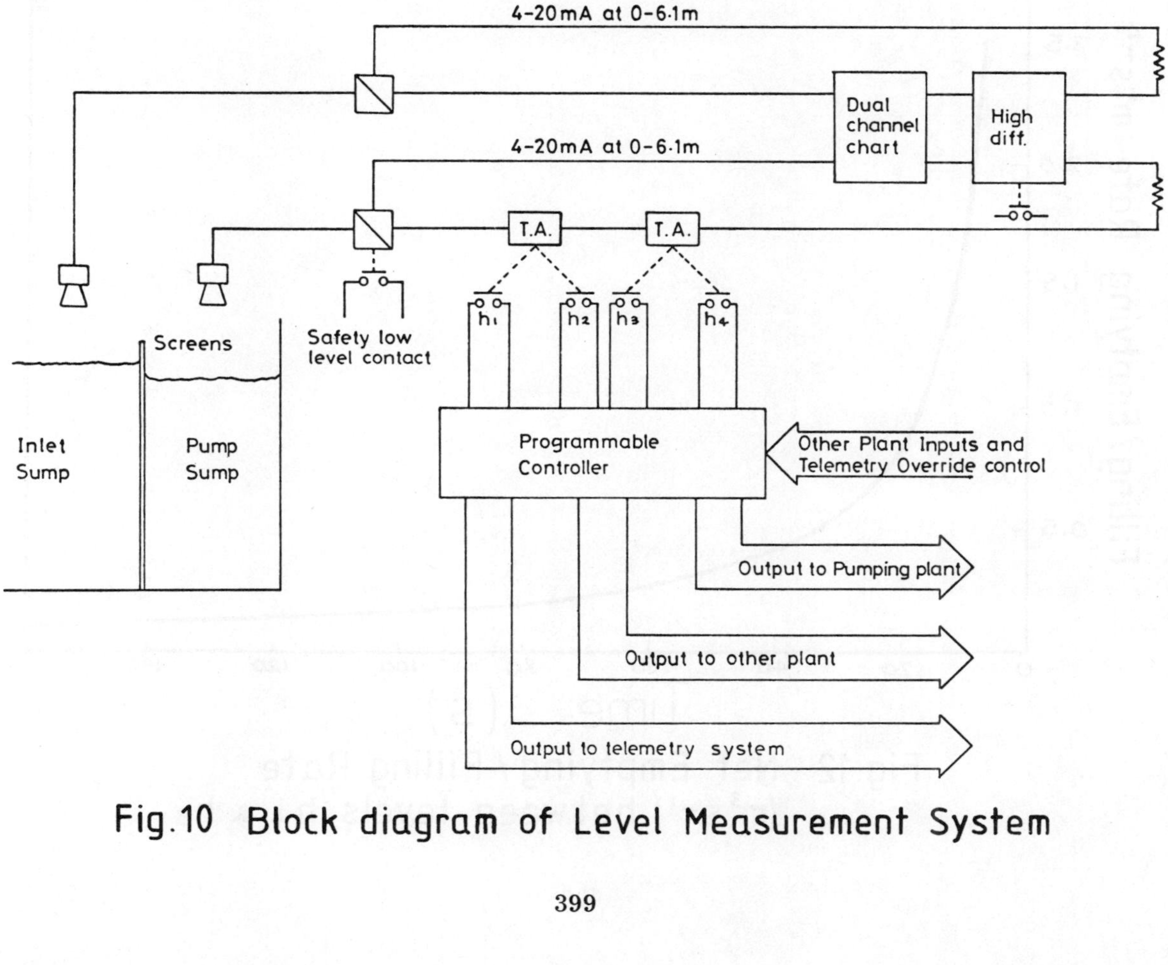

Fig. 10 Block diagram of Level Measurement System

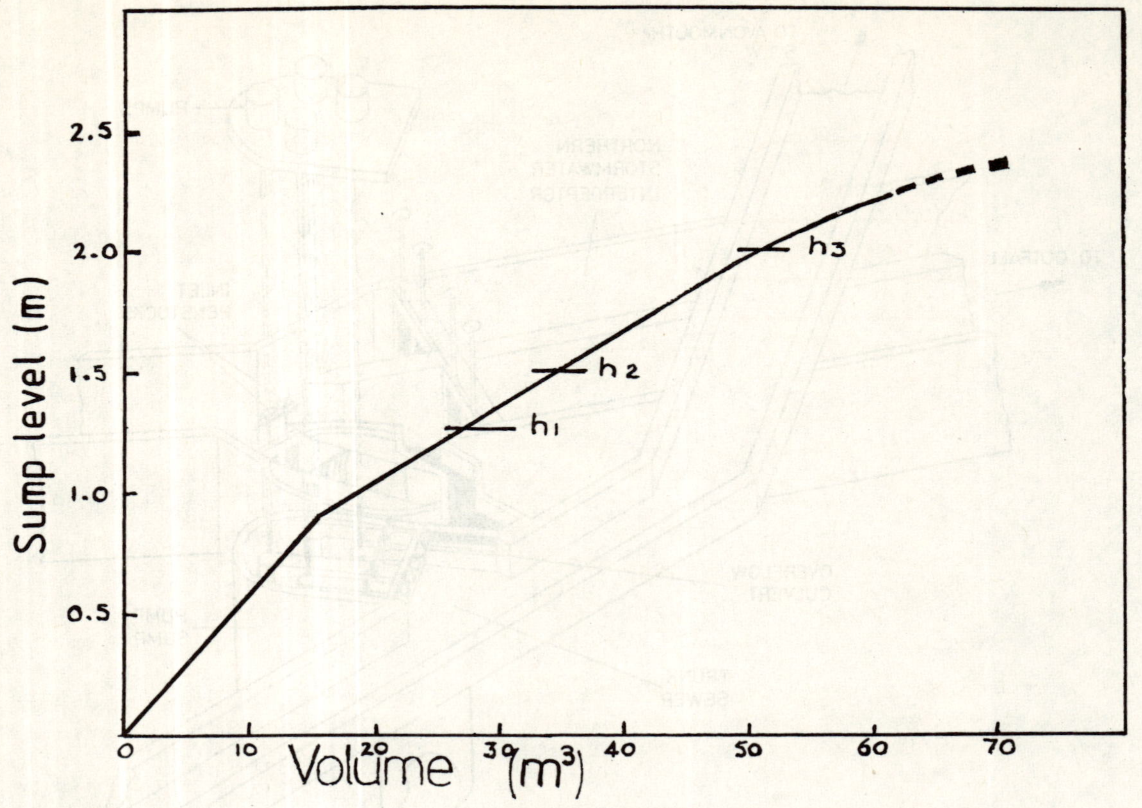

Fig. 11 Depth / Storage Curve

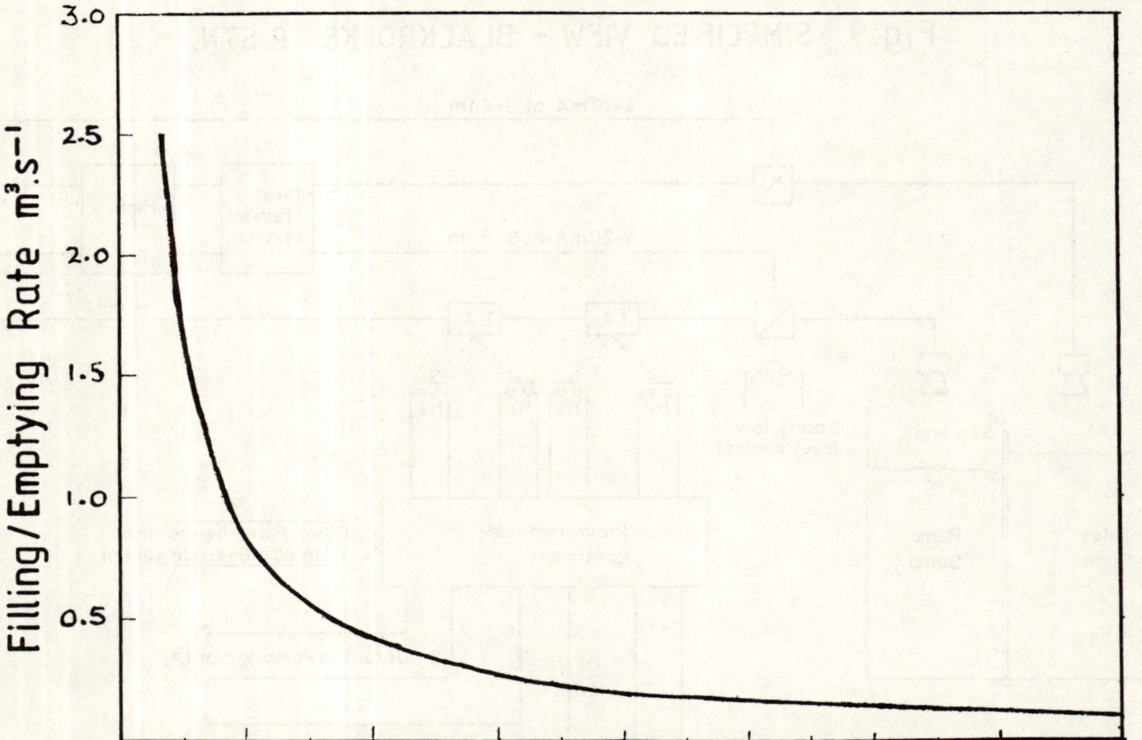

Fig. 12 Net Emptying / Filling Rate ($m^3.s^{-1}$) between levels h_2 & h_3

Sht. 1

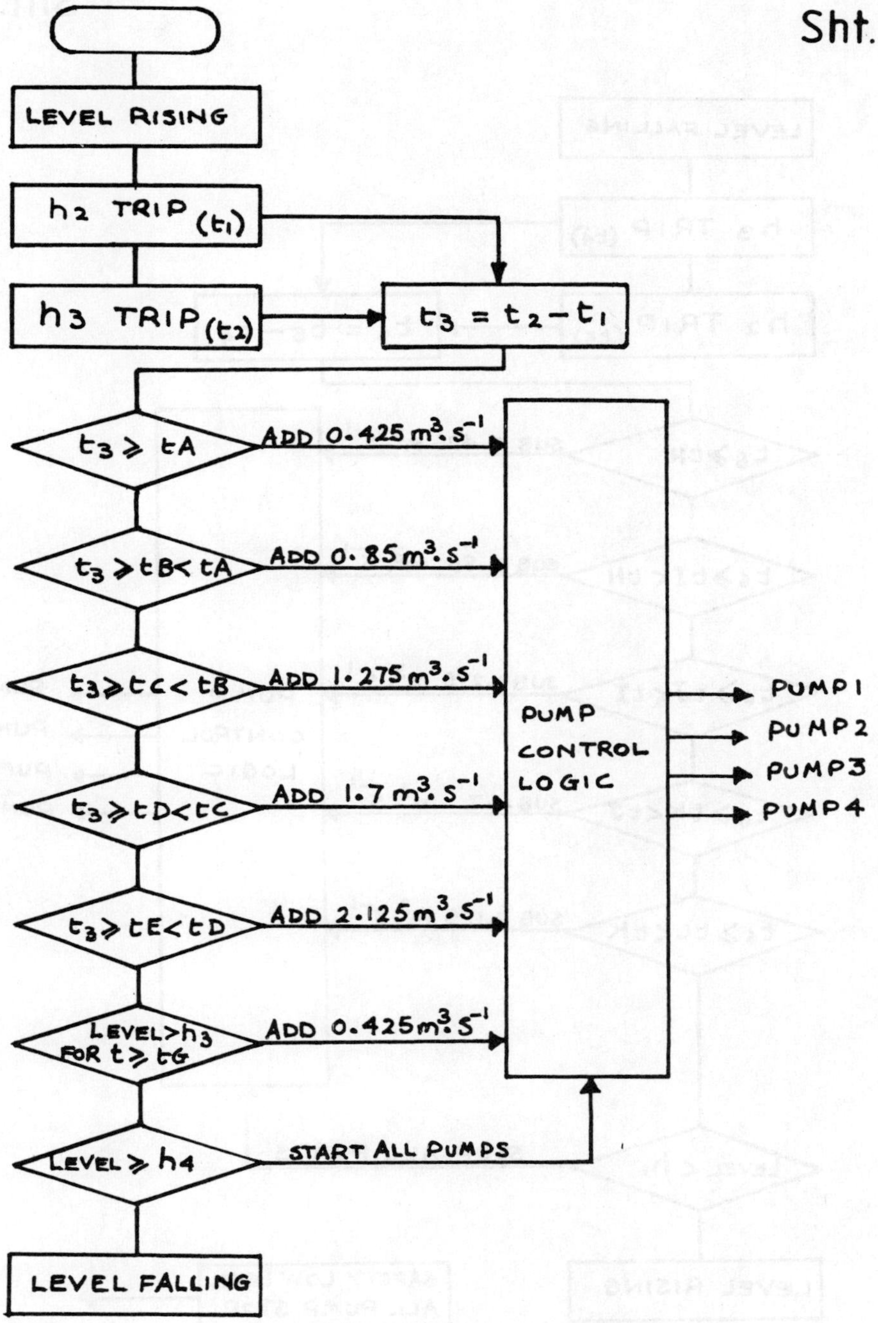

Fig. 13 PUMP CONTROL FLOW DIAG.

Sht. 2

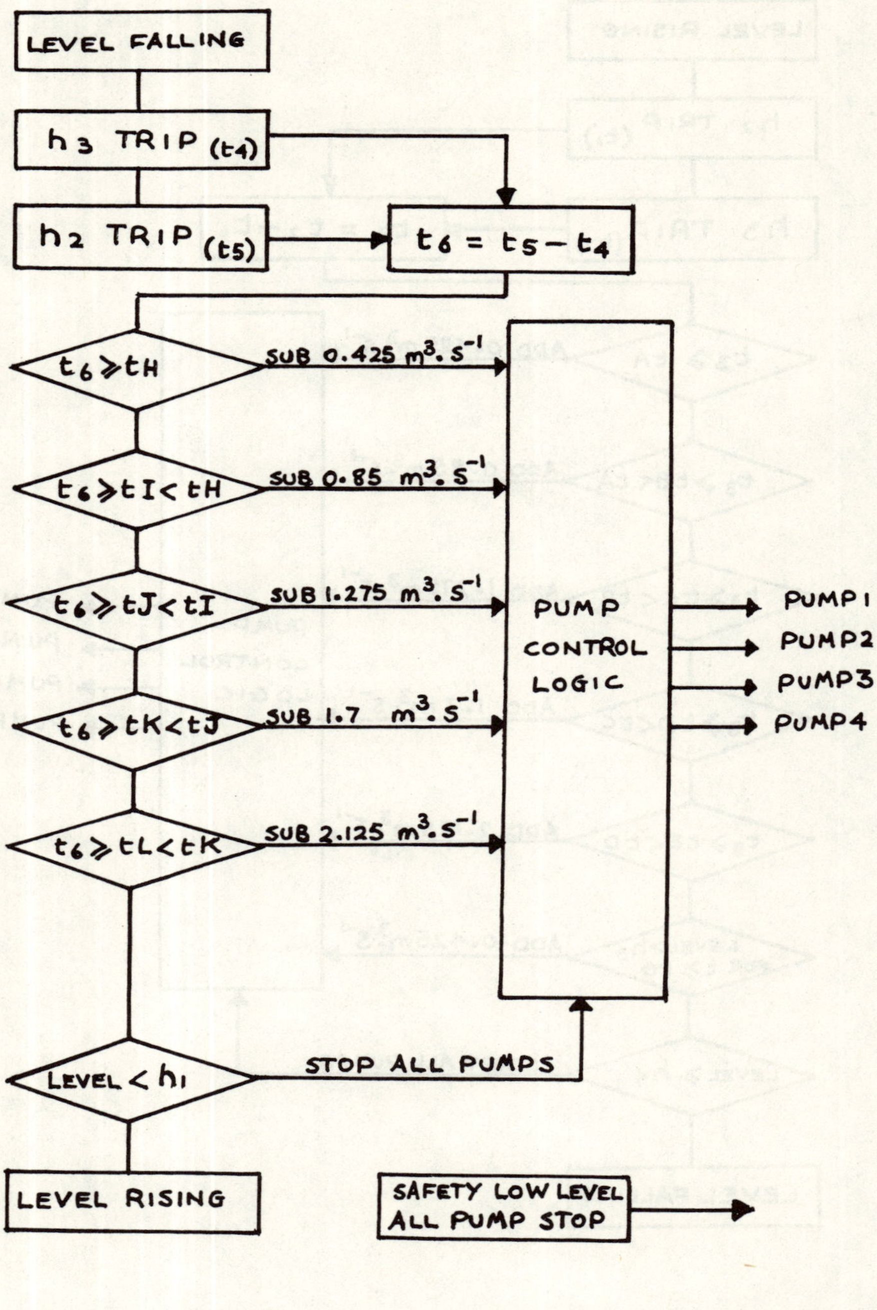

Fig.13 PUMP CONTROL FLOW DIAG.

International Conference on the
PLANNING, CONSTRUCTION, MAINTENANCE & OPERATION OF SEWERAGE SYSTEMS
12-14 September, 1984

PAPER J4

CONTROL AND OPERATION OF SEWERAGE SYSTEMS
PROBLEMS OF INSTRUMENTATION AND MEASUREMENT

Y. ZECH and C. DELMEE-WILLEMSENS, Department of Civil Engineering
J. BATARDY and E. PERSOONS, Department of Agricultural Engineering

Catholic University of Louvain
B-1348 Louvain-la-Neuve
Belgium

Summary

The construction of the new town of Louvain-la-Neuve (catchment area : 200 ha) gave the opportunity to develop a measurement network in the field of urban hydrology and hydraulics. The aim of this measurement network was to improve the knowledge of storm sewer flow behaviour. It rapidly appeared that the measurement itself was an interesting topic for research.

Hydrological measurement consists mainly of a network of eleven tipping-bucket raingages connected to a central computer which verifies and records changes in bucket position at one second intervals. The high measurement density permits a study of the spatial distribution of rainfall and its evolution.

For hydraulic aspects a 16.5 ha district of the town was selected. Some stormwater sewer sections were equipped for water depth and discharge measurement. It appeared that classical measurements made in one manhole or in one pipe section are generally not significant because of varied flow and transient phenomena. Some solutions are suggested.

Mathematical modelling was also tested with the aim of forecasting flood routing in the network from local measurement.

Held at The University of Reading, England. Organised and sponsored by BHRA, The Fluid Engineering Centre, Cranfield, Bedford in conjunction with the Water Research Centre.

NOMENCLATURE

g	= acceleration of gravity (9.81 m s^{-1})
n	= Manning's roughness coefficient ($m^{-1/3}$ s)
x	= longitudinal abscissa (m)
y	= water depth (m)
A	= wetted area (m^2)
Ic	= computed rainfall intensity (mm/h)
Ie	= calibration rainfall intensity (mm/h)
Ir	= recorded rainfall intensity (mm/h)
Iv	= actual rainfall intensity (mm/h)
L	= surface width (m)
Q	= discharge (m^3/s or l/s)
R	= hydraulic radius (m)
So	= bottom slope

1. INTRODUCTION

In Belgium, as in several other countries, the problem of operation of urban drainage system is more and more difficult. Urbanization has modified the storm water inflows so that overflows are more frequent. Also sewer networks have become more complex, due to underground works (underground railways, parkings, etc...), development of new urban districts and rehabilitation of old ones. Sometimes the exact lay-out of the network is not completely known.

On the other hand, the creation of many new wastewater treatment plants needs better knowledge of the hydraulic behaviour of sewerage systems. The optimal operation of these plants is based on a suitable forecasting of the network outflow.

These two exigencies - avoiding overflows and optimazing water treatment plants - lead to the need of appropriate measurements and correct interpretation of these in such a way that a numerical modelling may be used to determine the outflow from the measured data.

The construction of the new town of Louvain-la-Neuve gave the opportunity to study problems of measurements and modelling sewer flows. The new town, planned for a population of 30000 inhabitants has now reached 14000. The catchment was devoted to agriculture before the creation of the town. The outlet of the water shed is a small river with a natural discharge not exceeding 1 m3/s. The urbanization will strongly increase this outflow (to 20 m3/s for a 10-year storm), leading to the necessity of designing a storage lake at the outlet of the watershed. This artificial reservoir will be used as a permanent recreation area. The sewerage system of the new town is thus a separate one : a storm water network supplies the lake with rain-water while waste water is collected to a treatment plant.

The creation of a separate system in a new urban area, together with the development of a dense telephonic network, gives the opportunity of studying storm water runoff in the best conditions.

The goal of this paper is to present our experience in the design and operation of a measurement network in storm sewers, and to suggest possible ways of using these measures for improving the drainage system operation. The measurement facilities designed for research about storm water flow were in themselves a very interesting research topic.

2. RAINFALL MEASUREMENT

2.1 Network installation

The rainfall measurement network installed in the watershed of Louvain-la-Neuve was conceived in order to estimate the influence of the spatial variability of heterogeneous storms on the characteristics of the hydrographs recorded on a small urban catchment.

The installation conditions of the raingages were defined in order to make an objective comparison between the data of the different measurement stations. These conditions are especially important in this kind of basin as the distance between the raingages and the response time are short.

In addition to the classical criteria involved in the installation of the raingages, other constraints have been considered to define the different measurement sites (distance between the raingages and the buildings, proximity of telephone lines for the information transmission, uniform repartition in the watershed, protection to avoid voluntary damages).
Fig. 1 represents a scheme of the urban catchment of Louvain-la-Neuve and its rainfall measurement network which includes 11 raingages.

2.2 The raingages

2.2.1 Description

The raingages are all of the tipping-bucket type. The receiving surface is 200 cm2 and the nominal volume required to tip the bucket is 2 cm3 (raingage EDF) (Ref. 1). One information can therefore be recorded for each 0.1 mm of rainfall.

2.2.2 Calibration of the tipping bucket raingages

The calibration of this kind of raingage can be made in such a way as to eliminate the error due to the loss of water during the tipping of the bucket (Ref. 2 and 3). Within the framework of our study, a specific calibration curve for each raingage has been defined through the relationship between the actual intensity (Iv) and the recorded intensity (Ir) for a given calibration intensity (Ie). Several forms of adjustments have been tested in order to describe parametrically these curves. Below a recorded intensity Ir=45mm/h, the function Iv=F(Ir) remains linear (Ref. 4). Above this value, a loss occurs. For this second part of the curve, a polynomial adjustment (second order)is preferred to other functions on the base of the analysis of the correlation coefficients and standard deviation values obtained between Iv and Ic (calculated intensity from the function parameters).

2.3 Transmission of the informations

Each raingage is connected to the control center through an open telephone line. A control panel allows the engineer to see the functioning of the system in real time on an aerial photograph. The tipping of the bucket induces, through a magnet and a relay, an impulse at the extremity of the line. The impulse is sent to the computer and to the control panel where a Light Emitting Diode for each raingage simulates each tipping.It is so possible to control the working of the raingages during a rainfall event. Another LED for each station allows the engineer to verify the state of the transmission lines when it is not raining.

The computer contains an interface for 16 inputs. Each of them corresponds to one bit of its memory. The raingages are connected to the different inputs so that each tipping modifies the state of the bit corresponding to its raingage.

2.4 Storage of the informations

A time base of one second is generated in the control panel and is introduced in the computer through one of the inputs of the interface. The program stores the initialization date and time close to one second. At each second, the program scans the state of the bits corresponding to the different raingages and compares it with the state of the preceding second. If one of the bits is modified, the program stores the data on the following form : the new state of all the bits (two bytes) and the number of seconds elapsed since the last change. When the value of the time counter reaches 255 (no tipping during 255 second), the state of the raingages (2 bytes) and the value of the time counter (255 s, 1 byte) are stored. Then, the time counter in initialized again. The state of all the raingages must be stored because several buckets may tip during the same second : this procedure of storage allows then to record data in the same format.

When one page of the memory is filled, the data are transfered on magnetic disk.

Several information concerning the state of the raingages and the raingages and the number of stored data may be visualized on the screen of the computer.

2.5 Example of result

Concerning the rainfall-runoff relation problem of an urban catchment, which is characterized by a short concentration time, Fig. 2 illustrates the importance of the rainfall spatial variability on short time scale, especially for the convective storms. This example represents two hyetographs recorded during the same storm (time scale identical in relative value) by two raingages located in the rainfall progression axis and distant of 1.5 km. The total amount of water for the whole rainfall was similar (raingage 1 : 11.2 mm - raingage 7 : 11.8 mm). Nevertheless, the internal structure of the two hyetographs is appreciably different : indeed, the two peaks of the rainfall were recorded at raingage 1 with a delay of three minutes as compared to raingage 7.

3. HYDRAULIC MEASUREMENT

The Lauzelle district of Louvain-la-Neuve was chosen for particular study of storm flow and its correlation with rainfall intensity. This district presents a water shed of 19.5 ha for a future population of 3000 inhabitants. Fig. 4 shows the storm sewers network. Three typical sections are selected for measurement : sections P84' - P23 (circular pipe, Ø 50 cm), P27 - P28 (Ø 80 cm) and P28 - P 29 (Ø 90 cm). The measured parameters are the water depth and the discharge.

3.1 Problems of measurement

Two ways may be used to measuring the discharge in a sewer pipe :

- Indirect measuring from the water depth
- Direct measuring from the water depth and the velocity

3.1.1 Measuring discharge from water depth

For the measurement of the water level, various instruments may be used :

- Resistance gauge : the electrical conductivity between a pair of stainless steel wires is related to their depth of submersion. Such a gauge is not suitable for sewer water because of problems of corrosion.

- Capacitance gauges : the electric capacitance also depends on the depth of submersion. These instruments are better protected against corrosion and are able to measure rapidly fluctuating water levels. The measurement may be located in a manhole or in a pipe.

- Ultrasonic system measuring the return time of an ultrasonic signal reflected by the water surface. This kind of instrument is easy to use but a minimum distance is required between the transmitter-recorder and the water surface : the measurement is thus located in a manhole.

- Bubbler pressure gauges measuring the air pressure in a air tube outlet at the bottom of the sewer pipe. These instruments are selfcleansing if the measurement is continuous and the air outlet always submerged. But the auxiliary air pump needs electric power and maintenance.

- Membrane manometers where the elastic deformation of the membrane is transformed to changes in electric resistance, capacitance or induction. Sewer water may cause some corrosion problems to such a tranducer but constant improvement is obtained in this way.

- For large collectors or manholes, it is possible to adapt the geometrical shape to form a kind of Venturi or Parshall flume. These special sections must be located in such a way that the conveyance of the sewer pipe does not too much decrease.

The principle of such a measurement is based on a permanent relation between local water depth and discharge. However, this steady relation is seldom effective. Fig. 5 shows the computed water profile in the sewer network for an exceptional rainfall intensity of 100 mm/h (10-year intensity corresponding to the water shed concentration time is 45 mm/h). The computation is based on simplified hypotheses (Ref. 5 and 6) : steady flow neglecting concentration time, storage effects and manhole head losses. In spite of these simplifications, it clearly appears that uniform flow seldom occurs and that control sections also are not frequent in an ordinary sewer network, not especially designed for measurement.

Another problem is the occurrence of hydraulic jumps which may introduce a double possibility of water depth (sub- or supercritical flow) for the same discharge. Water level measurement in the manholes is still a worse solution : the velocity field is very irregular into a manhole and a part of the kinetic energy may be transformed to level rising due to local expansion.

The measurement of local water depth for determining the discharge is thus not really suitable although often used. Except in the case of new network where control section may be designed, a constant relation between water depth and discharge may seldom be found.

The above objections are still stronger when transient phenomena are taken into account. Fig. 6 summarizes experimental results of laboratory investigations on a physical model of sewer with three pipe sections separated by two manholes (Ref. 7). The aim of this investigation was the study of hydraulic instabilities. The diagram of Fig. 6 gives possible variations of headwater depth and tailwater depth for a given discharge. In this particular example, discharges till 15 l/s ($Q/D^{5/2} < 1.90$) present no special behaviour. However for greater discharges the curves present two branches depending on the evolution of tailwater depth : if tailwater depth is increasing we successively observe the following situations : no jump (part A'B' of the curves), jump in the middle section (part B'C'), jump in the upstream section (part C'D') which progressively moves in the upstream direction leading to full flow (part D'G) ; for decreasing tailwater level no jump can appear due to a lack of upstream aeration (part D'H). In fact presence of incidental obstacle in sewer may cause a change in type of flow leading to instabilities with pressure fluctuation and appearance of air pockets. These pulsating phenomena make a suitable discharge measurement impossible.

3.1.2 Measuring discharge from water depth and velocity

In the frequent case where no control section may be found in the sewer network, it is more accurate to have a simultaneous measurement of depth and velocity.

A functional block diagram of a typical combined flowmeter is shown in Fig. 7. This instrument consists in a bubbler pressure gauge coupled with an electromagnetic velocity sensor (Marsch - Mc Birney model 265). As can be seen this sensor measures the local velocity close to the bottom of the pipe. If low water level occurs, a predetermined velocity is assumed, otherwise the local velocity is multiplied by a correction factor to represent the mean velocity used in the continuity equation for computing the discharge.

Some problems are met with this kind of measurement. The first is the analogous computation of the wetted area from the water depth and of the discharge from this area and the mean velocity : this procedure is relatively slow compared with numerical computation, so that it is often more accurate to pick up the primary electrical signals before treating them numerically. A more difficult problem is the correction of the local velocity for computing the mean velocity. It is suggested to calibrate the instrument by explorating the velocity field for some typical discharges. But this calibration is only possible in a manhole when the instrument is located into the pipe far enough from the manhole to avoid excessive turbulence.

Some of the former objections described in the case of water depth measurement are also valid for the simultaneous measurement of depth and velocity. So the transient phenomena, and above all the flow instabilities strongly disturbe the velocity field and thus the relation between local and mean velocities.

3.2 Hydraulic instrumentation of the Lauzelle district

According to above considerations, following solutions were adopted for the three measurement sections :

- For the 50 cm diameter pipe P84' - P23, measurement of water depth in two sections, using a capacitance gauge.

- For the 80 cm diameter pipe P27 - P18, measurement of water depth in three sections, using a capacitance gauge.

- For the 90 cm diameter pipe P28 - P29, just downstream the pipe P27 - P28 and presenting the same discharge because there is no lateral junction in manhole P28, measurement of water detph in one section, using a capacitance gauge, and measurement of discharge by the above described flowmeter.

The used capacitance gauge is a prototype designed by the civil engineering laboratory (Laboratoire du Génie Civil LGC) of the University. The classical single capacitance is replaced by a sequence of small capacitances corresponding to a level increment of 2 cm. Each of these cells are connected to a processor which may detect whether a cell is dry, submerged or only wet. Permanent scanning allows not only determining water level but also detecting eventual misworking of the instrument. Fig. 8 shows a comparison between theoretical and measured water level for relatively fast variations : for rising water level the measure is almost perfect ; for decreasing water level, a small delay appears due to temporary persistence of water film on the instrument surface.

The choice of a double (pipe P84'-P23) or triple (pipe P27-P28) measurement of water depth is a trial to avoid the difficulty of finding control sections in ordinary sewer networks. In principe, the discharge may be derived from two water levels ; discharge and roughness coefficient from three water depths according to water profile equation (Ref. 5) :

$$\frac{dy}{dx} = \frac{So - \dfrac{Q^2 n^2}{R^{4/3} A^2}}{1 - \dfrac{Q^2 L}{g A^3}}$$

where the wetted area A, the hydraulic radius R and the surface width L depend on the water depth y.

Till now our experience cannot be considered as decisive about the accuracy of such a measurement. It seems that an accurate knowledge of the water profile slope dy/dx requires more caution than a level measurement. Anyhow a double measurement of the water depth permits the verfication of whether the flow is not too different from a uniform one and also suppresses uncertain measurments.

The above equation is written for gradually varied flow in steady conditions. Derivation of the discharge from transient water depth is a not solved problem. Although immediate knowledge of the discharge variation is seldom needed for practical use, our investigation have shown that sudden change in water level may occur, of which recording and interpreting is useful for scientific purpose.

To complete the measurement network, a tringular weir and a Parshall flume are set at the outlet of the stormwater network of the town. This measurement is required to study rainfall-runoff relation at a greater scale (Fig. 1)

Another problem is that of power supply. As it can be seen in the sheme of Fig. 3, although 220 V alternating current was available close to the measurement site, it was preferred to insert a battery in the electric circuit for avoiding power failure which may occur during thunder-storm which are precisely the most interesting measurement time. However such a design needs a complementary device to transform direct to alternating current.

3.3 Signals transmission and recording

Signals transmission is realized by private telephonic network. Distances being relatively important (2,... 3 km), modem are used at the ends of the transmission lines.

Duplex lines are foreseen in such a way that the central computer is able to drive each measurement processor. So the hydraulic measurement is switched off as long as no rainfall is observed by the hydrological network. In this way, an electrical connection is actually set between the two central computers driving respectively hydrological and hydraulic measurement.

The transmission celerity is 1200 bauds, allowing a complete measurement cycle all five seconds.

Collected informations are treated and recorded into disk units and magnetic tapes.

3.4 Examples

Fig. 9 and 10 give examples of typical measurements. For a given pipe section, the following data may be recorded

1. Local velocity close to the bottom of the pipe (electromagnetic velocity sensor of the Marsch - Mc Birney flowmeter)
2. Mean velocity or "corrected velocity" computed from multiplying the local velocity by a correction factor. If the water depth is too small the measurement is not accurate and the "corrected velocity" is recorded as zero
3. Water depth from the bubbler gauge of the flowmeter
4. Water depth from the capacitante gauge
5. Discharge directly from the flowmeter
6. Discharge computed from the mean velocity (2) and the bubbler gauge water depth (3)
7. Discharge computed from the mean velocity (2) and the capacitance gauge water depth (4)
8. Discharge computed only from the capacitance gauge water depth, assuming a uniform flow

These eight data per section may be combined with rainfall data for graphical presentation.

Figure 9 depicts the measurements of a typical shower. Water depth from bubbler gauge (3) and from capacitance gauge (4) were sketched. The velocity curve represents corrected values (2) : velocities corresponding to small water depth are not taken into account. Discharge curves are computed from bubbler gauge water depth (type 6) and from capacitance gauge water depth (type 7)

It can be observed significant differences in water depth depending on the kind of instrument. Bubbler pressure gauge presents a certain delay above all after a dry period during which air pump is automatically stopped with possible appearance of condensation in the air outlet, leading to pressure alteration. The first peak of a rainfall may be so damped. The capacitance gauge seems to be more accurate above all for non continuous measurement. Nevertheless small discharge are not recorded because of the lack of suitable velocity measurement.

Figure 10 shows a trial to solve this problem. In this case besides the computation from the bubbler gauge water depth and the velocity (type 6), the discharge is also calculated from the water depth only, considering this as uniform depth (type 8).

Comparison between the two curves shows that the assumption of uniform depth leads to undervaluation of the actual discharge. Nevertheless no beter way was found for taking into account the smallest discharges. Anyhow these cases of small discharges are the least interesting for operation of sewerage systems.

On the other hand the end of the velocity curve shows that for small values of velocity the measurement is not more suitable.

The problems of small discharges will decrease when the urbanization of Lauzelle district is completed because discharge and water depth will have more significant value compared with possible conveyance of the pipe of which only a little part is now used.

3.5 Conclusion about hydraulic measurement

The necessity of ensuring suitable measures for scientific interpretation has led to examine thoroughly the principles of hydraulic measurement in a storm sewer network. The so-gained experience may be useful for future design of such a measurement network for operation and control of treatment plant.

4. FLOOD ROUTING FORECASTING FROM MEASUREMENT

If a sewer network has to be equipped for real time operation, measure interpretation is as important as measuring self.

Our research about hydraulic phenomena into a storm sewer network is also the occasion for testing suitable numerical simulation of flood routing.

Besides classical methods based on hydrological hypotheses such as Muskingum method or RERAM program (Ref. 8), hydraulic method of integrating the Saint-Venant equations was developed (Ref. 6). This way using characteristic method is more tedious but also more accurate and thus easier for further comparisons.

The following hypotheses were used in the model :

- Selection of main branches for first computation step, the secondary ones being treated in a second step.
- Use of characteristics method in each of the main pipes and assembling by continuity equations at the nodes.
- Introduction of lateral discharges in the manholes with a distribution on several nodes for avoiding difficulties due to sudden discharge increasing. Both rectangular or triangular distribution may be used for lateral discharges (Ref. 9).
- Use of steady flow formulae for evaluating local head losses.
- Taking into account of local head losses in manholes, junctions...

Taking into account of mobile jumps and local full flow (with positive or negative pressure) has to be introduced in the model.

Till now our data base is not extended enough for testing significantly the validity of our model.

5. CONCLUSION

The paper describes some problems met in the design of a measurement network for urban strom sewer. Although the aim of this research is above all studying hydrological and hydraulic phenomena, this experience may be extended to the problem of urban drainage operation.

The greatest difficulty is correctly evaluating the consequences of approximations in measurement and modelling. Often measurement is reduced to one local water depth estimate without verifying errors due to non-uniform or transient flow. Moreover, simplifications in the mathematical modelling of flood routing may lead to non-realistic forecasting with further difficulties for operating the sewer network or the treatment plant.

It is wished that constant progresses in measurement design and flow modelling will contribute to improving operation of urban drainage.

6. REFERENCES

1. Guillot, P. : "Tipping-bucket mini-raingages" (Mini-pluviomètre à auget basculant) Technical note EDF, A3 DTG/TM, 1981 (in French)

2. Massalek, J. : "Calibration of the tipping-bucket raingage" J. Hydrol., 1981, 53, pp. 343-354

3. Oberlin, G.; Galea, G. and Rosique, J.B. : "Raingages calibration" (Le tarage des pluviomètres). La météorologie, 1980, 6th Serie, 20-21, pp. 23-32 (in French)

4. Oberlin, G.; Galea, G. and Rosique, J.B. : "Calibration of tipping-bucket raingages" (Etalonnage des pluviographes à augets basculeurs). Agence de Bassin Seine-Normandie, 1974, CTGREF Hydrologie, 25 pp. (in French)

5. Chow, V.T. : "Open-channel hydraulics". New-York, Mc Graw-Hill, 1959

6. Yevjevich, V. and Barnes, A.H. : "Flood Routing through storm Drains". Colorado State University, Fort-Collins, 1970, Hydrology papers 43-44-45-46

7. Zech, Y.; Delmée-Willemsens, C.; Dewandre, J.F. and Jacquemart, J.L. : "Instability Problems in urban storm sewers". Third International Conference on Urban Storm Drainage, Göteborg, 1984 (submitted for inclusion)

8. Desbordes, M. : "Applying models to studies of urban runoff : RERAM program" (Application des modèles aus études de ruissellement urbain : le programme RERAM). Conference "Rainfall - Runoff Relation" (Relation Pluie-Débit), Société Hydrotechnique de France, Paris, november 1979 (in French)

9. Sivaloganathan, K. : "Flow Computations for Storm Sewer Systems". First International Seminar on Urban Drainage Systems, Southamptan, 1982, pp. 2/77-91.

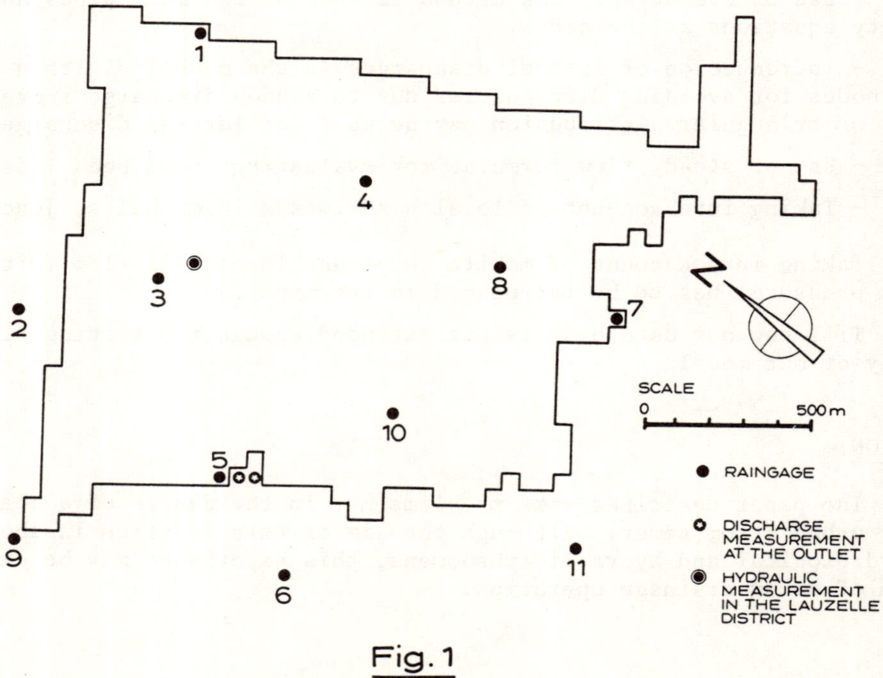

Fig. 1

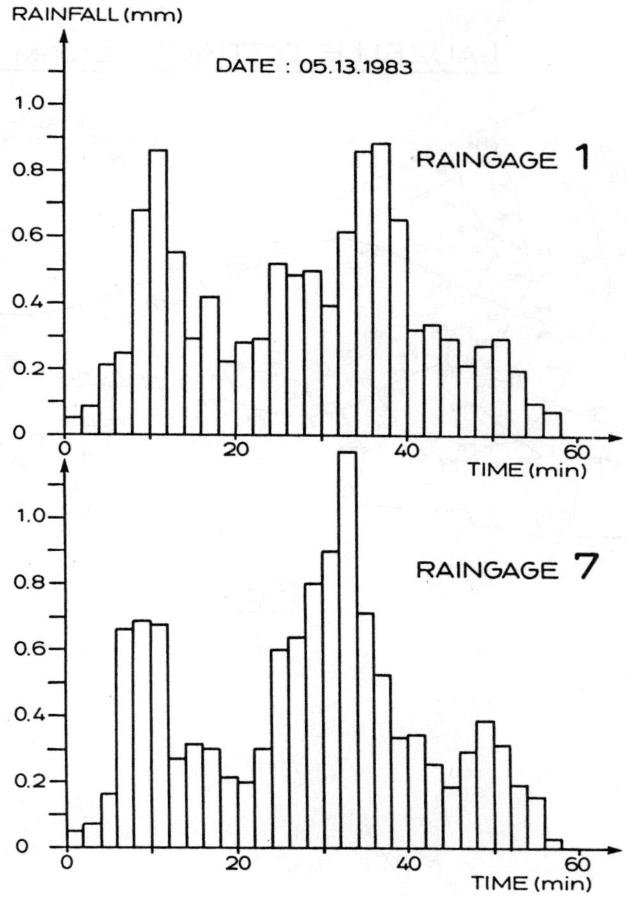
RAINFALL (mm)
DATE : 05.13.1983
RAINGAGE 1
RAINGAGE 7
1.0
0.8
0.6
0.4
0.2
0
20
40
60
TIME (min)

Fig.2

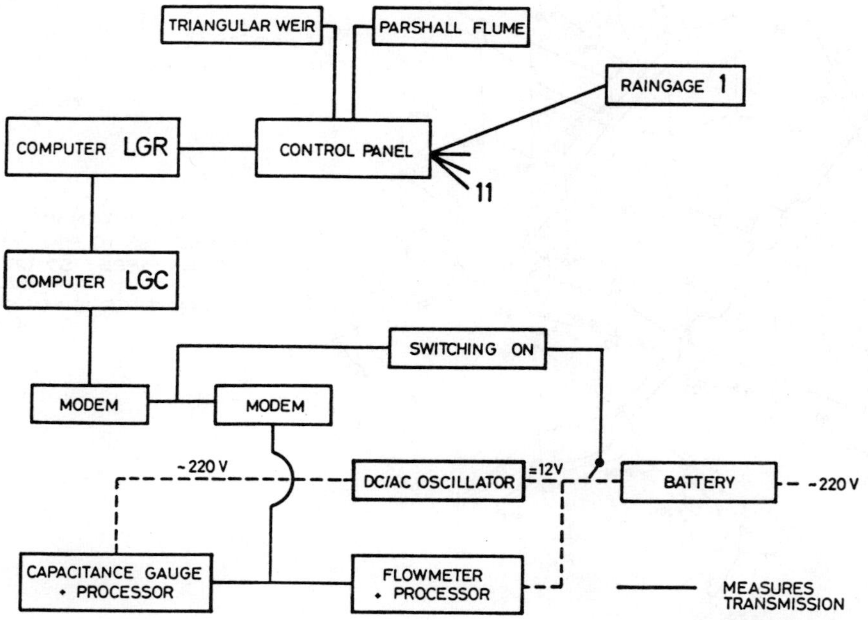
TRIANGULAR WEIR
PARSHALL FLUME
RAINGAGE 1
COMPUTER LGR
CONTROL PANEL
11
COMPUTER LGC
SWITCHING ON
MODEM
MODEM
~220 V
DC/AC OSCILLATOR
=12V
BATTERY
~220 V
CAPACITANCE GAUGE + PROCESSOR
FLOWMETER + PROCESSOR
MEASURES TRANSMISSION
POWER SUPPLY

Fig.3

LAUZELLE DISTRICT - STORM SEWER NETWORK

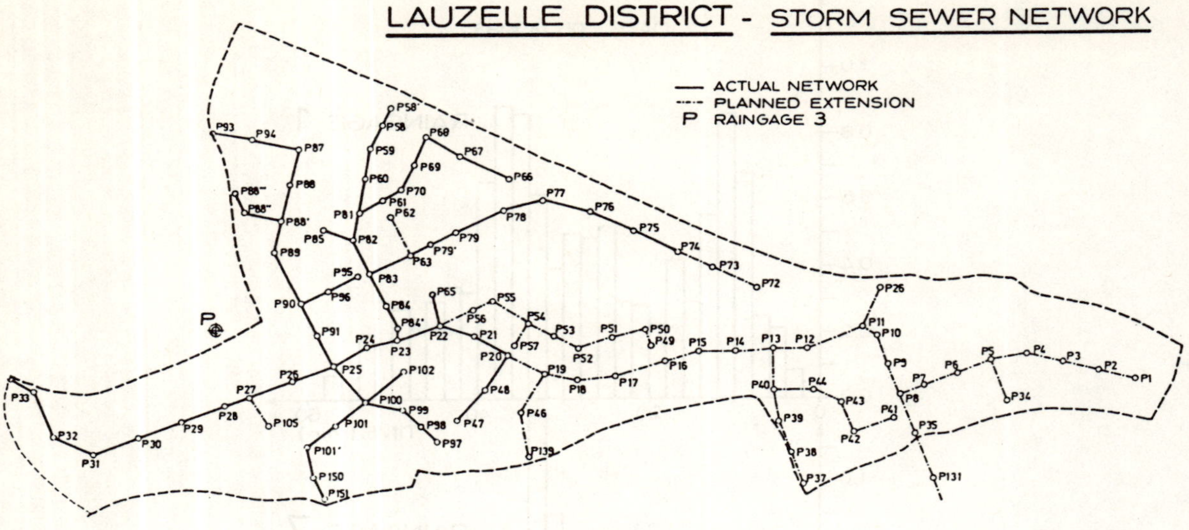

Fig.4

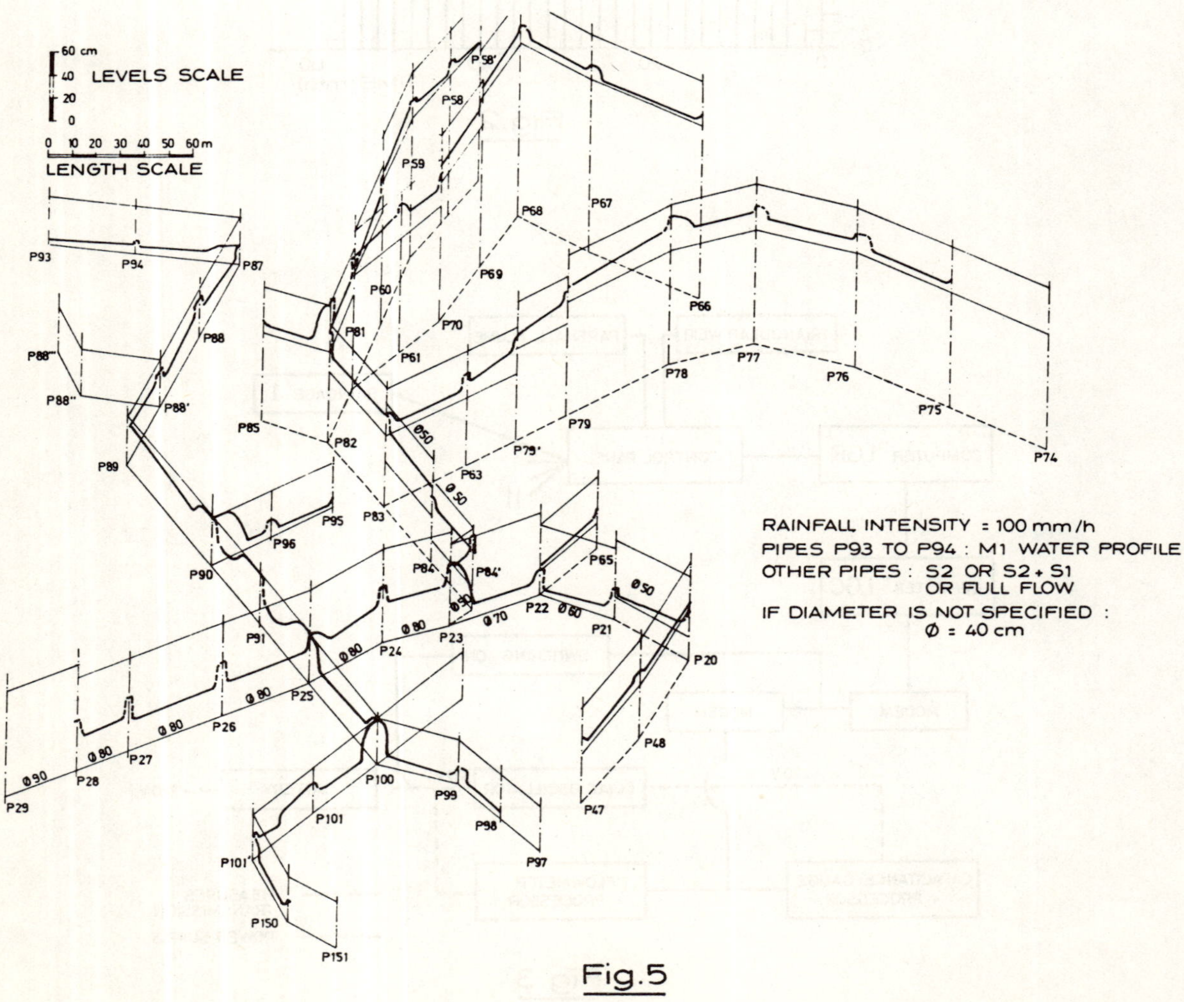

Fig.5

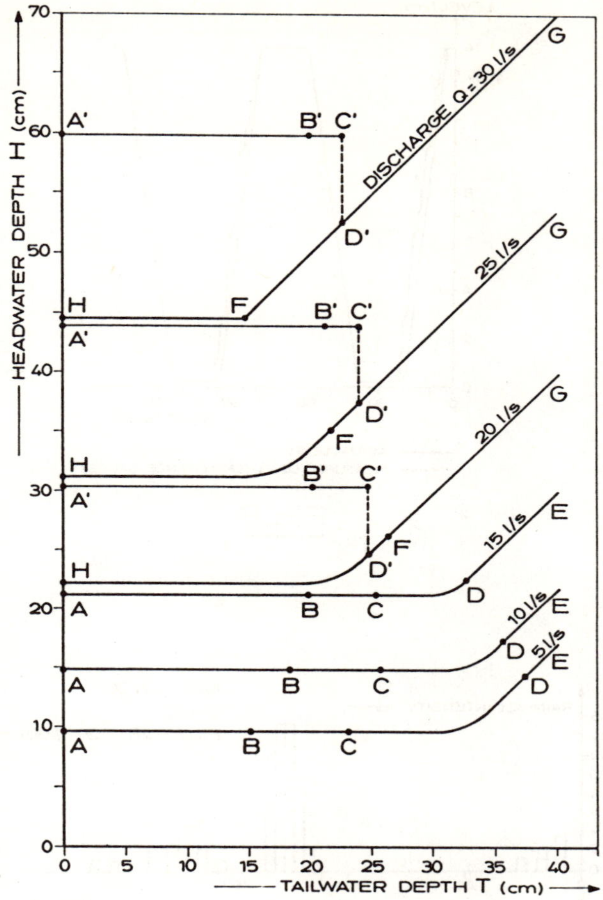

HEADWATER DEPTH H (cm)
TAILWATER DEPTH T (cm)
DISCHARGE Q = 30 l/s
25 l/s
20 l/s
15 l/s
10 l/s
5 l/s
A'
B'
C'
D'
H
F
G
A
B
C
D
E

Fig. 6

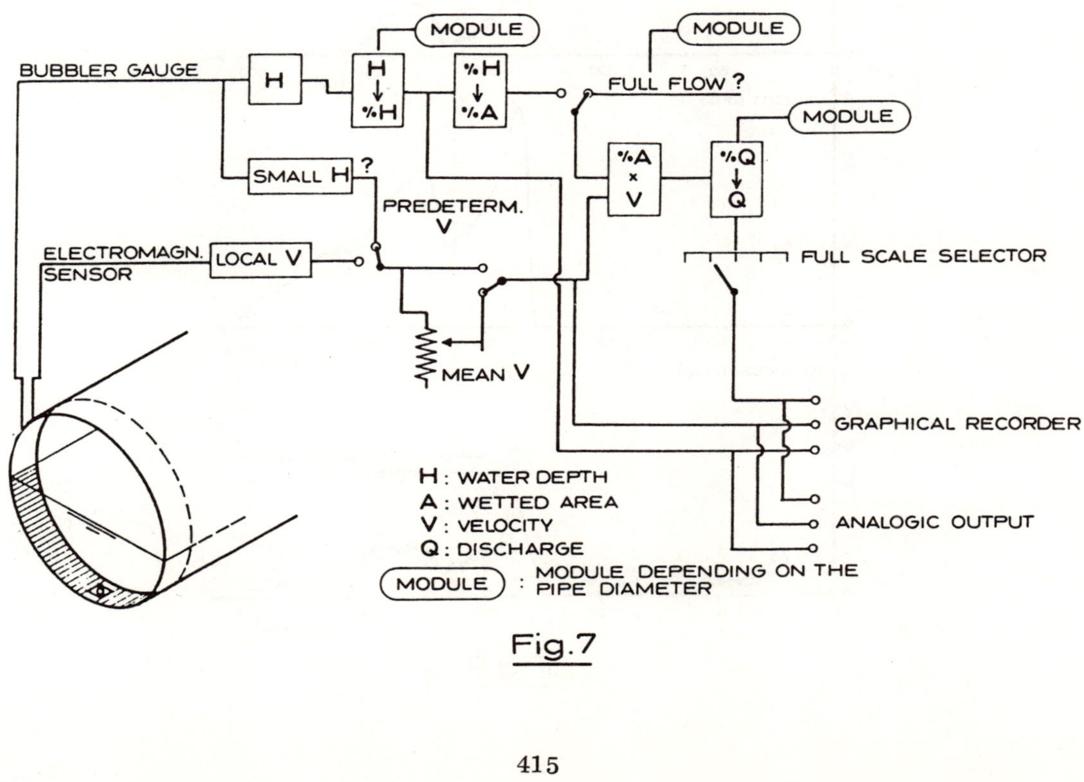

BUBBLER GAUGE
H
MODULE
H ↓ %H
%H ↓ %A
FULL FLOW ?
MODULE
MODULE
SMALL H ?
%A × V
%Q ↓ Q
PREDETERM. V
ELECTROMAGN. SENSOR
LOCAL V
FULL SCALE SELECTOR
MEAN V
GRAPHICAL RECORDER
H : WATER DEPTH
A : WETTED AREA
V : VELOCITY
Q : DISCHARGE
ANALOGIC OUTPUT
MODULE : MODULE DEPENDING ON THE PIPE DIAMETER

Fig. 7

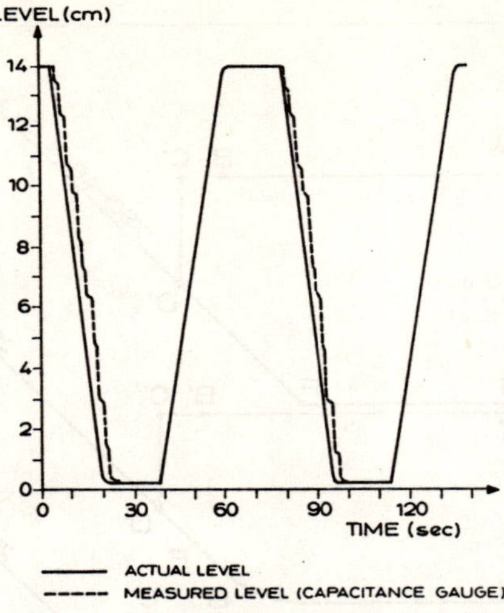
LEVEL (cm)
TIME (sec)
ACTUAL LEVEL
MEASURED LEVEL (CAPACITANCE GAUGE)

Fig.8

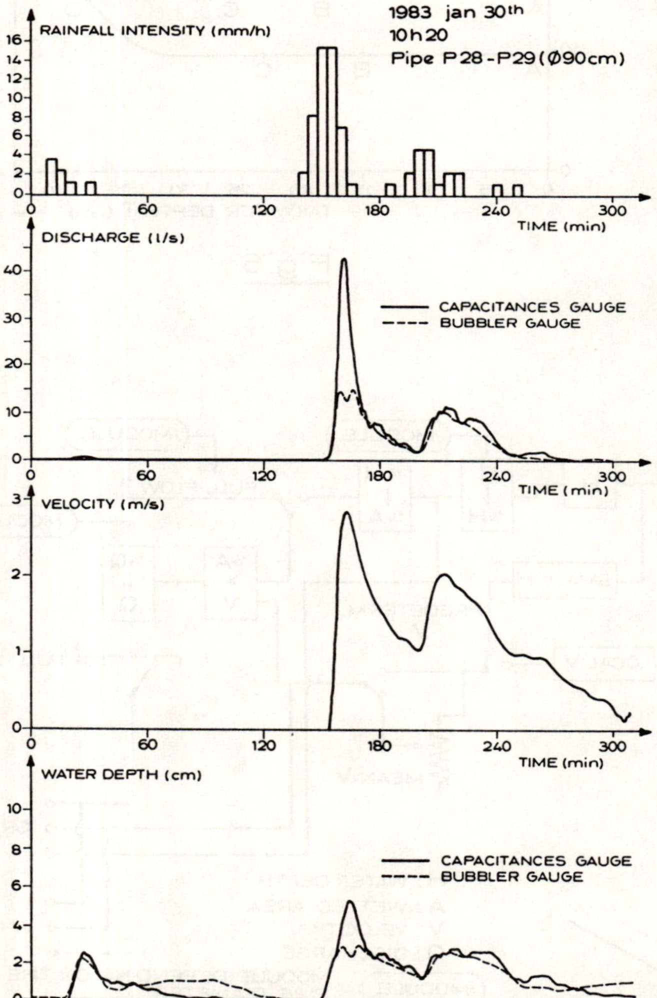
1983 jan 30th
10h20
Pipe P28-P29(Ø90cm)
RAINFALL INTENSITY (mm/h)
TIME (min)
DISCHARGE (l/s)
CAPACITANCES GAUGE
BUBBLER GAUGE
VELOCITY (m/s)
WATER DEPTH (cm)

Fig.9

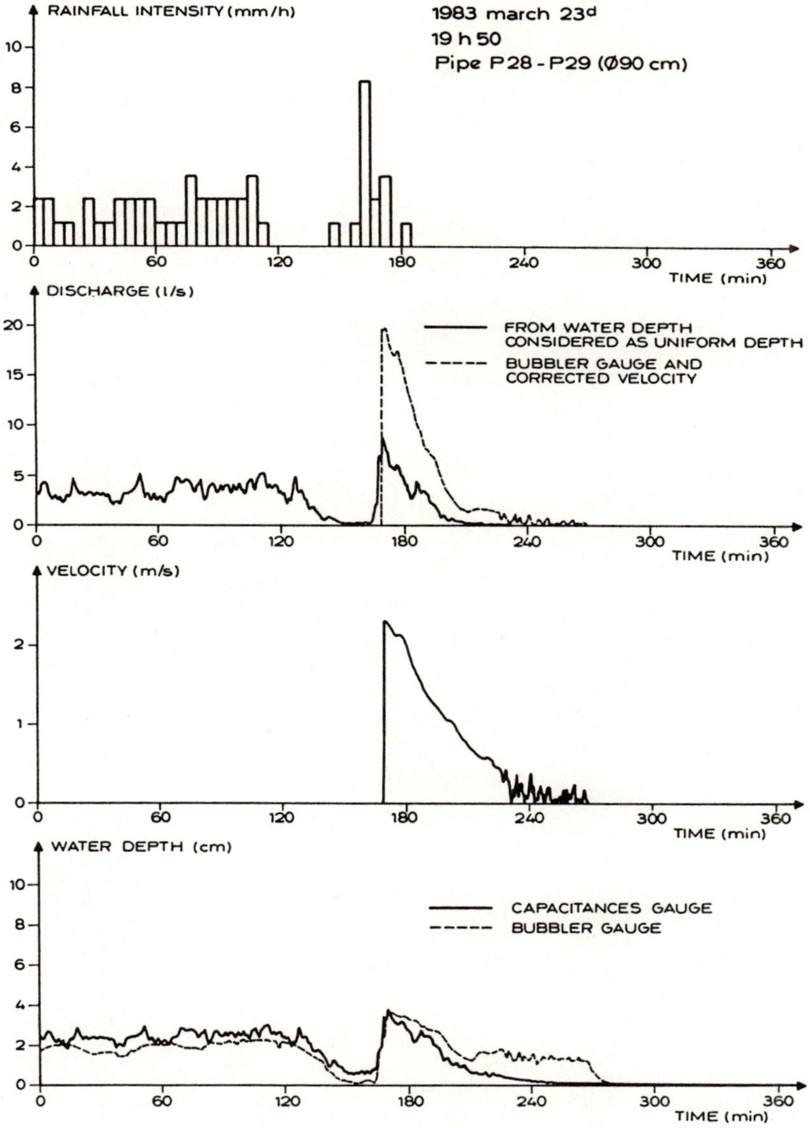
RAINFALL INTENSITY (mm/h)
1983 march 23d
19 h 50
Pipe P28 - P29 (Ø90 cm)
DISCHARGE (l/s)
FROM WATER DEPTH CONSIDERED AS UNIFORM DEPTH
BUBBLER GAUGE AND CORRECTED VELOCITY
VELOCITY (m/s)
WATER DEPTH (cm)
CAPACITANCES GAUGE
BUBBLER GAUGE
TIME (min)

Fig. 10

International Conference on the
PLANNING, CONSTRUCTION, MAINTENANCE & OPERATION OF SEWERAGE SYSTEMS
12-14 September, 1984

MEASUREMENT OF FLOW IN PART FILLED SEWER PIPES USING THE ELECTROMAGNETIC TECHNIQUE

A.A. Rowse

Sarasota Automation Ltd.,
King's Worthy, Winchester,
Hampshire SO23 7QA
England

Summary

The electromagnetic technique has been used for measuring flow in full pipes for many years. The need to measure flow in foul and storm sewers has instigated the development of the electromagnetic flowmeter for use in part filled pipes.

The main differences in the application of the technique include measuring lower velocities, measuring the level of the water surface and using different electrode and coil configurations. These differences have necessitated the development of more sensitive instrumentation to cover the tiny electrode voltages which vary with velocity. The measurement of water level has been achieved using a non-contact ultrasonic technique and this measurement combines with the voltages to infer flow.

The electromagnetic technique basically indicates the average velocity of the flow, but, due to practical limitations in positioning the coil, a non-uniform magnetic field is imposed on the pipe. This non-uniformity is compensated to some extent by the positioning of the electrodes.

The main source of electrical interference is a variable electro-chemical effect on the electrode/water/air interface and precautions are outlined to minimise this interference. Results are presented from laboratory trials at the National Engineering Laboratory from operational installations in the Severn-Trent Water Authority region.

Held at The University of Reading, England. Organised and sponsored by BHRA, The Fluid Engineering Centre, Cranfield, Bedford in conjunction with the Water Research Centre.

NOMENCLATURE

K_1 = overall calibration constant

K_3 = depth term constant

K_4 = $depth^2$ term constant

N = number of turns in the coil

h = water depth in millimetres

D = pipe diameter in millimetres

1. INTRODUCTION

There is a well established need in the water industry to measure flows in existing storm and foul sewers, both for long term storm flow studies, and, in some cases, for operational control. Conventional electromagnetic pipe flowmeters have been in use for over 20 years but they have not given encouraging results in dirty water or where air may get entrained in the pipe. However, the technology of an open channel electromagnetic flowmeter developed in the early 1970s (Ref. 3) seemed to offer a possible fruitful starting point.

This meter, Fig. 1., operates on a principle similar to that of an electric dynamo. If a length of conductor moves through a magnetic field, a voltage is generated between the ends of the conductor. In the meter, the magnetic field is generated by the coil which is wound either above or below the river. The conductor is the water moving through the magnetic field and the ends of that conductor are the banks. The very small voltage generated is sensed by the electrodes on the banks, and these are connected to the input of the electronic console. The faster the water moves, the greater is the voltage generated.

To overcome the problems caused by the presence of the Earth's magnetic field, battery effects of the probes in the water, and external electric noise, the magnetic field produced by the coil is reversed at regular intervals. Measurements are taken with the field first one way and then the other.

To provide a measure of discharge, water level must be measured. The basic equation relating the various variables is of the form

$$\text{Discharge} = \frac{\text{Probe Voltage} \times \text{Depth}}{\text{Magnetic field}} \times \text{Calibration Constant}$$

For a particular site, with a fixed size of coil and number of turns, the equation becomes of the form

$$\text{Discharge} = \frac{\text{K1} \times \text{Probe Voltage} \times (\text{K3} + \text{Depth} + \text{K4 Depth}^2)}{\text{Coil Current}}$$

The non linear depth function (K3 + Depth + K4 Depth 2) takes account of the change of magnetic field with distance from the coil.

The strength of the magnetic field created by the coil is about 100 amp turns per metre, compared with about 35 amp turns/metre for the Earth's field.

The voltage generated across the river as a result of the water moving through the magnetic field generated by the coil will be between a few tens of, and a couple of hundreds of, micro volts.

Interfering signals are picked up by the probes, whose magnitude is many thousands of times greater than the required signal. Generally,

1.1 BBC Radio 4: magnitude up to 4 milli volts between the probes.

1.2 Electric mains leakage: magnitude up to 1.5 volts between a probe and earth and 2 milli volts between the probes.

1.3 Slowly changing bias voltage generated by the electrode/water/air interface: up to 0.5 volts

The electronic system uses sophisticated analogue and digital techniques to detect measure and average the required signal.

This type of meter has been further developed over the past four years for use in insulated channels, and its reliability significantly improved by the use of microprocessor technology. With an insulated channel, the electrodes are not short circuited by the river or channel bed. Such a short circuit greatly reduces the measured signal by an ill defined amount, and also injects electrical noise into the system. The technique has now been successfully applied to some 13 river sites, (one in the U.S.A. and the rest in the U.K.). Widths range from 3 to 22 metres, the larger channels being trapezoidal in cross section, and with water velocities as low as a few millimetres/second. Following this success, the problem of applying the technique to part filled pipes containing raw sewage was investigated.

2. DESIGN PROBLEMS

A flowmeter which is to be used for long periods in sewers, has ideally to meet the following criteria:-

2.1 It must be capable of measuring flow in part full pipes with depths ranging from about 15% of pipe diameter to surcharge; where there is no relationship between level and flow.

2.2 It must be capable of measuring reverse flow.

2.3 It must not be excessively affected by deposition of solid material or grease.

2.4 It must operate in raw sewage, containing paper, rag and solids, and the surface may be covered in foam.

2.5 The sensors must not create a significant head loss, nor act as a trap for catching debris, rag, paper, etc.

2.6 The sensors must not require regular maintenance which may involve the attendance of a sewer gang.

2.7 The sensors must meet the requirements of Zone 1 flammable atmospheres.

2.8 The traditional requirements of electronic systems for accuracy and speed of response may be relaxed. Uncertainties of ±5 to ±10% and response times of up to 5 minutes are in many cases acceptable.

Various potential problem areas were identified and initially experiments centred on the following were conducted:-

a) The effect of grease and other contaminants on the electrodes.

b) The measurement of depth using acoustic transducers mounted above the water surface.

c) The calibration of prototype pipe flowmeters using clean water and settled sewage.

d) The obtaining of BASEEFA (British Approvals Service for Electrical Equipment in Flammable Atmospheres) Approval.

From these initial studies, prototype meters of the configuration shown in Fig. 2 were constructed.

The trials on the electrodes showed that no fouling problems emerged with stainless steel electrodes encased in a porous polythene membrane. However, on sites where surface waves of 20 mm or so existed, some development was required to reduce the electrical noise generated by the wave action. The use of narrower electrodes and better packing to keep putrid water out of the electrode housing proved beneficial.

The measurement of depth could be made reliably using acoustic devices, providing the water did not cover the transducer, or leave large deposits of debris on it.

The transducers had to be mounted well above the pipe if ambiguous results at high levels were to be avoided. Some further development is still needed to tailor the existing designs of depth gauge to the particular needs of the sewer environment. In particular, in the handling of signals from water surfaces which contain standing waves and in the prevention of solids being deposited in the depth gauge standpipe.

Calibration trials were conducted at the National Engineering Laboratory,

East Kilbride, in November 1981 and October 1982, and also at the Finham Works of the Severn-Trent Water Authority, using different configurations of coil and electrodes. The results were very encouraging and produced empirical calibrations which appeared to be consistent to within about ±5% and which followed anticipated calibrations.

No serious problems were encountered in meeting BASEEFA Zone 1 approval, except that at present the availability of the depth gauges is limited. Inevitably, some cost penalties exist in providing approved electrical and atmospheric barriers between the sewer and the electronic console, and also in burying the coil in concrete.

3. RESULTS FROM FIELD TRIALS

Tests have been carried out at five separate sites, one in the Thames and the remainder in Severn-Trent Water Authority regions, all with raw sewage flowing through the meters.

Pipe sizes ranged from 0.5 to 1.6 metres diameter, some on inlets to works and others in trunk sewers.

The problems which have been highlighted have been:-

3.1 Effect of rough water surface and foam on the depth gauge

3.2 Effect of waves generating voltages on the electrodes

3.3 Effect of sudden changes in level and in sewage quality causing the bias voltage of the probes to change faster than the electronic equipment can accommodate.

3.4 Possible injection of noise and the generation of electric potentials from metalwork in the channels both upstream and downstream of the meters.

The problem of the depth gauge has been overcome by using peak sensing circuits to remember the water level, whenever a satisfactory echo is achieved, and to maintain this reading for perhaps several tens of seconds between echos.

The electrode problems appear at some sites and not others. They are particularly noticeable at one site where the water enters the meter from two different sources, having different water qualities. One of them joins at right angles to the meter a distance of only about 2 diameters upstream.

Also at this site, the meter section is surrounded by a steel channel. These two site peculiarities are considered to contribute to the generation of occasional large step changes in measured electrode potential. A calibration test on this site in which the meter was compared with the rate of filling of a storm tank indicated a good relationship (±5%) punctuated by sudden low readings from the gauge which gave an overall discrepancy of some 25%.

Results from the other four sites are more acceptable and those with a smoother water surface have given very encouraging results. Laboratory trials on two different pipes with different coil sizes and electrode positions have been reported by the Hydraulics Research Station (Refs. 1 and 2). These trials were carried out at the National Engineering Laboratory.

The results of the latest tests are summarised in Fig. 3. The meter used had the same electrode and coil geometry as that now employed on the operational sites. The graph shows actual results for various flows and levels plotted as the ratio, 'flow/signal' versus depth. It indicates that the same law applies for flows over a 6 : 1 range and that a non linear relationship occurs between signal (= electrode voltage/coil current) and depth for the same volumetric flow.

The relationship between the variables was:-

$$\text{Flow} = \frac{\text{Electrode Voltage}}{\text{Coil Current}} \times \frac{765}{N}\left(\frac{h}{D} - 0.38\frac{h^2}{D^2}\right) \times D^2 \times 10^{-5}$$

where electrode voltage is in microvolts
coil current is in amps
flow is in litres/second

Generally, agreement between meter and actual flow was within $\pm5\%$.

There are obvious practical benefits especially during installation of having a simple coil placed above the pipe, of a diameter no more than that of the pipe. However, this means that the magnetic field is non uniform and is much weaker towards the bottom of the pipe. Tests were carried out with electrodes placed only in the lower part of the pipe. Although this gives a generally more linear result, the readings were variable, especially when the water level was about half pipe full.

The effect of abnormal vertical velocity gradients on a meter having a non uniform magnetic field has not yet been studied. However, it is generally believed that velocity profiles are quickly established in sewers, which, in general, do not contain sharp level changes within pipe runs.

4. CONCLUSIONS

The studies and field experiments have shown that the electromagnetic technique offers a solution to the problem of long term flow measurement in foul sewers and other part filled conduits. The many problems of working in potentially hazardous atmospheres, dirty water, foam laden surfaces, and of inserting a meter into an existing pipe, can be overcome.

In many cases they have been inserted 'in the wet', without the need to stop the flow.

5. REFERENCES:

(1) Hydraulics Research Station Report No. DE 55 March 1982

(2) " " " Report No. DE 60 February 1983

(3) Newman, J.D. "Electromagnetic Gauging in Effluent Channels" Water Research Centre Conference on Opportunities for Innovation in Sewerage, Reading March 1977

ELECTROMAGNETIC OPEN CHANNEL FLOWMETER

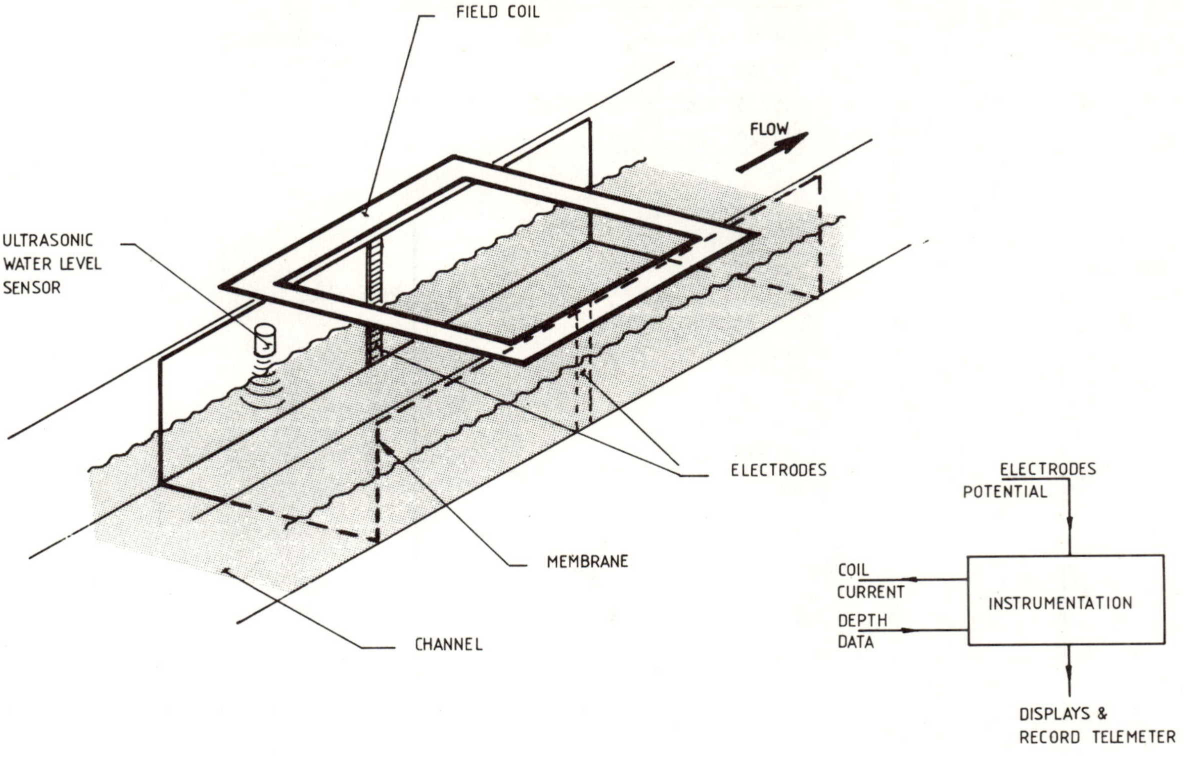

FIGURE .1.

ELECTROMAGNETIC PART FILLED PIPE FLOWMETER

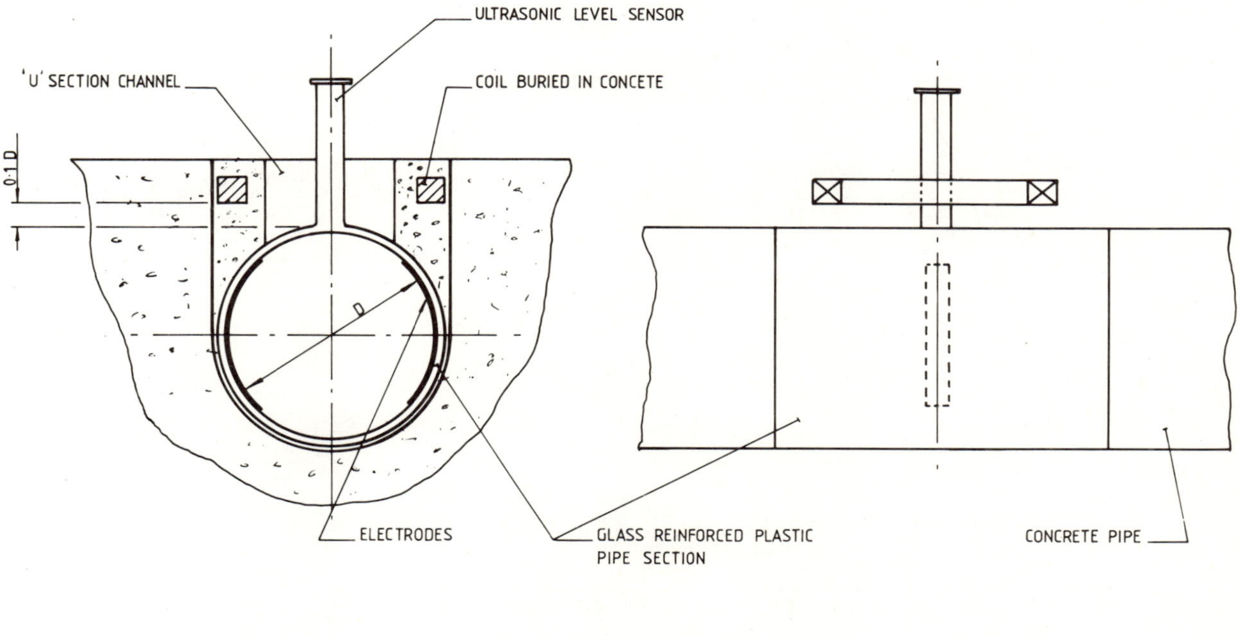

FIGURE. 2.

FIG. 3.

Data for 450mm I/D PIPE

Practical data NEL Trials October '82

- • 9·5 l/s on 18-10-82
- + 60 l/s } 19-10-82
- △ 36 l/s } 19-10-82
- □ 20 l/s } 19-10-82

Electrodes from 0·15 to 0·85 h/D

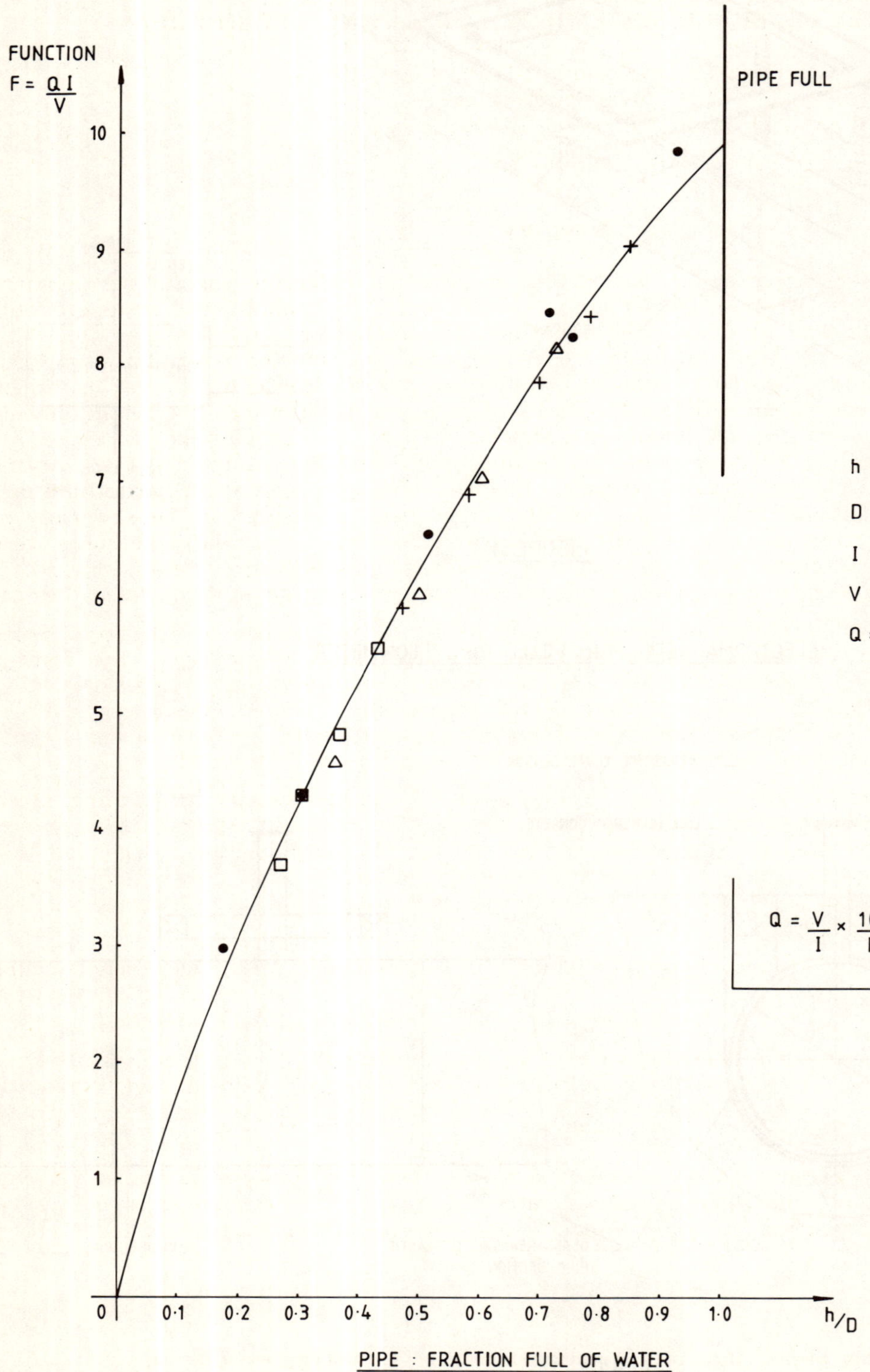

PIPE : FRACTION FULL OF WATER

h = Variable water depth mm.

D = 450 mm.

I = Coil current Amps

V = Electrode voltage μ Volts

Q = Flow litres / sec.

$$Q = \frac{V}{I} \times \frac{10^{-4}}{N} \times 76{\cdot}5 \left(\frac{h}{D} - \frac{0{\cdot}38h^2}{D^2} \right) D^2$$

International Conference on the
PLANNING, CONSTRUCTION, MAINTENANCE & OPERATION OF SEWERAGE SYSTEMS
12-14 September, 1984

PLANNING, CONSTRUCTION AND OPERATION OF A MULTILEVEL SEWERAGE SYSTEM FOR LOW LAYING URBAN AREAS

Prof. V. de Kosinsky, DSc CEng FICE
University of Liege
6, Quai Banning, B-4000 Liège
Belgium

Summary

Cost effective sewerage of low laying areas often calls for careful development of appropriate strategies. When an urban area is sufficiently high above the sewerage outlets the most economic way of discharging the flow is usually by gravity. Unfortunately this is not always the case, and there might be several isolated or combined reasons making gravity discharge impossible. The most common difficulty is the existence of low laying areas, being occasionally or permanently below the level of a nearby river where the sewers should discharge. Another concern for the municipal engineer can be river regulation works resulting in the water levels of the river being artificially raised and maintained too high for the satisfactory operation of the gravity outlets. Embankments, on one hand protecting low laying areas from inundation from a river, on the other hand also cut off the sewerage of these areas from the river, causing indirect flooding from storms if adequate measures were not taken. In the Liege region both heavy mining subsidence and high flood levels due to river works called for the development of a special multilayer sewerage system, consisting of:

1. networks of collecting sewers intercepting all the water coming from areas located permanently above the highest water stage in the river. The flow is evacuated into the river by gravity through pressure pipes, having no connection with the sewerage system or the groundwater table of the lower districts they cross to reach the river. These pipes work under pressure during floods. This high level system also contains detention basins.
2. networks of large shallow sewers, called "upper networks", collecting storm water, sewage and industrial waste water. Most of these sewers discharge normally by gravity into the river, but during floods their flow is diverted into pumping stations. The design and management of this low level shallow sewerage system must take into account the particular technical difficulties in the subsidence areas.
3. networks of relatively small diameter drains placed at great depths and called "lower networks" collecting the water seeping into the cellars in the valley. This flow is being continuously pumped by automatic pumping stations.

The paper first sets forth the various sewerage problems met in the Liege region. After this introduction the organizational and financial background is described. Next the planning and operation of the sewerage system is presented, followed by the description of the design and construction.

Although the present paper describes a special sewerage system successfully developed in the Liege region, the same principles can be applied to any region located in a river valley, having urban areas laying occasionally or permanently below the level of a river.

Held at The University of Reading, England. Organised and sponsored by BHRA, The Fluid Engineering Centre, Cranfield, Bedford in conjunction with the Water Research Centre.

1. INTRODUCTION.

1.1. Sewerage Problems

A very typical example of a combined set of sewerage problems low laying urban areas can meet exists in and around Liege, one of the largest towns in Belgium. The origins of the problems are fourfold :

- heavy mining subsidences resulting in more and more areas settling below the level of the river;
- very high flood discharges in the river Meuse crossing the region;
- river regulating works artificially upholding the level of the river for navigation purposes;
- flood-walls protecting low laying areas also impeding the flow of storm water into the river.

1.2. Mining subsidences

Coal-mining has always been a tradition in the Liege basin, and it became really intensive in the nineteenth century. However, the same mining from which originated the whole industrial development of the region resulted also in a deadly threat for the entire industrial valley.

An 1810 law authorized mining along about 23 km of the river valley, upstream and downstream of Liege, with the exception of the city itself.

In the mines it was found uneconomic to backfill systematically all the cavities left by the coal seams, and as a result, very heavy mining subsidences have taken place. Upstream of Liege, for instance, the ground has for many years been settling at a rate of 0,05m to 0.10m a year. It has been found that the subsidence at the surface equals or exceeds half of the thickness of the mined coal seams.

As the ground does not settle in the areas where there has been no mining, i.e. at the limits of the basins upstream and downstream of the mines, and under the city itself, two large depressions formed :

- one above Liege, where the subsidence is nil at the edges, but very quickly reaches 2m to 6m, and even more in several low spots,
- one below Liege, with two low spots where the subsidence is also more than 2m to 6m.

The real threat for the urban and industrial areas came from the river as the water stages remained independent of the mining subsidences, because the river-bed did not settle at either end, upstream and downstream of the mines, neither in the middle under the city itself. As a result, many areas of the valley had been settling, often several metres, with respect to the river, and flooding became more and more frequent and serious.

As an exemple, the main street of one of the industrial suburbs would have never been flooded without mining subsidence. In 1872, that is after 57 years of mining, the water rose there to 1.10m. Without protection works, nowadays the street would permanently be flooded and a 100-year flood, like the one in 1872, would drown the lowest part of the street under 5.40m of water (see Fig. 1).

1.3. Floods

The discharge of the river Meuse at Liege is extremely variable. In summer it can fall below 30 m^3/s, but in winter it can reach 3000 m^3/s. In the past catastrophic floods occurred regularly. The first historical flood mentioned is that of 1036. The oldest recording is that of 1571, and the highest one is from 1643.

A flood in 1880 inundated only a few low spots, but in 1925-26 an only 0.50m higher flood inundated 2500 hectares in the valley : industry and commerce came to a standstill for several months and there was a danger of epidemy. 6000 houses were flooded, more than 1000 of them up to the first floor.

1.4. River Regulation

The great floods of 1925-26 called for important flood defence works. Levees and embankments were rapidly constructed, designed to resist to the highest floods, even after all the mining subsidences had taken place. Other important works, such as dredging and removing islands and weirs helped to regulate the river. In addition, modern movable weirs, which can completely be lifted out of the river during floods, maintain a constant minimum level to make navigation possible even during droughts, round the year fluvial transport being vital to the heavy industries of Liege.

1.5. Storm Drainage

However, containing the river between flood-walls and facilitating flood discharges were not sufficient to save the region from flooding. As a matter of fact, when the river is at flood levels, storm precipitation falling in the valley and on the surrounding hills and plateaux cannot flow freely into the river, and would soon flood the lowest areas. And to the volume of rainwater, we must also add the sewage and industrial waste waters.

During floods, and in some places even during normal or low-water periods, the sewerage system must also be cut off from the river, to stop the river back-flooding the valley through these pipes.

In addition, if no special measures were taken, water seeping from the river under the embankments would also raise the groundwater table, flooding first the cellars, then the lower parts of the valley, inundating finally the whole urban and industrial area.

2. ORGANIZATION AND FINANCE

2.1. Organization

To protect themselves from all the dangers described in the introduction, the municipalities of the valley created a co-operative association charged with the design, construction and operation of a special sewerage system, called by an old French word "Demergement". First the municipalities upstream of Liege joined in 1928 in an "Association Intercommunale pour le Demergement". Others followed in 1941, and since 1947 all the "communes" concerned are members of it.

The protected urban area in the valley is 1000 hectares (10 km²) and the drainage basins of the hills and plateaux have a surface of 7500 hectares (75 km²).

The association of the municipalities is organized like a public agency. It has departments to look after the design, construction and operation of the sewerage system :

- A team of 21 engineers, architects and technicians carries out the design and site supervision of the construction works;
- Another team of 68 operates the system, including the maintenance of the pumping stations described later on;
- A third team of 6 looks after the maintenance of the sewers and outlets.

The technical departments are backed up by the usual administrative, social and financial services.

The association is supervised by a committee of three university professors, specialized in civil and hydraulic engineering.

2.2. Finance

As far as the capital investment is concerned for the construction and equipment it is far beyond the means of the municipalities and is financed to :

- 83% by the State through the Ministry of Public Works,
- 12,5% by the Province of Liege,
- 4,5% by the concerned municipalities.

On the other hand, the municipalities finance the operational costs, being helped however by the State. As a matter of fact they receive a subsidy amounting to 75% of their sewerage expenses. It must be pointed out that these 75% were borne up to 1966 by the collieries and it is only since their gradual closing down that the State took over the subsidizing.

3. PLANNING AND OPERATION

3.1. The principles of the multilevel sewerage system

The water stage in the river being artificially maintained, large industrial and built-up areas in the valley find themselves, due to mining subsidence, below high, normal or even low water level. As it has been pointed out, such a situation presents special sewerage problems. The most simple solution would have been to envisage the pumping of all the flow coming from the sewers as well as the storm drainage. But this solution would have meant very heavy investment and operational costs, so a special sewerage system has been developed.

As shown in Fig. 2, this special sewerage system consists of networks on three levels :

a. It was to avoid the expensive construction and operation of high capacity pumping stations, that the decision was taken to lay collecting "belt" sewers on both sides of the river, high enough (taking into account future mining subsidences) on the hillsides to have enough head to discharge into the river by gravity even under flood conditions. The flow is evacuated into the river by gravity through pressure pipes called "exutories" having connection neither with the sewerage system nor with the water-table of the lower districts they cross to reach the river. These pipes work under pressure during floods. This high level system also contains detention basins

b. In the valley, networks of large shallow sewers, called "upper networks", collect storm water, sewage and industrial waste water. Most of these sewers discharge normally by gravity into the river, but during floods their flow is diverted into pumping stations. The design and management of this low level shallow sewerage system must take into account the particular technical difficulties in the subsidence areas.

c. Also in the valley, networks of small diameter drains (usually ø 0.4m, sometimes ø 0.5m), placed at great depths (between 3m and 7.50m depth) and called "lower networks" collect the water seeping into the cellars in the valley. This flow is being constantly pumped by automatic pumping stations.

The reasons for choosing a double system of networks in the valley are twofold :

- as long as the ground level permits, the "upper networks" can discharge into the river by gravity most of the time, except during floods. A combined system would have meant deep networks and consequently continuous pumping.
- as far as construction is concerned, the laying of deep large section sewers of a combined system would have been practically impossible in most of the very narrow streets carrying heavy traffic and full of water and other service mains and all sorts of cables. Furthermore, as most of the nearby houses have most likely suffered already settlement damage and are cracked and weakened, deep excavations have to be supported by special linings, which must often be abandoned in the trench. Alternatively, costly thrust-driving techniques must be used.

At the design stage, a comparative study is carried out for every urban area to be drained, based on the construction and operation costs of both solutions, keeping in mind the difficulties of the construction. These studies led to the adoption of a double network of "upper" and "lower" sewers for most of the valley upstream of the city.

In the valley downstream of the city, the situation is rather different and here the economic and technical studies concluded in favour of a combined system of sewerage.

3.2. Operation

When the river stage is normal, most of the "upper networks" can discharge into the river by gravity. During floods, the networks are cut off from the river by floodgates, and their discharge is diverted to pumping stations, which pump the water into the river. These stations are called "principal pumping stations". Out of the 28 such stations planned, there are 21 working and one under construction. Ten of them pump 24 hours a day, because of the heavy mining subsidence, and several others will have to switch to continuous operation with the subsidence progressing.

The flow of the "lower networks" is discharged to "secondary pumping stations". These stations work continuously, pumping from the "lower networks" into the nearest "upper network". There are 13 "secondary pumping stations" planned upstream of the city, and all but one are built and working.

3.2.1. The Principal Pumping Stations

In such a special urban sewerage system the reliability must be utmost. A breakdown of a pumping station, even only during a short storm, may mean the inundation of large urban areas.

In the same time, the running expenses (about 200,000 US $ a year per station) consisting for 90% of the salaries and social charges, must be kept as low as possible.

In every station, water from the main sewers is discharged through thrash-racks into two independent suction chambers. They are normally connected to each other, but either of them can be disconnected for maintenance, one chamber being sufficient for the good working of the pumping station.

The mixed-flow horizontal-shaft (for easy maintenance) pumps turn at a constant speed of 485 rpm. They have several inspection ports making cleaning and maintenance an easy and quick operation. The stations are equipped with three (one on stand-by) to six (two on stand-by) pumps, discharging between 0.6 m^3/s and 1.5 m^3/s each.

In the older stations the suction chambers are located below the pumps, the priming being done by vacuum-pumps. In the newer stations pumps and suction chambers are side by side, so the system is self-priming, making unnecessary the permanent staffing of these telecommanded stations.

The automatization of the older station has also been undertaken, the final aim being to have only two stations with permanent staff, one upstream and one downstream of the city. In addition there are naturally breakdown and maintenance teams available at short notice.

Every station has power supplies from two independent networks, with several connections to each one.

3.2.2. Secondary Pumping Stations

These stations are smaller but often deeper than the principal ones.

In these secondary pumping stations, there are also two suction chambers, either of which can also be disconnected for maintenance. Above each suction chamber there is a 0.15 m^3/s to 0.25 m^3/s pump unit, which turns at a speed of 750 rpm, the pumped water being more clean. One of the pumps is always on stand-by. The suction chambers being pressurized, this system is also self-priming. Power supply is by twin cables from the nearest principal pumping station, connected to two independent networks again.

The secondary pumping stations are automatic, their working being controlled from the main ones. They are only visited for maintenance and repairs.

4. DESIGN AND CONSTRUCTION

4.1. Sewers

The sewers laid in the ground subject to mining subsidence must follow all the movements without suffering any damage.

For this reason, from the beginning, the sewers have been made up of short lengths (3m) of circular reinforced concrete or asbestos pipes, with flexible joints. In some cases pipes are concreted in situ in 6m lengths.

The shallow combined sewers of the "upper networks" must resist to heavy traffic loads (trains, lorries, etc...). When the Association was formed, there were no prefabricated pipes available on the market to such specifications and during more than 50 years the Association designed and cast their own concrete pipes. It is only recently, that this side-activity has been abandoned.

The slopes of the laid sewers are designed to remain sufficient, even after differential mining subsidences, to evacuate always the maximum discharge, which means that for every length of sewer, the future settlements must be computed.

As far as the "exutories" are concerned, discharging the flow of the "belt" or collector sewers of the hillsides and plateaux, they are also made up of short sections with flexible joints. Wherever the "exutories" must work under pressure, they are split up into two parallel conduits, one of them being sufficient to discharge the full flow. Any one of them can be disconnected for maintenance.

As shown in Fig. 3, upstream of the city there are five "exutories" working on each side of the river, the biggest one having a capacity of 35 m^3/s. Downstream, on each side there are three "exutories" under construction. Here the major difficulties are the construction of a siphon under a 120m wide navigation canal, with three conduits of 1.80m by 2.30m, and the thrust-driving of a 490m long ø 2.50m diameter reinforced concrete collecting sewer.

The first sewers date from 1928, and in many places the subsidences are well over 2m. In spite of that, all the sewers (there are more than 200 km of them) are in good working order, with the exception of two short lengths of 150m and 100m, where they had to be relaid. And one of these lengths crosses a geological fault, which was impossible to avoid.

4.2. Pumping Stations

All the pumping stations must also be designed to resist to mining subsidence. In addition, as they immerse deep in the water-table, cracks cannot be tolerated as the walls must be watertight to protect from the groundwater the pumps and electro-mechanical equipment.

To take into account the effects of the settlements, the pumping stations are constructed monolithic in reinforced concrete, with the following two design hypotheses :

a. the station is supported in the middle with one quarter cantilevering at each side;

b. the station is simply supported on both sides on a quarter of its length, i.e. it is not supported on half of its length in the middle.

Out of the 33 pumping stations built to date, several have been subjected to 2m and more of settlement and are leaning in all directions, but none of them suffered the least damage from these movements.

As shown in Fig. 4 and Fig. 5, for further security, the watertight walls of the stations rise above the ground, higher than the level of a 100-year flood, just in case one of the embankments bursting.

5. CONCLUSIONS

In the areas upstream of the city, most of the works are finished and there has been no inundation since 1926 the sewerage functioning perfectly even during flood conditions. Even the cellars are dry, but in a few streets. People forget slowly the formerly damp groundfloors, and the new generation will never see the high "duck-trams" built on large wheels to cope with flood water in the streets.

In the downstream municipalities, which joined the Association after the Second World War only, there is still much to do, and not all areas are yet safe from the flooding of the river or from heavy storms.

At the time of writing this paper, early February 1984, a flood event occurred, nearly comparable to that of 1925-26 (3000 m^3/s). But as the flood protection and sewerage system was designed with a 20% safety factor, it has proved to be perfectly able to cope with the situation in the Liege region.

The "Association Intercommunale pour le Démergement et l'Epuration des Communes de la Région Liégeoise" (AIDE) has today 56 years of specialized experience in urban sewerage and its works and management have been most successful in keeping the region free from flooding. In the first 50 years the actualized investment amounted to 120 million US Dollars, and the "Demergement" system discharges into the river an average of 28 million m^3 of water per year.

Figure 1. Flood Levels and Mining Subsidence.

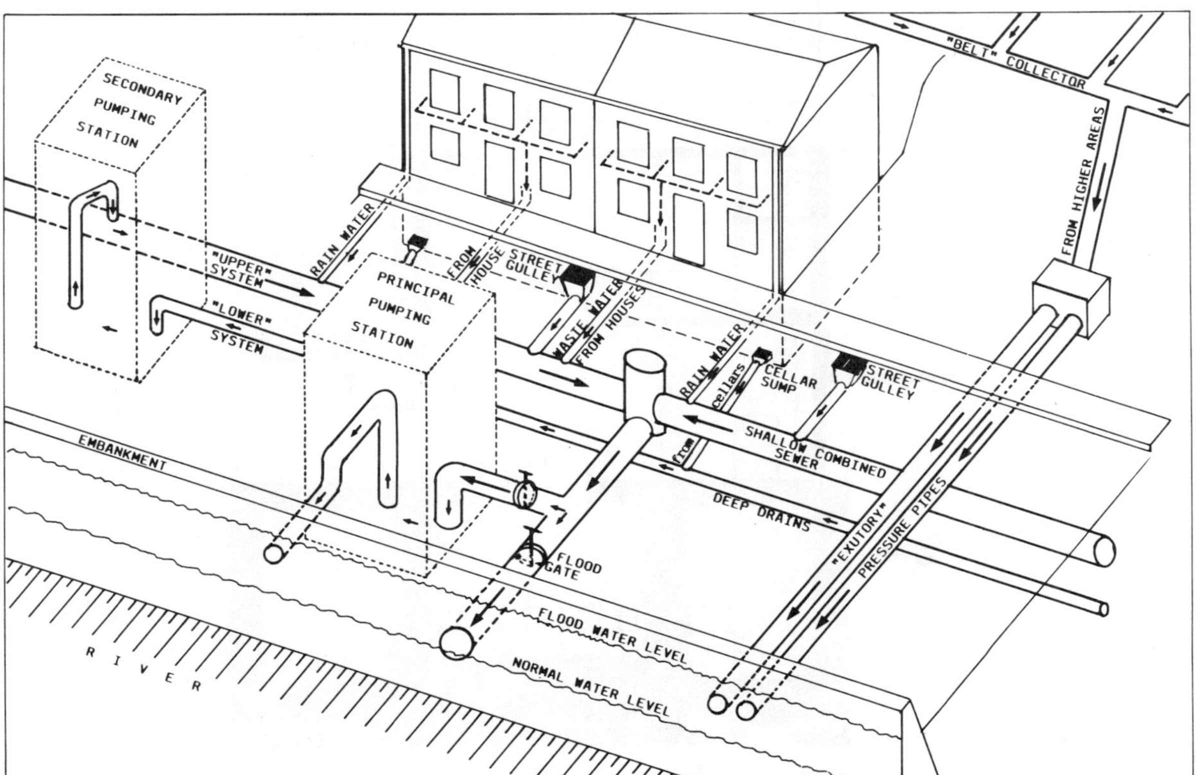

Figure 2. The Principle of the Multilevel Sewerage System.

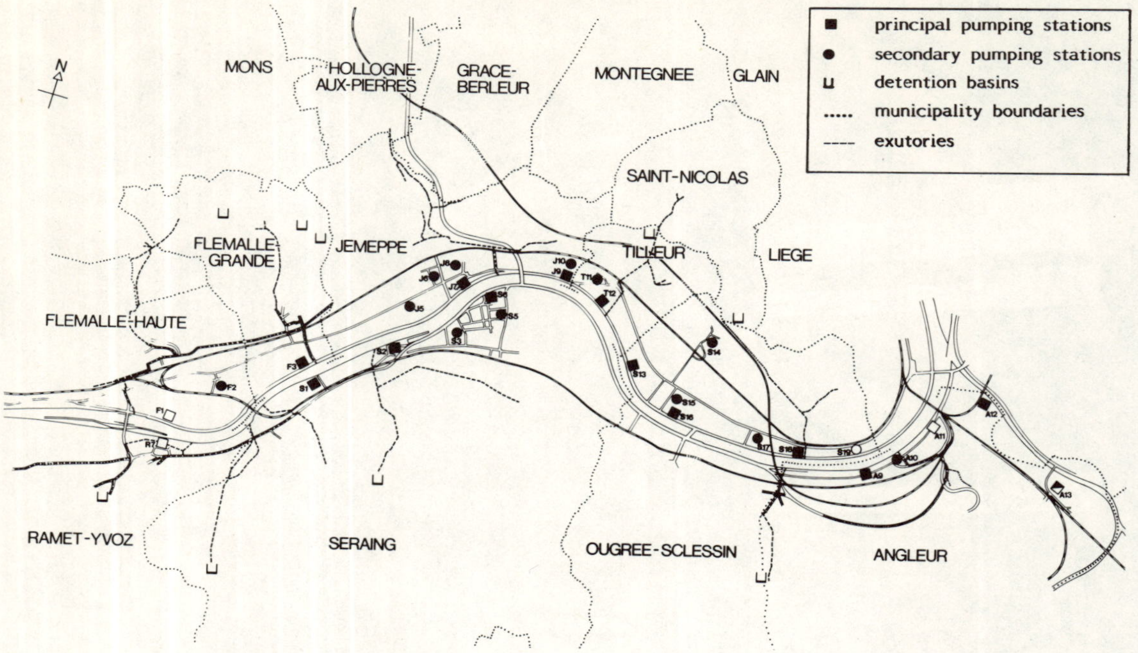

Figure 3. The Sewerage System upstream of Liege.

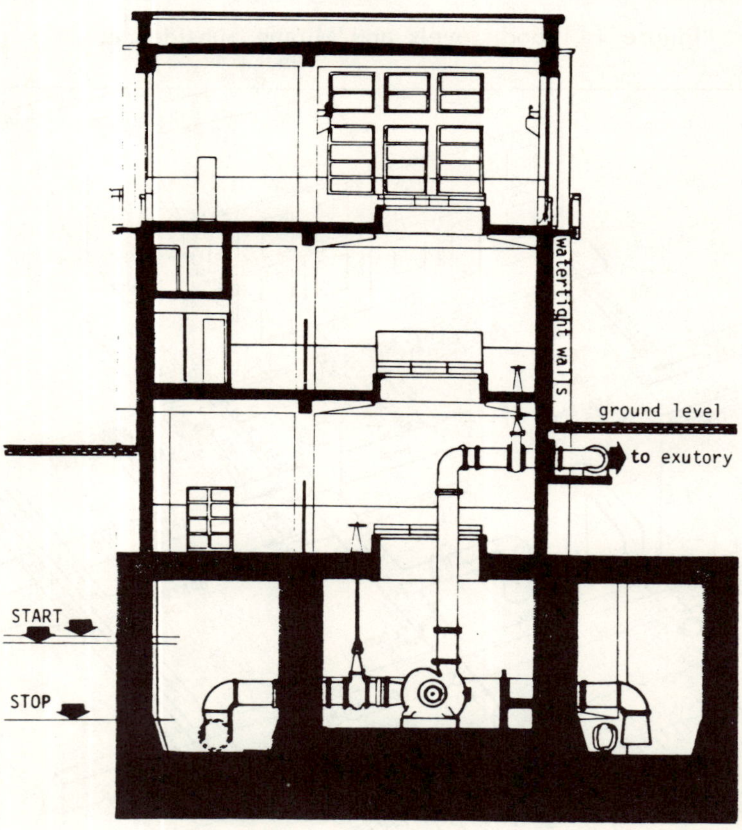

Figure 4. Typical Principal Pumping Station.

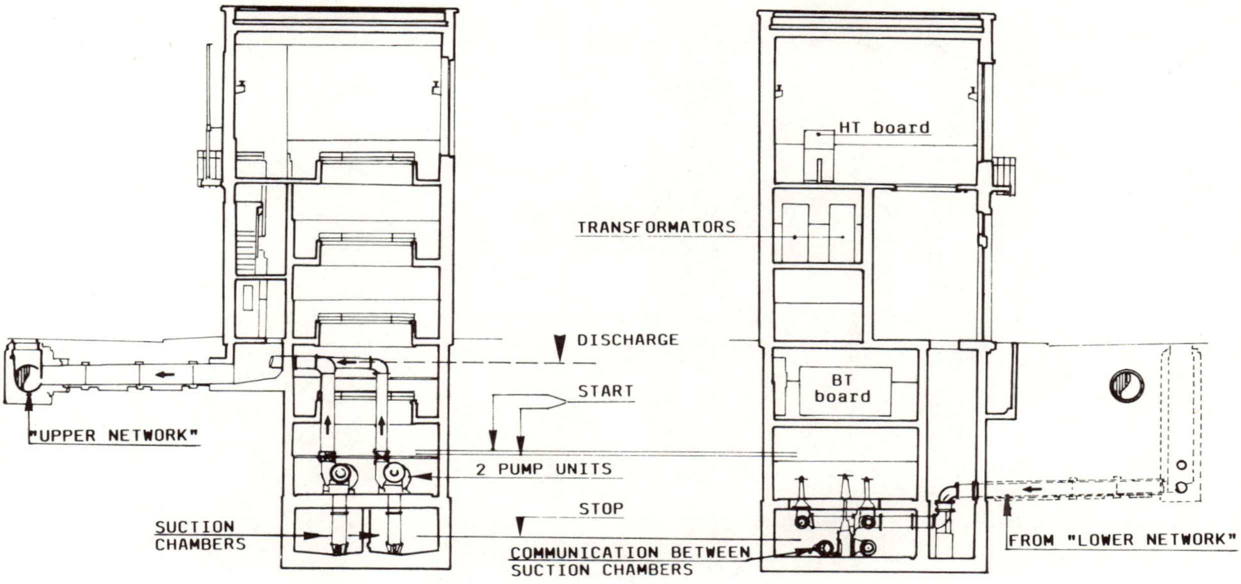

Figure 5. Typical Secondary Pumping Station.

International Conference on the
PLANNING, CONSTRUCTION, MAINTENANCE & OPERATION OF SEWERAGE SYSTEMS
12-14 September, 1984

PAPER K2

KING'S LYNN SEWERAGE

PLANNING FOR TOMORROW'S USE WITH YESTERDAY'S HERITAGE

M.W. Child B.Sc(Hons), C.Eng.,M.I.C.E.,M.I.W.P.C.

Anglian Water,
Cambridge Division,
ELY,
Cambs

I.H. Roebuck M.Sc., C.Eng.,M.I.C.E.,M.I.P.H.E.

Binnie and Partners
Artillery Row,
LONDON SW1P 1RX

Summary

The conflict between the expectations of the customer who requires a higher level of service, and the cost he is prepared to pay has been a catalyst in directing engineers to seek better value from our underground sewer heritage. In order to improve sewer performance it is necessary to know how the system works and to be able to test and cost different options under varying conditions. The Wallingford Procedure, by linking field data and computer simulation enables such investigations to be carried out with confidence. Much has already been written about the application of the Wallingford Procedure. This paper, provides an observation of one study, its objectives and strategies, and examines some of the practical aspects of data collection and analysis. The paper stresses the need for data accuracy and also describes a method of modifying hydraulic analysis to make allowances for sediment.

Held at The University of Reading, England. Organised and sponsored by BHRA, The Fluid Engineering Centre, Cranfield, Bedford in conjunction with the Water Research Centre.

NOMENCLATURE

Q	=	discharge
Qf	=	full bore discharge
Ds	=	depth of sediment
Dw	=	depth of water
D	=	diameter
Ks	=	pipe roughness factor

1. INTRODUCTION

King's Lynn in common with many urban centres in the U.K. has an underground sewer heritage that has been developed in a totally piecemeal fashion. Many of the older sewers in the town date back to our Victorian forefathers efforts to rid the town of disease and human squalor. Several of the larger sewers were constructed in the mid 1800's with subsequent major additions between the wars and in the 1950's. Since the second world war Lynn has seen a rapid development with the population rising from 20,000 to over 37,000 persons mainly as a result of an overspill agreement with the former G.L.C. This overspill development has also seen a re-development of the town centre. Whilst the activity of providing shopping precincts, pedestrianised areas, rear vehicular accesses and car-parks progressed on the surface, little or no effort was put into looking at the sewers underground or planning for tomorrow's use.

The Anglian Water Authority took over direct responsibility for the sewers in 1978 and as a result of a major collapse commenced a survey of the older man entry parts of the system. Several of the older brick sewers have now been successfully relined. In addition the Authority started strategic CCTV and locational surveys of the whole sewerage system in 1981/82. As more information became available and as a result of the frequency of flooding, the Authority decided that a full-scale study of the system was justified. Binnie and Partners were appointed to produce a report with the following objectives:-

(i) Ascertain the structural and hydraulic performance of the system.

(ii) Provide a detailed investment plan of renovation, augmentation and renewal for the next decade or so.

This paper discusses briefly the early development of the sewerage system and considers in detail some of the practical aspects of data collection and analysis completed to date. One of the problems encountered was the amount of sediment in the system. Details of a method of modifying data input for the computer simulation to make allowances for sediment in sewers is included.

Submission of the final report is expected in June 1984.

2. HISTORICAL ASPECTS

Whilst in many towns the problem was the need for channels and pipes for the removal of waste from the areas of population, in Lynn a ready made sewerage system was available by using the numerous watercourses or "fleets" (a channel filled by the tide and left shallow at low water). Not only did these channels or fleets receive a flush from the effects of the tide, but at low tide, freshwater from the surrounding higher ground further cleansed the system. Unfortunately the fleets became over-utilised, not only for the transport and transfer of goods to the neighbouring warehouses, but also for the tipping of domestic refuse from the notorious "muckhills". This over use unfortunately coincided with a reduction in freshwater flows because of the diversion of upstream water for town fortifications and water power for the town corn mills. The insanitary condition of the town, mainly as a result of the over use of fleets, reached its peak in the 16th century. The Fleets were no longer an asset, but had become a public health liability. The following comments illustrate some of the problems:-

1557

Three mayors required in one year because of deaths due to disease.

1597

Death rate highest in County 153/1,000 population.

16th Century Saying

"Ye rotten Borough of King's Lynn,
dead in trade but alive in sin,
famous for unmarried daughters and filthy tidal waters".

By the 19th century public health problems remained critical and cholera and typhoid were rampant. The public pressure for better sanitary conditions for the town made it necessary to fill in and culvert the insanitary fleets. (As shown in Fig. 1.). The following comments illustrate the environmental problems at that time:-

Early 19th Century Saying

"A nasty stinking sinkhole of sin which the map of the County denominates Lynn".

July 1832

Cholera outbreak. Many families extinguished. One such family of the name Ormiston is recorded:-

"The mother was buried on the Monday afternoon,
Tuesday at 6 a daughter was buried,
Wednesday at 9 another daughter,
Thursday at 8, the father,
.... in ten days, nine individuals of the same family buried".

1841

The death rate was the highest in the County of Norfolk.

1865

Reported by the "Times" a typhoid outbreak causing 40 deaths. "The people of Lynn have little taste for pure water.....they evidently prefer sewage and they like it neat!".

The following features of today's inherited system are note worthy:-

(i) The sewers are below high tide levels in the Gt. Ouse and the system is tide locked if a storm occurs during high water.

(ii) The sewers have been laid at flat gradients and in certain cases backfalls are present.

The combination of flat gradients and tide locking has resulted in sedimentation thereby reducing available capacity. Despite the expansion of the town no new outfalls have been constructed in over 40 years.

3. THE INITIAL STUDY

To meet the objectives of the study it was agreed at the outset that an initial report should be prepared to collate all existing data and make recommendations for the final study. It was considered fundamental to the management of the project to locate engineering staff responsible for data collection at King's Lynn.

The initial report identified five drainage areas subject to flooding which required more detailed study. This study area covered 15 square kilometres and included 190 kilometres of sewer, 27 sewage pumping stations, 2 stormwater pumping stations, 17 storm overflows and 6 interconnections between drainage areas.

A total of 150 flooding events in 56 different streets occurring between 1977-83 were grouped into the respective drainage areas. Each flooding event, its location, rainfall and high tide level were tabulated to determine the areas most prone to flooding and the conditions required to produce it.

On the basis of information collected, a detailed study period of eight months was recommended for additional data collection, analysis and completion of the final report.

Additional data collection was recommended as follows:-

Site Inspections.

(i) Additional locational sewer surveys to provide data for hydraulic analysis and for determination of drainage area boundaries.

(ii) Inspection of areas to determine permeable and impermeable surfaces.

(iii) Determination of the extent to which surface water connections have been made to separate foul sewers.

(iv) Inspection of overflows and tidal outfalls to ascertain their mode of operation, capacity and periods of closure.

(v) Measurement of sediment depths at regular intervals throughout the system.

(vi) Analysis of sediment samples.

Instrumentation and Tests

(i) Flow measurement at major collectors to validate the computer flow simulation model.

(ii) Flow gauging at interconnections between drainage areas and at minor overflows using white washed boards.

(iii) Pump performance tests at major pumping stations and pump hours run monitoring at minor pumping stations.

(iv) Hydrogen sulphide monitoring in areas where malodours had been reported.

(v) Rainfall recording at two sites to correlate rainfall intensity with measured flows and levels within the system.

CCTV Surveys.

(i) CCTV survey of a minimum of 5% of the sewers within the study area with location chosen to include principal and minor sewers, sewers of a variety of ages, and sewers of different forms of construction.

(ii) Sewer cleaning prior to CCTV survey of lengths known to contain substantial depths of sediment.

Methods of data analysis were recommended as follows:-

(i) Hydraulic analysis using the Wallingford Storm Water Package (WASSP program).

(ii) Computer analysis of CCTV survey results to include selective information retrieval and production of summary reports.

(iii) Computer production of longitudinal sections of main sewers and those in areas of flooding.

4. LOCATIONAL SURVEYS

In common with all Water Authorities Anglian Water has devoted a proportion of its annual budget to collection of information on the location and condition of our underground heritage. Sound validated data on locational surveys is a pre-requisite to both hydraulic analysis and reliable examination of subsequent options and costs.

The manpower to carry out the necessary locational survey work could not be provided from "in-house" resources and therefore contracts have been placed with specialist survey firms. It became apparent that the accuracy of much of the data was suspect; probably the result of contractors tendering for the work at too low a price.

Based on the experience of these early contracts new contract clauses (see Appendix I) were drawn up to ensure that the accuracy of the results could be monitored during the progress of the survey work. The first contract in Anglian Water using these new clauses was let in 1983 at King's Lynn. Performance monitoring under the new contract clauses showed that inaccurate surveys in line, level, and layout were being produced.

The majority of the errors have been attributed to the contractor's method of working. However, discrepencies between Ordnance Survey Benchmarks were also identified. To overcome this Anglian Water now provide all survey contractors with validated benchmark information.

The extent of the level errors found during the initial performance checking at Lynn is shown in Appendix II. Appendix II also includes recent information on level errors found on a survey carried out before the inclusion of the performance checking clause. In addition to the level errors illustrated errors in layout and line have been identified on both surveys

The tolerances included in the performance clause are considered to be those that can be achieved by good surveying practice. They are not necessarily those tolerances that should be applied because of the future use of the data.

5. PLOTTING LOCATIONAL SURVEY DATA

From locational survey data 1/1250 scale wall plans were produced and acted as a repository for information collected in the field. Drainage area boundaries and information on combined and separate systems, permeable and impermeable areas and those areas not draining to the system were plotted.

Field data sheets from locational survey data were also prepared for hydraulic analysis using WASSP. A typical sewer field data sheet is included in Appendix III. Care must be taken when translating field information to the computer code sheets; WASSP uses pipe numbers rather than manhole numbers and does not include all manholes. A cross reference pipe code number is therefore included on the field data sheets.

The field data sheet provided the means for the production of longitudinal sewer sections of major sewers and those sewers in areas of flooding. A computer programe called PEEWIT has been developed to rapidly produce finished drawings on a drum plotter.

The program has the following features:-

(i) Automatic generation of Al size drawings to any scale.

(ii) Plotting of services crossed by the sewer.

(iii) Plotting of description of areas crossed.

(iv) Calculation of sewer gradients and invert levels of intermediate manholes for which levels are not given in the data.

(v) Automatic splitting of the sewer branch if the length exceeds that of a standard sheet.

(vi) Automatic calculation of datum level for each branch.

The sewer sections will indicate visually the sewer gradients, highlight potential causes of problems, provide a further basis for the recommendation of solutions and provide a working record for Operations personnel.

6. FLOW MEASUREMENT

A contract was placed with a specialist contractor for sewer flow monitoring over a five week period using ten depth and velocity monitors with data loggers and two tilting bucket rain gauges.

The monitors were sited at major outfalls prior to overflows and at selected points along principal sewer runs. Readings were taken at two minute intervals, the data being retrieved on a weekly basis. Depth and velocity checks were made by independent means to ensure proper calibration of the instruments.

One problem was encountered at a number of sites with sediment covering the sensor head, and causing loss of velocity readings. Upstream cleaning of sediment and/or relocation of the sensor to a point of higher flow velocity was the solution in some cases. At other sites sediment was a continual problem and was dealt with as follows:-

(i) in the case of coarse granular sediment the sensor retaining ring was carefully placed on the sediment.

(ii) in the case of finer sediment a steel wedge, the same depth as the sediment was placed inside the retaining ring under the sensor head.

Both methods proved successful, but particular attention was required to record the profile of the sediment and make the necessary depth corrections.

Two tilting bucket rain gauges with a tip discharge of 0.5mm were used; one was located at the centre of the study area, the other at the extremity. One gauge proved defective for part of the survey period therefore limiting the rainfall information to one set of readings. It is suggested that on future surveys of similar size, three such gauges should be used to allow for the possibility of breakdown.

In order to supplement the flow monitoring data additional measurements were collected as follows:-

Type of Measurement	Reason for Measurement	Instrument
Level monitoring	To indicate levels of flow at minor overflow chambers and interconnections between drainage areas.	White washed gauge boards.
Pump output test at major pumping stations.	(a) To ascertain the true output of the pumps for use with hydraulic model. (b) To ascertain the efficiency of the pumps for comparison with pump manufacturers' curves.	Stop watch, water level monitor, pressure gauge ammeter.
Pump hours run monitoring at minor pumping stations.	To ascertain the period of pumping over weekly period to identify infiltration and the effects of storm water.	Pump event recorders.

Such tests were easy to conduct and have provided valuable information in understanding how the system operates and why problems have arisen. An example of this is highlighted at one pumping station where it was found that the pump was operating for approximately 40% of the time throughout the night. Night inspection traced the source to a burst watermain which was contributing a base flow of approximately 4 litres/second.

7. SEDIMENT MEASUREMENT

Despite the recently introduced cleaning programme, substantial quantities of sediment are present in the sewers which seriously affect the hydraulic operation of the system in two ways:-

(i) The cross sectional area of flow is reduced.
(ii) The hydraulic roughness is increased.

Fig. 2. shows the effect of sediment in a pipe assuming the sediment invert is ten times as rough as the sewer walls. It can be seen that a sediment depth of 10% pipe diameter reduces the full pipe discharge by approximately 20%.

Depths of sediment were measured throughout the system at regular intervals. Initially measurements were taken in the manhole from ground level by the use of a steel tube with rounded end to penetrate to the invert and another tube of comparable length with a flat plate at the end to rest on top of the sediment. This was found to be inaccurate due to turbulence in the manholes and the relative smoothness of channel inverts. Therefore, a series of hand held tools were developed for measurement in the pipe.

Sediment depths recorded for one drainage area are indicated below:-

Proportional depth of sediment Ds/D	Number of Measurements	Percentage
0 - 0.1	40	48%
0.1 - 0.2	14	17%
0.2 - 0.3	9	11%
0.3 - 0.4	14	17%
0.4 - 0.5	5	6%
0.5 and above	1	1%
	83	100%

8. HYDRAULIC ANALYSIS

The results of the sediment depth measurements and the knowledge built up about how the system functions indicated that the Lynn sewers had inherent sediment problems. The depths of sediment encountered were sufficient to influence the pipe discharge and therefore it was decided that the effect of sediment should be included within the computer simulation. A method was proposed to modify the input data and introduce equivalent pipes to model the true sedimented conditions. Two options were considered as follows:-

(i) Modify roughness only to give the equivalent full bore discharge of the sedimented sewer.

(ii) Modify roughness and diameter to give an equivalent full bore discharge and pipe section.

Adjustment of roughness only has the advantages that only one parameter is changed and all levels and diameters truly represent physical points in the system. Good representation of the depth/discharge curve for an equivalent pipe with modified roughness for sediment depths of 20% can be obtained. Above this figure the representation becomes increasingly inaccurate and in addition errors in pipe volume become significant.

Therefore, a combination of the above methods is being used. For sediment depths below 20% roughness alone is adjusted. For greater depths, roughness and diameter are adjusted.

Fig. 3 . indicates the accuracy of this method. The actual depth/discharge curve of a pipe sedimented to 15% of diameter compares favourably with the curve for the same diameter pipe with roughness increased to give the same full bore discharge. At 50% sediment depth, the actual depth/discharge curve again compares favourably with the curve for the equivalent pipe when diameter and roughness are modified to give the same full bore discharge.

The adjustments are derived directly from the Colebrook White equation and are dependent only on roughness hydraulic radius and cross sectional area. Changes of pipe slope and viscosity have little effect.

It is important that these adjustments are made to the WASSP data files in a consistent and controlled manner, to avoid errors being introduced into the model over a period of time, perhaps by confusion of true and adjusted values.

A pre-processor program has been written for the WASSP pipe data file. The program reads the pipe data file that contains actual pipe diameter and pipe wall roughness and combines this with the sediment file information. It then calculates new roughnesses and diameters and creates a new WASSP pipe data file containing these values. It also creates a file containing a record of all changes to roughness and diameter. This method of data preparation which is shown in Fig. 4. has the following advantages:-

(i) No editing is done on the file containing the pipe system data.

(ii) The record file plus the WASSP output file will give a complete record of all adjustments and results from a run.

(iii) Different sediment conditions are easily represented by new sediment data files.

To validate the model the largest rainstorm recorded will be translated into the form of an input hydrograph and fed into the WASSP program. At the point of flow measurement the simulated flow hydrograph and recorded flow hydrograph will be compared. The model will be acceptable if hydrograph peak flows are within 20% of each other, flow volumes are similar and hydrograph shapes are similar. If these criteria are not satisfied three main reasons for differences will be examined:-

(i) Incorrect sewer data eg. unknown overflows.
(ii) Errors in calculation of subcatchments, permeable areas etc.
(iii) Mathematical error in the model; some pipe configurations can cause mathematical instability resulting in oscillating simulated flow. These can be eliminated by minor adjustments to the pipe data.

Having validated the model for the existing system further runs will be made with a design storm event of 1, 2 and 5 years. The selected design event will be reconsidered with a cleaned out system to check the sensitivity of flooding to sediment deposition. Further runs will be required to establish the influence of tide locking, improvements to be gained by relief sewers, scope for storm water pumping and sensitivity to additional storage. Conclusions and recommendations will also cover the benefit of cleaning some or all lengths and future maintenance.

9. STUDY COSTS

The study and surveys will have taken ten months to complete at an overall cost of about £200,000.

The costs at first sight may appear high. However, the costs need putting in perspective as follows:-

(i) The Authority employs capital assets in the study area valued at £100m. at current price.

(ii) The study costs are equivalent to the cost of replacing about 0.5 kilometres of sewer in an area containing 190 kilometres of sewer.

(iii) An unacceptable level of service is being provided for the area (150 flooding occurrences at 56 locations in 5 years).

10. CONCLUSIONS

The demands of the earlier inhabitants of Lynn for a healthier town, improved sanitary conditions and the prevention of endemic diseases resulted in the filling of the obnoxious fleets and the conveying of all wastewater by piped systems to the tidal river. Today the customer requires a level of service such that flooding, malodours and nuisances are totally removed, and expects this provision to continue for tomorrow's use.

How much needs to be spent on our ageing underground heritage and how should the limited money available be put to the best use? These are the over-riding questions being faced by all Water Authorities; answers to such questions can only be formulated after careful investigation. The need to properly plan a strategy to achieve the objectives is of fundamental importance.

The use of new techniques and computer programs such as PEEWIT and WASSP allow the modelling of systems to be carried out with confidence. However, the soundness of the solutions is only as reliable as the accuracy of the data input. Proper emphasis should be placed on the validation of data and value for money considerations rather than speed of data collection at lowest possible cost.

The use of contractors for CCTV, sewer locational and flow monitoring surveys is seen as a natural extension of the continuing use of contractors in the Water Industry. However, such contractors need careful monitoring and checking in the field if consistent standards are to be achieved. The lowest tender is not necessarily the best value for money. Many of the inaccuracies of sewer locational surveys stem from contractors submitting too low a price. A similar situation may arise with regard to sewer flow monitoring. At the present time sewer flow monitoring contractors offer a high level of service at a price acceptable to both parties. However, as more contractors move into the market the tendency for price cutting may occur with consequential loss of service and reliability of data.

To find the correct solutions to problems in old sewerage systems, sewers should not be investigated in a piecemeal fashion, limited to trouble spots only, but whole areas should be analysed. It is equally important to ensure that practical engineering staff are located in the field to supplement by investigation and field studies the sewer locational and flow monitoring data.

The sensitivity of pipe discharges to inherent sediment problems is such that sediment depths of as little as 10% of pipe diameter will seriously affect the hydraulic analysis. Where known sediment problems exist consideration should be given to including in the hydraulic model an allowance to offset the sediment in the sewers.

The costs of the study and subsequent analysis may appear at first sight to be high (in this case £200,000). Such costs need viewing, however, in the light of benefits, levels of service and the replacement costs of assets employed. It can then be seen that the cost of planning for tomorrow's use with yesterday's heritage is a sound commercial investment and good value for money.

ACKNOWLEDGEMENTS

The authors wish to thank Peter F. Bullock, General Manager, Cambridge Division, Anglian Water and Ted Flaxman of Binnie & Partners for permission to publish this paper. Thanks are also due to numerous staff of both Binnie & Partners and Anglian Water for advice and assistance.

The views expressed in this paper are those of the authors and do not necessarily reflect the views of Anglian Water or Binnie & Partners.

APPENDIX I

SEWER LOCATIONAL CONTRACT CLAUSES

PERFORMANCE CLAUSE

The following measurement tolerances shall apply:-

Parameter	Tolerance	Application
Level	± 25 mm	Applies to all levels excluding manhole covers.
	± 50 mm	Applies to manhole covers only.
Location	± 300 mm	Applies to the location (using the measured distances on the record cards provided by the Contractor) of the centre of the manhole cover, rodding eye cover, lamphole cover etc
All other dimensions	± 50 mm	

On completion of the survey of each 100 manholes the Contractor will supply the relevant manhole record cards or drafts thereof and notify the Engineer who will then carry out a performance test within 4 weeks of that notification. The Contractor shall supply a survey team to carry out the performance test, under the supervision of the Engineer, and a person appointed by the Contractor as his representative.

The performance test will consist of a re-survey of up to 5 manholes randomly selected by the Engineer and checked against the results obtained by the Contractor in the initial survey. The test shall include the reproving of all associated pipework lengths.

The survey will have been deemed to have failed the performance test if any item of manhole measurement falls outside the tolerances stated in the Performance Clause. Notwithstanding any of the foregoing if in the opinion of the Engineer the survey of the other data on the manhole record card is sufficiently defective then the survey may be deemed to have failed the performance test.

If the survey satisfies the performance test the Contractor shall be entitled to payment for that survey and the performance test at the rates included in the Bill of Quantities.

If the survey fails to meet the performance test, the Contractor shall resurvey that portion of the work at his own expense. He will not be paid for the work involved in the performance test. When he is satisfied that the previously failed work has been corrected he should inform the Engineer and a further performance test will be carried out in accordance with the above procedure on the remaining 95 manholes. Perforamce tests will be repeated at the Contractor's expense until the Engineer is satisfied that this portion of the work has met the requirements of the performance test as stated in sub-clause 3.14.3. The Contractor will then be paid in accordance with Clause 2.26 for that portion of the survey and the final performance test.

APPENDIX II

SEWER LOCATION INVERT LEVEL ERRORS

KING'S LYNN

Of the 5% random sample about 30% failed the initial performance check for ± 25mm.

Majority of errors in the range ± 50mm. Maximum error + 1.04m.

VALIDITY CHECK OF PRE PERFORMANCE CLAUSE CONTRACT

Validity check carried out in late 1983 on survey work carried out in 1982. Of the 5% random sample 20% failed the validity check for ± 50mm. Selection of errors:-

M.H. No.	
821	+ 94mm
903	+ 132mm
331	+ 268mm
084	- 453mm
381	- 130mm
871	+ 76mm

Majority of errors in the range ± 75mm to ± 150mm. Maximum error + 2.256m.

APPENDIX III

TYPICAL SEWER NETWORK FIELD DATA SHEET

SEWER NETWORK FIELD DATA SHEET										JOB TITLE DRAINAGE AREA							SHEET NO.
Manhole No	Cover Level (m)	Exit Invert Level (m)	Main Entry Invert Level (m)	Side Entry Invert Level (m)	MH area (sq m)	Pipe dia (mm)	Pipe length (m)	Gradient	Pipe material	Pipe Code No	Residential properties No	Residential properties Flow	Business properties No	Business properties Flow	DWF (l/s)	Sediment (mm)	Notes

Figure 1. Culverting operations in 1865

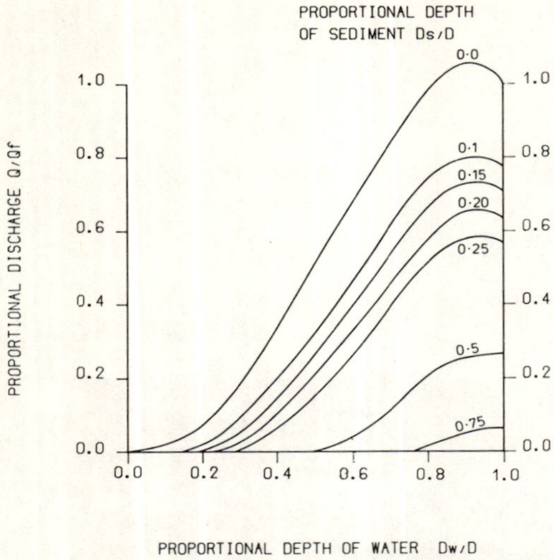

- 0.5m diameter pipe
- Ks walls = 3mm Ks sediment = 30mm
- Pipe gradient = 5m/km (not critical)

Figure 2.
Proportional depth/discharge curves for various depths of sediment

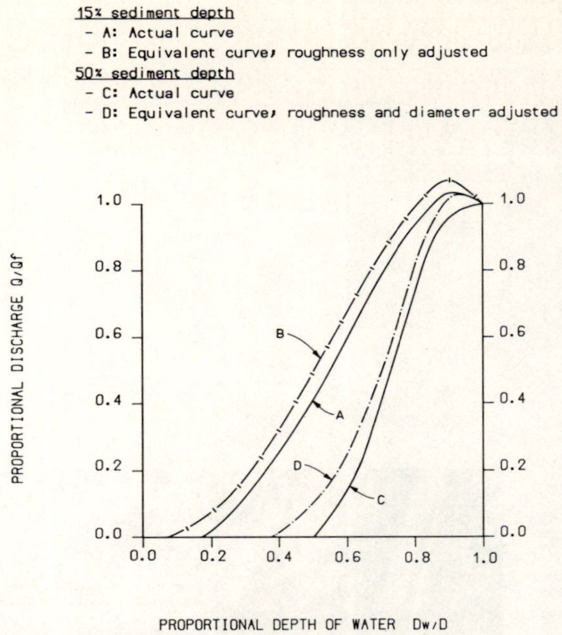

- 0.5m diameter pipe
- Ks walls = 3mm Ks sediment = 30mm
- Pipe gradient = 5m/km (not critical)

Figure 3.
Proportional depth discharge curves for actual pipes and equivalent modified pipes

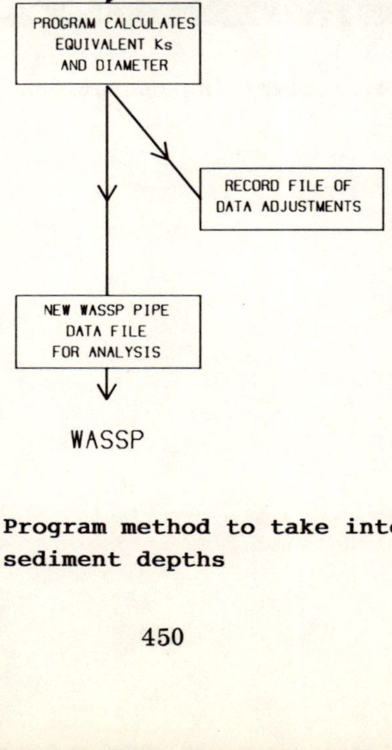

Figure 4. Program method to take into account sediment depths

International Conference on the
PLANNING, CONSTRUCTION, MAINTENANCE & OPERATION OF SEWERAGE SYSTEMS
12-14 September, 1984

PAPER K3

ASSESSING A SEWER NETWORK IN THE MIDDLE EAST

Mr. E. Walton, BSc, MSc, MICE, MIWPC, MIWES.

Miss J.M. Denner, MA, MSc, DIC, MICE, MIPHE.

Mr. J. Gay, BSc, MS, MICE, MIWPC, MIWES

BALFOURS Consulting Engineers
P.O. Box 1913,
Safat,
Kuwait.

BALFOURS Consulting Engineers
Yeoman House, 63 Croydon Road,
London SE20 7TW
England.

SUMMARY

Anticipating potential corrosion and other problems of the extensive asbestos cement sewer network in Kuwait, the Government commissioned Balfours in association with the WRC and others to undertake a 12-month field study to assess the extent of the problem and to make proposals for appropriate remedial measures.

The paper describes the following:-

1) CCTV inspection of 52km of pipe (ranging from 200mm to 800mm dia.), in-sewer still photography; examination of manholes; topographical surveys; measurement of sewer gas concentrations; flow recording programme and analysis of sewage samples.

2) Methods of data classification using micro-computer/word processing systems to assist in assignment of priorities, together with the use of computer programmes for rapid analysis of existing and proposed system hydraulics.

3) An evaluation of the applicability in Kuwait of non-man-entry sewer renovation techniques and manhole protection systems.

In conclusion the paper discusses the findings of the survey and recommendations for upgrading the sewer system.

Held at The University of Reading, England. Organised and sponsored by BHRA, The Fluid Engineering Centre, Cranfield, Bedford in conjunction with the Water Research Centre.

1. INTRODUCTION

In May 1982, the Kuwait Ministry of Public Works appointed Balfours, in association with Al Marzouk and Abi Hanna, to undertake a survey of a representative sample of about 10% of the sewerage system serving roughly half of Kuwait. The survey was to include CCTV (closed circuit television) inspection of the sample sewer lengths, together with flow and quality measurement at relevant locations and a topographical survey. A local sub-contractor was appointed to undertake the CCTV field work and associated sewer cleaning, and the UK Water Research Centre (WRC) was included in the consulting group as specialist advisors.

In addition to the survey, the Consultants were required to recommend a strategy for the renovation or replacement of those parts of the sewerage system (excluding pumping stations and rising mains) where remedial works were required to ensure that the system would achieve a reasonable working life. Tender documents were to be prepared for any proposed works.

Table 1 presents details of the sewerage system serving the study area: the sewers have generally been in commission for less than 12 years. Figure 1 shows the location of the study area.

Nominal Diameter (in)	(mm)	Length Installed (m)	% of total length
8	200	180,550	28.9
9	225	200	0.1
10	250	383,910	61.5
12	300	13,610	2.2
14	350	11,360	1.8
16	400	1,200	0.2
18	450	14,440	2.3
20	500	4,400	0.7
24	600	9,480	1.5
26	650	1,410	0.2
28	700	1,840	0.3
32	800	2,040	0.3

Table 1 : Size distribution of trunk and reticulation sewers in the Study Areas.

This paper summarises the scope and findings of the survey, and discusses sewer renovation options in the Kuwait context.

2. FIELD SURVEY WORK

The CCTV and level survey of the selected sewers in the study area and the manhole inspection were carried out over the period July 1982 to February 1983. Flow monitoring and assessment of sewage quality was commenced in November 1982 and completed in May 1983.

2.1 Topographical Survey

All manholes in the lengths of sewers inspected and also at other critical points in the system were checked against the available record drawings for level, depth and position. Revised record drawings and sections of the sewers were prepared during the study.

2.2 CCTV Survey

In total over 52km of sewers ranging in size from 200mm to 700mm in diameter were inspected by CCTV camera. Approximately 13km of trunk sewers were surveyed, the remainder being reticulation sewers. An average of 2800m of sewers were surveyed per week with sewer cleaning gangs working in advance of each CCTV unit.

The amount of sewer cleaning required was extensive and every effort was made to ensure that this was as unobtrusive as possible. The amount of sand and debris present in the sewers was far greater than expected and caused considerable cleaning problems and delays. Furthermore, many almost permanently surcharged sewer lengths made progress slower than expected.

The self-propelled BCS 85 system used for the CCTV work consists of a remote controlled camera with a near 100% distortionless wide angle lens and a fully automatic colour-still camera both mounted on a tractor with an automatic alignment system. The camera housing is lightweight, corrosion resistant and waterproof. The data console contained in the monitor van permits inspection information to be logged on the screen together with gradients and distances automatically calculated from sensor readings.

For survey work in pipe lengths where deposits of sand or debris could not be totally removed despite repeated cleaning, a camera mounted on skids was winched through.

2.3 Data management

Data records were prepared during the survey by manual charting of defects and significant features in standard forms based on recommendations by the WRC. These records were stored on floppy disc together with other field survey and inspection information and analytical results to provide a comprehensive data base with easy access for classification and interpretation. Indexed video tapes for each drainage area complemented these records and assisted with quality control checking of the information recorded by the field teams.

2.4 Flow measurement

120 flow measurement stations were selected along trunk sewers to determine the flow characteristics from each drainage area. The hydraulic performance of the system was assessed by a programme of flow measurement and hydraulic analysis. Two types of flow recorders were used: Marsh-McBirney chart recorders and the new Golden River Flow Survey Instrument Package.

(i) The Marsh-McBirney Type 265M Velocity Modified Flow Meter designed for location within manholes can measure both velocity and depth, compute discharge and record the data on a circular chart for either 24 hours or 7 days duration.

It monitors sewage velocity and sewage depth in open channel flow and combines these with the pipe diameter from a plug-in module to calculate the discharge. Average velocity is derived from a point velocity measurement made by a small streamlined velocity sensor using the electromagnetic effect and the sewage depth measurement is made by an air reaction gauge connected to an air line with its outlet adjacent to the velocity sensor. The velocity sensor and the air tube outlet are attached to an expanding circular band to form the sensor assembly and is placed inside the sewer and clamped against the wall. The configuration of the assembly is designed to shed debris and cause minimum disturbance to the flow.

(ii) The Golden River Flow Survey Instrument Package monitors the depth and velocity of flow and stores the data in the solid state memory of the Dataman component installed in the manhole.

The point velocity measurement is made using an ultra-sonic velocity sensor based on the Doppler effect and the measurement of flow depth is made by a silicon strain gauge type pressure transducer. These signals are processed in the Dataman prior to storage. After fastening the hydrodynamically shaped detector head to the appropriate size of mounting band, the band is inserted into the sewer and expanded against the pipe wall. The Dataman is programmed using the Retriever component with information relating to the site location, date, time, pipe diameter and interval of recording. Data is abstracted from the Dataman by interrogation using the Retriever and then transferred to the micro-computer for storage on disc.

2.5 Hydraulic Analysis

From the results of the flow measurements and population data derived from government census and planning consultants figures, the base flows, per capita sewage discharges and peaking factors were established. An hydraulic analysis programme was written for the micro-computer and used to derive the present and future flows and to assess the hydraulic capacity of both the existing system and renovated sewers or to determine the requirements for a new sewer. The results were presented in tabular form to highlight the hydraulic performance considerations for use in the assessment of remedial measures.

2.6 Sewage quality investigations

Sewage sampling was undertaken using automatic portable samplers at key locations in each drainage area. Each station was monitored for seven days with samples being collected daily and analysed in the Consultants laboratory established in Kuwait City. Individual and composite samples were analysed for biological oxygen demand (BOD), solids, chemical oxygen demand (COD), ammonia, chloride, sulphates, oil and grease and hydrogen sulphide.

3. A SYNOPSIS OF THE SURVEY FINDINGS

All sewers were constructed using asbestos cement pipes: some pipes have had both internal and external coatings of bituminous paint, often hand applied, others have had no protection at all. The distribution of sizes and lengths of gravity sewers in the project area are such that 92.7% are 300mm diameter or less while only 0.8% are larger than 600mm diameter.

3.1 Internal Structural Deterioration of Sewers

During the course of the survey, four distinct corrosion types were observed:

(i) Extensive exfoliation of asbestos fibre sheets. (Plate 1)

This appeared to be related to the method of manufacture. During manufacture, a cement impregnated asbestos fibre sheet is continuously wrapped around a cylindrical mandrel until the required pipe wall thickness is obtained. It appears that acid formed by the oxidation of hydrogen sulphide penetrates the first layer of fibre along a specific line of attack, which is generally either along the soffit of the pipe or just above the water level. The attack then proceeds into the next layer of fibre and behind the first one, causing de-bonding of the laminar structure. Where corrosion and exfoliation have occurred the fibre sheet has little residual strength and separates from the pipe wall, eventually dropping into the flow.

(ii) Nodules formed over large areas of certain lengths of sewer. (Plate 2)

In severe cases nodules were over 20mm in diameter with severe local disruption of the asbestos cement within the pipe wall. The method of attack is not known but is possibly caused by expansion of the cementitious content of the pipe material as it undergoes chemical change during corrosion. Susceptibility to attach may be related to the method of manufacture of the pipe or pinholes in the bitumen coating.

(iii) General corrosion which appeared as white bands along the soffit and/or the walls of the asbestos cement pipe. (Plate 3)

In some places there was no exfoliation and corrosion seemed to penetrate uniformly into the pipe wall over the area concerned. In other sewers, it appeared that corrosion penetrated behind the asbestos cement, and ultimately removes a layer from the whole of the affected area leaving a fresh surface which appeared to be undamaged. It is probable that the attack takes place along parts of the pipe wall either where the fluctuating water level occasionally wets the pipe (or where capillary action maintains damp conditions) or where condensation occurs on the roof of the pipe. Frequent wetting by immersion in sewage would wash away any acid formed before it could cause damage.

(iv) A pronounced ridge of corrosion product at, or just above, the normal water line where damp conditions may exist for long periods. (Plate 4)

The disruption and expansion of the pipe wall structure appeared to be uniform and the laminar structure of the asbestos cement clearly persisted in the corroded material. Generally this material is not washed off the sewer pipe, possibly because of the low flows pertaining. In some instances the capacity of the pipe was significantly reduced by this type of corrosion.

In cases where corrosion has progressed up the wall to the crown of the pipe the corrosion product may have become detached exposing fresh asbestos cement. In such cases the severity of attack may not be apparent.

Generally the corrosion products were bulkier than the original pipe wall. Measurements of these corrosion products indicated a dry bulk density of 0.84 g/cm^3 compared with a dry density of 2.03 g/cm^3 found for a piece of good asbestos cement pipe which had been in service for some years. A comparison of these indicated a volume increase of 142%.

A number of connections had been made to sewers between manholes using special side connection pieces. Visual evidence from the survey showed that these cast iron junction pieces were more susceptible to attack than the main pipe: all junction pieces had expanded greatly and some were exfoliated. The increased turbulence created at junctions strips off hydrogen sulphide in greater quantities than elsewhere and the pipe walls at such locations will also be subject to a greater extent of wetting thus producing conditions which are ideal for sulphide induced corrosion. The corrosion was often limited to the cast iron special junction pieces and there was a definite demarcation at each end of these specials.

Structural deterioration of sewers was found to be extensive. Table 2 shows the degree of corrosion for the lengths of trunk and reticulation sewer surveyed in each drainage area.

	Length Surveyed	Length of Trunk Sewer	Extent of Corrosion				Length of Ret'n. Sewer	Extent of Corrosion			
			S	M	L	OK		S	M	L	OK
TOTAL	52055	12870	6713	4528	1270	359	39185	9808	19006	5773	4598
%			52%	35%	10%	3%		25%	48%	15%	12%

Table 2 : Sewer Condition Summary

Note: All lengths are in metres.
S - severely corroded
M - moderately corroded
L - lightly corroded
OK- satisfactory

Displaced joints were numerous in many areas, indicative of pipe movement subsequent to the laying of the sewers. This was most probably a result of inadequate support to the pipe, caused by a lack of bedding to the pipe or lack of compaction to the bedding. Open joints were observed in several areas: these may have been the result of pipe movement but it is probable that poor workmanship during laying was the cause.

Cracked pipes were not common, but there were several in the industrial area. They were all severely corroded but it is not considered that this was the cause of failure as more severely corroded pipes without cracks were observed in similar locations at the same or greater depths. It is more probable that either the pipes were already damaged before laying or pipe support was inadequate due to poor workmanship during construction.

The CCTV survey did not show connection pipes protruding into the main sewers or other intrusions, such as tree roots.

3.2 Condition of Manholes

Throughout the study area there was evidence of corrosive attack in manholes. In the 2520 manholes inspected, walls were inspected below the cover slab and above the benching and the extent of corrosion noted. The findings are presented below with the percentage of observations of each corrosion classification being indicated.

State of Corrosion	Manhole top	Manhole base
Not significantly corroded	21%	15%
Lightly corroded	56%	53%
Moderately corroded	17%	21%
Severely corroded	6%	11%

Notes: Light corrosion was defined as step irons stripped of their galvanising and rusted, and aggregate exposed over large areas of concrete.

Severe corrosion was defined as a 50 per cent reduction in the step iron section in places or a loss of 50 mm of concrete over large areas.

Aggregate had been exposed on the benching of most manholes and in many cases parts of the benching had crumbled away. Corrosion of manhole covers and frames was not a serious problem. Trash baskets, slung below the manhole cover, were provided by the original designers in order to catch solids entering through the ventilation holes.

Problems associated with the maintenance of manholes were widespread. The following is a summary of observations made during the survey and Plates 5 and 6 illustrate the condition of manholes.

Manhole Defect	in category
Obstruction of flow	23%
fallen trash basket	6%
no evidence of trash basket	32%
surcharged	25%

Notes: Sand deposits were classed as obstructing the flow if they filled the channel to over half the pipe height.

It is not known how many surcharged manholes also contained obstructions to flow.

3.3 External Structural Deterioration of Sewers

Ground water attack of the asbestos cement sewers was considered to be a possible problem in Kuwait. It was known that superficial attack of asbestos cement water mains had been recorded. As relatively few of the water mains would have been below groundwater level, it was considered that sewers, which are generally deeper than water mains, could be expected to be more susceptible to ground water attack. Consequently, several sewers were exposed and three removed for inspection. In addition over thirty samples of asbestos cement pipe taken from burst water mains were inspected.

The water mains inspected were all unprotected pipes. With the exception of one example they all exhibited longitudinal fissuring 0.5 to 1.5mm deep.

Many fissures exhibited what appeared to be crystalline deposits running along their length. In one extreme case delamination had progressed from the longitudinal fissures and secondary lateral fissures had developed.

Although the fissuring of the water mains may have been stress assisted, there is no doubt that ground water attack of unprotected asbestos cement pipes does occur to some extent. The attack observed was generally only superficial but would probably have progressed gradually in time.

All three sewer pipes removed for examination had been protected by an external coating of bituminous paint as had one large diameter water main which was also examined. In all cases the coatings were still intact. Two of the pipes were approximately 700mm below ground water level. The other two were above ground water level. There were no signs of external attack on any of these pipes.

Sewers appeared to be adequately protected by the external bituminous coating, and if they can be renovated internally external deterioration should not limit their life.

3.4 Capacity of Sewers

The results of the flow monitoring programme showed that the majority of reticulation sewers were oversized. It was also apparent that some of the existing trunk sewers would not provide adequate capacity for future requirements. In some cases renovation would improve the capacity of sewers by reducing hydraulic resistance created by debris and the roughness of corrosion, but the vast majority of large diameter trunk sewers would require replacement.

For reticulation sewers the reduction in pipe diameter effected by slip lining or in-situ lining could be tolerated in most cases. In those instances where the diameter would be reduced excessively by sliplining an in-situ lining process was recommended.

Infiltration was detected in several trunk sewers and in isolated cases was large in relation to the sewage flow. The deeper sewers were more susceptible and therefore trunk sewers are more likely to be affected than reticulation sewers. Where infiltration was found to be a problem, and the pipes themselves are structurally sound, an in-situ type of lining would be most appropriate as some sealing of joints is achieved in the lining process.

In general the per capita domestic sewage flow contribution was lower than anticipated, averaging approximately 120 l/h.d for the areas studied. By reducing infiltration, renovation will extend the working life of some trunk sewers and pumping stations by reducing the hydraulic load. However where the trunk sewers are severely corroded and have only marginal capacity for future flows it would be cost-effective to replace them.

3.5 Sewage Characteristics

The sewage strength measured during the study varied widely with biological oxygen demand (BOD) ranging from 150 mg/l to over 700mg/l. However the strength of sewage is influenced by the water consumption, the amount of infiltration and socio-economic level of the area. From the results the per capita BOD discharge varied between approximately 50 to 65 g/h.d. Such values are typical of sewage elsewhere and analyses at the main sewage treatment works at Ardiyah gave an overall per capita BOD (including industry and commerce) of 60 g/h.d.

The chemical oxygen demand (COD) in most residential areas varied from 2 to 3 times the sewage BOD generally being between 450 and 700mg/l. This ratio is typical of purely domestic sewage. In the Industrial areas much higher COD values were found averaging over 1000mg/l, with peaks of over 3000mg/l.

Large amounts of oil and hydrocarbons were frequently observed in many sewers in the industrial area and occasionally in residential areas. The oil and grease content of the sewage in the industrial areas ranged from 440 to 2400mg/l, being significantly higher than the typical range of 50 to 150mg/l found in residential areas.

The sulphide concentration in the sewage was found to vary with the age of the sewage. At locations less than 2km from the head of a sewer the concentration of hydrogen sulphide occasionally exceeded 50 mg/l. The sulphide concentration in the atmosphere in manholes rarely exceeded 4ppm as the manholes are generally well ventilated. Occasionally, however, where there was turbulence in the manhole, the concentration in the atmosphere exceeded 50 ppm.

Ammonia concentrations generally varied with the flow from 30 to 100 mg/l. In small catchment areas the night-time concentration was significantly higher indicating that the majority of water was used for the flushing of WCs during this period.

Large quantities of sand and silt were measured in the sewage although accurate sampling was made difficult by the tendency for solids to settle out along the sewers as a result of surcharging.

Chloride concentrations were found to vary widely throughout the day in most residential areas, related to the use of the separate brackish water supply for gardens, yard hosing etc., and to infiltration of groundwater into the sewerage system.

3.6 Maintenance Problems

There was widespread evidence to indicate that the current level of the Ministry's considerable maintenance operations were still inadequate to cope with the problems which occur. Certain original design aspects have resulted in a higher maintenance requirement than could have been expected:

(i) the selection of 250mm diameter as the minimum size for reticulation sewers in most areas has led to low flow velocities in the majority of sewers;

(ii) the gradients of sewers in most areas are insufficient to achieve a self cleaning velocity at full or half full flow. Field measurements showed velocities in some trunk sewers to be less than 0.4m/s;

(iii) reticulation sewers have been laid on both sides of most roads in residential areas with low flows in both sewers;

(iv) the adoption of the ventilated manhole covers, which are in many cases below the surrounding ground level, has enabled copious quantities of sand to enter the sewerage system.

(v) galvanised mild steel trash baskets installed below to manhole covers to collect solids have become corroded extensively and many have subsequently collapsed into manholes creating obstructions to flow and permitting the sand passing through the ventilated covers to enter the sewer.

Low per capita sewage discharge found in many parts of the survey area has exacerbated the low velocities. As a result debris has been deposited throughout the sewerage network and it was apparent that the current scale of maintenance operations has been unable to cope.

Obstructions in manholes were common. Although most appear to have been initiated by the unfortunate combination of galvanised steel trash baskets and ventilated manhole covers coupled with a lack of routine inspection, many obstructions were caused by the deliberate dumping of rubbish into manholes. This problem was particularly noticeable in the industrial and commercial districts.

Surcharging of sewers and manholes was also widespread, much of it being caused by obstructions. However, in several areas temporary non-operation of pumping stations resulted in persistent surcharging over large areas.

4. SEWER RENOVATION

As 99% of the sewers are 600mm diameter or less, any renovation technologies to be considered must be based on non man-entry systems. None of the alternative systems available for small pipes has substantial proven applications, of say 10 years or more, particularly in an aggressive sewer environment such as that found in Kuwait. Therefore the review of renovation options and subsequent recommendations were based on an engineering interpretation of the best available information and not upon an evaluation of proven methods.

4.1 Functional Requirements of Renovation Systems

When considering renovation options it is important to establish whether renovation is required to:

(a) restore the structural integrity of the sewer pipes affected by the corrosive environment;

(b) prevent further deterioration of the pipes (i.e. protect those existing pipes which are corroded but still structurally stable);

(c) control infiltration.

(d) improve flow characteristics

The findings of the sewer survey work supplemented by information obtained in discussions with Ministry staff, resulted in the following observations:

- no cases of structural collapse or failure of the existing sewer system were observed in the surveyed sewer lengths and it is understood that collapses are very infrequent;
- there was substantial visual evidence of internal corrosion of the pipes ranging from mild to very severe;
- infiltration was observed in several locations and primarily occurred at joints in deeper sewers and at the joints between manhole bases and rings;
- the exteriors of sewer pipes coated with bituminous based paint have been relatively unaffected by groundwater attack;

The findings showed that the prime reason for renovation in Kuwait was protection of the system against internal corrosion.

4.2 Technical Evaluations of Options

For the reasons already given, the renovation systems which were to be considered relate to the continuous lining of non man-entry asbestos cement sewers and included:

- Pipeline insertion (sliplining);
- In-situ linings (Insituform type);
- Spray linings and coatings.

In reviewing sewer renovation in the Kuwait context, certain important points emerge:

(i) Large-scale capital expenditure will be necessary for sewer renovation in Kuwait and it would be ideal to carry out a comparative testing programme of the options over a number of years in order to select the preferred technique, but the condition of many of the existing sewers is such that the time is not available.

(ii) It was not possible in 1983 to recommend spray linings as a sewer renovation technique for Kuwait because the state-of-the-art of the remote spray lining of irregular surfaces, especially those which are difficult to prepare effectively, was not considered sufficiently advanced. It is, however, a technology that should be continually reviewed.

(iii) In several areas of Kuwait the sewerage reserve is fairly narrow with other services in close proximity and in order to avoid damage to other services excavation would have to be undertaken very carefully, particularly where dewatering is necessary.

(iv) In many cases the sewers to be renovated are relatively short lengths under major roads where disruption must be kept to a minimum.

(v) Provided the sewer network is properly cleaned on a planned programme, renovation by Insituform or sliplining will be possible where hydraulic capacity is not limiting. Where hydraulic capacity is limiting, the sewers will require relaying unless some of the flow upstream can be diverted into another main sewer which has spare capacity.

(vi) The CCTV survey has shown incidences of displaced joints to be common and consequently any sliplined pipe will have to be smaller than the existing sewer diameter and extensive grouting operations may prove necessary.

(vii) To avoid the need for existing connections to be re-made to the renovated sewer, diversion of the connection to the nearest manhole should be considered.

(viii) Very few manholes have been inspected externally, but corrosion of the buried concrete of other exposed structures has not been observed. It is not thought that manholes will deteriorate due to ground water attack.

It is not within the scope of this paper to discuss the technical aspects of alternative systems, but Appendix 1 summarises their relative merits in the Kuwait context. As can be seen in the Appendix tables, Insituform type systems have certain advantages for much of the Kuwait application.

Final selection of the renovation system as an alternative to relaying, will be governed to a large extent by a comparative economic evaluation. The tender documents subsequently issued for the first renovation contract requested prices for both systems to ensure commercial competitiveness. However the study cost estimates showed that Insituform is marginally cheaper than sliplining in Kuwait particularly for the larger sewer sizes. In general, renovation cost were estimated to be between about 50 and 80% of the cost of a new sewer depending on size and depth and whether the sewer was beneath a road or a verge (see Table 3). These costs do not reflect the additional benefits of lack of interference to traffic, amenity and other services.

	Renovation as a percentage of new cost			
	250mm dia		450mm dia	
Depth	slip lining	In-situ form	slip lining	In-situ form
2 - 3	74	77	96	90
3 - 4	64	65	91	84
4 - 5	54	49	86	74
5 - 6	49	43	80	67

Table 3 : Renovation cost as a percentage of new sewer cost at same depth.

Note: The above figures do not include for road reinstatement as most sewers are in verges, or for traffic diversion or other intangible costs. If road reinstatement were included the <u>relative</u> cost would be lower for slip-lining and further reduced for Insituform.

5. CONCLUSIONS

The Kuwait Ministry of Public Works showed considerable foresight in commissioning a comprehensive survey of the older parts of the sewerage system. The separate sewer system in Kuwait is typical of several in the Middle East i.e. asbestos cement sewers operating in a hot climate.

The survey highlighted the advanced state of corrosion of large parts of the system particularly towards the downstream sections of the system and also downstream of small pumping station discharges. This corrosion results from acid attack due to the formation of hydrogen sulphide in the septic sewage. Design features and maintenance problems exacerbate the problems of septicity; particularly since they result in the presence of large quantities of sand and other solids.

Whilst a CCTV survey of between 5 and 10% of sewers is considered to be a reasonably representative sample, thorough cleaning and inspection of each sewer length will be required (to evaluate its condition) prior to any renovation.

Continuous flow measurement in sewers carrying heavy silt loads and having partially corroded sections is difficult and it is preferable to use equipment which has previously undergone extensive field trials. It is however, an important exercise in order to provide information for hydraulic analysis of the system.

Sewer renovation is a relatively new field of technology which is developing rapidly. Whilst renovation by Insituform and slip lining has been proposed in Kuwait for those sewers which will not by hydraulically overloaded in the future, each case should be considered separately as technology advances. Ideally it is preferable for sewerage agencies to anticipate deterioration at a sufficiently early date to permit local pilot trials of alternative methods of renovation before placing large contracts.

In conclusion, it is considered important that information on renovation studies and contracts, both successes and failures, should be published to help engineers and scientists tackle the problem of deteriorating sewerage assets in a wide range of local circumstances.

Acknowledgement

The Authors wish to thank the Ministry of Public Works, Kuwait, for their permission to publish the results of the survey and for the information and extensive assistance provided by their staff.

APPENDIX 1 - Sheet 1

Relative technical merits of sliplining as a technique for sewer renovation

Advantages	Disadvantages
+ Liner material can be inspected and tested prior to insertion.	- Displaced joints may require liner to be significantly smaller than existing pipe diameter.
+ Less interference to traffic and public amenity than relaying of sewers.	- If HDPE liner is used it has significant wall thickness thereby reducing the diameter of renovated sewer.
+ Developments in slip-lining techniques permit short lengths to be slip-lined with less disruption than with Insituform.	- Substantial excavation, reinstatement and temporary works required particularly in areas of high groundwater.
+ Smoothness of liner may improve flow (and solids transport) characteristics.	- It is necessary to grout the annulus between liner and existing pipe at manholes to prevent ingress of ground-water and complete filling of the annulus between liner and existing pipe will be necessary if long term structural strength of existing pipe is in doubt.
+ Several competitive contractors to undertake work.	- Jointing of liner lengths requires close attention.
+ Can undertake longer single lengths than Insituform where conditions permit	- Excavation necessary to form side connections.
	- Problems can be encountered if towed liner becomes jammed for any reason.
	- Where significant infiltration exists all joints require to be pressure tested and sealed prior to grouting operations.

APPENDIX 1 - Sheet 2

Relative technical merits of Insituform type system as a technique for sewer renovation	
Advantages	**Disadvantages**
+ Minimum temporary works and interference to traffic and public amenity	- A patented technique with a small number of licensed contractors thereby limiting competition.
+ Minimum installation time	- Long term durability and strength of composite liner unknown in Kuwait conditions.
+ Liner follows the cross-sectional profile of the existing sewer pipe thereby maximising the diameter of the renovated pipe.	- No standard specification currently available for Insituform liner material.
+ Not significantly affected by displaced pipe joints and can accommodate bends with only minor wrinkling.	- Problems can be encountered if setting of the lining occurs out of correct alignment for any reason.
+ Liner and existing pipe are contiguous with any localised depressions or pockets in pipe wall capable of being filled by excess resin being provided in the liner.	
+ Smoothness of the liner may improve flow (and solids transport) characteristics.	
+ Excavation may not be necessary for side connections on larger pipes as in-pipe cutting systems have been developed.	
+ Does not require the sealing of leaking joints prior to lining.	

APPENDIX 1 - Sheet 3

Spray Linings

In-situ spray lining of small diameter water mains using cementitious materials has been undertaken widely for many years. More recently, thermosetting resin sprays have been used to line iron or steel water mains where the objective has been to prevent corrosion rather than to contribute to any structural strength.

Important factors in spray lining are the condition of the surface to be coated and its preparation so that the coat can adhere effectively. If the completed spray linings are not totally free from small defects they will eventually deteriorate and peel off as has happened with the bituminous lining of AC pipes. In existing small diameter sewers it would not be possible to check thoroughly the quality of the finished surface.

The difficulty of adequate preparation of the surface of moderately or severely corroded sewers for the acceptance of a spray lining, presently renders the process unsuitable for all but lightly corroded sewers.

Spray-on lining would almost certainly require all major sources of infiltration to be sealed, by means of remote control devices, prior to commencement of spraying. It would, however, have the advantage that side-connections may be more readily dealt with.

The ability of currently available spray-on linings to withstand any sewer cleaning operations in the future is questionable.

The Water Research Centre are investigating thicker spray coatings, incorporating fibres, which would also have structural strength but this technique is still in a development stage.

Plate 1. Severe corrosion with extensive exfoliation of asbestos fibre.

Plate 2. Lightly corroded sewer with nodular corrosion over most of pipe surface.

Plate 3. Medium corrosion progressing from general corrosion.

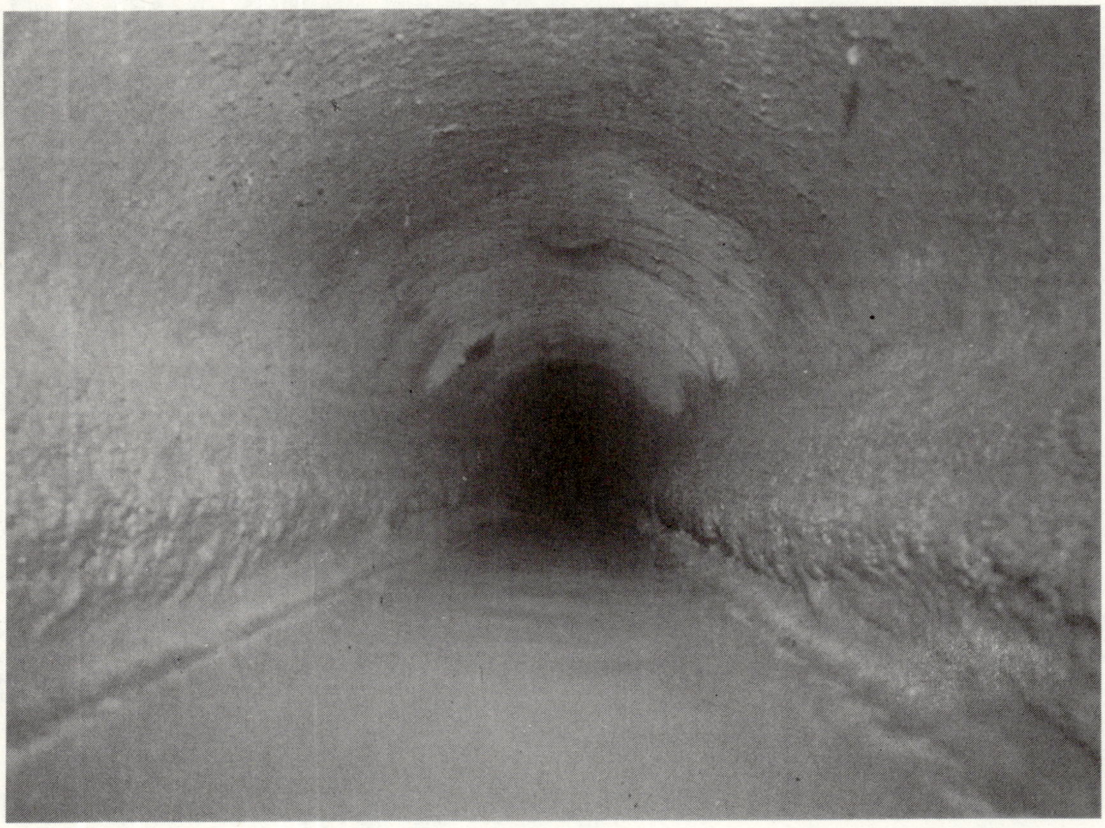

Plate 4. Medium severity corrosion with pronounced ridge of corrosion product just above the water line.

Plate 5. Typical fallen trash basket and blockage in manhole.

Plate 6. Debris removed from manhole in industrial area. The cover of this manhole was in place and intact.

International Conference on the
PLANNING, CONSTRUCTION, MAINTENANCE & OPERATION OF SEWERAGE SYSTEMS
12-14 September, 1984

PAPER K4

THE IMPROVEMENT OF SMITH STREET SEWER, LONDON

K.J. Riddell and M.J. Marriott

C.H. Dobbie and Partners
154-160 Croydon Road
Beckenham
Kent
U.K.

Summary

The Smith Street Sewer in London is a combined sewer serving a population of around 19,000 in a built up catchment area of 130 hectares.

The decision to reline the existing sewer and construct a deep stormwater relief sewer is described. The adopted Scheme provides both structural improvement, and reduction in storm surcharge levels. Design considerations related to these two main aspects of the improvement works are discussed.

The choice of a precast gunite segmental lining for this man entry, brick built sewer is presented, with description of alternative materials considered for relining. Details are given of the design and construction of the selected system.

Emphasis is made of the importance in sewer renovation of considering both hydraulic and structural aspects. The overall improvement of Smith Street Sewer provides an example of this approach.

Held at The University of Reading, England. Organised and sponsored by BHRA, The Fluid Engineering Centre, Cranfield, Bedford in conjunction with the Water Research Centre.

1. Introduction

The Smith Street sewer serves an intensely built-up area comprising parts of the Royal Borough of Kensington and Chelsea and of the City of Westminster. The sewer is served by Kensington Road and Brompton Road branch sewers and drains an area of some 130 hectares which includes the Victoria and Albert Museum, Brompton Oratory, Harrods department store and part of the Royal Hospital. Although there is a lot of movement of population to and from this area during the day, these various movements are thought to balance and the total population based on the 1971 census, is estimated to be 19,000. A plan of the area is shown in figure 1.

Besides acting as main sewers for the area, the sewers also serve as local drains for the streets along which they pass, and there are numerous small drainage connections from buildings and road gullies. The system is completely combined, the main sewers carrying both foul sewage and surface water. It has not been possible to establish exactly when the sewer was constructed, but its route can be followed on a drawing signed by Bazalgette dated 1856 and also on some older, but undated, drawings. Bazalgette was Engineer to the Metropolitan Board of Works, and has been widely acclaimed for his work in planning and implementing London's comprehensive sewerage scheme. (Ref. 1).

The sewer starts in Kensington Road some 70 m east of Exhibition Road and runs east before it turns south along Rutland Gate to enter the Brompton Road about 120 m. east of the Brompton Oratory. Here, it is joined by the Brompton Road branch, and continues south-west, along the Brompton Road, to Thurloe Place where there is an offtake to Low Level Sewer no. 2. The sewer continues beneath Brompton Road, crosses over the London Transport Underground Railway in a specially designed cast iron culvert and continues, via a number of roads (including Smith Street) and the grounds of the Royal Hospital, to outfall into the River Thames almost opposite Battersea Park Pier. The last 100 metres or so, before the outfall, was constructed in the second half of the nineteenth century when the Royal Hospital Creek was filled and the Chelsea Embankment was constructed. This length contains an offtake to Low Level Sewer No. 1, running under the Chelsea Embankment. The outfall is flap-gated as a significant length of the sewer is affected by the tide level in the River Thames.

The Smith Street Sewer is constructed in brick of varying cross-sectional shapes and sizes. A number of these are shown in figure 2. The original construction has served remarkably well for over a century and a quarter, but, close inspection showed that the structure was cracked and fractured in places, particularly at the crown, and a number of bricks were missing completely. There was also evidence of significant infiltration, deformation of the sides and crown and wear of the invert.

The overall size of the sewer seems to have been dictated by the requirements of man-entry rather than hydraulic capacity, although the sewer does, generally, increase in size in the downstream direction. However, in view of the tidal influence on the outfall and the limited offtake to the Low Level Sewer No. 1, the Smith Street Sewer can surcharge to ground level during times of heavy rainfall. This exceeds the levels of many basements in the drainage area and incidence of flooding has been recorded. It is therefore from both hydraulic and structural points of view that the decision was made to renew the Smith Street Sewer.

2. The Choice of a Scheme

The initiative for the improvement of Smith Street Sewer was generated in 1973 by the Public Engineering Services Committee of the Greater London Council, who at that time were the controlling Authority. A detailed investigation of the condition of the existing sewer was undertaken and, at this time, a hydraulic analysis was carried out for steady flows only.

From such information as was obtainable from water supply records etc, dry weather flow was estimated at 432 litres per head per day. Of this, 20% of the total flow was caused by infiltration, which was to be expected in an old brick sewer of this type. The analysis was carried out for a combination of twice dry weather flow coupled with a Ministry of Health storm which was a standard design criterion for storm sewers at this time. Any new works were to be designed to convey the above flows against a Thames level of 4.57 ODN.

The scheme proposed at this stage involved the complete reconstruction of Smith Street Sewer, with the exception of the short length of cast iron culvert, at an estimated cost of £2.65 m. It was recommended that this reconstruction should, more or less, follow the line of the existing sewer and be constructed wholly in tunnel, using bolted concrete segments varying from 1476 to 2086 mm internal diameter, to ameliorate the inevitable problems that would be caused by existing services and traffic disruption if the job were to be carried out in open cut.

Shortly after presentation of the scheme outlined above, responsibility for sewerage in London passed to the Metropolitan Public Health Division of Thames Water. They were mindful of the high estimated cost of a reconstruction scheme and also the high degree of uncertainty in such an estimate due to the anticipated problems in maintaining flows, encountering services, traffic disruption and the making of over six hundred side connections. For the above reasons, it was decided to investigate an alternative solution to the hydraulic and structural problems of the Smith Street Sewer.

The alternative considered was to separate the hydraulic and structural components of the scheme by providing additional flow and storage capacity by means of a new relief sewer constructed in deep tunnel and to reline the existing Smith Street Sewer in order to prevent its further deterioration and maintain its safety and efficiency in carrying low flows. The route of the proposed Relief Sewer is shown in figure 1.

In order to assess the performance of this alternative, a more detailed hydraulic analysis was undertaken. A schematic diagram of the network analysed is shown in figure 3. The tunnel proposed was to be 2.54 m in diameter and 1475 m in length yielding a total storage volume of 7,474 m^3. It was to be fed by an air-entraining siphon designed such as to allow levels to build up in the Smith Street Sewer almost to soffit level at the offtake point, and then to maintain this level within a tight margin irrespective of draw-off rate into the relief tunnel. Rainfall frequency analysis was based on the recommendations contained in Vol II of the 'Flood Studies Report' (Ref 2) and the storm profile used for the generation of storm hydrographs was the '50% Summer' defined by the Meteorological Office in the same volume. A range of durations and return periods were analysed in order to determine the onset of various operating states of the system.

The relief sewer is in deep tunnel and, when full, operates as an inverted siphon providing additional flow capacity for the system as a whole. Prior to this it merely provides additional off stream storage and, after each occassion it is called into use, requires to be evacuated by pumping. A simplified routing analysis showed that the most critical storm duration for storage volume requirment, during periods of high Thames level, was 3 hours and that the volume available would be equalled or exceeded an average of once every two years. Flow hydrographs for this situation are shown in figure 4 and it will be seen that, for a considerable distance downstream of the offtake, the flow in the Smith Street Sewer reverses during the peak of the storm.

At higher return periods, with the Relief Sewer acting as an inverted siphon, flood routing effects do not significantly reduce peak flows. Figure 5 shows the surcharge levels that may be tolerated along the Smith Street Sewer. Under these conditions, the total outflow from the combined system is 6.07 m^3/sec, which may be expected to occur slightly more frequently than once every 2 years. However it must be borne in mind that this occurence must be preceded by an event capable of filling the tunnel and be simulaneous with a high Thames level. It is estimated that the return period of overall 'system failure' is about once in ten years.

The cost of this alternative scheme, comprising tunnel and relining work, was estimated in 1978 to be £2.71 m. The original scheme estimate, updated to 1978, was now approximately £5 m. and therefore the alternative scheme, in addition to other advantages, showed a significant saving in capital cost.

3. Relining of Smith Street Sewer

With the decision described above to renovate the Smith Street Sewer over a length exceeding 3 kilometres consideration was given to a number of different techniques.

The main materials considered were Glass Reinforced Cement (GRC), Glass Reinforced Plastic (GRP), Gunite, and Polyester Resin Concrete (PRC). These were all considered in the context of a preformed lining with a grouted annular surround acting compositely with the original structure, as shown in figure 6.

With the need during construction to maintain flows, which would at times fill the sewer completely at short notice, it was considered advisable to limit the necessity of internal strutting during the grouting operation. With the existing cross section typically having high straight vertical sides, linings with significant inherent bending strength were preferred. This pointed to either Gunite or PRC of 40 mm thickness. Of these two materials, preformed gunite was preferred, being more amenable to local irregularities by use in conjunction with some insitu guniting, and therefore considered more economic in this context.

The sewer relined with precast gunite panels shows a loss in volume or cross section of around 30%, with a reduction in conveyance when flowing full of between 20% and nil, depending on how pessimistic an assumption is taken for the roughness of the existing brickwork. In this case, a reduction in conveyance approaching 20% is considered to apply.

However, as the sewer is more than adequately sized for flows during dry weather, and in view of the planned construction of the Smith Street Relief Sewer, this reduction in conveyance when flowing full was considered acceptable. The constructional reasons in favour of a thicker lining therefore prevailed.

Factors which may have altered the choice of lining material, had circumstances been different, include:

(i) a need to maximise conveyance, which would have favoured thinner linings than the one selected.

(ii) presence of aggresive industrial effluent, which may have required a non cement based material rather than the gunite selected.

(iii) changes in the relative costs of materials. In this connection, it is wise to keep in mind the relative costs of moulds for the different materials, and the number of sources of supply which are available.

The arrangement of panels is important, for a combination of constructional and structural reasons, as will be described.

In a combined sewer, it is advantageous to be able to carry the flow during dry weather within the invert panels. In this way, overpumping need only be employed while invert panels are placed, if the invert panel is suitably sized. Figure 10 shows overpumping in operation during invert laying.

Often it is not practicable to form a joint capable of transmitting the full stress that can be taken by the lining. Joints can therefore be points of weakness, and are best limited in number. Certainly a two piece lining is to be preferred, comprising an invert and a crown, but the dimensions of the sewer may preclude this. Alternatives are to use either a three or four piece lining. In the case shown in figure 7 a four piece lining was preferred, comprising an invert, crown and two side panels. The main reason against the three piece lining is the joint at the crown, which is at the most likely position for a tension crack to form.

The relined sewer can be analysed structurally by a collapse mechanism approach. Following the approach used in recent Water Research Centre publciations (Ref 3 and 4), the existing sewer fabric is deemed to have zero tensile strength, and a relined sewer acting compositely is assumed to derive its strength from the lining acting in tension and the original sewer in compression. These forces separated by a lever arm related to the thickness of the sewer wall, give the bending moment capacity to resist applied loads.

For a typical cross section, two main collapse mechanisms exist as shown in Figure 8. Subsequent analysis by this method of the design employed at Smith Street showed that for the cross section in figure 7, mechanism A dominates, and based on an estimated tensile strength of gunite, a satisfactory safety factor in excess of 3 is obtained.

In considering collapse mechanisms, the thickness of the original structure gives added safety. Indeed, it is this which contributes to the stability of the existing structure, although this should not be relied on indefinitely as the costs of rehabilitation following extreme structural failure are likely to be much higher than those for preemptive renovation.

It can be seen that by suitable location of panel joints at or near to pin joints in the likely failure mechanisms, it is possible to minimise the effect of these points of weakness. However it is still advisable to maximise the tensile strength at joints to resist alternative mechanisms of failure. With preformed gunite, continuity of reinforcement across joints will act in this way.

The length of Smith Street Sewer to be relined is joined by over 600 side connections, of which over 50 are of man entry size. The smaller connections often have flap valves at the outlet. These are accommodated within a recess in the lining. This is not a significant problem in a man entry sewer. One aspect concerning side connections which does warrant comment is the level at which the larger Borough Sewers enter. As relining of the main sewer raises the invert by about 75 mm, a step could be left unless the incoming sewer is regraded. Figure 9 shows how this step was avoided; the length of regrading depended on the slope of the incoming sewer.

For construction purposes, a number of temporary access shafts were to be sunk at an approximate maximum spacing of 200 metres. An ingenious trolley system for transporting and handling units enabled consideration to be given in one instance on site to working with a longer distance between shafts exceeding 250 metres.

In contrast to pipelaying, it is often recommended that relining proceeds in a downstream direction. We would endorse this opinion, although this was not necessarily the chosen direction of construction at Smith Street, and produced no insuperable adverse effects.

Disadvantages in working upstream include the tendency for increased settlement of suspended solids upstream of the work, in the location which will next require to be cleaned and relined. Dealing with infiltration and other local flows during invert laying can also be affected by the direction of working. In particular it was found by experience that to work concurrently toward a point in both directions could prove a problem in areas subject to groundwater infiltration.

The relining work described above was let for construction as two separate contracts, of approximately equal lengths. Relining of the upstream half of the sewer network was successfully completed in 1981 by Rees Seergun Limited. At the time of writing, the second contract, comprising the length of Smith Street Sewer downstream of the siphon, is being relined by Miller Buckley Civil Engineering Limited.

4. Thoughts for the future

Renovation work poses the question of how in the following century the system will be futher renovated.

It is inevitable that new methods and materials will be developed, and we do not intend to speculate on these. However, it is likely that economic means of sewer renovation will continue to avoid reconstruction and will involve some reduction in size by relining methods. In view of this a 'whole system' approach to renovation is essential. Structural and hydraulic problems must be considered in conjunction. A key aspect in this respect is the use of storage to improve hydraulic performance.

There are means of analysis of growing sophistication available to take account of the existing storage within systems, and to analyse surcharged networks. Devices to maximise this 'on stream' storage have been used, which involve the choking of flows at suitable locations.

An 'off stream' storage approach, as used on the Smith Street Sewer, usually involves a degree of pumping. Operational experience will have to guide to what extent an approach of this sort is to serve as a model for future schemes. A major factor in planning must be to incorporate flexibility in use, to allow not only for present day operation, but for future changes.

Schemes involving on and off stream storage or relief sewers have particular relevance to combined systems, but there is still a considerable variation in domestic flows, with peaking factors of up to 6 times the median dry weather flow in common use for design purposes. A 'whole system' approach could include measures to reduce both the total and peak values of these flows, to keep the maximum benefit from existing systems even after relining.

If the installation of water meters to domestic properties does gain widespread acceptance, will we see this followed by the introduction of a similar charging structure for sewage flows? A policy of charging households on the basis of use of the system may have the effect of reducing total flows. Would the imposition of off peak rates, as used for various other services, achieve a further end of more balanced flow in separate sewerage systems, with less diurnal variation and lower peaks?

The possibility of this extent of metering may seem elaborate, and maybe similar ends could be achieved by other means such as education and persuasion. It does however serve to illustrate the fact that in a 'whole system' approach to improvement and maintenance of the nation's sewerage system, there are aspects which lie outside the direct influence of the Engineering Authorities concerned. This fact is more readily recognised of Public Health Engineering in the less developed countries, but is always present and is likely to assume increasing importance here to maintain present standards if levels of funding are not significantly increased.

Acknowledgements

This paper is published with the permission of Thames Water Authority, but the view expressed are those of the Authors and not necessarily of the Authority.

References

1. Doxat, J : 'The Living Thames', Hutchinson Benham, London, 1977.

2. Natural Environment Research Council : 'Flood Studies Report', N.E.R.C., London, 1975

3. Glennie, E.B : 'Structural design of renovated sewer systems' External Report 56E, Water Research Centre, March 1982.

4. Water Research Centre : 'Sewerage rehabilitation manual' 1984.

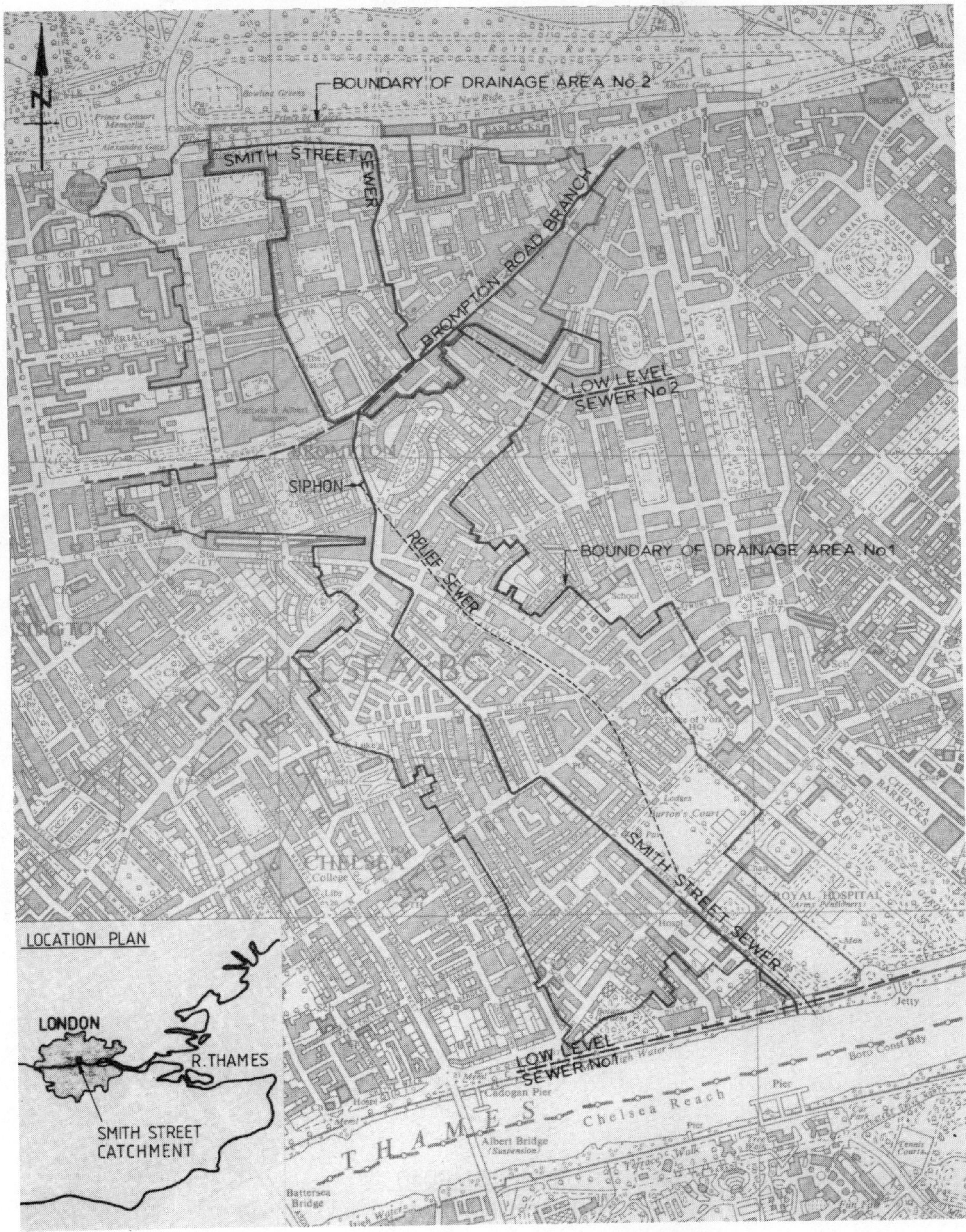

Figure 1: PLAN OF SMITH STREET SEWER DRAINAGE AREA

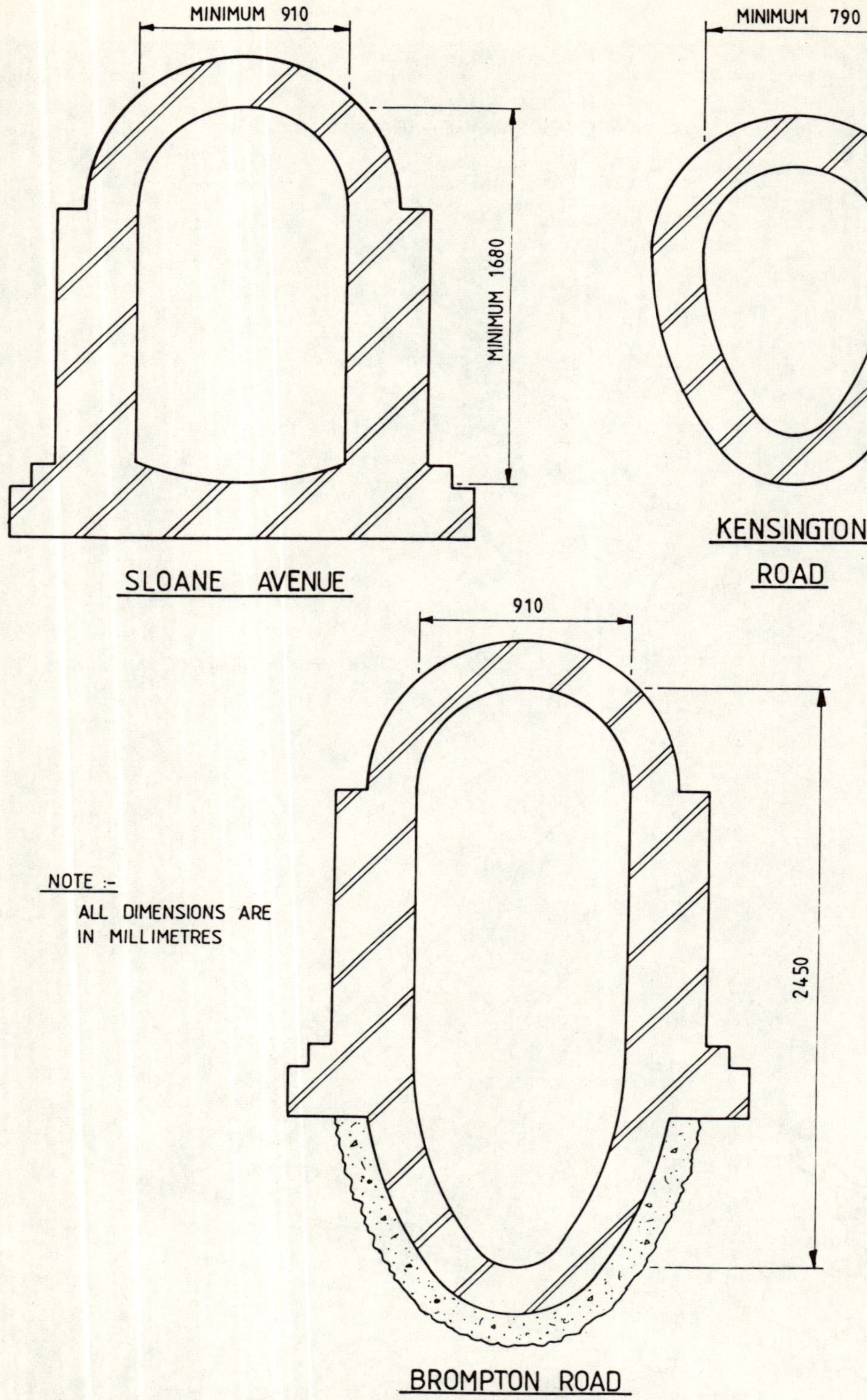

Figure 2: TYPICAL CROSS SECTIONS OF SMITH STREET SEWER

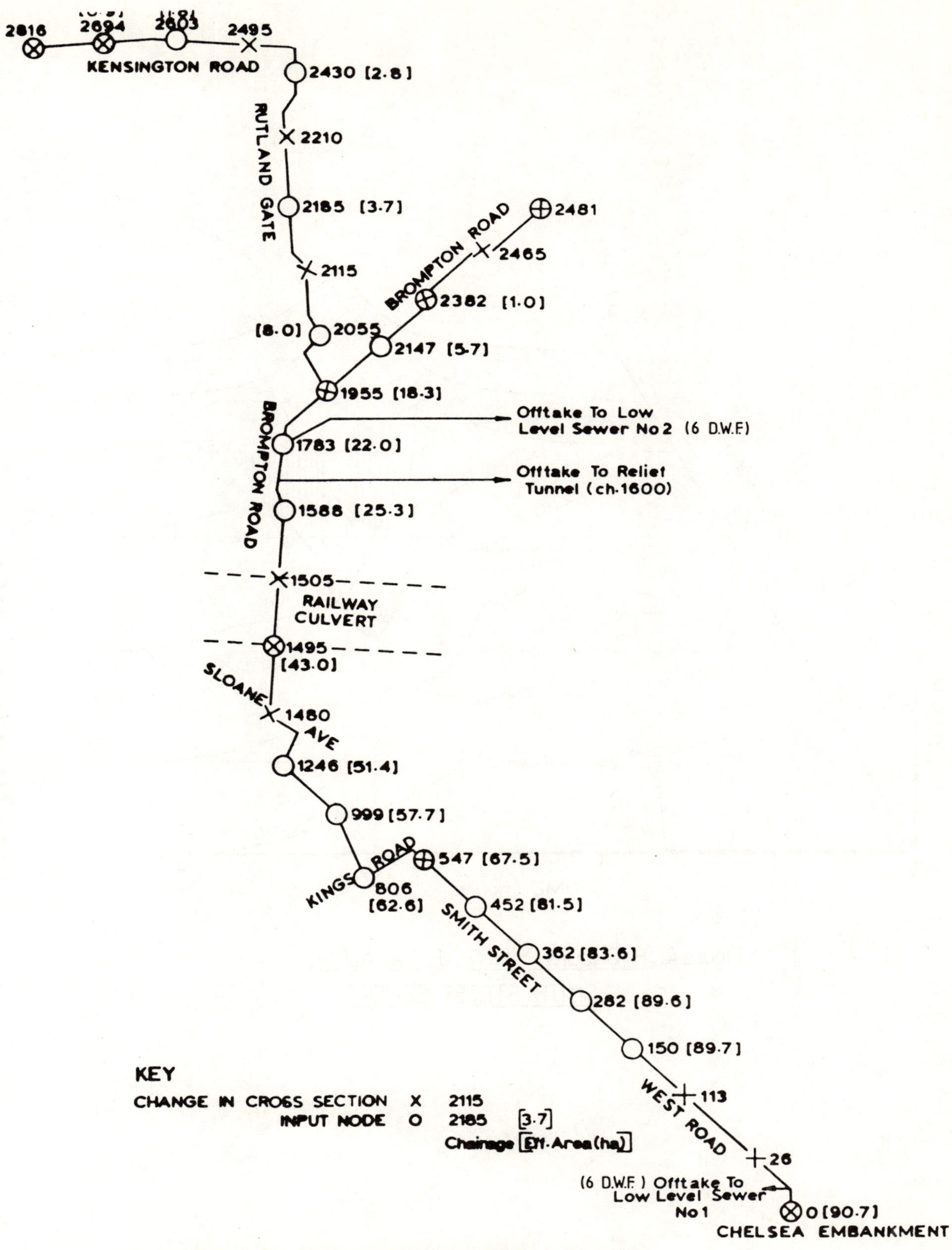

Figure 3: SCHEMATIC DIAGRAM OF SMITH STREET SEWER

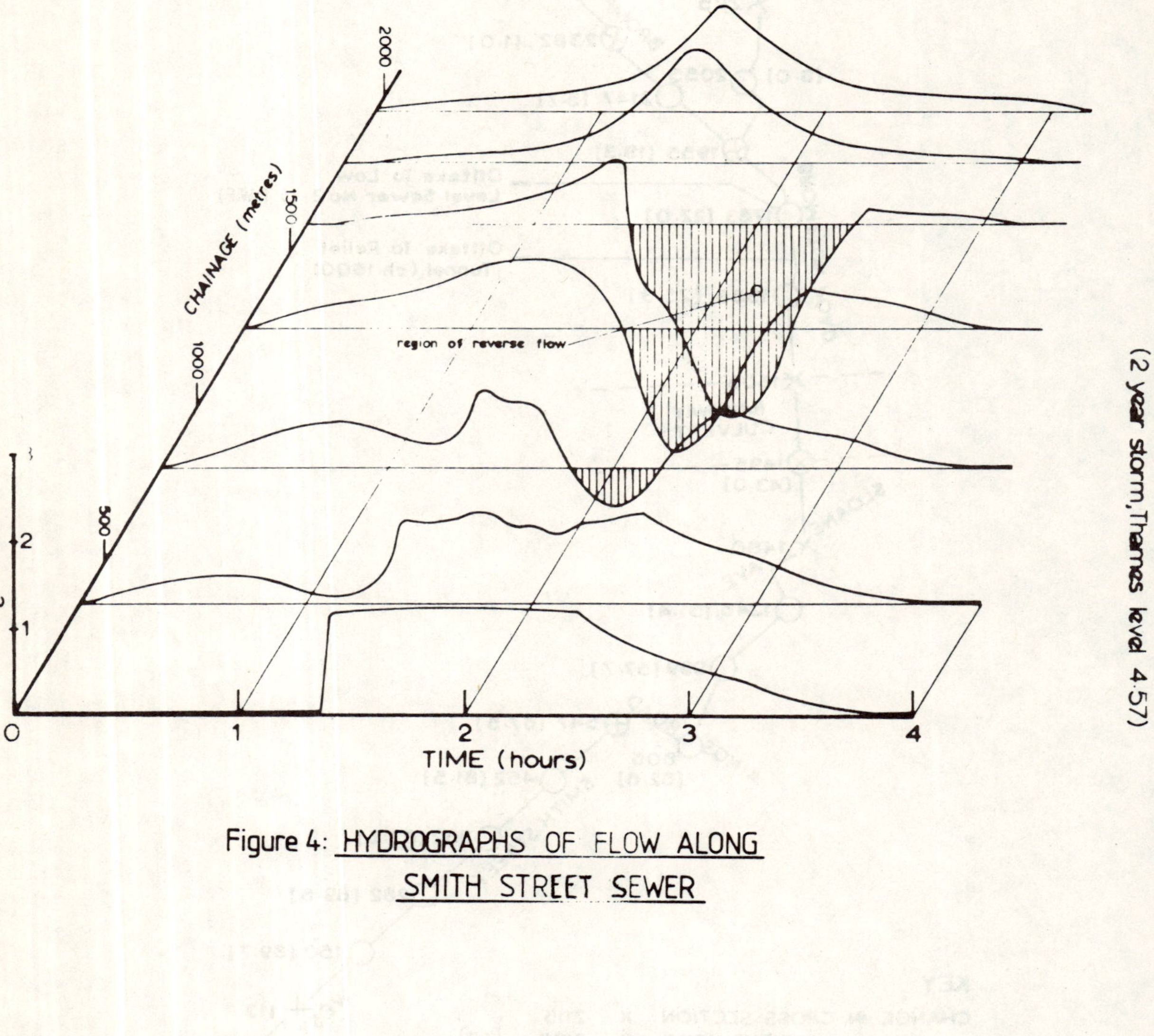

Figure 4: HYDROGRAPHS OF FLOW ALONG SMITH STREET SEWER

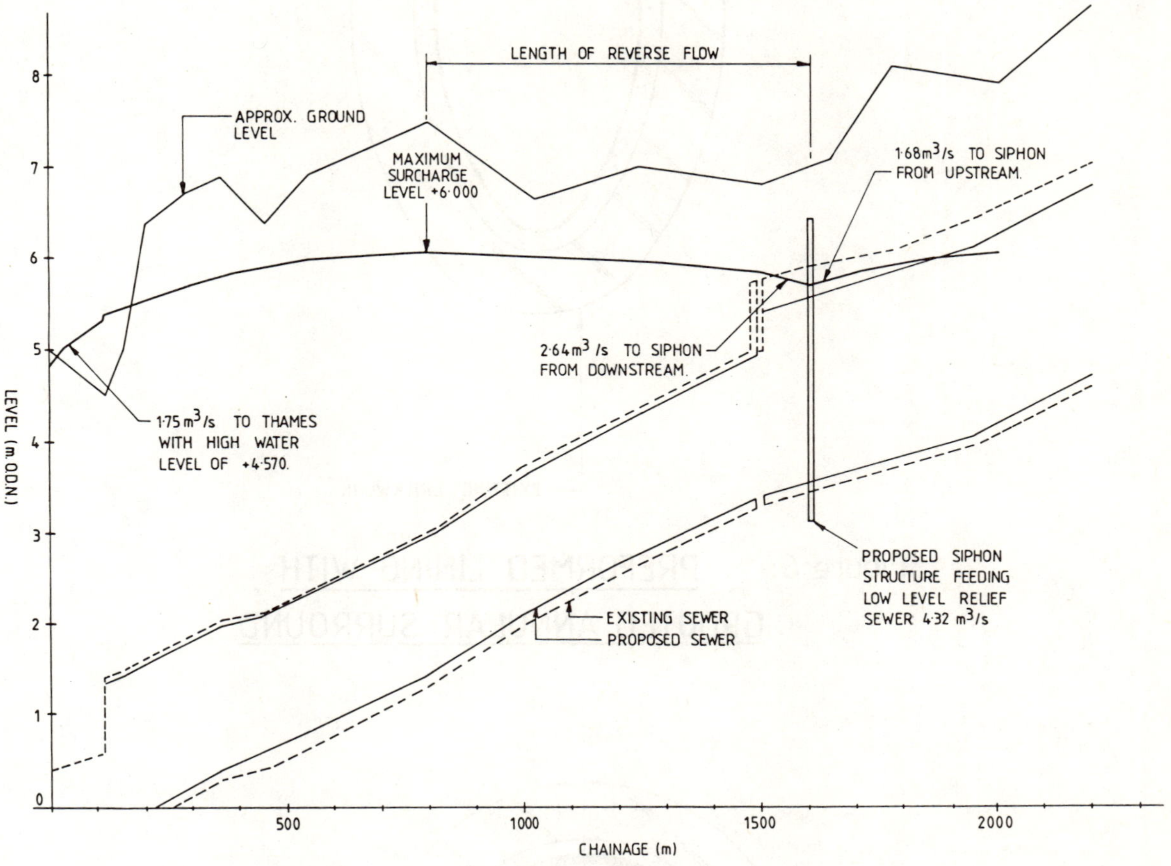

Figure 5: SURCHARGE LEVELS ALONG RELINED SMITH STREET SEWER

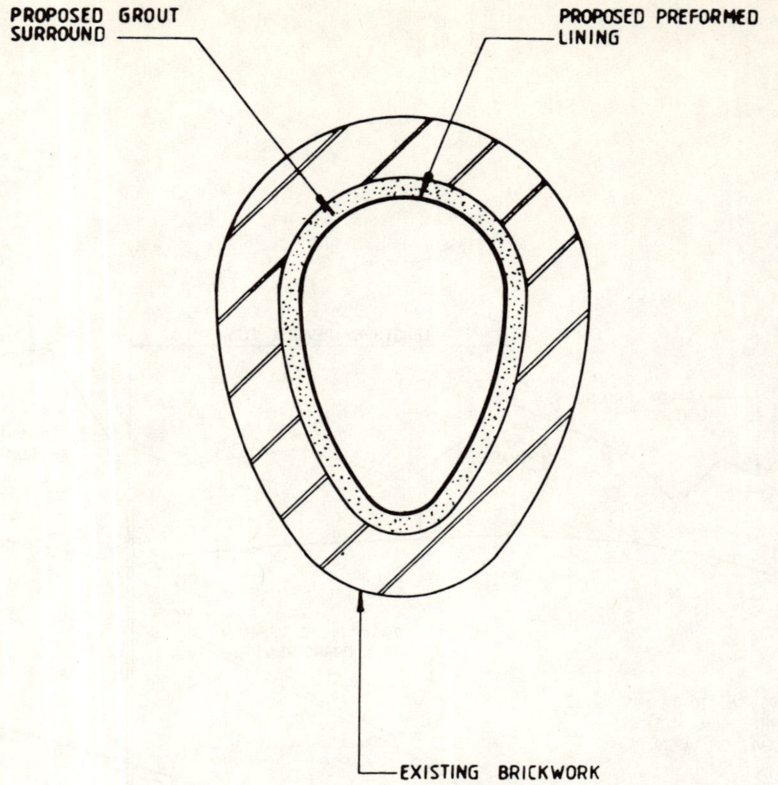

Figure 6: PREFORMED LINING WITH GROUTED ANNULAR SURROUND

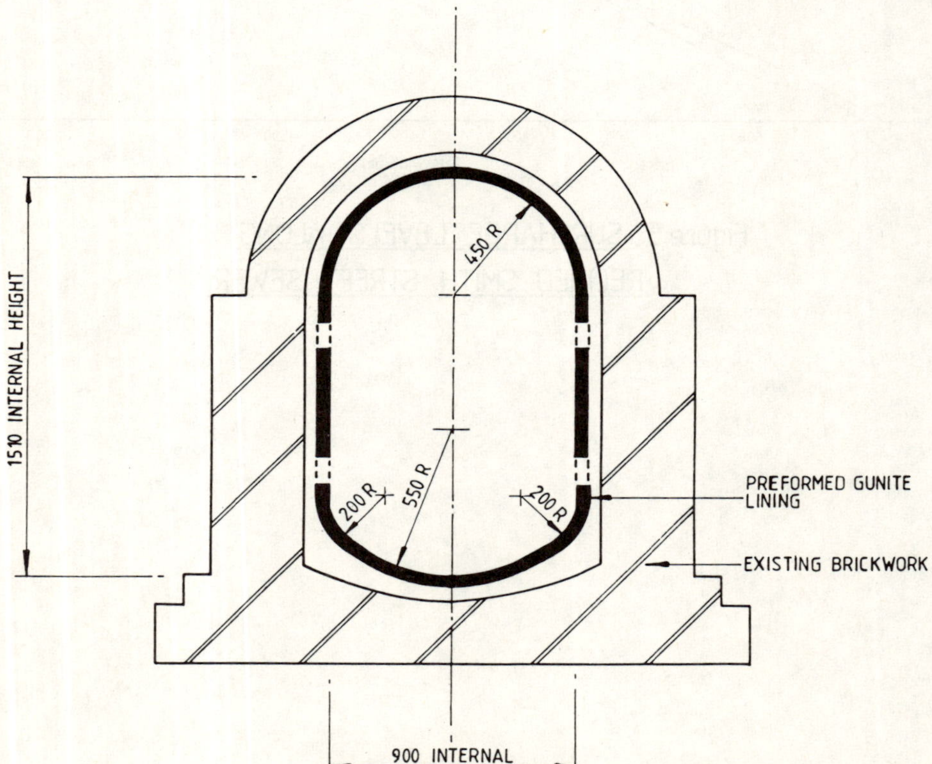

Figure 7: RELINED SEWER SECTION TYPE A2

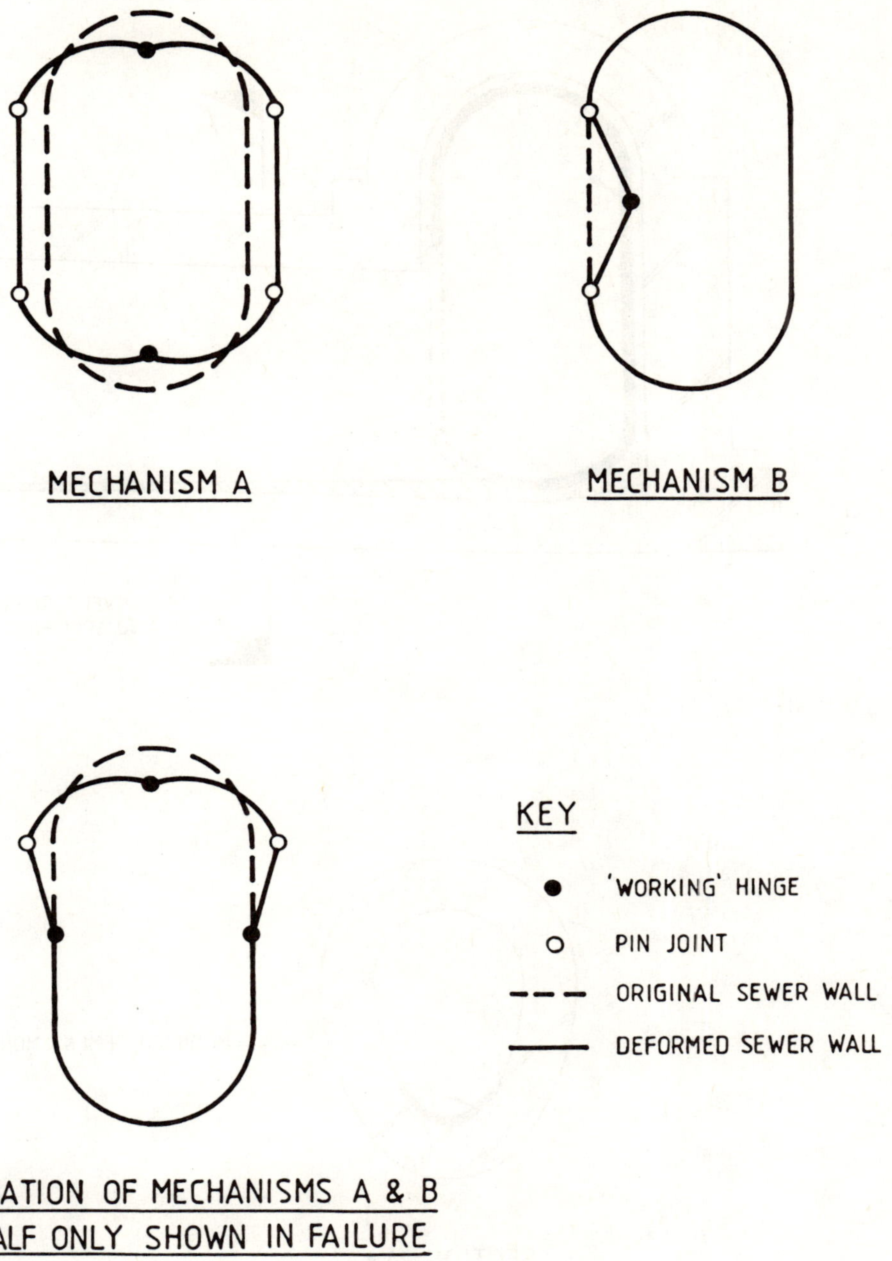

Figure 8: POSSIBLE FAILURE MECHANISMS

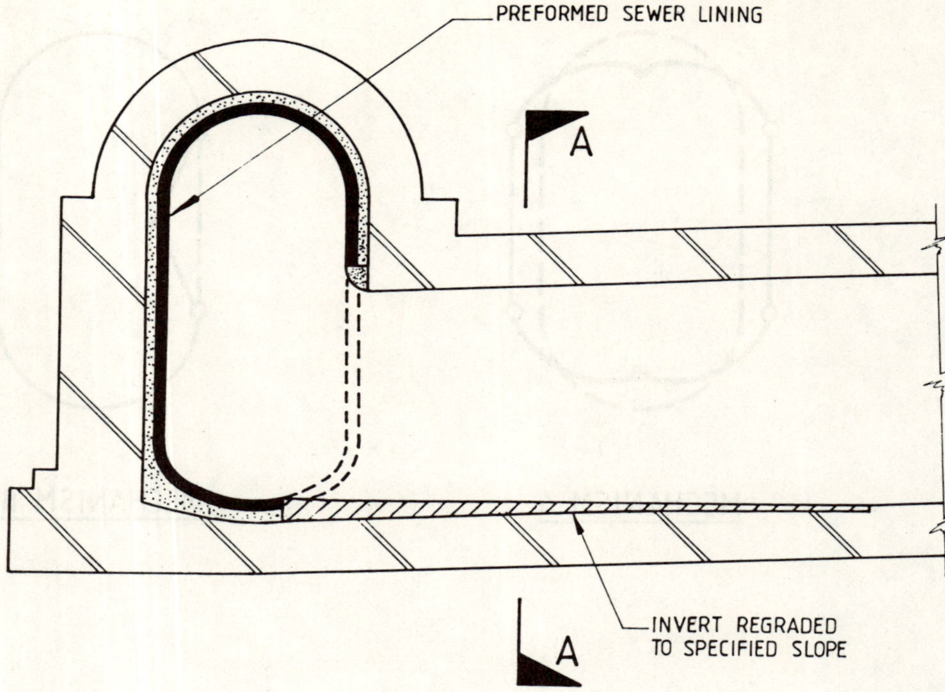

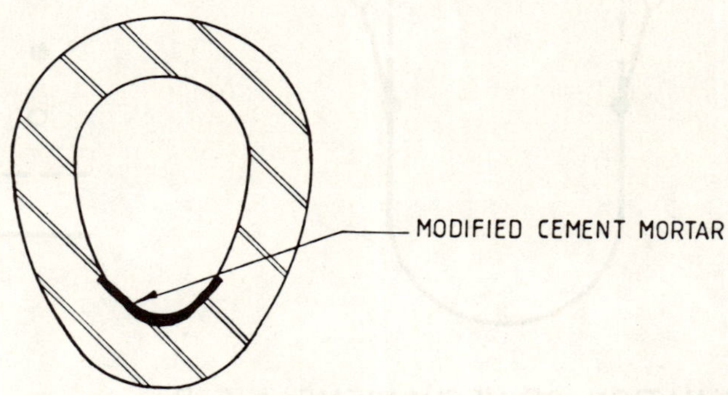

SECTION A-A

Figure 9: DETAIL SHOWING REGRADING INVERTS OF BOROUGH SEWERS

Figure 10:

Levelling invert units during relining of Smith Street Sewer beneath the Royal Hospital, Chelsea. The photograph shows an overpumping main supported within the sewer on Acrow struts.

International Conference on the
PLANNING, CONSTRUCTION, MAINTENANCE & OPERATION OF SEWERAGE SYSTEMS
12-14 September, 1984

PIPELINE REHABILITATION TECHNIQUES

B JAY SCHROCK PE

J S C International Engineering
2261 Grove Avenue
Sacramento
CA 95815
USA

SYNOPSES

The US Environmental Protection Agency (EPA), the Research Foundation of the American Public Works Association (APWA), and the National Association of Sewer Service Companies (BASSCO) individually and in concert are redefining and re-evaluating inflow/infiltration (I/I) reduction criteria and methodology. Various accepted parameters will be addressed and discussed.

The engineering consulting community has significant experience in the planning, design. analysis. and structural assessment of existing sewerage systems. When these various engineering parameters have been evaluated, it has historically been commonplace to suggest pipeline replacement. This normally necessitates relocating along various alternate routes and increasing pipe diameters. Relatively new pipeline renovation or rehabilitation techniques have been developed and are now being utilised.

The proper and complete rehabilitation of sewer service laterals on sanitary sewers reduces I/I levels often permitting reduction in size of mainline or trunk sewers. Proper hydraulic model evaluation often reduces the size requirements for combined sewers. I/I reduction normally reduces certain structural duress in the components of the sewerage system. Also, early prioritised rehabilitation of pipelines is cost-effective when compared to crisis response or later rehabilitation. The Engineering Consultant´s "comfort zone" is replacement in lieu of rehabilitation. The multitude of available rehabilitation techniques should be evaluated.

Modern day sewerage pipelines have been installed during the past 150 years. Funds are not available for replacing these failing pipelines. Therefore, it is paramount that a systemised renovation program be initiated in all communities around the world.

Held at The University of Reading, England. Organised and sponsored by BHRA, The Fluid Engineering Centre, Cranfield, Bedford in conjunction with the Water Research Centre.

1. PAST EFFORTS

Sewer system evaluation and rehabilitation was an important part of the US Environmental Protection Agency (EPA) Construction Grants Program, beginning in 1974, in an effort to control excessive infiltration/inflow (I/I). In 1980, EPA conducted a study to determine the effectiveness of this program, and concluded that the program, on the whole, was not getting its money's worth. Minor adjustments were made in the Program; however, not until this year have any significant modifications been incorporated.

One of the primary reasons given was the non-funding of repair for side sewers and service laterals that, in many cases, were responsible for the greatest infiltration volumes. Although large diameter interceptors were typically corrected, the extraneous flows from tributary laterals continued and systems ended up with only partial correction. Also, migration of infiltration locations seemed to develop after remedial work was completed. In at least one case, extraneous flows actually increased. The migration of infiltration locations probably occurred as correction work eliminated groundwater drainage into sewer systems, allowing groundwater tables to rise and subject other defective pipes in the systems to submergence. The failure to make all necessary corrections made rework a poor investment of public funds.

There are several other reasons for failures after remedial construction. Grouting of pipe joints is a common area of rehabilitation failure when it is not properly done. Other rehabilitation failure problems stem from overlooking or failing to understand the complex mechanisms of structural failures during the rehabilitation evaluation. Often, little is done in rehabilitation programs to correct external pipeline deficiencies where groundwater movement has affected the sewer, or where soil movement, erosion, and voids have been prominent causes of the structural and joint failures being repaired.

Disappointing experiences in the past have been attributable to incomplete correction of the entire problem, or failure to deal with the causes. When a thorough evaluation is performed and appropriate rehabilitation measures are employed based on competent design and construction, the results are sound.

The Research Foundation of the American Public Works Association (APWA) during 1983 conducted workshop seminars at each of the EPA Regional Offices located throughout the United States. The results of these meetings were used as input for establishing new EPA guidelines for pipeline system evaluation and rehabilitation. Member organisations of the National Association of Sewer Service Companies (NASSCO) at various geographical locations have taken a more firm line in guaranteeing the sewer agency authorities certain reductions in I/I. With these various efforts by organisational bodies, there seems to be an increase in the comfort zone by the engineering community.

2. FUTURE EFFORTS

EPA has recently released their CG-84/85 criteria on qualification for grant funds. The funded projects that qualify for sewer rehabilitation now require that anticipated flows in excess of the following must be verified and confirmed:

1. 120 gallons per capita per day and less is considered non-excessive infiltration.

2. 275 gallons per capita per day and less is considered non-excessive inflow.

As part of the post-construction project evaluation and prior to final certification, the grantee will be required to determine whether I/I reduction efforts have been cost-effective. An essential part of this determination is the identification of the actual amount of I/I reduction that has been achieved through the corrective program. This requires an analysis and evaluation of system flow records during the wet-weather season immediately following system upgrading. A flow analysis will be needed to determine the overall reduction of I/I as reflected at the treatment plant. Accordingly, unless the trunk lines to the treatment plant are surcharged during the storm periods, I/I flow data should be collected at the plant rather than at multiple points throughout the collector system. The Sewer System Evaluation Survey (SSES) should not be repeated. and information developed should be limited to that required to develop a post-construction I/I flow figure. This flow figure compared with the original I/I flow will provide the actual percentage reduction achieved.

It will also be necessary to identify the actual costs of system rehabilitation associated with the I/I reduction. These costs should include the cost of design and SSES costs expended to identify acceptable corrective action.

The construction grant regulations require that the engineer or engineering firm provide the following services during the one-year performance period:

- Direct operation of the project
- Train or provide for training of operating personnel
- Prepare curricula and training materials for operating personnel
- Review the operations and maintenance (O&M) manual to accommodate actual operating experience
- Advise the grantee whether the project is meeting its performance standards
- An analysis of the cause of the project's inability to meet the performance standards
- A schedule for undertaking in a timely manner the corrective action necessary to bring the project into compliance
- The scheduled date for certifying to the reviewing agency that the project is meeting its performance standards.

3. INTRODUCTION OF METHODS

Rehabilitation of deteriorated sewer pipelines can be justified when the total evaluation of all existing data of the investigation and analysis process has been completed. The effects of groundwater play an important part in the analysis of sewer pipeline rehabilitation. Groundwater can, at times, rise and fall creating a soil pumping action adjacent to the failing pipe. The movement of groundwater can affect and loosen the soil above, adjacent to, and below the pipes.

Several distinct features are quite apparent in failing sewers. The most common failures are either structural or corrosive, or both. These basic failures lead to other problems in sewers such as root intrusion, infiltration, exfiltration, stoppages, and/or collapse.

The best rehabilitation method chosen for any sewer pipeline renewal is the one that meets the parameters for solving the entire problem which includes external as well as internal conditions.

The various internal and external rehabilitation methods are discussed individually in this paper. For the purposes of this discussion, the following rehabilitation methods and categories of sewer shapes and sizes will apply.

Sewer Shapes	Old Pipe, mm
Circular Shape	
1. Small diameter	400 mm and smaller
2. Intermediate diameter	450 mm to 850 mm
3. Large diameter	900 mm and larger
Egg, Elliptical, Oval, or Ovoid Shape	
4. Small size circumference	2,000 mm and less
5. Intermediate size circumference	2,001 mm to 3,000 mm
6. Large size circumference	3,001 mm and larger
Irregular Shape	
7. Small size circumference	1,500 mm and less
8. Intermediate size circumference	1.501 to 2,500 mm
9. Large size circumference	2,501 mm and larger

4. EXTERNAL REHABILITATION

Most external rehabilitation methods are performed above the ground or by excavating next to the pipe. Two common methods are chemical grouting and cement grouting. Both methods are appropriate for solving problems of significant groundwater movement, washouts, soil settlement, and soil voids.

5. REHABILITATION METHODS

A EXTERNAL

1. Chemical Grouting

2. Cement Grouting

B INTERNAL

1. Chemical Grouting

2. Cement Mortar Lining
   (a) Centrifugal Process
   (b) Reinforced Centrifugal Process
   (c) Mandrel Process

3. Coatings

4. Reinforced Shotcrete Placement

5. Concrete Placement
   (a) Slip Form or Fixed Form
   (b) PVC Sheet Liners (T-lock)
   (c) PE Sheet Liners (shot bolted)

6. Fibreglass Reinforced Cement Liners (GRC)

7. Fibreglass Reinforced Plastic Liners (GRP)

8. Resin-Impregnated Polyester
   (a) Epoxy Resin
   (b) Polyester Resin

9. Sliplining or Pipe Insertion Renewal
   (a) PE Pipe (extruded)
   (b) PE Pipe (spiral-welded)
   (c) Pb Pipe
   (d) ABS Pipe
   (e) PVC Pipe
   (f) RPM Pipe
   (g) RTR Pipe
   (h) D.I Pipe (cement-lined)
   (i) D.I Pipe (poly-lined)

10. Flexible Pipe Rerounder

11. Root Control and Removal

12. Excavation and Replacement or Repair

5.1 Chemical Grouting

Chemical grouts consist of solutions of two or more chemicals that react to form a gel or a solid precipitate as opposed to cement or clay grouts that consist of suspensions of solid particles in a fluid. The reaction in the solution may be either chemical or physiochemical and may involve only the constituents of the solution of these constituents reacting with other substances encountered. The reaction causes a decrease in fluidity and a tendency to solidify and form occlusions in channels or fill voids in the material into which the grout has been injected.

Most applications of chemical grouts are made into saturated or partially saturated materials. The majority of chemical grouting applications are related to water shutoff rather than strength increase. For sewer rehabilitation purposes, grouting would be effective in reducing or eliminating infiltration but would not significantly improve the structural condition.

The basic method of external chemical grouting is to place a relatively impermeable barrier called a grout curtain. It consists of one or more interlocking rows of grouted soil cyclinders. Each of the individual cylinders is formed by the injection of grout through a pipe or drilled hole. These cylinders should have relatively uniform cross-sections throughout their depth. For uniformity, it is important to grout short sections or stages of any individual hole so as to minimise the opportunity for grout to flow preferentially throughout the hole depth. The time required for the grout to solidify or gel time is directly related to pumping rates and to depth. The initial injections of the actual grout curtains are generally used to arrive at values for stage length, pumping pressure, and gel time. The diameter of the sewer pipeline, the size of problem area, the depth of sewer, and the groundwater zones help to determine the spacing between the grouting holes and the required number f rows. Economics usually limit the spacing between holes to 1.5 meters or less.

One of the most widely used chemical grouts for curtain walls, especially in the fine soils, is the acrylamide-type. Acrylamide grouts are mixed in proportions that produce stiff gels from dilute water solutions when properly reacted. Several reactants and mixtures of reactants may be used. The gel is essentially impermeable to water. A beneficial property of the acrylamide chemical grout is that pumping times in excess of gel times can be utilised due to the low shear strength of the gelled grout. Gel times can be controlled by varying the temperature and the catalyst concentration. This property is effective in minimising dilution with groundwater and subsequent displacement of the grout. Chemical grouting with acrylamides in dry soils presents problems, and it is not recommended for external rehabilitation adjacent to water pipelines. Acrylamide is a toxic chemical that can be absorbed into the body through broken skin, inhalation, and swallowing. Because of this toxicity, potential hazards in handling and usage can occur if not supervised by technically qualified personnel.

5.2 Cement Grouting

Portland cement grouts can be used to form an impermeable subsurface barrier but are restricted in application to medium sands or coarser materials because of the larger size of cement particles. For purposes of filling voids and washouts adjacent to sewer pipelines, various Portland cements are ideal. Unlike most chemical grouts, Portland cement grouts are relatively inexpensive. Several types of Portland cements have been used successfully. Type III cement is often selected because of its small particle size. A variety of water to cement ratios can be used depending on subsurface conditions. Strength characteristics are generally not important where grouting is primarily intended to fill voids surrounding a buried sewer pipeline. Proper design of the water to cement ratio results in grout mixtures which are easy to mix and inject, and have strengths of 3.45 to 6.9 Mega-Newtons per square meter (MN/mn^2). Clays can be added to cement for grouting, to form gels. and to prevent settlement of the cement from suspension. They do. however, have the problem of no well-defined setting time and they have a slow strength development. Because of this, they are normally not utilised in sewer rehabilitation void grouting.

For extremely large void filling applications, other cements such as pozzolan and fly ash mixes can be used and are more economical than straight Portland cement and soil-cement mixtures. In addition to the above, Portland cement grouts may be used as a filler and accelerator in silicate grouts and may be compatibly mixed with acrylamide to improve water shutoff capabilities and injection characteristics.

When external grouting is done on pipelines, it is absolutely imperative that bedding and lateral support be accomplished initial to any above pipe void filling. This is especially true when rehabilitating flexible pipes such as brick, etc.

The need for external rehabilitation is often not practical on small-sized pipelines due to the nature of certain failures, and the conditions of the ground above the pipe. Internal and external rehabilitation is usually completed on larger-sized pipelines where external problems are more readily determined. Virtually, all rehabilitation requires some form of internal work.

6. INTERNAL REHABILITATION

The following rehabilitation analysis will be related to methods and materials used for internal sewer pipeline renewal.

6.1 Chemical Grouting

Chemical grouting is most commonly used to seal deteriorated and leaking pipe joints, service connections, and open joints. Although chemical grouting is normally used in small-diameter pipes, large pipes can also be successfully grouted at the joints with special equipment. There are four basic grouting materials used for internal grouting of sewer pipelines; acrylamide gel, acrylate gel, urethane gel, and polyurethane foam are the main types of chemical grouts currently available.

The chemicals necessary to form acrylamide or acrylate gels are usually mixed in two separate tanks and pumped through separate hoses to the pipeline joint to be sealed. One tank is used to mix and dispense a solution of water and the acrylamide or acrylate grouting material. The other tank is used to mix and dispense a solution of water and a catalyst. The water and catalyst solution initiates the chemical reaction when mixed with the acrylamide solution. Additives can be included in either solution to help control shrinkage, reaction or "gel" time, and other variables.

Urethane grout materials may be used to form either an elastomeric gel, much like the acrylamide and acrylate gels, or a rubber-like foam. Urethane gel is different from the acrylamide or acrylate gels in that water is the catalyst for the urethane gel material. The properties of the reacted gels are also different. Urethane gel seals pipeline joints by forming an elastomeric collar within the pipe joint as well as by consolidating soils and filling voids outside the joint.

Urethane gel is applied in essentially the same manner as the acrylamide and acrylate gels; however, the equipment used for the polyurethane foam grout, which also is activated by water is normally used.

Polyurethane foam differs from the gel grouts in that the foam is used to form an in-place pipeline gasket. The application techniques and equipment used for the placement of the foam grout do not fill voids or stabilise the soil outside of the pipe joint.

The test-and-seal technique is used on most pipeline segments selected for sealing. This technique minimises the potential for groundwater migration to other, unsealed joints in a manhole-to-manhole sewer segment. With this method, each joint is first tested to determine if it is tight. After the joint has been located with a closed-circuit TV inspection camera, the rubber sleeves at each end of the packer are inflated until they form a seal against the inside of the pipe and isolate the joint to be tested. If the joint fails the air or water test, the grout and catalyst solutions are pumped under pressure into the void between the two inflated sleeves and out into the surrounding soil.

The amount of grout needed to seal a defect depends upon the size of the leak. The acrylamide, acrylate, and urethane gels usually are pumped until the gel solidifies; the back pressure will then indicate to the operator that the leak has been successfully sealed. The operator then deflates the rubber sleeves and moves to the next joint for testing and possible sealing and continues until all the joints and radial cracks in the manhole-to-manhole segment have been tested and, where necessary, sealed.

The success of grouting using a packer unit can be assessed by conducting a low-pressure test of the sealed defect once the setting time has elapsed. Pressures must be controlled to avoid rupturing the grout seal.

In large-diameter sewers where worker access is adequate, leaking joints of concrete and clay pipes can be injected with chemical grouts using a nozzle-type applicator.

The initial low cost and speed of chemical grouting lends itself well to nonstructural rehabilitation of sewer pipelines. The need for dewatering of the line to be grouted, groundwater movement, toxicity problems, and lack of accuracy in the determination of set time have been deterrents to the use of chemical grouting.

6.2 Cement Mortar Lining

Deteriorated internal surfaces can be renewed in sewer pipes made of concrete, steel, and iron by applying cement mortar. Lining material is usually a 1:2 Portland cement mortar, using cement highly resistant to sulphates. Lining thicknesses are usually 6 mm or greater. The thickness depends upon the condition of the existing pipe. In pipe diameters below 600 mm, reinforcing is not normally required; however, in diameters above 600 mm that are badly deteriorated, it is desirable to reinforce the cement mortar lining with a spirally-wound reinforcing rod.

6.2.1 Application Methods

There are three methods of applying cement mortar linings to pipelines in place: centrifugal process, reinforced centrifugal process, and mandrel process. All three methods require thorough cleaning and dewatering. The methods are briefly described below.

1. **Centrifugal Process.** A variable-speed winch pulls a revolving mortar-dispensing head through the pipe at a specified rate to control lining thickness. Access openings are required every 150 to 200 meters depending on the diameters. Large-diameter pipes can increase these distances when working areas inside the pipe become greater. Electric generators are used to supply power to the winch and the mortar mixer. In sizes above 250 mm, troweling is usually done to provide a smoother finish and ultimately a greater carrying capacity. In large sewers, service laterals are usually plugged prior to lining.

   The cost of centrifugal in-place lining depends on a number of factors, including pipe diameter, pipe length, condition of the line, depth and profile, access, traffic control, and thickness requirement. Unit costs generally decrease significantly with increasing length of line. Centrifugal in-place lining is applicable to pipe sizes from 250 to 3,700 mm. The current "state-of-the-art" for this method is usually most economical above 600 mm in diameter. One of the advantages to this cement lining method is that the line can be placed in service 24 hours after the lining process. This process is not as speedy as other renewal methods, but it has proven to be effective and in many cases competitive.

2. **Reinforced Centrifugal Process.** Severely deteriorated sewer pipes above 600 mm in diameter may require additional reinforcement based upon structural design considerations. This reinforcing process consists of three steps: (1) a course of mortar one half the final lining thickness is placed by the centrifugal machine without troweling; (2) a spirally-wound reinforcing rod is placed, and the size and spacing of the rod are determined by structural design requirements, and (3) after the steel rod has been placed, a second course of mortar is spun into place to the final desired thickness. The spiral rod has two advantages over prefabricated cage steel: (1) it requires less steel and (2) it conforms to the inside contour of the pipeline.

The addition of reinforcing and increased installation time raises the cost of this method significantly; however, when compared to shotcrete and reinforced concrete methods, it is competitive.

3. **Mandrel Process.** The mandrel process is applicable to sewer pipes from 100 to 400 mm in diameter with few service connections. This process uses a pressurised extrusion technique as follows: (1) a tight-fitting baffle is threaded onto a cable in the pipeline, (2) a two-part sand and one-part cement mortar mix is loaded into the pipe against the baffle, and (3) a conically-shaped mandrel is then secured to the cable. The mandrel is composed of centering springs, metering springs, a perforated skirt, and a troweling edge.

   The assembly is pulled through the pipe usually for 100 to 150 metres. Back pressure of the baffle forces the mortar over the cone of the mandrel and against the pipe wall. The perforated skirt squeezes out excess moisture, and the troweling edge produces a smooth, dense coating. Because of the pressure created during the lining process, service connections must be plugged.

   This method has found minimal application in the sewer rehabilitation field due to the closely spaced excavations required: slow speed, high cost, and diameter limitations.

   It is recommended that when the cement mortar linings is employed where high sewage temperatures and/or acidic corrosion are prevalent that certain sulphide control measures be incorporated in the sewer system.

6.3 Coatings

Generally, the use of chemical-resistant linings and coatings have had marginal success. Most problems can be related to permeability, pinholes, adhesion, elongation, and flexibility characteristics. Some of these have apparently been overcome by material improvements and/or a thicker application of these materials.

A number of in-depth investigative exposure testings have been completed under various governmental authorities. A large number of coatings on steel and concrete surfaces generated a significant amount of data. Proprietary coatings composed of urethane and phenolic epoxy performed quite well and were classified as highly resistant.

6.4 Reinforced Shotcrete (Gunite) Placement)

Gunite is defined as a mixture of a fine aggregate, cement, and water applied with air pressure by a cement ejector. The mixed material, in a relatively dry state, is transported through a hose from the cement ejector to the nozzle, which is a special mixing chamber where water is added. Here, hydration takes place, and the resulting mixture is shot into place under air pressure.

Concrete that can be placed with not more than 11.5 to 13 litres of water per sack of cement will possess high density and have less shrinkage. Ordinary poured or hand-placed structural concrete with twice the water to cement ratio of gunite usually provides half the in-place density or strength values.

Due to the extremely high density and low permeability of well-placed gunite, the penetration of water is held to a minimum. Gunite is usually more chemically resistant to acids than ordinary concrete. Well-placed gunite shows water or sewage absorption to be somewhat less than 5 percent. The percentage of voids can be less than one-half of that of poured concrete; therefore, normal sewer sediment buildup usually occurs at a lower rate. Also, special aggregates and cements with high alumina contents (Class II or V) can be incorporated in the mix.

Tests show that well-placed gunite possesses a modulus of elasticity (E) of over 27,500 MN/m^2 and a coefficient of thermal expansion similar to that of low-carbon steel.

Experiments and actual use in the field have shown that the adhesion or bonding between gunite and properly cleaned old surfaces is usually stronger than the base material. Normally, the more deteriorated and/or irregular the old pipeline, the better the adhesion. Oils, greases, foreign materials, and loose mortar or cement must be removed prior to application of gunite.

6.5 Concrete Placement (Slip Form or Fixed Form)

The placement of reinforced or nonreinforced concrete for rehabilitation is an effective renewal method for a variety of sewer conduit shapes. Normally, this method is used in large-diameter sewers with adequate access where materials can be handled effectively. Highly sulphate-resistant Portant cement is normally required.

1. **Slip Form or Fixed Form.** In large sewer pipes and tunnels, band-built forms or prefabricated forms are built and/or installed in a section of the sewer and concrete is poured around them. The forms are removed and re-used until the entire sewer renewal is completed. The structural condition of the existing conduit determines whether or not steel reinforcement is required. Reinforcing steel can be single or multiple layers of welded wire mesh or hand layed-up cages attached to the deteriorated conduit wall by threaded inserts.

   One noteable installation of this type is the Davern Street Storm Sewer in Saint Paul, Minnesota. A horseshoe sewer tunnel was rehabilitated by inserting prefabricated corrugated aluminium sleeves 4.5 metres long into the tunnel and pouring concrete around them. This proved to be less costly than total replacement, complete clay-tile replacement, or other methods investigated.

   Structurally reinforced concrete lining rehabilitation is very credible and simple to design. The sewer must be thoroughly cleaned and dewatered prior to rehabilitation.

   In many rehabilitation cases, concrete placement is the only solution. Seldom is it applicable below 1,200 mm in diameter. It does, however, have renewal applications in virtually any shape of sewer conduit.

2. **Polyvinylchloride (PVC) Sheet Liners.** For additional corrosion protection against aggressive sewage when reinforced or non-reinforced concrete is placed, protection can be obtained by the use of mechanically bonded PVC sheet liners. These liners contain interlocking ribs of T´s to hold them in place after the concrete is cured. These linings are placed on the forms prior to pouring, and heat fusion or solvent welding is used to join each section of liner. After the forms are removed, the next section is joined by heat fusing or solvent welding at the joint. The entire PVC plastic liner is tested with a spark device to check for holes, defects, and damage. Hydraulic capacities can be greatly improved, and corrosion of the conrete is virtually eliminated.

   The cost of the pliable plastic liners is reasonable; however, their installation is expensive, but extremely effective if done properly.

3. **Polyethylene (PE) Sheet Liners.** As with the PVC sheet liners, polyethylene provides excellent corrosion protection. After a thorough hydraulic cleaning of the deteriorated concrete down to the level of sound concrete, the pipe is cement mortar troweled filling in significant irregularities. The PE sheet liner is then placed directly against the relatively smooth pipe. The sheets are held in place by the use of forms, stulls, or blocking and are physically connected to the pipe with stainless steel bolts or impact nails. Adjacent sheets are bonded by heat fusing.

6.6 Fibreglass Reinforced Cement Liners (Segmented)

Fibreglass reinforced cement liners are thin panels developed to rehabilitate sewers above 1,100 mm. These linings are segmented to fit the diameter required. Circular, oval, rectangular, elliptical, ovoid or V-shaped sewers can be lined.

The linings are manufactured in a composite of cement and glass fibres. The normal wall thickness is 10 mm; however, this is occasionally modified. The linings have high mechanical and impact strength with good acid and alkalinity resistance. They are also highly resistant to abrasion with negligible absorption and permeability.

After a thorough line cleaning and dewatering, the segments are installed in normally 1.25 metre lengths which overlap at the joints, and the flanges on the segments may be predrilled for fixing by screw or impact nail gun. The segmented rings are anchored on spacers; and upon final assembly. the entire section is cement-pressure grouted. Side laterals are cut-in and grouted prior to final erection.

This method provides flexibility to accommodate variations in grades, slopes, cross-section, and deterioration. The linings are not designed to support earth loads; however, they are effective in providing new lining for a structurally sound sewer. The interior smooth surface can improve hydraulic capacities.

Although the segmented sections are light and easy to handle. the installation of these linings is labour-intensive and slow. It is not a recommended method for structurally inadequate sewers. These linings (invented in England) are available on a custom-design basis in the United States. At this time, it is difficult to determine the competitive comparability of glass-reinforced cement linings with other methods.

6.7 Fibreglass Reinforced Plastic Liners (Segmented)

Similar in nature to glass-rienforced cement liners, these liners are used predominantly in sewers above 1,100 mm. Formed in segments with a range of available wall thicknesses, these linings are not structural components, but they do provide an adequate corrosive barrier and smooth lining for structurally sound sewers.

These segmented liners are manufactured from fibrelgass and a choice of resins from polyester to vinylester. Choices of resins can be made based on the corrosion resistance required. This composite of fibreglass and resin can provide excellent hydraulic characteristics due to the smooth inner surface. The material has high mechanical and impact strength and good abrasion resistance. The linings have little absorption and negligible permeability. Wall thickness and laminate designs can be varied easily for custom installations.

The installation of segmented glass-reinforced plastic (GRP) liners is not difficult due to the light weight of the segments. After a thorough cleaning and dewatering, the segments are placed in the sewer with mechanically locking joints, bolts, or impact nails at each joining section. They are blocked in place and after the entire section is complete. the space between the linings and the existing sewer is cement pressure grouted in place to prevent sagging and deformation.

6.8 Resin-Impregnated Polyester

This method utilises a polyester needle felt tubing, manufactured to the exact dimensions of the pipe to be relined. The tubing is saturated with a thermosetting or epoxy resin and is called "Insituform" by the manufacturer. Its purpose is to create a new pipe within the damaged original pipe.

6.8.1 Application Method.

This tubing is tailored to the exact circumference and length of the pipe to be renewed. After saturation with resin, the tube is inserted through an attached angle inversion tube into the existing sewer from a manhole or other access point without the need for special excavations. The tube is then extended with cold water under pressure, inverting it through the existing pipe and pressing the special felt material against the inner walls. The water is then heated, curing the resin-saturated material, and forming a hard impermeable pipe within the damaged original pipe.

The new lining thickness usually is 9 to 25 mm and due to its smooth interior. flow capacities can be improved. This method allows for the filling of cracks, bridging gaps, articulation through pipe defects, and angles up to 45 degrees depending on diameter. "Insituform" is normally not designed to carry total external earth loads as is the pipe. Inspection is required to determine those areas where, for structural reasons, this method should not be used. The chemical resistance and abrasion of the resistance liner is excellent for normal applications in sewers. Chemically resistant resins are available for specific applications. A special cutting device is available commercially to remotely cut in service connections with the aid of a television camera. Proper curing of the lining is important to ensure successful attachment to the pipe wall at service connections so that sewage cannot penetrate the area behind the lining at the connection.

This method results in direct savings due to no excavation, little traffic control. low labour requirements, and rapid installation. To date. the largest installation in the United States has been 54 inches in diameter. This method requires that the sewer be dewatered and cleaned prior to installation.

Comparatively, the "Insituform" method is economically competitive although this will vary broadly with the size and length of the pipe to be rehabilitated.

6.9 **Slip Lining or Pipe Insertion Renewal**

Insertion renewal. that is, inserting a new pipe material into the old deteriorated sewer pipeline has become an economical and popular method of rehabiltation. A variety of pipe materials may be used, and selection can be based on certain features, benefits, and economics of materials and installation. Slip lining accomplished to date has been from 150 to 2,700 mm in diameter. At present. the following materials have been used in the insertion renewal process of rehabilitation.

1. **Polyethylene (PE) Pipe (Extruded).** PE pipe is manufactured in three types: low. medium. and high density. The most commonly used for sewer slip lining is high density (HPDE). High-density compounds are rigid and hard, strong, tough, corrosion-resistant, and well suited for sewer applications. Structural characteristics are a function of wall thickness design as in most plastics and other homogenous materials. Due to the low modulus of elasticity. careful design consideration should be given to buckling (ring buckling). PE pipes are categorised by the Standard Dimension Ratio (SDR), and pipe grades are also classified by stress-cracking resistance. Two national specifications are available for reference and design. They are: ASTM D-2447 and ASTM F-714. These standards apply generally to circular cross-section extruded PE pipe from 200 to 1,600 mm.

   Generally. PE pipe is pulled through the existing sewer in long lengths as shown in ASTM F-585. It is butt-fuse welded above the ground prior to pulling. The larger the diameter. the thicker the wall, the longer it takes to weld the joints together. The excavation required for this method is dependent upon the depth of the sewer, the diameter, and the wall thickness of the insertion pipe.

   Large diameter, thick-walled pipes have extremely large entrance radii which require large excavations for installation purposes. The pipe is attached to a pulling head and connected to a winch. usually electrical. at the other end of the sewer. Rollers are used above ground to facilitate the movement of the pipe during the pull. The pipe is pulled through the entire section at once. In some cases, stresses in the pipe wall are too high to make a single pull and multiple pulls must be considered. Maximum pulling lengths vary depending on the size, condition, and alignment of the old pipeline. Service connections can be connected generally by excavation and using a heat fusion saddle, or tapping saddle. In most cases, except for some larger diameter sewers, service connections must be excavated, the connection made. and then backfilled. Occasionally. when greater structural integrity is desired, the annular space between the PE liner and the existing sewer may be grouted for the entire length. Grouting process also prevents movement due to temperature changes, negative bouyance (flotation), and collapse of the inserted pipe in high groundwater locations. Additional pipe design considerations should be investigated. (When grouting of the annular space is required, the pipe must be designed to withstand the pressures associated with the installation of the grout).

PE pipe insertion provides an excellent corrosion and abrasion-resistant sewer conduit with potentially very good hydraulic characteristics due to its smooth inner surface. The flexibility of PE pipe allows it to articulate some offset in joints, manholes, and pipe size changes. The fuse-welded joints, if properly fused and cured, are virtually leak-free. In most cases, dewatering of the existing sewers is required. However, in some cases of low flows, the annular space may be adequate to convey the sewage.

2. **Polyethylene Pipe (Spiral-Welded).** This PE product is generally the same as the previous one with the following exceptions: the manufacturing process, size range, and design flexibility. Spiral-welded PE pipe can be manufactured in sizes from 300 to 3,600 mm in diameter, and in virtually any length depending upon the handling criteria. The load-bearing strength of this pipe is a function of the wall thickness and rib spacing. The rib spacing can be adjusted in thickness and length to meet particular structural needs. Especially in large diameters, this allows for excellent pipe economics. This type of high-density PE pipe is joined with a bell and spigot and sealed with an elastomeric gasket or heat-fused joints. Physical and chemical properties are similar to circular extruded PE pipe. Due to the method of manufacturing, any diameter can be made to meet the specific needs of a rehabilitation project. This results in maximum diameter and minimum annular space in the rehabilitated sewer.

   The installation methods for this product depend on the diameter and length specified. Economics are obtained by the excavation of a small installation pit when using bell and spigot pipe. In many instances, the pipe may be inserted while the sewer is flowing.

3. **Polybutylene (PB) Pipe (Extruded).** PB pipe resembles medium-density PE pipe in stiffness and chemical resistance but has higher strength under sustained stress. It retains this strength with increasing temperature. Its upper temperature limit is about 80 degrees C for gravity sewer applications. As a liner for the sewer renewal process, it is extremely effective where extra protection is required for high temperatures and aggressive industrial waste effluents. Thinner walls may be than used for PE pipe under the same structural load. Standard specifications regarding PB pipe are ASTM F-809 and F-809M.

   The installation of PB pipe is virtually the same as PE pipe. However, the maximum diameter currently available is 1.100 mm.

4. **Acrylonitrile-Butadiene-Styrene (ABS) Pipe.** ABS pipe is manufactured in SDR-pressure rated (ASTM D-2282) and gravity sewer (ASTM D-2751) types. These standards incorporates sizes up to and including 300 mm. The compounding of these polymers provides a hard, tough, strong and chemically resistant plastic. This plastic has been utilised often for small diameter, short lengths of pipe slip lining.

5. **Polyvinylchloride (PVC) Pipe.** PVC is manufactured as a pipe with a minimum amount of compounding ingredients for both pressure (ASTM D-2241) and non-pressure (ASTM D-3034 and F-679) applications which provide high strength, good chemical resistance, good toughness, and dimensional stability. These standards incorporate sizes up to and including 900 mm. This plastic has been utilised often for small diameter, short lengths of pipeline slip lining.

6. **Reinforced Plastic Mortar (RPM) Pipe.** The use of RPM pipe for rehabilitation of sewers and pipelines has been used extensively. RPM pipe can be used to reline cracked, corroded, and severely deteriorated existing sewers of all types. RPM

pipe is inserted into the existing deteriorated sewer using 6-metre lengths and a variety of installation methods. At present, diameters, ranging from 450 to 1,700 mm are available.

RPM pipe is a thermosetting composite of fibreglass, polyester resin combined with other resins, and a sand matrix. Commonly, each diameter has a specified minimum wall thickness based upon stiffness factors developed from ring compression, ring buckling, and longitudinal compression criteria. RPM pipe possesses very good chemical resistance and is corrosion-resistant for sewer applications as required in ASTM D-3681. Hydraulic friction can be low due to the very smooth interior surface, and the pipe has good abrasion and erosion resistance. RPM pipe has high strength and a high modulus of elasticity when compared to thermoplastics.

One manufacturer has developed an inverted bell and spigot joint using an elastomeric gasket for sealing. The flow restriction with this joint is inconsequential, and it allows for a constant outside diameter in addition to a large cross-sectional area to push or jack against. Other proprietory jointing systems are available. National specifications on design, materials, and testing for RPM pipe for ASTM D-3262, ASTM D-3517 and ASTM D-3754.

The installation of RPM pipe for relining purposes requires an excavated pit approximately 9 metres in length and 1.25 metres wider than the old pipe diameter. The existing sewer must be cleaned prior to the insertion and restrictions and alignment parameters determined in order to properly size the liner. After the pit is dug, the existing sewer is exposed and the top half of pipe is removed in the pit area. Generally, RPM pipe as with other segmented pipe liners can be inserted while the sewage is flowing. A half-full sewer, for most diameters, provides adequate flotation for lining distances up to 1.500 metres from one installation pit. Lower flows reduce the insertion distance. RPM pipes are usually pulled or pushed upstream against the sewage flow. In many cases, the installation pit can be set in the centre of the pipeline reach to be relined, and the pipe pulled or pushed in two directions. Pushing by jacking devices and pulling with cables and winches provides for rapid installation of RPM pipe. Additional installation pits may be required where angular offsets occur. Pipe joint pull to small and relining curvatures is limited to very small allowable angles.

Grouting the annular space between the pipes at the manhole connections is required, but grouting of the annular space for the entire pipeline is usually not required if the liner is strong enough to withstand the anticipated loads in event of collapse. High groundwater and excessive structural deterioration usually determines whether annular space grouting is required.

Minimum traffic obstructions, light weight, high rates of installation, and reasonable costs are the apparent advantages of RPM pipe used for rehabilitation. Limitations in its use include the difficulty in making service connections and the minimum allowable curvature criteria.

7. **Reinforced Thermosetting Resin (RTR) Pipe.** Similar in many respects to RPM pipe, RTR pipe has been used for rehabilitating sewers. It is usually inserted by pulling it through the existing sewer. RTR pipe is manufactured in diameters from 100 to 4,900 mm and in lengths up to 24 metres depending on diameter. A variety of joining systems are available, including bell and spigot with an elastomeric gasket, field wrapped, and locking tension types. Most types may be used in sewer relining.

RTR pipe is manufactured from polyester resins, other corrosion-resistant resins, and fibreglass filaments wound on a mandrel or applied inside a mandrel as a composite. It exhibits high axial and longitudinal strength in addition to good abrasion and erosion resistance. A smooth interior normally provides low friction factors.

8. **Ductile Iron (DI) Pipe.** DI pipe, Class 50, has been successfully used as a slip lining pipe. Where the existing pipe has experienced corrosion problems, polyvinyl-lined DI pipe should be used. For pipes experiencing only structural problems, the standard DI pipe with cement lining is normally sufficient.

   Due to the high modulus of elasticity, the normal concern for collapse (ring buckling) resistance is at a minimum. If necessary, a heavier class pipe may be utilised with the same outside diameter. The longitudinal compressive strength is very high, permitting the pipe to be pushed relatively long distances. The installation of DI pipe for relining purposes requires an excavated pit approximately 9 metres in length and about 1.25 metres wider than the old pipe diameter. Noteworthy of mention is that some reduction in normal Bell O.D. has been facilitated.

6.10 Flexible Pipe Rerounder

A method and apparatus is available for substantially reconstituting the rounding of flexible pipe after being installed. The apparatus is comprised of a moveable housing, sized to at least 95 percent of the pipe inside diameter. The housing is pulled through the pipeline while inducing vibration; causing rerounding of the pipe. The vibration induces a reallocation of the surrounding soil medium. The structural bearing capacity of the flexible pipe has been densified to the level of carrying six times the original vertical load. The end result is that the pipe zone envelope can be densified to 360 degrees around the pipe.

One of the primary benefits is that any pipe deflecting beyond allowable limits can be rerounded without new excavations and installations.

6.11 Root Control and Removal

Removal and control of roots that intrude into sanitary sewer systems is an ongoing maintenance task for most system operators. An effective root-control program requires an understanding of the basics of root control.

Roots grow toward moisture and nutrients by a continuous process that adds cells, one at a time, at the end of the root. This cell-by-cell growth enables roots to penetrate small openings in sewer pipes. Because roots seek moisture and nutrients, root intrusion tends to be a problem only in sewers installed in soils where moisture is limited. This includes pipelines that are continuously or seasonally above the groundwater table. Root intrusion is a more common problem in service connections because they are normally shallower and frequently not as well constructed as the sewer mains.

Root intrusion not only may increase the rate of infiltration into the sewer pipe or joint by expanding the opening, it may also substantially increase the potential for pipeline blockage. Once a root enters a sewer pipe, it expands to increase the mass available for moisture absorption. This not only screens out some of the solids in the wastewater but decreases upstream velocities, intensifying problems with deposition of solids in the sewer.

Various herbicides are available for chemical treatment. Techniques for applying herbicides generally consist of flood treatment (soaking), spraying, or foaming.

The use of foam herbicides is increasing because they are relatively easy to apply. In addition, foam is effective in treating root intrusion in building services without the problems generally associated with the flood treatment technique.

6.12 Excavation and Replacement

This technique involves the removal of the existing pipes from the ground and replacing them with new ones. When suitable materials and construction methods are utilised, the technique may produce the most effective rehabilitation results. The cost of this technique, however, is normally much higher than other rehabilitation techniques and the time requirement is usually much longer. Therefore, the cost of construction, administration, inspection, etc., must be evaluated when determining the most cost-effective alternate.

This technique is normally considered for application under one or more of the following conditions:

1. In locations where pipes have lost their structural integrity and are collapsed, crushed, broken, or badly deteriorated and cracked.

2. In cases where pipe size enlargement, change in grade, and/or pipe realignment are needed in addition to pipe deficiency corrections.

3. In cases where the causes of damage to the existing pipes (such as corrosion, soil movement, increasing traffic load, etc) have been identified and it is desirable to prevent the reoccurrence of these damages by replacing the existing pipes with new ones having higher quality and greater strength.

Just as for new sewer construction, this rehabilitation technique may require the removal of pavement, disruption of traffic, dewatering, well-pointing, shoring, interference with utilities and structures, often old pipe removal, and maintenance of existing service. In addition, during the construction period, the sewage flows in the various sewer reaches are usually bypassed. All the cost involved in these tasks should be considered when comparing this technique with other rehabilitation techniques.

7. GENERAL CONCLUSIONS

In most cases, the rehabilitation of sewer pipelines can be accomplished effectively and at a lower cost than excavation and replacement.

It is extremely important to evaluate the physical condition of the existing sewer to adequately determine the best rehabilitation method and the most apparent cost-effective solution.

Many unique materials and techniques are available for sewer pipeline rehabilitation, and consideration should be given to external problems as well as internal problems. The diameter, cross-section, variation in flow, number of connections, dewatering requirements, access, location, depth, nature of the wastewater, movement of groundwater, amount of extraneous flow, and soil settlement conditions are additional factors of concern in a rehabilitation evaluation.

External grouting and void filling help to solve groundwater infiltration and sewage exfiltration problems.

Internal chemical grouting, sliplining with various pipes, and placement of several mechanical and cementious materials are the techniques currently utilised in sewer pipeline rehabilitation.